Suzuki–Miyaura Cross-Coupling Reaction and Potential Applications

Special Issue Editor
Ioannis D. Kostas

MDPI • Basel • Beijing • Wuhan • Barcelona • Belgrade

MDPI

Special Issue Editor
Ioannis D. Kostas
National Hellenic Research Foundation
Greece

Editorial Office
MDPI AG
St. Alban-Anlage 66
Basel, Switzerland

This edition is a reprint of the Special Issue published online in the open access journal *Catalysts* (ISSN 2073-4344) from 2016–2017 (available at: http://www.mdpi.com/journal/catalysts/special_issues/suzuki_miyaura).

For citation purposes, cite each article independently as indicated on the article page online and as indicated below:

Author 1; Author 2. Article title. *Journal Name* **Year**, *Article number*, page range.

First Edition 2017

ISBN 978-3-03842-556-4 (Pbk)
ISBN 978-3-03842-557-1 (PDF)

Table of Contents

About the Special Issue Editor v

Preface to "Suzuki–Miyaura Cross-Coupling Reaction and Potential Applications" vii

Christophe Len, Sophie Bruniaux, Frederic Delbecq and Virinder S. Parmar
Palladium-Catalyzed Suzuki–Miyaura Cross-Coupling in Continuous Flow
Reprinted from: *Catalysts* 2017, 7(5), 146; doi: 10.3390/catal7050146......................................1

Katia Martina, Maela Manzoli, Emanuela Calcio Gaudino and Giancarlo Cravotto
Eco-Friendly Physical Activation Methods for Suzuki–Miyaura Reactions
Reprinted from: *Catalysts* 2017, 7(4), 98; doi: 10.3390/catal7040098......................................24

Tom Willemse, Wim Schepens, Herman W. T. van Vlijmen, Bert U. W. Maes and Steven Ballet
The Suzuki–Miyaura Cross-Coupling as a Versatile Tool for Peptide Diversification and
Cyclization
Reprinted from: *Catalysts* 2017, 7(3), 74; doi: 10.3390/catal7030074......................................104

Magne O. Sydnes
The Use of Palladium on Magnetic Support as Catalyst for Suzuki–Miyaura
Cross-Coupling Reactions
Reprinted from: *Catalysts* 2017, 7(1), 35; doi: 10.3390/catal7010035......................................136

Takeru Kamigawara, Hajime Sugita, Koichiro Mikami, Yoshihiro Ohta and Tsutomu Yokozawa
Intramolecular Transfer of Pd Catalyst on Carbon–Carbon Triple Bond
and Nitrogen–Nitrogen Double Bond in Suzuki–Miyaura Coupling Reaction
Reprinted from: *Catalysts* 2017, 7(7), 195; doi: 10.3390/catal7070195......................................150

**Hamza Boufroura, Benjamin Large, Talia Bsaibess, Serge Perato, Vincent Terrasson,
Anne Gaucher and Damien Prim**
Aziridine- and Azetidine-Pd Catalytic Combinations. Synthesis and Evaluation of the
Ligand Ring Size Impact on Suzuki-Miyaura Reaction Issues
Reprinted from: *Catalysts* 2017, 7(1), 27; doi: 10.3390/catal7010027......................................160

**Roghayeh Sadeghi Erami, Diana Díaz-García, Sanjiv Prashar, Antonio Rodríguez-Diéguez,
Mariano Fajardo, Mehdi Amirnasr and Santiago Gómez-Ruiz**
Suzuki-Miyaura C-C Coupling Reactions Catalyzed by Supported Pd Nanoparticles for the
Preparation of Fluorinated Biphenyl Derivatives
Reprinted from: *Catalysts* 2017, 7(3), 76; doi: 10.3390/catal7030076......................................169

Melania Gómez-Martínez, Alejandro Baeza and Diego A. Alonso
Graphene Oxide-Supported Oxime Palladacycles as Efficient Catalysts for the Suzuki–Miyaura
Cross-Coupling Reaction of Aryl Bromides at Room Temperature under Aqueous Conditions
Reprinted from: *Catalysts* 2017, 7(3), 94; doi: 10.3390/catal7030094......................................185

**Mujeeb Khan, Mufsir Kuniyil, Mohammed Rafi Shaik, Merajuddin Khan,
Syed Farooq Adil, Abdulrahman Al-Warthan, Hamad Z. Alkhathlan,
Wolfgang Tremel, Muhammad Nawaz Tahir and Mohammed Rafiq H. Siddiqui**
Plant Extract Mediated Eco-Friendly Synthesis of Pd@Graphene Nanocatalyst: An Efficient and
Reusable Catalyst for the Suzuki-Miyaura Coupling
Reprinted from: *Catalysts* **2017**, *7*(1), 20; doi: 10.3390/catal7010020..201

Jiahuan Yu, An Shen, Yucai Cao and Guanzhong Lu
Preparation of Pd-Diimine@SBA-15 and Its Catalytic Performance for the Suzuki
Coupling Reaction
Reprinted from: *Catalysts* **2016**, *6*(12), 181; doi: 10.3390/catal6120181..215

Arindam Modak, Jing Sun, Wenjun Qiu and Xiao Liu
Palladium Nanoparticles Tethered in Amine-Functionalized Hypercrosslinked Organic Tubes
as an Efficient Catalyst for Suzuki Coupling in Water
Reprinted from: *Catalysts* **2016**, *6*(10), 161; doi: 10.3390/catal6100161..232

Jiří Schulz, Filip Horký, Ivana Císařová and Petr Štěpnička
Synthesis, Structural Characterization and Catalytic Evaluation of Anionic Phosphinoferrocene
Amidosulfonate Ligands
Reprinted from: *Catalysts* **2017**, *7*(6), 167; doi: 10.3390/catal7060167...243

About the Special Issue Editor

Ioannis D. Kostas with a Degree in Chemistry (University of Thessaloniki, 1986) and a PhD in Organometallic Chemistry (University of Athens, 1991) worked a Post-doctoral Fellow at the Vrije Universiteit in Amsterdam (1994–1995) with F. Bickelhaupt, and then at the Max-Planck-Institut für Kohlenforschung in Mülheim an der Ruhr (1995-1996) with M.T. Reetz. In 1996, he jointed the National Hellenic Research Foundation in Athens as a Researcher, in which he is "Research Director" since 2007. Since 2015, he is also Visiting Professor at the University of Thessaly (Department of Biochemistry and Biotechnology). His research interests are focused on: (a) transition metal homogeneous catalysis, including asymmetric catalysis and catalysis in aqueous medium, (b) catalysis by metal nanoparticles, (c) organic synthesis of bioactive compounds. He is pioneer on the use of thiosemicarbazones as ligands in Pd-catalyzed coupling reactions, and also metalloporphyrins as catalysts in the Suzuki reaction and the hydrogenation of unsaturated aldehydes. One of the catalysts (CAS no. 219954-63-9) published by his group for the Suzuki-Miyaura coupling is commercially available by numerous companies. He has co-authored 43 journal articles and four book chapters. He has been coordinator or participant in 21 European and National competitive programs. He has been invited as a reviewer for proposals and fellowships within National, European and American programs, and for a number of papers (as well as Editor in Chief) by 43 scientific journals. He has also been member of the Editorial Advisory Board in 6 scientific journals. He has supervised 20 pre-graduate, MSc, PhD students and post-docs, and he is teaching at a post-graduate level.

Preface to "Suzuki–Miyaura Cross-Coupling Reaction and Potential Applications"

Suzuki–Miyaura cross-coupling remains a powerful tool in organic synthesis for C–C bond formation and has various industrial applications, for example, the synthesis of pharmaceuticals and materials. Intensive research efforts are being made into finding ways of improving and expanding the scope of this process, and the development of more efficient catalytic systems for this extremely important reaction is still a hot research topic of enormous academic and industrial interest.

This Special Issue, consisting of four reviews, two communications and six articles, focuses on recent promising research and novel trends in the broad field of Suzuki–Miyaura cross-coupling employing a range of different palladium catalysts. Homogeneous or heterogeneous catalysis in organic or aqueous medium, using conventional conditions or non-conventional techniques such as microwave and ultrasound irradiation, grinding and photo-activated processes as green chemistry approaches, as well as continuous flow technology are included. The catalysts described herein are unsupported metal complexes, catalysts immobilized on solid supports, ligand-free catalytic systems or metal nanoparticles. Studies on catalyst recycling, coupling of non-activated substrates, mechanistic insights, as well as potential applications for the synthesis of fine chemicals and intermediates used in the manufacture of bioactive compounds and materials are also of great interest.

The Catalysts editorial team would like to thank the Editor-in-Chief, Professor Keith Hohn, for arranging this Special Issue. Professor Hohn has also expressed his gratitude to the Senior Assistant Editors, Jiuyu Guo and Shelly Liu, and the Editorial Office staff of the journal for their assistance and understanding during this issue's preparation. Special thanks are, of course, also due to all the contributing authors and to the reviewers for their assessments and recommendations concerning the submitted manuscripts. The authors are also to be thanked for their cooperation with the subsequent revisions and I believe that these efforts are reflected in the high quality of the work presented in this Special Issue.

Ioannis D. Kostas
Special Issue Editor

catalysts
MDPI

Review

Palladium-Catalyzed Suzuki–Miyaura Cross-Coupling in Continuous Flow

Christophe Len [1,*], Sophie Bruniaux [1], Frederic Delbecq [2] and Virinder S. Parmar [3]

[1] Centre de Recherche Royallieu, Université de Technologie de Compiègne (UTC), Sorbonne Universités, CS 60319, F-60203 Compiègne CEDEX, France; sophie.bruniaux@utc.fr
[2] Ecole Supérieure de Chimie Organique et Minérale (ESCOM), 1 rue du Réseau Jean-Marie Buckmaster, F-60200 Compiègne, France; f.delbecq@escom.fr
[3] Institute of Advanced Sciences, 86-410 Faunce Corner Mall Road, Dartmouth, MA 02747, USA; virparmar@gmail.com
* Correspondence: christophe.len@utc.fr; Tel.: +33-344-234-323

Academic Editor: Ioannis D. Kostas
Received: 15 March 2017; Accepted: 25 April 2017; Published: 9 May 2017

Abstract: Carbon–carbon cross-coupling reactions are among the most important processes in organic chemistry and Suzuki–Miyaura reactions are the most widely used protocols. For a decade, green chemistry and particularly catalysis and continuous flow, have shown immense potential in achieving the goals of "greener synthesis". To date, it seems difficult to conceive the chemistry of the 21st century without the industrialization of continuous flow process in the area of pharmaceuticals, drugs, agrochemicals, polymers, etc. A large variety of palladium Suzuki–Miyaura cross-coupling reactions have been developed using a continuous flow sequence for preparing the desired biaryl derivatives. Our objective is to focus this review on the continuous flow Suzuki–Miyaura cross-coupling using homogeneous and heterogeneous catalysts.

Keywords: palladium; continuous flow; Suzuki–Miyaura; cross-coupling

---

1. Introduction

Among the main reactions in organic chemistry, C–C bond formation via a cross-coupling reaction catalyzed by transition metals is undoubtedly the most important and has been exploited very widely in the recent years. Palladium, the most widely used metal, enables the synthesis of complex and functionalized organic molecules and its chemistry possesses different interesting facets such as heterogeneous and homogeneous catalysis under mild experimental conditions compatible with many functional groups [1–5]. Several palladium catalyzed cross-coupling reactions such as Heck [6–11], Suzuki [12–16], Sonogashira [17–21], Stille [22–25], Hiyama [26], Negishi [27], Kumada [28], Murahashi [29] and Buchwald–Hartwig [30,31] have been developed over the years.

Due to current impetus in promoting green chemistry for sustainable development, both for academic and industrial research, chemists have recently established catalytic reactions based on renewable resources, atom economy, less hazardous chemical steps, safer (least toxic) solvents, auxiliaries and alternative technologies such as continuous flow, microwave irradiation, ultrasound irradiation, etc. In the context of green chemistry, catalysis and alternative media, different cross-coupling reactions such as Suzuki–Miyaura in batch reactors have been developed in aqueous media or in water as sole green safer solvent via conventional heating or microwave irradiation [32–43]. Continuous flow chemistry as alternative technology offers significant processing advantages including improved thermal management, mixing control, application to a wider range of reaction conditions, scalability, energy efficiency, waste reduction, safety, use of heterogeneous catalysis, multistep synthesis and much more [44–49]. Two different reactors, micro and meso (or flow) reactors, exist and the devices depend on the channel dimensions, from 10 to 300 µm for the micro reactor (also called milli or mini) and from 300 µm to more than 5 mm for the meso reactor. Several advantages and disadvantages are associated with the micro and meso reactors. The main advantages for the micro reactor are the low material input, low waste output, excellent mass transfer properties, fast diffusive mixing and the disadvantages are the low throughput, tendency to channel blockage and high pressure drop. In the case of solid handling due to confined conditions and increasing of the concentration to have a better productivity, the use of continuous sonication could prevent clogging [50]. For the meso reactor, the advantages are the high throughput, low pressure drop and possibility to handle solids for heterogeneous catalysis. Few disadvantages for meso reactors are poor mass transfer property, slower mixing, etc. Different studies have described the theory and practicalities of scaled-out micro and meso reactors but no practical examples of large-scale production have been described. Palladium-catalyzed cross-coupling reactions in continuous flow reactors have been reported in the literature at temperatures higher than 60 °C [51–63], while only few studies have described micro and meso reactors for the C–C bond formation at temperature lower than 60 °C. In parallel with the synthesis of low molecular weight compounds, this technique has been applied by academic and industrial groups for the production of polymers [64–67]. For the sake of clarity, this review describes continuous flow selective palladium-catalyzed cross-coupling reactions having a good energy efficiency at temperatures ranging between 0 °C and 80 °C.

2. Accepted Mechanism of Suzuki Cross-Coupling

The Suzuki–Miyaura cross-coupling reaction [12–16] is one of the most versatile and frequently employed method for C–C bond formation. It consists of the coupling of organoboron compounds (organoborane, organoboronic acid, organoboronate ester and potassium trifluoroborate) with aryl, alkenyl and alkynyl halides. Nowadays, a large variety of boronic acids are commercially available. The general Suzuki–Miyaura catalytic cycle occurs through oxidative addition, transmetallation and reductive elimination [13–15,68–70]. After formation of the catalytic species Pd(0), generated in situ starting from palladium Pd(II) or directly from Pd(0) derivatives, oxidative addition of the aryl halide ArX furnishes the palladium complex (ArPdXLn). The transmetallation step occurs by conversion of the palladium halide (ArPdXLn) in the presence of the base RO$^-$ to a nucleophilic palladium alkoxy complex (ArPdORLn). This complex subsequently reacts with a neutral organoboron compound Ar'B(OH)$_2$ to afford the diaryl complex (ArPdAr'Ln) in a *cis–trans* equilibrium. Then, reductive elimination of the *cis* form gives the biaryl derivative Ar–Ar' and Pd(0) (Scheme 1) [15].

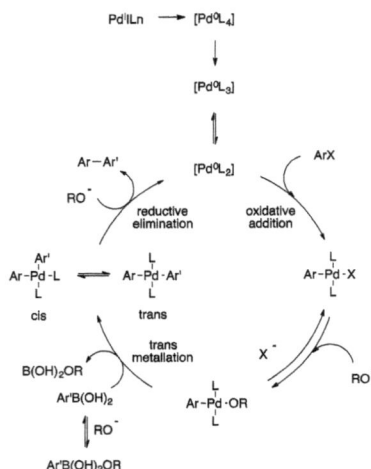

Scheme 1. Mechanism of the homogeneous Suzuki–Miyaura reaction.

Using supported palladium catalysts, Suzuki–Miyaura cross-coupling reaction is a heterogeneous catalysis [71]. During the reaction, the palladium Pd(II) could be release from the surface of the solid support and this leaching palladium could be responsible for the catalysis as a (quasi)homogeneous mechanism (Scheme 2) [72–83].

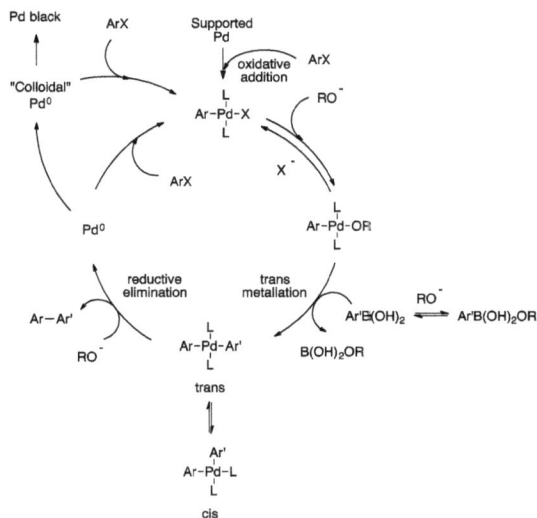

Scheme 2. Mechanism of the heterogeneous Suzuki–Miyaura reaction.

3. Homogeneous Suzuki–Miyaura Cross-Coupling Reaction in Continuous Flow

Buchwald reported an efficient synthesis of biaryls from aryl halide substrates using a successive lithiation/borylation/Suzuki–Miyaura cross-coupling sequence in three successive mesoreactors [84]. Starting from aryl bromide, the bromine-lithium exchange afforded the corresponding aryllithium which reacted with borate to form the boronate agent. Conventional Suzuki–Miyaura cross-coupling reaction using homogeneous second-generation palladium let precatalyst **a** furnished the target biaryl derivatives (Scheme 3). One of the main drawbacks of this nice concept was the formation of solids such as lithium triisopropylarylborate during the process; optimization of the nature of the solvent (THF and H_2O), the concentration of reagents and the use of acoustic irradiation have been reported to avoid the formation of such solids.

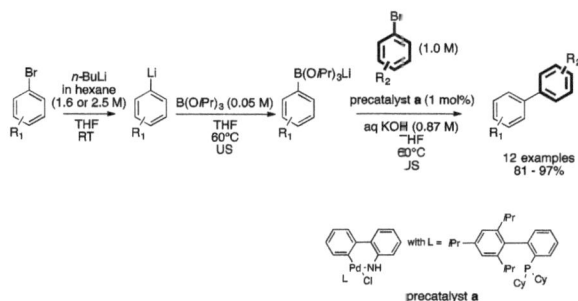

Scheme 3. Lithiation/borylation/Suzuki–Miyaura cross-coupling sequence for the synthesis of biaryl derivatives.

In this report, the development of different reactors made with a perfluoroalkoxyalkane (PFA) tube having the inner diameter of 1 mm has been described [84]. A solution of arylbromide in THF and a solution of *n*-butyllithium in hexane (1.6 M or 2.5 M) were injected simultaneously, then mixed at a T-shaped mixer and delivered to the first reactor (*reactor 1*) at room temperature with a flow rate of 50–78 µL min^{-1} and a varying residence time (2–120 s). A solution of diluted B(OiPr)$_3$ in THF was injected with a flow rate of 1 µL min^{-1} and mixed with the exiting stream of aryllithium derivative at a T-shape mixer. The mixed stream was introduced to the second reactor (*reactor 2*) at 60 °C under acoustic irradiation with a residence time of 1 min. Then, a solution of aqueous KOH (0.87 M) and a solution of aryl halide (1.00 M) and XPhos precatalyst (**a**, 1 mol %) in THF were successively injected into the exiting stream with a flow rate of 100 µL min^{-1} and 21–40 µL min^{-1}, respectively. The combined mixture was introduced to the third reactor (reactor 3) at 60 °C under acoustic irradiation with a residence time of 10 min (Scheme 4). Ultrasound chemistry was used for reactors 2 and 3 to avoid reactor clogging and ensure a good mixing of reagents during the formation of the borate and the Suzuki–Miyaura cross-coupling reaction.

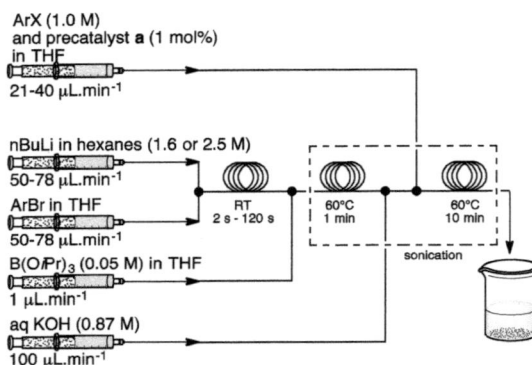

Scheme 4. Lithiation/borylation/Suzuki–Miyaura cross-coupling sequence for the synthesis of biaryl derivatives in a microflow system.

Application of the above methodology was realized with a range of various aryl halides (Figure 1), the limiting step of the process being the lithiation of aryl halides. In their hands, Buchwald described that the aryl bromide could be lithiated at room temperature. Whatever the nature of the starting aryl bromide having different electronic and steric demands in *para*, *meta* and *ortho* positions, the aryllithium and then the corresponding lithium arylborate were obtained in good yields. For the third step, the Suzuki–Miyaura cross-coupling reaction with aryl bromide or chloride with both electron-withdrawing and electron-donating substituents, afforded the target compounds in good yields. It was noteworthy that non-canonic heteroatomic halides such as quinoline, isoquinoline, pyrimidine and benzothiophene were good reagents for the continuous flow reaction.

It is noteworthy that five-membered 2-heteroaromatic boronic acids are unstable at room temperature and consequently give low yields in the Suzuki–Miyaura cross-coupling reaction [85–90]. Consequently, Buchwald turned attention to the lithiation/borylation/Suzuki–Miyaura cross-coupling of heteroarenes such as thiophene and furan derivatives; starting from furanic derivatives, selective deprotonation of the hydrogen atom in position 2 at room temperature afforded the corresponding lithium analog which reacted with borate to form the boronate agent. Then, conventional homogeneous Suzuki–Miyaura cross-coupling reaction furnished the target biaryl derivatives. After optimization of the first continuous flow process (Scheme 4), the borylation was made at room temperature with a reduced time (6 s vs. 60 s) and acoustic irradiation was not needed for this step in reactor 2 (Scheme 5) [84].

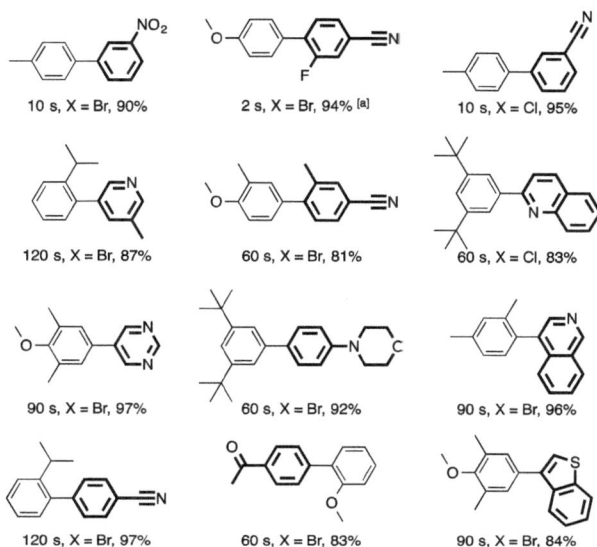

10 s, X = Br, 90%	2 s, X = Br, 94% [a]	10 s, X = Cl, 95%
120 s, X = Br, 87%	60 s, X = Br, 81%	60 s, X = Cl, 83%
90 s, X = Br, 97%	60 s, X = Br, 92%	90 s, X = Br, 96%
120 s, X = Br, 97%	60 s, X = Br, 83%	90 s, X = Br, 84%

Figure 1. Substrate scope of continuous flow lithiation/borylation/Suzuki–Miyaura cross-coupling sequence starting from aryl bromides.

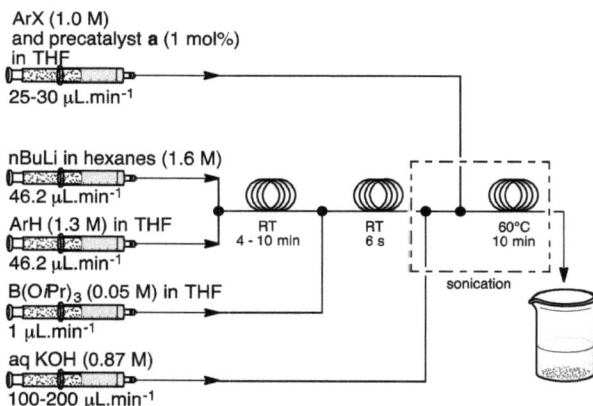

ArX (1.0 M)
and precatalyst **a** (1 mol%)
in THF
25-30 µL.min^{-1}

nBuLi in hexanes (1.6 M)
46.2 µL.min^{-1}
ArH (1.3 M) in THF
46.2 µL.min^{-1}
B(O*i*Pr)$_3$ (0.05 M) in THF
1 µL.min^{-1}
aq KOH (0.87 M)
100-200 µL.min^{-1}

RT
4 - 10 min

RT
6 s

60°C
10 min

sonication

Scheme 5. Lithiation/borylation/Suzuki–Miyaura cross-coupling sequence of heteroarenes with aryl halides in a flow system.

Application of the method was realized to show the scope of the reaction [84]. Starting from thiophene, 2-alkylthiophene and 2-alkylfuran, borylation in two steps was efficient: the coupling with different substituted aryls and heteroaromatic halides afforded the target compounds in good yields (Scheme 6). This novel process allows the use of low-cost heteroarenes instead of more expensive and unstable 2-heteroaromatic boronic acids and 2-heteroaromatic bromides.

In order to illustrate the synthetic potential of this methodology, Diflunisal [91,92] was obtained in a multi-step sequence [84]. Starting from 4-bromoanisole, the lithiation/borylation followed by Suzuki–Miyaura cross-coupling with 1-bromo-2, 4-difluorobenzene permitted the synthesis of the key intermediate in the production of Diflunisal (Scheme 7).

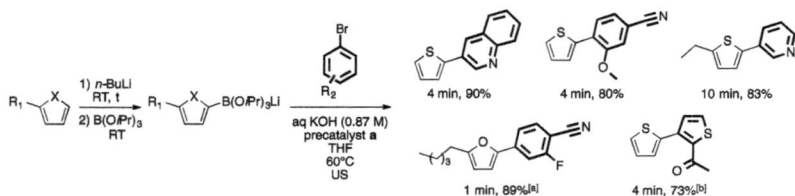

Scheme 6. Substrate scope of continuous flow lithiation/borylation/Suzuki–Miyaura cross-coupling sequence starting from furan derivatives: [a] 0.44 M NaF aqueous solution was used instead of KOH; and [b] 0.87 M KF aqueous solution was used instead of KOH.

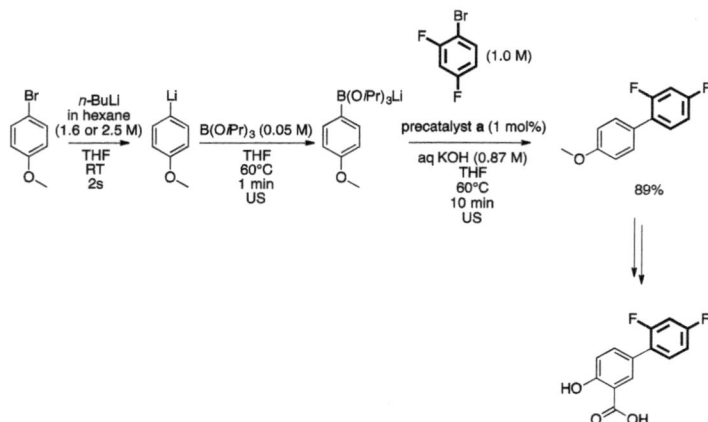

Scheme 7. Total synthesis of Diflunisal via lithiation/borylation/Suzuki–Miyaura cross-coupling in a microflow system.

In order to develop an automated, droplet-flow microfluidic system applied to Suzuki–Miyaura cross-coupling reaction, Buchwald and Jensen reported a systematic methodology including key mechanistic insights [93].

A three-step flow diazotization, iododediazotization and Suzuki–Miyaura cross coupling reaction has been reported by Organ starting from aniline derivatives [94]. Starting from the arylamine, the diazotation followed by the introduction of iodide atom furnished the iodobenzene derivatives. Then conventional Suzuki–Miyaura cross coupling afforded the biphenyl derivatives (Scheme 8).

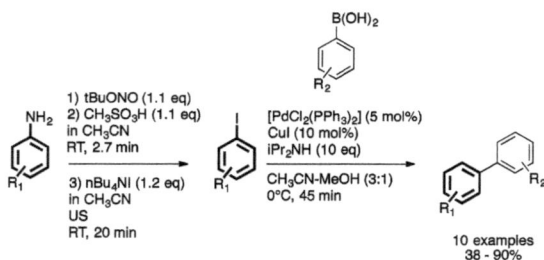

Scheme 8. Diazotization/iododediazotization/Suzuki–Miyaura cross coupling sequence for the synthesis of biaryl derivatives.

Three reactors were made with PFA capillary tubing with an inner diameter of 1.52 mm and different volumes. The residence time in reactors was adjusting the length of the reactor tubing. A solution of aniline derivative in CH_3CN and a solution of tBuONO in CH_3CN were injected simultaneously, followed by mixing with a T-mixer and injection of a solution of methanesulfonic acid in CH_3CN. The three solutions were used with a flow rate of 22 μL min^{-1}. The mixed stream was introduced to the first reactor at room temperature with a residence time of 2.7 min. Then, a solution of nBuNI in CH_3CN was injected into the stream with the same flow. The combined mixture was introduced to the second reactor for which is immersed in an ultrasonic bath. The residence time was 20 min at room temperature and then the segmented effluent was temporarily collected in an intermediate reservoir. Due to the used reservoir a continuous flow unit (CFU) was accommodated. A solution of [$PdCl_2(PPh_3)_2$], CuI, *i*Pr$_2$NH in CH_3CN and a solution of boronic acid in MeOH were injected simultaneously to the main stream and introduced to the third reactor at 60 °C for 45 min (Scheme 9) [94].

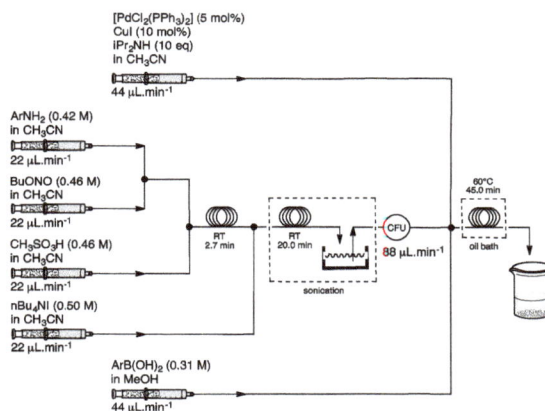

Scheme 9. Diazotization/iododediazotization/Suzuki–Miyaura cross-coupling sequence of aniline derivative with aryl halides in a flow system.

Application of the above protocol with little variations to the production of biphenyl compounds was reported (Figure 2). In function of the different steric and electric demands in the aromatic core, the coupling gave satisfactory yields [94].

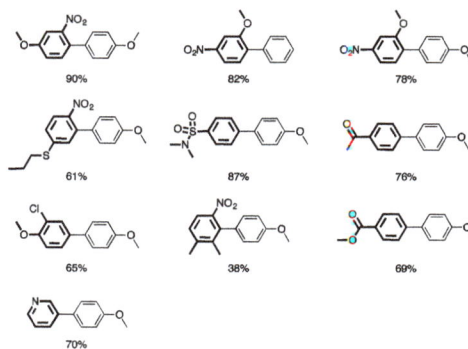

Figure 2. Substrate scope of continuous flow diazotization/iododediazotization/Suzuki–Miyaura cross-coupling sequence starting from aniline derivative.

Application to continuous flow process on a large scale was reported recently and could open new way for industrial use [95,96].

4. Heterogeneous Suzuki–Miyaura Cross-Coupling Reaction in Continuous Flow

A suitable solid support having Pd(II) species precursors to Pd(0) catalysts are now commercially available but different groups prefer to design their home-made catalysts. In the first part the use of Pd(0) reagent is reported and in the second part Pd(II) is described.

Monguchi and Sajiki reported a palladium on carbon-catalyzed Suzuki–Miyaura coupling reaction using an efficient and continuous flow system (Scheme 10) [97]. To investigate the scope of the reaction, a range of arylboronic acids and halogenobenzene derivatives were tested in mild conditions for 20 s during a single-pass (Figure 3). The authors have reported the detection of little leaching (<1 ppm).

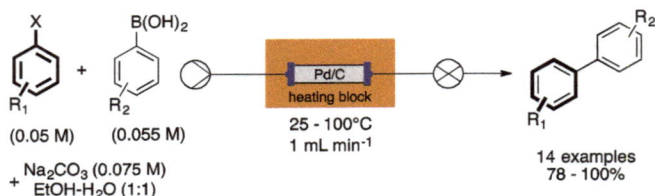

Scheme 10. Suzuki–Miyaura cross coupling in a flow system using H-Cube®.

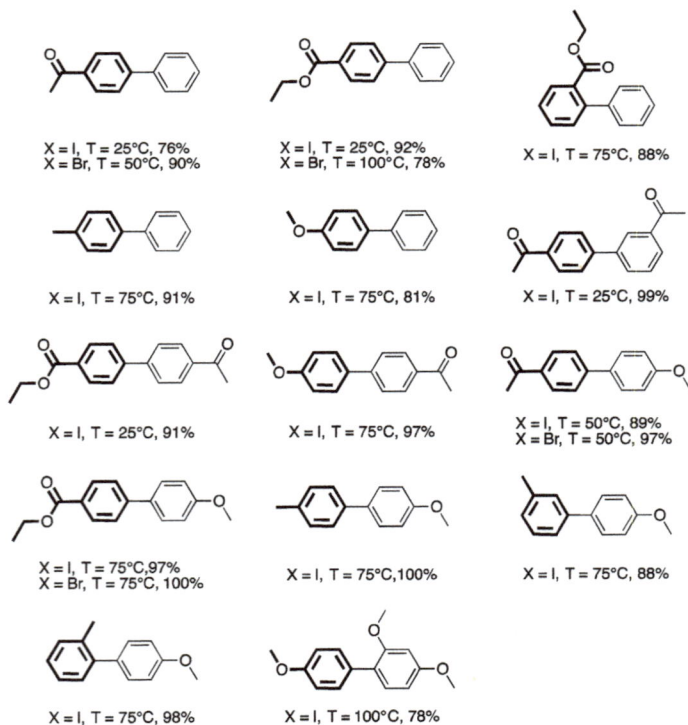

Figure 3. Substrate scope of continuous flow Suzuki–Miyaura cross-coupling sequence in a flow system using H-Cube®.

Using the same apparatus H-Cube®, another group reported the Suzuki–Miyaura cross coupling in the presence of graphene supported palladium nanoparticles [98]. A solution of 4-bromobenzaldehyde and phenylboronic acid dissolved in H_2O-EtOH-THF (1:1:1) was injected with a flow rate of 0.2 mL min^{-1} resulting in a contact time of less than 1 min at 135 °C. The target biaryl compound was obtained in good yield with a conversion of 96% (Scheme 11).

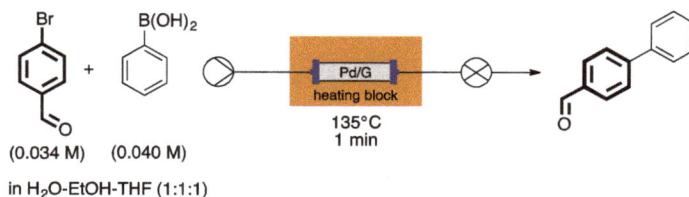

(0.034 M)   (0.040 M)

in H_2O-EtOH-THF (1:1:1)

Scheme 11. Suzuki–Miyaura cross coupling in a flow system using H-Cube®.

In 2006, Canty was the first to develop a macroporous monolith support as a suitable substrate for anchoring a palladium complex for Suzuki–Miyaura cross-coupling continuous flow capillary microreactors [99]. Ten years after, Nagaki reported an efficient three-step flow sequence using Pd catalyst [100]. The aryllithium obtained from arylbromide reacted with B(OMe)$_3$; after the borylation reaction, the Suzuki–Miyaura cross coupling reaction in the presence of immobilized Pd(0) on the polymer afforded the target biaryl derivatives (Scheme 12).

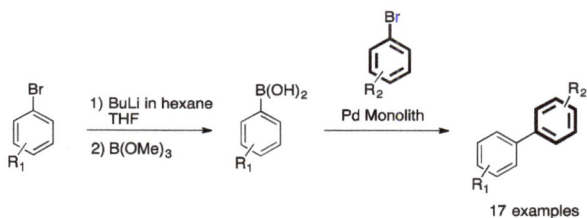

17 examples

Scheme 12. Halogen/lithium exchange/borylation/Suzuki–Miyaura cross-coupling sequence for the synthesis of biaryl derivatives.

In this report, a solution of bromobenzene in THF (0.10 M) and a solution n-BuLi (0.6 M in hexane) were injected simultaneously with a flow rate of 6.0 mL min^{-1} and 1 mL min^{-1}, respectively, in a micromixer (ID = 500 μm) and then in a reactor (ID = 1000 μm) for a residence time of 1.7 s. A solution of diluted B(OMe)$_3$ (0.12 M) in THF was injected with a flow rate of 6.0 mL min^{-1} and the main stream was introduced to a micromixer (ID = 500 μm) and the second reactor (ID = 1000 μm) for a residence time of 2.0 s (Scheme 13). After producing the boronic acid solution, iodoaryl derivative (0.33 M) in methanol was added and the mixture was passed through the palladium catalyst at 100 °C with a residence time of 4.7 min or at 120 °C with a residence time of 9.4 min.

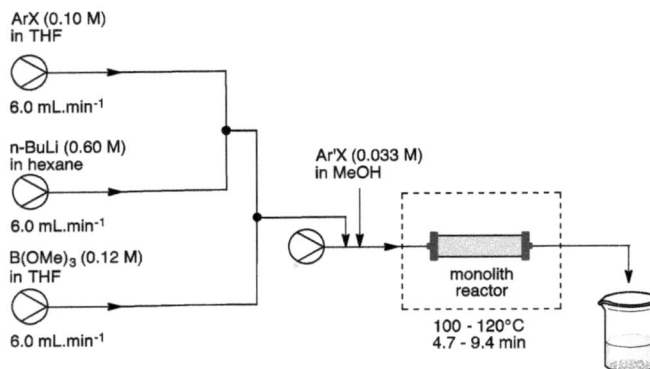

Scheme 13. Halogen/lithium exchange/borylation/Suzuki–Miyaura cross-coupling sequence for the synthesis of biaryl derivatives in a microflow system.

Application of the above methodology was successfully applied to the cross-coupling of various functional aryl and heteroaryl iodides (Figure 4) [100]. It was noticeable that cyano derivatives in this process tolerated the experimental conditions. Adapalene, a drug used for the treatment of acne, was produced in 86% yield by applying this methodology.

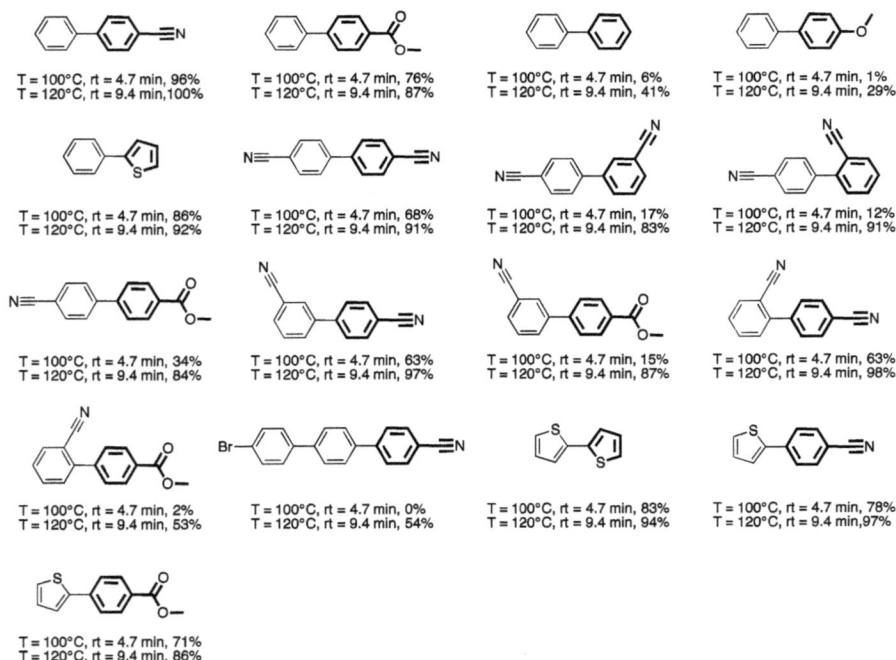

Figure 4. Substrate scope of continuous flow lithiation/borylation/Suzuki–Miyaura cross-coupling sequence in a flow system using immobilized Pd on polymer monolith.

A large-scale Suzuki–Miyaura cross-coupling reaction using solid supported palladium Pd^0 nano/microparticles and ultrasound irradiation was reported in continuous flow by Das [101]. The continuous flow technique used by the authors required a syringe, a reservoir, a pump and a reaction vessel. After the introduction of the aryl bromide, phenylboronic acid and potassium carbonate in $MeOH-H_2O$ in the reservoir via the syringe **a**, the reagents were pumped (127 mL min^{-1}) to the reaction vessel **e** where the solid supported palladium (0) nano/microparticles (SS-Pd) as heterogeneous catalyst had been charged. Ultrasonication of the mixture (20 kHz) was realized and the reaction product was poured through **b** into the reservoir. Two exits (**d** and **g**) were present to recover the mixture after completion of the reaction (Scheme 14). In comparison with microreactor and mesoreactor, this process permitted to furnish biaryl derivatives on a gram scale in continuous flow.

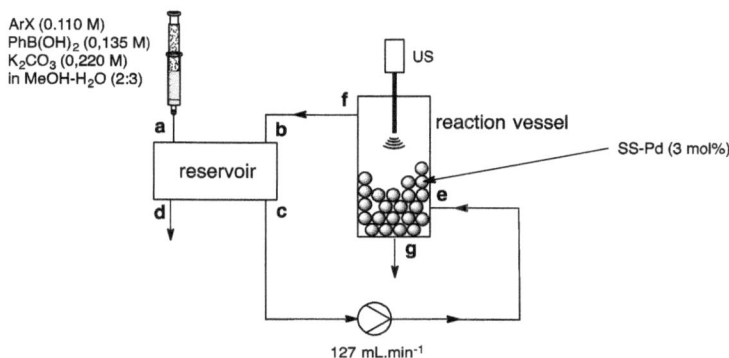

Scheme 14. Continuous flow Suzuki–Miyaura cross-coupling on a gram scale of the substrate.

Reactions of various aryl iodides with phenylboronic acids gave excellent yields. Aryl iodides having different substituents were explored without significant change in their reactivity. The activation of the aryl chloride was more difficult than expected, but gave good yields using this methodology (Scheme 15) [101].

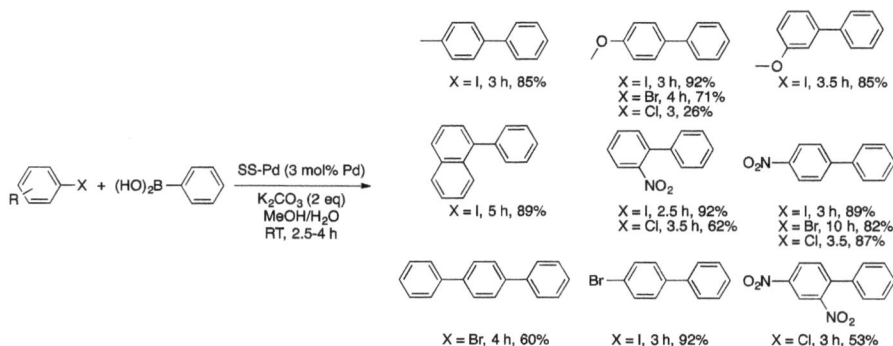

Scheme 15. Continuous flow Suzuki–Miyaura cross-coupling reaction of phenyl boronic acid with aryl halides using SS-Pd as heterogeneous catalyst.

The catalytic stability in MeOH-H_2O under flow conditions was studied by Das and a mechanism was proposed (Scheme 16) [101]. No significant loss of activity was observed after recycling five times. In their hands, the SEM analysis of SS-Pd showed the presence of Pd(0) nano-microparticulates on the solid support which implies its reusability and minimum leaching of Pd from the solid surface.

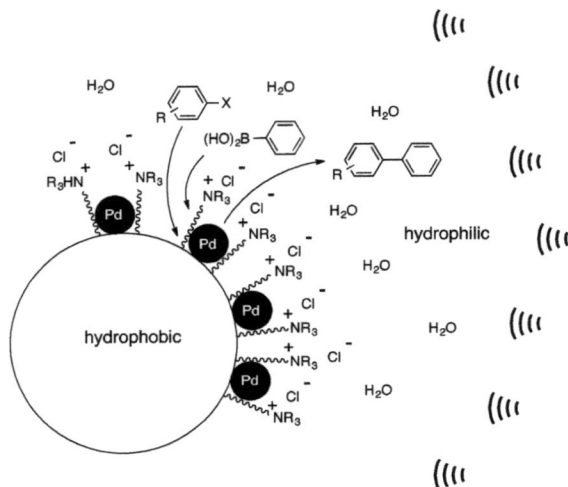

Scheme 16. A schematic diagram of the Suzuki–Miyaura cross-coupling reaction of phenyl boronic acid with aryl halides using SS-Pd as heterogeneous catalyst.

Martin-Matute developed a novel strategy using Pd nanoparticles supported in a functionalized mesoporous Metal-Organic Frameworks (MOFs) [102]. To the best of our knowledge, it was the first report on the use of metallic nanoparticles supported on MOFs in flow chemistry for catalytic applications. A mixture of aryl halide, boronic acid/ester and K_2CO_3 in water and ethanol was passed through a column of homemade 8 wt % Pd@MIL-101-NH_2 at room temperature for a residence time of 35–40 min (Scheme 17).

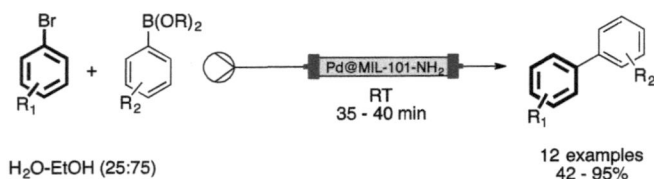

Scheme 17. Suzuki–Miyaura cross coupling in a flow system using Pd MOF.

Variations of the aryl halides and boronic acid derivatives permitted the production of a mini-library (Figure 5). The authors always used the same cartridge for the preparation of the mini-library of biaryl derivatives. It is noticeable that after the reaction, the recovery catalyst was found to be partially crystalline with a remaining Pf content of 6.81 wt % (initially 7.29 wt %) [102].

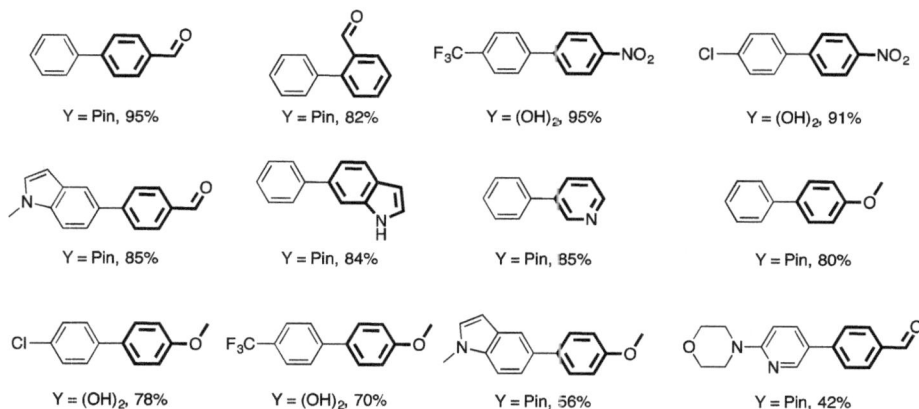

Figure 5. Substrate scope of continuous flow Suzuki–Miyaura cross-coupling sequence using immobilized Pd on MOFs.

A very nice strategy based on dendrimer-encapsulated Pd nanoparticles as catalyst in flow reactor was developed by Verboom [103,104]. In contrast with the conventional heterogeneous Suzuki–Miyaura cross-coupling reaction using cartridge filled with solid catalysts, Verboom's method anchored Pd nanoparticles onto the inner walls of the flow reactor. Aryl halides (10 mM) were mixed with boronic acid derivatives (15 mM) in ethanol at 80 °C using n-Bu$_4$NOH (20 mM) as base at 80 °C. The solution was passed through the catalytic microreactor with a residence time of 13 min (Scheme 18).

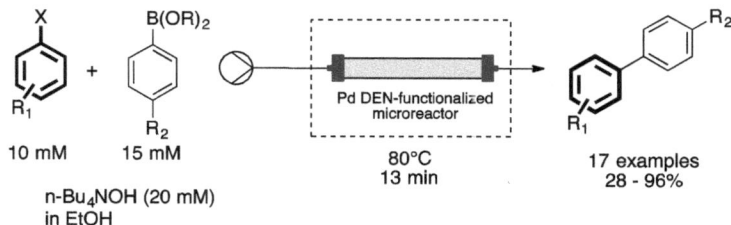

Scheme 18. Suzuki–Miyaura cross coupling in a flow system using dendrimer-encapsulated Pd nanoparticles.

The electronic substituents effects have been studied and different biaryl compounds have been obtained (Figure 6). This strategy demonstrated the influence of dendrimers in the stabilization of the Pd NPs with low metal leaching [103,104].

Another group developed dendrimers for continuous flow Suzuki–Miyaura cross-coupling reaction [105]. Variation was noticeable since the authors described magnetic Fe$_3$O$_4$ fixation of dendron-functionalized iron oxide nanoparticles containing Pd nanoparticles. In this process, the non-covalent magnetic fixation of solid material inside the glass reactor microstructures was applied using external magnetic forces and the reversible immobilization of catalyst materials onto the wall of microchannels was possible. Application of this methodology was realized to produce only one compound using 4-methoxy-1-bromobenzene and boronic acid.

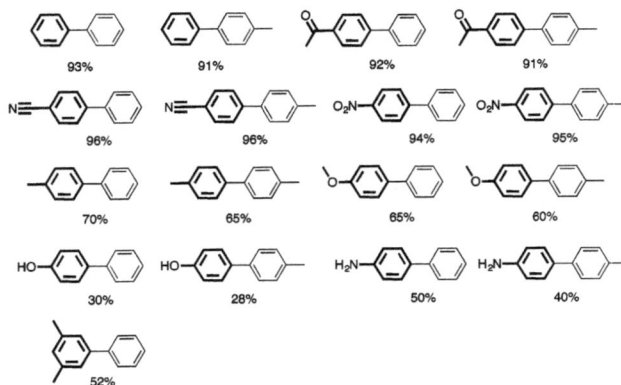

Figure 6. Substrate scope of continuous flow Suzuki–Miyaura cross-coupling sequence using Pd dendrimers encapsulated microreactor.

Another strategy used palladium nanoparticles immobilized in a polymer membrane for the Suzuki–Miyaura cross-coupling reaction but this particular area has not been developed in this review. As examples, some works have been written recently in this field and are just quoted in this review for interested researchers wishing to gain deeper knowledge of the field [106–108].

Alcazar reported an efficient cross-coupling reaction using commercial heterogeneous silica-supported palladium catalyst and a mesoreactor [109]. The authors used a simple and efficient experimental set-up using a 6.6 mm (internal diameter) Omnifit column containing 1 g of heterogeneous catalyst and commercial boronic acids and aryl halides (Scheme 19). A solution of aryl halide in THF and a solution of boronic acid and base in water were pumped at 0.2 mL min^{-1} with two independent pumps. The flow streams met at a T-shaped mixer and then passed through a column containing SiliaCat DPP-Pd as diphenylphosphine palladium (II) heterogeneous catalyst at 60 °C with a residence time of 5 min. A biphasic solvent system such as THF-H$_2$O was used to ensure complete dissolution of any solid and avoiding any subsequent clogging.

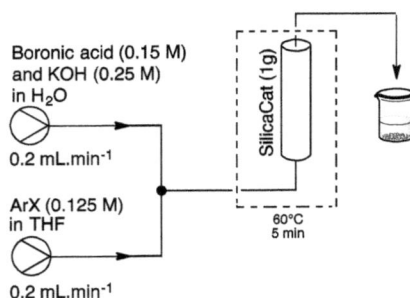

Scheme 19. Continuous flow Suzuki–Miyaura cross-coupling sequence using SiliaCat DPPP-Pd as supported catalyst.

Application of this strategy permitted the synthesis of the biaryl derivatives starting from halides/pseudohalides and (4-methoxyphenyl) boronic acid in excellent yields (Scheme 20) [109]. Whatever the leaving group on the benzene ring, the target biphenyl derivatives were obtained in high yields. Of course, the use of aromatic ring bearing electron-donor groups such as 2,4-dimethoxy analogs gave lower yields (50%). It was notable that bromo- and chloropyridines provided good yields and the ester functionality was tolerated despite the use of KOH as strong base. Using this process, the authors claimed that the crude products are clean and free of phosphine ligand avoiding the need of chromatographic purification. Moreover, low leaching of palladium from the support and the stability of the catalyst after more than 30 cycles was observed.

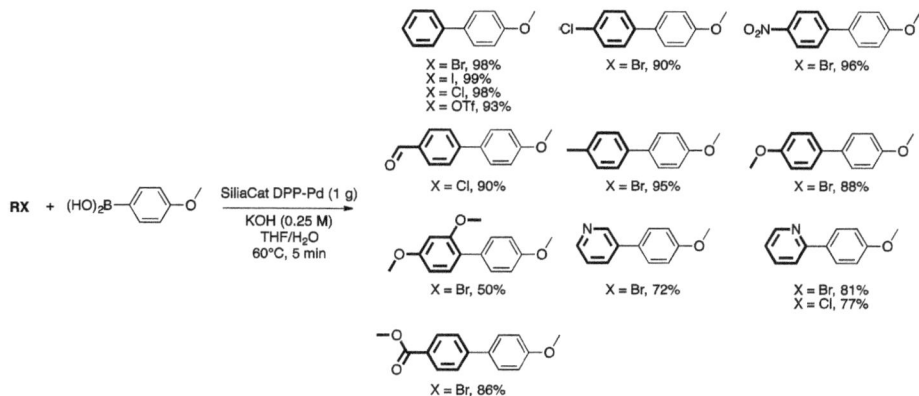

Scheme 20. Substrate scope of continuous flow Suzuki–Miyaura cross-coupling sequence with different aryl bromides and 4-methoxyphenylboronic acid.

In order to further explore the scope of the reaction, Alcazar reported the use of bromobenzene and phenyltriflate as starting materials with different boronic acid derivatives (Scheme 21) [109]. Excellent yields were obtained with commercial boronic acids and boronic ester, borane and borate freshly prepared from the corresponding bromo derivatives by metalation [110].

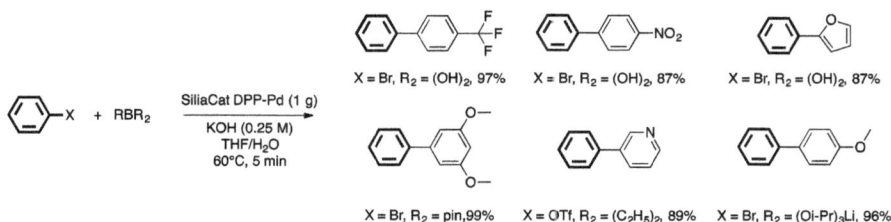

Scheme 21. Substrate scope of continuous flow Suzuki–Miyaura cross-coupling reaction with different aryl bromides/triflate and phenylboronic acid derivatives.

Recently, Kappe presented a comparative investigation of four commercial immobilized phosphine-based Pd catalyst [111]. One of them was SiliaCat DPP-Pd as diphenylphosphine palladium (II) heterogeneous catalyst developed by Pagliaro [112–114]. In this work, the best process used two stock solutions. The first solution contained aryl halide (0.83 M) in THF and the second one phenylboronic acid (0.45 M) and K_2CO_3 (0.55 M) in a mixture of H_2O-EtOH (1:1). These two solutions were pumped in different feeds, 0.055 mL min^{-1} and 0.155 mL min^{-1}, respectively, and mixed in a T-mixer and then introduced to the catalyst cartridge of the X-cube flow reactor at 80 °C (Scheme 22). Under these conditions, full conversion was obtained in less than 20 min and almost quantitative yield of the biaryl target compound was reported.

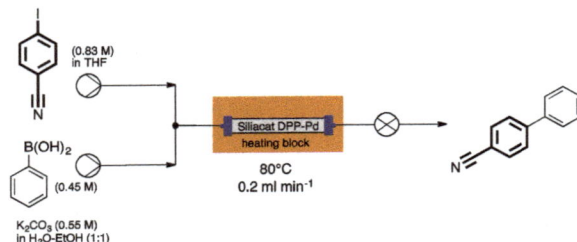

Scheme 22. Suzuki–Miyaura cross coupling in a flow system X-Cube using commercial Siliacat DPP-Pd.

Buarque and Esteves reported an interesting work using heterogeneous catalyst from Covalent Organic Frameworks (COFs) [115], COFs are different than MOFs since COFs do not contain metallic ions or heavy elements as part of their structures. In this study, the authors developed Pd(OAc)$_2$@COF-300 for the Suzuki–Miyaura cross-coupling reaction in continuous flow (Scheme 23). The mixture of bromobenzene, phenylboronic acid in a solution of MeONa (2 M) in MeOH was injected on to a glass column (Omnifit column with a volume of 6.3 mL), which was filled with glass beads (2 mm) and Pd(OAc)$_2$@COF-300 (100 mg). The residence time was 20 min and the temperature was maintained at 60 °C; under these conditions, the maximal conversion was obtained between 20 and 40 min with a very high degree of selectivity.

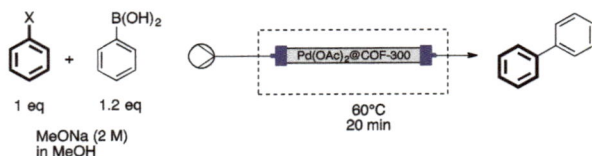

Scheme 23. Suzuki–Miyaura cross coupling in a flow system using Pd(OAc)$_2$@COF-300.

An efficient approach was reported for the production of furan-based biaryls [116]. A mixture of aryl halide, boronic acid derivative and TBAF in methanol (0.37 M) was injected through an X-cube fitted with a FC1032 catalyst at flow rate of 0.5 mL min^{-1} at 120 °C for 2 h (Scheme 24).

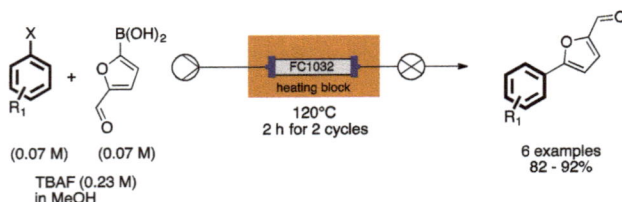

Scheme 24. Suzuki–Miyaura cross coupling in a flow system X-Cube using FC1032 catalyst.

Over to catalyst cycles, the furan derivatives were obtained in good yields (82–92%) using FC1032 catalyst as *t*-butyl based palladium polymer (Figure 7) [116].

The same process was developed after substitution of FC1032 catalyst by PdCl$_2$(PPh$_3$)$_2$ DVB catalyst at flow rate of 0.3 mL min^{-1} at 120 °C for 3 h. It is well known that PdCl$_2$(PPh$_3$)$_2$ DVB catalyst is a more efficient catalyst than FC1032 catalyst. In this regard; starting with the deactivated aryl bromides or aryl chlorides in the presence of PdCl$_2$(PPh$_3$)$_2$ DVB catalyst afforded the target furan-based biaryls in 83–92% yields (Figure 8) [116].

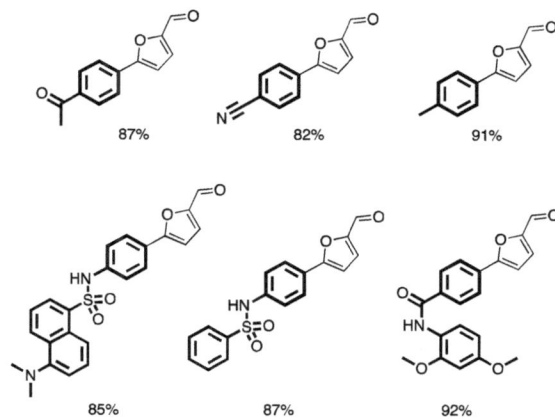

Figure 7. Substrate scope of continuous flow Suzuki–Miyaura cross-coupling sequence using FC1032 catalyst.

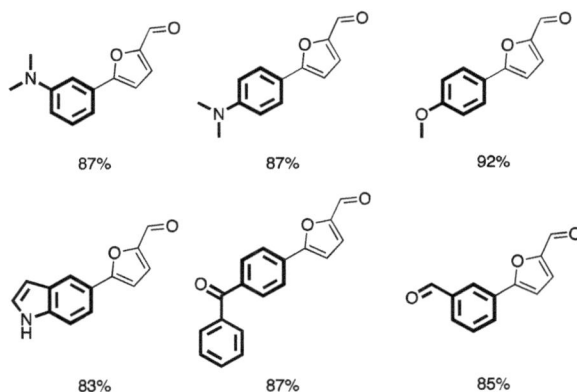

Figure 8. Substrate scope of continuous flow Suzuki–Miyaura cross-coupling sequence using PdCl$_2$(PPh$_3$)$_2$ DVB catalyst.

5. Concluding Remarks

The main focus of this review has been the observance of the continuous flow chemistry and Suzuki–Miyaura cross-coupling reactions. Homogeneous Suzuki–Miyaura cross-coupling reactions have been reported in two different elegant and modular strategies: (i) lithiation/borylation/homogeneous Suzuki–Miyaura sequence using a three-step triphasic flow system; and (ii) diazotization/iododediazotization/homogeneous Suzuki–Miyaura sequence using a three-step triphasic flow system. More examples have been reported in heterogeneous Suzuki–Miyaura cross-coupling reactions. Some groups used Pd(0) as the active catalyst and some groups preferred to start with Pd(II) as precursor of Pd(0). Whatever the type of catalyst, homogeneous, heterogeneous, Pd(II) or Pd(0), the residence times were less than one hour and the Pd loading were low compared with the conversion, yield and selectivity. As mentioned by Kappe, "*palladium which is leached from the support is most likely responsible for the catalysis, thus suggesting a (quasi)homogeneous mechanism*". In this regard, homogeneous metal catalyst/ligand system should probably be more efficient if the recycling of the catalyst could be improved.

Depending on the parameters used (concentrations, temperature, pressure, etc.), the lifetime of all the elements of the process, pumps, pipes and reactors, is longer or shorter. To date, no realistic study has been published on this aspect. Varying the nature of the materials, and the designs of the reactor with the microfluidic system, the possibilities to work in high concentrations are new avenues to explore in the future. Chemists and chemical engineers have the means to pave the way to a more widespread implementation of continuous flow strategies for the production of industrially relevant products in the future. Importantly, we hope that these demonstrated advantages of combining Suzuki–Miyaura cross-coupling reaction and flow processes can stimulate further advances in the field from the younger generations for the benefit of the chemical industry in the future.

Author Contributions: S.B. and F.D. analyzed the data; C.L. and V.S.P. wrote the paper.

Conflicts of Interest: The authors declare no conflict of interest.

References

1.   Yin, L.; Liebscher, J. Carbon–carbon coupling reactions catalyzed by heterogeneous palladium catalysts. *Chem. Rev.* **2007**, *107*, 133–173. [CrossRef] [PubMed]
2.   Barnard, C. Palladium-catalyzed C–C coupling: Then and now. *Platin. Met. Rev.* **2008**, *52*, 38–45. [CrossRef]
3.   Johansson Seechurn, C.C.C.; Kitching, M.O.; Colacot, T.J.; Sniekus, V. Palladium-catalyzed cross-coupling: A historical contextual perspective to the 2010 Nobel Prize. *Angew. Chem. Int. Ed.* **2012**, *51*, 5062–5085. [CrossRef] [PubMed]
4.   Agrofoglio, L.A.; Gillaizeau, I.; Saito, Y. Palladium-assisted routes to nucleosides. *Chem. Rev.* **2003**, *103*, 1875–1916. [CrossRef] [PubMed]
5.   Polshettiwar, V.; Len, C.; Fihri, A. Silica-supported palladium: Sustainable catalysts for cross-coupling reactions. *Coord. Chem. Rev.* **2009**, *253*, 2599–2626. [CrossRef]
6.   Heck, R.F. Acylation, methylation, and carboxyalkylation of olefins by group VIII metal derivatives. *J. Am. Chem. Soc.* **1968**, *90*, 5518–5526. [CrossRef]
7.   Mizoroki, T.; Mori, K.; Ozaki, A. Arylation of olefin with aryl iodide catalyzed by palladium. *Bull. Chem. Soc. Jpn.* **1971**, *44*, 581. [CrossRef]
8.   Heck, R.F.; Nolley, J.P. Palladium-catalyzed vinylic hydrogen substitution reactions with aryls benzyl, and styryl halides. *J. Org. Chem.* **1972**, *37*, 2320–2322. [CrossRef]
9.   Heck, R.F. Palladium-catalyzed reactions of organic halides with olefins. *Acc. Chem. Res.* **1979**, *12*, 146–151. [CrossRef]
10. Dieck, H.A.; Heck, R.F. Organophosphinepalladium complexes as catalysts for vinylic hydrogen substitution reactions. *J. Am. Chem. Soc.* **1974**, *96*, 1133–1136. [CrossRef]
11. Beletskaya, I.P.; Cheprakov, A.V. The Heck reaction as a sharpening stone of palladium catalysis. *Chem. Rev.* **2000**, *100*, 3009–3066. [CrossRef] [PubMed]
12. Miyaura, N.; Yamada, K.; Suzuki, A. A new stereospecific cross-coupling by the palladium-catalyzed reaction of 1-alkenylboranes with 1-alkenyl or 1-alkynyl halides. *Tetrahedron Lett.* **1979**, *20*, 3437–3440. [CrossRef]
13. Miyaura, N.; Yanagi, T.; Suzuki, A. The palladium-catalyzed cross-coupling reaction of phenylboronic acid with haloarenes in the presence of bases. *Synt. Commun.* **1981**, *11*, 513–519. [CrossRef]
14. Miyaura, N.; Suzuki, A. Palladium-catalyzed cross-coupling reactions of organoboron compounds. *Chem. Rev.* **1995**, *95*, 2457–2483. [CrossRef]
15. Suzuki, A. Recent advances in the cross-coupling reactions of organoboron derivatives with organic electrophiles, 1995–1998. *J. Organomet. Chem.* **1999**, *576*, 147–168. [CrossRef]
16. Suzuki, A. Cross-coupling reactions of organoboranes: An easy way to construct C–C bonds. *Angew. Chem. Int. Ed.* **2011**, *50*, 6722–6764. [CrossRef] [PubMed]
17. Sonogashira, K.; Tohda, Y.; Hagihara, N. A convenient synthesis of acetylenes: Catalytic substitutions of acetylenic hydrogen with bromoalkenes, iodoarenes and brompyridines. *Tetrahedron Lett.* **1975**, *16*, 4467–4470. [CrossRef]
18. Paterson, I.; Davies, R.D.; Marquez, R. Total synthesis of the Callipeltoside aglycon. *Angew. Chem. Int. Ed.* **2001**, *40*, 603–607. [CrossRef]

19. Toyota, M.; Komori, C.; Ihara, M. A concise formal total synthesis of Mappicine and Nothapodytine B via an intramolecular hetero Diels-Alder reaction. *J. Org. Chem.* **2000**, *65*, 7110–7113. [CrossRef] [PubMed]

20. Nicolaou, K.C.; Dai, W.M. Chemistry and biology of the enediyne anticancer antibiotics. *Angew. Chem. Int. Ed.* **1991**, *30*, 1387–1416. [CrossRef]

21. Wu, R.; Schumm, J.S.; Pearson, D.L.; Tour, J.M. Convergent synthetic routes to orthogonally fused conjugated oligomers directed toward molecular scale electronic device applications. *J. Org. Chem.* **1996**, *61*, 6906–6921. [CrossRef] [PubMed]

22. Milstein, D.; Stille, J.K. A general, selective, and facile method for ketone synthesis from acid chlorides and organotin compounds catalyzed by palladium. *J. Am. Chem. Soc.* **1978**, *100*, 3636–3638. [CrossRef]

23. Milstein, D.; Stille, J.K. Palladium-catalyzed coupling of tetraorganotin compounds with aryl and benzyl halides. Synthetic utility and mechanism. *J. Am. Chem. Soc.* **1979**, *101*, 4992–4998. [CrossRef]

24. Stille, J.K. The palladium-catalyzed cross-coupling reactions of organotin reagents with organic electrophiles. *Angew. Chem. Int. Ed.* **1986**, *25*, 508–524. [CrossRef]

25. Espinet, P.; Echavarren, A.M. The mechanisms of Stille reaction. *Angew. Chem. Ed. Int.* **2004**, *43*, 4704–4734.

26. Hiyama, T.; Hatanaka, Y. Palladium-catalyzed cross-coupling reaction of organometalloids through activation with fluoride ion. *Pure Appl. Chem.* **1994**, *66*, 1471–1478. [CrossRef]

27. King, A.O.; Okukado, N.; Negishi, E. Highly general stereo-, regio-, and chemo-selective synthesis of terminal and internal conjugated enynes by the Pd-catalyzed reaction of alkynylzinc reagents with alkenyl halides. *J. Chem. Soc. Chem. Commun.* **1977**, 683–684. [CrossRef]

28. Tamao, K.; Sumitani, K.; Kumada, M. Selective carbon-carbon bond formation by cross-coupling of Grignard reagents with organic halides. Catalysis by nickel-phosphine complexes. *J. Am. Chem. Soc.* **1972**, *94*, 4374–4376. [CrossRef]

29. Yamamura, M.; Moritani, I.; Murahashi, S.I. The reaction of σ-vinylpalladium complexes with alkyllithiums. Stereospecific syntheses of olefins from vinyl halides and alkyllithiums. *J. Organomet. Chem.* **1975**, *91*, C39–C42. [CrossRef]

30. Hartwig, J.F. *Handbook of Organopalladium Chemistry in Organic Synthesis*; Negishi, E., Ed.; Wiley-Interscience: New York, NY, USA, 2003; p. 1051.

31. Jiang, L.; Buchwald, S.L. *Metal-Catalyzed Cross-Coupling Reactions*; Meijere, A., Diederich, F., Eds.; Wiley-VCH: Weinheim, Germany, 2004; p. 699.

32. Polshettiwar, V.; Decottignies, A.; Len, C.; Fihri, A. Suzuki-Miyaura cross-coupling reactions in aqueous media: Green and sustainable syntheses of biaryls. *ChemSusChem* **2010**, *5*, 502–522. [CrossRef] [PubMed]

33. Fihri, A.; Luart, D.; Len, C.; Solhi, A.; Chevrin, C.; Polshettiwar, V. Suzuki-Miyaura cross-coupling reactions with low catalyst loading: A green and sustainable protocol in pure water. *Dalton Trans.* **2011**, *40*, 3116–3121. [CrossRef] [PubMed]

34. Sartori, G.; Enderlin, G.; Herve, G.; Len, C. Highly effective synthesis of C-5-substituted 2′-deoxyuridine using Suzuki-Miyaura cross-coupling in water. *Synthesis* **2012**, *44*, 767–772. [CrossRef]

35. Hassine, A.; Sebti, S.; Solhy, A.; Zahouily, M.; Len, C.; Hedhili, M.N.; Fihri, A. Palladium supported on natural phosphate: Catalyst for Suzuki coupling reactions in water. *Appl. Catal. A Gen.* **2013**, *450*, 13–18. [CrossRef]

36. Sartori, G.; Enderlin, G.; Herve, G.; Len, C. New efficient approach for the ligand-free Suzuki-Miyaura reaction of 5-iodo-2′-deoxyuridine in water. *Synthesis* **2013**, *45*, 330–333. [CrossRef]

37. Decottignies, A.; Fihri, A.; Azemar, G.; Djedaini-Pilard, F.; Len, C. Ligandless Suzuki-Miyaura reaction in neat water with or without native β-cyclodextrin as additive. *Catal. Commun.* **2013**, *32*, 101–107. [CrossRef]

38. Gallagher-Duval, S.; Herve, G.; Sartori, G.; Enderlin, G.; Len, C. Improved microwave-assisted ligand free Suzuki-Miyaura cross-coupling of 5-iodo-2′-deoxyuridine in pure water. *New J. Chem.* **2013**, *37*, 1989–1995. [CrossRef]

39. Enderlin, G.; Sartori, G.; Herve, G.; Len, C. Synthesis of 6-aryluridines via Suzuki-Miyaura cross-coupling reaction at room temperature under aerobic ligand-free conditions in neat water. *Tetrahedron Lett.* **2013**, *54*, 3374–3377. [CrossRef]

40. Herve, G.; Sartori, G.; Enderlin, G.; Mackenzie, G.; Len, C. Palladium-catalyzed Suzuki reaction in aqueous solvents applied to unprotected nucleosides and nucleotides. *RSC Adv.* **2014**, *4*, 18558–18594. [CrossRef]

41. Herve, G.; Len, C. First ligand-free, microwave-assisted, Heck cross-coupling reaction in pure water on a nucleoside—Application to the synthesis of antiviral BVDU. *RSC Adv.* **2014**, *4*, 46926–46929. [CrossRef]

42. Lussier, T.; Herve, G.; Enderlin, G.; Len, C. Original access to 5-aryluracils from 5-iodo-2′-deoxyuridine via a microwave-assisted Suzuki-Miyaura cross-coupling/deglycosylation sequence in pure water. *RSC Adv.* **2014**, *4*, 46218–46223. [CrossRef]

43. Hassine, A.; Bouhrara, M.; Sebti, S.; Solhy, A.; Luart, D.; Len, C.; Fihri, A. Natural phosphate-supported palladium: A highly efficient and recyclable catalyst for the Suzuki-Miyaura coupling under microwave irradiation. *Curr. Org. Chem.* **2014**, *18*, 3141–3148. [CrossRef]

44. Haswell, S.J.; Watts, P. Green chemistry: Synthesis in micro reactors. *Green Chem.* **2003**, *5*, 240–249. [CrossRef]

45. Frost, C.G.; Mutton, L. Heterogeneous catalytic syntheis using microreactor technology. *Green Chem.* **2010**, *12*, 1687–1703. [CrossRef]

46. Wiles, C.; Watts, P. Continuous flow reactors: A perspective. *Green Chem.* **2012**, *14*, 38–54. [CrossRef]

47. Newman, S.G.; Jensen, K.F. The role of flow in green chemistry and engineering. *Green Chem.* **2013**, *15*, 1456–1472. [CrossRef]

48. Wiles, C.; Watts, P. Continuous process technology: A tool for sustainable production. *Green Chem.* **2014**, *16*, 55–62. [CrossRef]

49. Vaccaro, L.; Lanari, D.; Marrochi, A.; Strappaveccia, G. Flow approaches towards sustainability. *Green Chem.* **2014**, *16*, 3680–3704. [CrossRef]

50. Falb, S.; Tomaiuolo, G.; Perazzo, A.; Hodgson, P.; Yaseneva, P.; Zakrzewski, J.; Guido, S.; Lapkin, A.; Woodward, R.; Meadows, R.E. A continuous process for Buchwald-Hartwig amination at micro-, lab-, and mesoscale using a novel reactor concept. *Org. Process Res. Dev.* **2016**, *20*, 558–565.

51. Gemoets, H.P.L.; Hessel, V.; Noel, T. Aerobic C–H olefination of indoles via a cross-dehydrogenative coupling in continuous flow. *Org. Lett.* **2014**, *16*, 5800–5803. [CrossRef] [PubMed]

52. Reynolds, W.R.; Plucinski, P.; Frost, C.G. Robust and reusable supported palladium catalysts for cross-coupling reactions in flow. *Catal. Sci. Technol.* **2014**, *4*, 948–954. [CrossRef]

53. Bourne, S.L.; O'Brien, M.; Kasinathan, S.; Koos, P.; Tolstoy, P.; Hu, D.X.; Bates, R.W.; Martin, B.; Schenkel, B.; Ley, S.V. Flow chemistry syntheses of styrenes, unsymmetrical stilbenes and branched aldehydes. *ChemCatChem* **2013**, *5*, 159–172. [CrossRef]

54. Peeva, L.; da Silva Burgal, J.; Vartak, S.; Livingston, A.G. Experimental strategies for increasing the catalyst turnover number in a continuous Heck coupling reaction. *J. Catal.* **2013**, *306*, 190–201. [CrossRef]

55. Sharma, S.; Basavaraju, K.C.; Singh, A.K.; Kim, D.P. Continuous recycling of homogeneous Pd/cu catalysts for cross-coupling reactions. *Org. Lett.* **2014**, *16*, 3974–3977. [CrossRef] [PubMed]

56. Peeva, L.; Arbour, J.; Livingston, A. On the potential of organic solvent nanofiltration in continuous Heck coupling reactions. *Org. Process. Res. Dev.* **2013**, *17*, 967–975. [CrossRef]

57. Domier, R.C.; Moore, J.N.; Shaughnessy, K.H.; Hartman, R.L. Kinetic analysis of aqueous-phase Pd-catalyzed, Cu-free direct arylation of terminal alkynes using a hydrophilic ligand. *Org. Process. Res. Dev.* **2013**, *17*, 1262–1271. [CrossRef]

58. Tukacs, J.M.; Jones, R.V.; Darvas, F.; Dibo, G.; Lezsak, G.; Mika, L.T. Synthesis of γ-valerolactone using a continuous flow reactor. *RSC Adv.* **2013**, *3*, 16283–16287. [CrossRef]

59. Yang, G.R.; Bae, G.; Choe, J.; Lee, S.; Song, K.H. Silica-supported palladium-catalyzed Hiyama cross-coupling reactions using continuous flow system. *Bull. Korean Chem. Soc.* **2010**, *31*, 250–252. [CrossRef]

60. Phan, N.T.S.; Brown, D.H.; Styring, P. A facile method for catalyst immobilization on silica: Nickel-catalyzed Kumada reactions in mini-continuous flow and batch reactors. *Green Chem.* **2004**, *6*, 526–532. [CrossRef]

61. Alonso, N.; Zane Miller, L.; Munoz, J.D.M.; Alcazar, J.; Tyler McQuade, D. Continuous synthesis of organozinc halides coupled to Negishi reactions. *Adv. Synth. Catal.* **2014**, *356*, 3737–3741. [CrossRef]

62. Tan, L.M.; Sem, Z.Y.; Chong, W.Y.; Liu, X.; Hendra; Kwan, W.L.; Ken Lee, C.L. Continuous flow Sonogashira C–C coupling using a heterogeneous palladium-copper dual reactor. *Org. Lett.* **2013**, *15*, 65–67. [CrossRef] [PubMed]

63. Zhang, H.H.; Xing, C.H.; Bouobda Tsemo, G.; Hu, Q.S. *t*-Bu$_3$P-coordinated 2-phenylaniline-based palladacycle complex as a precatalyst for the Suzuki cross-coupling polymerization of aryl dibromides with aryldiboronic acids. *ACS Macro Lett.* **2013**, *2*, 10–13. [CrossRef] [PubMed]

64. Schulte, N.; Breuning, E.; Spreitzer, H. Method for the Production of Polymers. U.S. Patent 20,080,207,851 A1, 28 December 2004.

65. Seyler, H.; Jones, D.J.; Holmes, A.B.; Wong, W.W.H. Continuous flow synthesis of conjugated polymers. *Chem. Commun.* **2012**, *48*, 1598–1600. [CrossRef] [PubMed]

66. Gao, M.; Subbiah, J.; Geraghty, P.B.; Chen, M.; Purushothamar, B.; Chen, X.; Qin, T.; Vak, D.; Scholes, F.H.; Watkins, S.E.; et al. Development of a high-performance doncr-acceptor conjugated polymer: Synergy in materials and device optimization. *Chem. Mater.* **2016**, *28*, 3481–3487. [CrossRef]

67. Mitchell, V.D.; Wong, W.W.H. *Synthetic Methods for Conjugated Polymer and Carbon Materials*; Leclerc, M., Morin, J.-F., Eds.; Wiley-VCH: Weinheim, Germany, 2017; p. 65.

68. Amatore, C.; Jutand, A.; Le Duc, G. Kinetic data for the transmetallation/reductive elimination in palladium-catalyzed Suzuki-Miyaura reactions: Unexpected triple role of hydroxide ions used as base. *Chem. Eur. J.* **2011**, *17*, 2492–2503. [CrossRef] [PubMed]

69. Amatore, C.; Jutand, A.; Le Duc, G. Mechanistic origin of antagonist effects of usual anionic bases (OH$^-$, CO$_3{}^{2-}$) as modulated by their countercations (Na$^+$, Cs$^+$, K$^+$) in palladium-catalyzed Suzuki-Miyaura reactions. *Chem. Eur. J.* **2012**, *18*, 6616–6625. [CrossRef] [PubMed]

70. Carrow, B.P.; Hartwig, J.F. Distinguishing between pathways for transmetalation in Suzuki-Miyara reactions. *J. Am. Chem. Soc.* **2011**, *133*, 2116–2119. [CrossRef] [PubMed]

71. Cantillo, D.; Kappe, C.O. Immobilized Transition Metals as Catalysts for Cross-Couplings in Continuous Flow—A Critical Assessment of the Reaction Mechanism and Metal Leaching. *ChemCatChem* **2014**, *6*, 3286–3305. [CrossRef]

72. Narayanan, R.; El-Sayed, M.A. Effect of catalysis on the stability of metallic nanoparticles: Suzuki reaction catalyzed by PVP-palladium nanoparticles. *J. Am. Chem. Soc.* **2003**, *125*, 8340–8347. [CrossRef] [PubMed]

73. Narayanan, R.; El-Sayed, M.A. Effect of catalytic activity on the metallic nanoparticle size distribution: Electron-transfer reaction between Fe(CN)$_6$ and thiosulfate ions catalyzed by PVP-platinium nanoparticles. *J. Phys. Chem. B* **2003**, *107*, 12416–12424. [CrossRef]

74. De Vries, A.H.M.; Mulders, J.; Mommers, J.H.M.; Henderickx, H.J.W.; De Vries, J.G. Homeopathic ligand-free palladium as a catalyst in the Heck reaction. A comparison with a palladacycle. *Org. Lett.* **2003**, *5*, 3285–3288. [CrossRef] [PubMed]

75. De Vries, J.G. A unifying mechanism for all high-temperature Heck reactions. The role of palladium colloids and anionic species. *Dalton Trans.* **2006**, *21*, 421–429. [CrossRef] [PubMed]

76. Zhao, F.; Bhanage, B.M.; Shirai, M.; Arai, M. Heck reactions of iodobenzene and methyl acrylate with conventional supported palladium catalysts in the presence of organic and/and inorganic bases without ligands. *Chem. Eur. J.* **2000**, *6*, 843–848. [CrossRef]

77. Bhanage, B.M.; Shirai, M.; Arai, M. Heterogeneous catalyst system for Heck reaction using supported ethylene glycol phase Pd/TPPTS catalyst with inorganic base. *J. Mol. Catal. A* **1999**, *145*, 69–74. [CrossRef]

78. Reetz, M.T.; Westermann, E. Phosphane-free palladium-catalyzed coupling reactions: The decisive role of Pd nanoparticles. *Angew. Chem. Int. Ed.* **2000**, *39*, 165–168. [CrossRef]

79. Reetz, M.T.; Helbig, W.; Quasier, S.A.; Stimming, U.; Breuer, N.; Vogel, R. Visualization of surfactants on nanostructured palladium clusters by a combination of STM and high-resolution TEM. *Science* **1995**, *267*, 367–369. [CrossRef] [PubMed]

80. Thathagar, M.B.; Ten Elshof, J.E.; Rothenberg, G. Pd nanoclusters in C–C coupling reactions: Proof of leaching. *Angew. Chem. Int. Ed.* **2006**, *45*, 2886–2890. [CrossRef] [PubMed]

81. Gaikwad, A.V.; Holuigue, A.; Thathagar, M.B.; Ten Elhof, J.E.; Rothenberg, G. Ion- and atom-leaching mechanisms from palladium nanoparticles in cross-coupling reactions. *Chem. Eur. J.* **2007**, *13*, 6908–6913. [CrossRef] [PubMed]

82. Ananikov, V.P.; Beletskaya, I.P. Toward the ideal catalyst: From atomic centers to a "cocktail" of catalysts. *Organometallics* **2012**, *31*, 1595–1604. [CrossRef]

83. Kashin, A.S.; Ananikov, V.P. Catalytic C–C and C–heteroatom bond formation reactions: In Situ generated of preformed catalysts? Complicated mechanistic picture behing well-known experimental procedures. *J. Org. Chem.* **2013**, *78*, 11117–11125. [CrossRef] [PubMed]

84. Shu, W.; Pellegatti, L.; Oberli, M.A.; Buchwald, S.L. Continuous-flow synthesis of biaryls enabled by multistep solid-handling in a lithiation/borylation/Suzuki-Miyaura cross-coupling sequence. *Angew. Chem. Int. Ed.* **2011**, *50*, 10665–10669. [CrossRef] [PubMed]

85. Kabri, Y.; Gellis, A.; Vanelle, P. Synthesis of original 2-substituted 4-arylquinazolines by microwave-irradiated Suzuki-Miyaura cross-coupling reactions. *Eur. J. Org. Chem.* **2009**, *2009*, 4059–4066. [CrossRef]

86. Gill, G.S.; Grobelny, D.W.; Chaplin, J.H.; Flynn, B.L. An efficient synthesis and substitution of 3-aroyl-2-bromobenzo[*b*]furans. *J. Org. Chem.* **2008**, *73*, 1131–1134. [CrossRef] [PubMed]

87. Organ, M.; Calimsiz, S.; Sayah, M.; Hoi, K.; Lough, A. Pd-PEPPSI-IPent: An active, sterically demanding cross-coupling catalyst and its application in the synthesis of tetra-ortho-substituted biaryls. *Angew. Chem. Int. Ed.* **2009**, *48*, 2383–2387. [CrossRef] [PubMed]

88. Dang, T.; Chen, Y. One-pot oxidation and bromination of 3,4-diaryl-2,5-dihydrothiophenes using Br_2: Synthesis and application of 3,4-diaryl-2,5-dibromothiophenes. *J. Org. Chem.* **2007**, *72*, 6901–6904. [CrossRef] [PubMed]

89. Maeda, H.; Haketa, Y.; Nakanishi, T. Aryl-substituted C3-bridged oligopyrroles as anion receptors for formation of supramolecular organogels. *J. Am. Chem. Soc.* **2007**, *129*, 13661–13674. [CrossRef] [PubMed]

90. James, C.A.; Coelho, A.L.; Gevaert, M.; Forgione, P.; Snieckus, V. Combined directed ortho and remote metalation-Suzuki cross-coupling strategies. Efficient synthesis of heteroaryl-fused benzopyrannones from biaryl O-carbamates. *J. Org. Chem.* **2009**, *74*, 4094–4103. [CrossRef] [PubMed]

91. Giodiano, C.; Coppi, L.; Minisci, F. Process for the Preparation of 5-(2,4-difluorophenyl)-salicylic Acid. U.S. Patent 5,312,975, 17 May 1994.

92. Hannah, J.; Ruyle, W.V.; Jones, H.; Matzuk, A.R.; Kelly, K.W.; Witzel, B.E.; Holtz, W.J.; Houser, R.A.; Shen, T.Y.; Sarett, L.H. Novel analgesic-antiinflammatory salicylates. *J. Med. Chem.* **1978**, *21*, 1093–1100. [CrossRef] [PubMed]

93. Reizman, B.J.; Wang, Y.M.; Buchwald, S.L.; Jensen, K.F. Suzuki-Miyaura cross-coupling optimization enabled by automated feedback. *React. Chem. Eng.* **2016**, *1*, 658–666. [CrossRef] [PubMed]

94. Teci, M.; Tilley, M.; McGuire, M.A.; Organ, M.G. Using anilines as masked cross-coupling partners: Design of a telescoped three-step flow diazotization, iododediazotization, cross-coupling process. *Chem. Eur. J.* **2016**, *22*, 1–10. [CrossRef] [PubMed]

95. Ormerod, D.; Lefevre, N.; Dorbec, M.; Eyskens, I.; Vloemans, P.; Duyssens, K.; Diez de la Torre, V.; Kaval, N.; Merkul, E.; Sergeyev, S.; et al. Potential of homogeneous Pd catalyst separation by ceramic membranes. Application to downstream and continuous flow processes. *Org. Process Res. Dev.* **2016**, *20*, 911–920. [CrossRef]

96. Cole, K.P.; Campbell, B.M.; Forst, M.B.; McClary Groh, J.; Hess, M.; Johnson, M.D.; Miller, R.D.; Mitchell, D.; Polster, C.S.; Reizman, B.J.; et al. An automated intermittent flow approach to continuous Suzuki coupling. *Org. Process Res. Dev.* **2016**, *20*, 820–830. [CrossRef]

97. Hattori, T.; Tsubone, A.; Sawama, Y.; Monguchi, Y.; Sajiki, H. Palladium on carbon-catalyzed Suzuki-Miyaura coupling reaction using an efficient and continuous flow system. *Catalysts* **2015**, *5*, 18–25. [CrossRef]

98. Brinkley, K.W.; Burkholder, M.; Siamaki, A.R.; Belecki, K.; Gupton, B.F. The continuous synthesis and application of graphene supported palladium nanoparticles: A highly effective catalyst for Suzuki-Miyaura cross-coupling reactions. *Green Process Synth.* **2015**, *4*, 241–246. [CrossRef]

99. Bolton, K.F.; Canty, A.J.; Deverell, J.A.; Guijt, R.M.; Hilder, E.F.; Rodemann, T.; Smith, J.A. Macroporous monolith supports for continuous flow capillary microreactors. *Tetrahedron Lett.* **2006**, *47*, 9321–9324. [CrossRef]

100. Nagaki, A.; Hirose, K.; Moriwaki, Y.; Mitamura, K.; Matsukawa, K.; Ishizuka, N.; Yoshida, J. Integration of borylation of aryllithiums and Suzuki-Miyaura coupling using monolithic Pd catalyst. *Catal. Sci. Technol.* **2016**, *6*, 4690–4694. [CrossRef]

101. Shil, A.K.; Guha, N.R.; Sharma, D.; Das, P. A solid supported palladium(0) nano/microparticle catalyzed ultrasound induced continuous flow technique for large scale Suzuki reactions. *RSC Adv.* **2013**, *3*, 13671–13676. [CrossRef]

102. Pascanu, V.; Hansen, P.R.; Bermejo-Gomez, A.; Ayats, C.; Platero-Prats, A.E.; Johansson, M.J.; Pericas, M.A.; Martin-Matute, B. Highly functionalized biaryls via Suzuki-Miyaura cross coupling catalyzed by Pd@MOF under batch and continuous flow regimes. *ChemSusChem* **2015**, *8*, 123–130. [CrossRef] [PubMed]

103. Ricciardi, R.; Huskens, J.; Verboom, W. Dendrimer-encapsulated Pd nanoparticles as catalysts for C–C cross-couplings in flow microreactors. *Org. Biomol. Chem.* **2015**, *13*, 4953–4959. [CrossRef] [PubMed]

104. Ricciardi, R.; Huskens, J.; Holtkamp, M.; Karst, U.; Verboom, W. Dendrimer-encapsulated palladium nanoparticles for continuous-flow Suzuki-Miyaura cross-coupling reactions. *ChemCatChem* **2015**, *7*, 936–942. [CrossRef]

105. Rehm, T.H.; Bogdan, A.; Hofmann, C.; Lob, P.; Shifrina, Z.; Morgan, D.G.; Bronstein, L.M. Proof of concept: Magnetic fixation of dendron-functionalized iron oxide nanoparticles containing Pd nanoparticles for continuous-flow Suzuki coupling reactions. *ACS Appl. Mater. Interfaces* **2015**, *7*, 27254–27261. [CrossRef] [PubMed]

106. Seto, H.; Yoneda, T.; Morii, T.; Hoshino, Y.; Miura, Y. Membrane reactor immobilized with palladium-moaded polymer nanogel for continuous-flow Suzuki coupling reaction. *AIChE J.* **2015**, *61*, 582–589. [CrossRef]

107. Gu, Y.; Favier, I.; Pradel, C.; Gin, D.L.; Lahitte, J.F.; Noble, R.D.; Gomez, M.; Remigny, J.C. High catalytic efficiency of palladium nanoparticles immobilized in a polymer membrane containing poly(ionic liquid) in Suzuki-Miyaura cross-coupling reaction. *J. Membr. Sci.* **2015**, *492*, 331–339. [CrossRef]

108. Dai, Y.; Formo, E.; Li, H.; Xue, J.; Xia, Y. Surface-functionalized electrospun titania nanofibers for the scavenging and recycling of precious metal ions. *ChemSusChem* **2016**, *9*, 2912–2916. [CrossRef] [PubMed]

109. Munoz, J.D.M.; Alcazar, J.; de la Hoz, A.; Diaz-Ortiz, A. Cross-coupling in flow using supported catalysts: Mild, clean, efficient and sustainable Suzuki-Miyaura coupling in a single pass. *Adv. Synth. Catal.* **2012**, *354*, 3456–3460. [CrossRef]

110. Noel, T.; Kuhn, S.; Musacchio, A.J.; Jensen, K.F.; Buchwald, S.L. Suzuki-Miyaura cross-coupling reactions in flow: Multistep synthesis enabled by a microfluidic extraction. *Angew. Chem. Int. Ed.* **2011**, *50*, 5943–5946. [CrossRef] [PubMed]

111. Greco, R.; Goessler, W.; Cantillo, D.; Kappe, C.O. Benchmarking immobilized Di- and Triaryphosphine palladium catalysts for continuous-flow cross-coupling reactions: Efficiency, durability, and metal leaching studies. *ACS Catal.* **2015**, *5*, 1303–1312. [CrossRef]

112. Pandarus, V.; Gingras, G.; Beland, F.; Ciriminna, R.; Pagliaro, M. Process intensification of the Suzuki-Miyaura reaction over sol-gel entrapped catalyst SiliaCat DPP-Pd under conditions of continuous flow. *Org. Process Res. Dev.* **2014**, *18*, 1550–1555. [CrossRef]

113. Pandarus, V.; Gingras, G.; Beland, F.; Ciriminnia, R.; Pagliaro, M. Fast and clean borylation of aryl halides under flow using sol-gel entrapped SiliaCat DPP-Pd. *Org. Process Res. Dev.* **2014**, *18*, 1556–1559. [CrossRef]

114. Pandarus, V.; Ciriminnia, R.; Gingras, G.; Beland, F.; Drobod, M.; Jina, O.; Pagliaro, M. Greening heterogeneous catalysis for fine chemicals. *Tetrahedron Lett.* **2013**, *54*, 1129–1132. [CrossRef]

115. Goncalves, R.S.B.; de Oliveira, A.B.V.; Sindra, H.C.; Archarjo, B.S.; Mendoza, M.E.; Carneiro, L.S.A.; Buarque, C.D.; Esteves, P.M. Heterogeneous catalysis by covalent organic frameworks (COF): Pd(OAc)$_2$@COF-300 in cross-coupling reactions. *ChemCatChem* **2016**, *8*, 743–750. [CrossRef]

116. Trinh, T.N.; Hizartzidis, L.; Lin, A.J.S.; Harman, D.G.; McCluskey, A.; Gordon, C.P. An efficient continuous flow approach to furnish furan-based biaryls. *Org. Biomol. Chem.* **2014**, *12*, 9562–9571. [CrossRef] [PubMed]

catalysts

MDPI

Review

Eco-Friendly Physical Activation Methods for Suzuki–Miyaura Reactions

Katia Martina, Maela Manzoli, Emanuela Calcio Gaudino and Giancarlo Cravotto *

Dipartimento di Scienza e Tecnologia del Farmaco and NIS—Centre for Nanostructured Interfaces and Surfaces, University of Turin, Via P. Giuria 9, Turin 10125, Italy; katia.martina@unito.it (K.M.); maela.manzoli@unito.it (M.M.); emanuela.calcio@unito.it (E.C.G.)
* Correspondence: giancarlo.cravotto@unito.it; Tel.: +39-011-670-7684

Academic Editor: Ioannis D. Kostas
Received: 31 December 2016; Accepted: 16 March 2017; Published: 23 March 2017

Abstract: Eco-compatible activation methods in Suzuki–Miyaura cross-coupling reactions offer challenging opportunities for the design of clean and efficient synthetic processes. The main enabling technologies described in the literature are microwaves, ultrasound, grinding (mechanochemistry) and light. These methods can be performed in water or other green solvents with phase-transfer catalysis or even in solventless conditions. In this review, the authors will summarize the progress in this field mainly from 2010 up to the present day.

Keywords: Suzuki–Miyaura cross-coupling reaction; Pd catalysts; microwaves; ultrasound; ball milling; light

1. Introduction

The present review deals with the use of non-conventional methods for Suzuki–Miyaura reactions (SMC) [1]. In particular, the results obtained with different unconventional techniques were described by following a Green Chemistry approach, and focusing on recent methodologies based on microwave (MW) and ultrasound (US) irradiations, grinding and photo-activated processes. We aimed at describing several original works published in the field after 2010; however, we also took on board previous papers which can be considered seminal works on the topic and which therefore cannot be ignored. Here, the main advantages and drawbacks of the abovementioned methodologies will be comprehensively covered, serving as a guide for further research on innovative SMC transformations under unconventional activation. Moreover, due to length limits, these topics have been briefly explored, and readers are encouraged to refer to the cited reviews for in depth dissertation.

In particular, MW-assisted chemistry relies on the ability of the reaction mixture to efficiently absorb MW energy, producing rapid internal heating (in-core volumetric heating) by the direct interaction of electromagnetic irradiation with the molecules in the reaction mixture. Although MW irradiation are currently applied as non conventional method to promote fast chemical transformations, there has been considerable speculation on their own effect. Indeed, the debate is still open and focused on discerning between thermal effects due to the rapid heating and high bulk reaction temperatures reached under MW dielectric heating, and other specific or nonthermal microwave effects. These effects, which are not linked to a macroscopic change in reaction temperature, have been hypothesized to originate from a direct interaction of the electromagnetic field with specific molecules, intermediates, or even transition states in the reaction medium [2].

Mechanochemistry is a branch of solid-state chemistry where intramolecular bonds are broken by mechanical action followed by further chemical reactions. At the same time, mechanochemistry can also generate radicals via the breaking of weak bonds and under extreme surface plasma conditions generated by mechanical impact. However, mechanochemistry is not usually be applied to liquid or to solid–liquid reactions but in this frame, sonication could represent a viable alterantive by virtue of the mechanical and chemical events occurring in liquid medium via the unique phenomenon of acoustic cavitation: the rapid nucleation, growth and collapse of micrometer-scale bubbles generate in the sonicated liquid.

Cavitation, that is a non linear phenomenon that generally depend on external parameters such as frequency, intensity and nature of sonicated solvents, could act by means of two classes of different effects: radical and mechanical ones. The former arises from the sonolysis of molecules which can occur mainly at the bubble interface as well as in the interior cavity. While the latter effect, that follow cavity collapse, originates from shear forces, microjets and shock waves that occur outside the bubble, resulting in profound physical changes when solids or metals are present. Both effects are responsible for speeding up organic transformations thanks to the so-called "mass transfer effect" induced by reaction mixture sonication [3].

Among the above mentioned innovative technologies, sunlight is an abundant and easily accessible energy supply and therefore it can display great potential to activate environmentally benign organic transformations. The use of light would favour chemical synthesis at room temperature, as well as it would avoid thermally induced side reactions. However, this approach suffers from two main drawbacks: the high energy ultraviolet (UV) component is often required to drive the reactions, and usually photochemical reactions give low selectivity to the desired products. Photocatalytic reactions, which can be performed under mild conditions to obtain high product selectivity can easily overcome these disadvantages [4].

This review encompasses four main sections specifically dedicated to the different activation methodology type: MW, US, grinding and light. Each section guides the readers through a series of reactions—from the most common reaction to the more complex one—while examining reaction conditions and peculiar features. The discussion is directed to the importance of combining the homogeneous or heterogeneous catalyst (mainly Pd-based catalyst) with the unconventional tecniques in order to overcome the current limitations of organochemistry.

Due to the variety of nanostructured materials and substrates, it has been very difficult to rationalize the order chosen to describe the reported examples. After a brief paragraph on the SMC reaction without any activation procedure, the discussion is focused on the MW-assisted reaction carried out in the presence of an homogeneous and then an heterogeneous catalyst, with the aim to give an overall description on the effects observed on the reactivity (in terms of yields, reaction time and temperature and so on) under MW irradiation. A very short paragraph has been dedicated to Ni catalyzed couplings. Moreover, some information on the effect of MW on the heterogeous catalysts have been also reported. A special section has been dedicated to the preparation of heterogeneous catalysts for SMC reaction by using MW. Then the discussion moves to the US-assisted SMC and to the use of US to synthesise heterogeneous catalysts for SMC reaction and for MW-assisted SMC reaction. At this point, mechanochemical activation is described and the overview finishes with the use of light to promote the SMC reaction in the presence of opportunely designed photocatalytic systems. The conclusion section gives a description of the main open questions and of the opportunities given by the innovative methodologies in the future scenario.

A Table of contents, that summarises how the review is organized, is here provided for the sake of clarity:

1. Introduction
Some Insights into the Suzuki–Miyaura reaction
2. Microwave-Assisted SMC Reactions
2.1. MW Irradiation in Homogeneous Catalyzed SMC
2.1.1. SMC from Arene to Heterocycle Decoration
2.1.2. Organotrifluoroborates in MW-Assisted SMC
2.1.3. MW-Assisted One-Pot Protocols

2.1.4. Ligands in MW-Assisted SMC

2.1.5. MW Promoted Ni Catalyzed SMC

2.2. Moving from Homogeneous to Heterogeneous Catalysis in MW Promoted SMC

2.3. Solid Supported Pd Catalyzed MW Promoted SMC

Use of MWs for the Synthesis of the Catalysts for the SMC Reaction

3. Ultrasound-Assisted SMC Reactions

3.1. Ultrasound-Assisted SMC Reactions

3.2. Ultrasound Assisted Heterogeneous Preparation of Catalysts Suitable for SMC Reactions

3.3. Ultrasound Assisted Heterogeneous Catalyst Preparation for MW Promoted SMC Reactions

4. Mechanochemical Activation of SMC

5. When the SMC Reaction is Driven by Light

6. Conclusions and Future Perspectives

Some Insights into the Suzuki–Miyaura Reaction

The Suzuki–Miyaura cross-coupling reaction (SMC hereafter) was first reported in 1979 and was rapidly adopted by most synthetic labs as the method of choice for the formation of aliphatic and aromatic C–C bonds [5–8]. The coupling of organoboron reagents with organic halides or pseudohalides in the presence of a palladium (or nickel) catalyst and a base is now a classic procedure, even at the industrial level, for the synthesis of fine chemicals, natural products, pharmaceuticals and polymers [9,10]. The literature bears witness to the enormous impact that this reaction has had, as documented by the 700 reviews published. Furthermore, every year of the last decade has seen around 50–60 surveys highlight the versatility and the peculiar advantages of SMC. Of these advantages, low toxicity, easy access to organoboron reagents, their chemoselectivity and mild reaction conditions are perhaps most significant [11]. Greener protocols have been widely described, several of which are carried out in aqueous solvents thanks to the stability of boronic acids in water [12–14], however, the poor solubility of substrates and the low stability of metal catalysts in water mean that phase-transfer catalysts and water-soluble phosphine ligands have been used [15]. Transition metal nanoparticles (e.g., Pd, Fe) can be deposited onto the surface of the mesoporous nanocomposites through a mechanochemical protocol recently SMC have been performed in the solid-state under mechanochemical activation and a green approach to protocol optimization evaluates the catalyst recycling [16]. In this review, the most important eco-compatible activation methods in SMC, including microwaves (MW), ultrasound, grinding (mechanochemistry), light, and the main advances made over last few years will be described. The use of non-conventional energy sources is a challenging topic that has paved the way for new opportunities in the design of clean and efficient synthetic processes.

2. Microwave-Assisted SMC Reactions

2.1. MW Irradiation in Homogeneous Catalyzed SMC

2.1.1. SMC from Arenes to Heterocycles

MW-assisted organic syntheses have recently been receiving ever increasing attention as they are seen as a viable alternative to conventional heating. It enables the rapid optimization of procedures to be carried out without the need for direct contact between chemical reactants and the energy source. The main advantage of MW irradiation is fast volumetric heating. Differences in solvent and reactant dielectric constants mean that selective dielectric heating can furnish notable enhancement in the directly transfer of energy to the reactants, which induces an instantaneous internal temperature increase [17]. Recent years have seen the use of MW-induced heating become one of the most efficient procedures for carrying out organic syntheses and, in the last 20 years, the scientific community has focused on the search for MW assisted procedures to carry out environmentally friendly SMC [18,19]. Moreover, the high versatility of SMC is becoming more prominent thanks to the number of commercially available reagents and several SMC procedures have been already reported in the *Journal of Chemical Education* [20–22]. A valuable example describes MW-assisted methods for the preparation of 5-phenyl-2-hydroxyacetophenone derivatives in water with K_2CO_3, tetrabutylammonium bromide (TBAB) and Pd(OAc)$_2$ (15 min irradiation at 400 W, 150 °C) [23].

The haloaromatic electrophilic species used in metal-catalyzed processes are commonly aryl iodides or bromides as the catalyst can undergo oxidative addition into fairly weak carbon-iodine or carbon-bromine bonds quite easily, while aryl chlorides are employed more rarely as the electrophilic coupling partner is much less reactive due to the higher carbon-chlorine bond strength. SMC can also be performed with aryl pseudohalides such as sulfonates (such as triflates, mesilates and tosilates) affording to biaryl under mild condition.

The low reactivity of chloro-arenes as reaction partners in cross-coupling reactions means that advances in these protocols have only been obtained in presence of various palladium complexes and using ingenious and dedicated ligands. The efficacy of MW irradiation in this protocol has been demonstrated by Bjørsvik et al. who developed a MW-promoted C–C coupling between highly substituted and congested 1-chloro-2-nitrobenzene and phenylboronic acid (Scheme 1) [24]. The reactions were performed in MeOH and water (4:1), using Pd(PPh$_3$)$_4$ as the catalyst, Na$_2$CO$_3$ and TBAB for 30 min under MW irradiation at 120 °C.

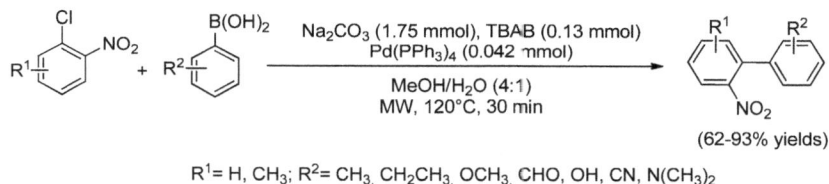

Scheme 1. Microwave (MW)-promoted C–C coupling between highly substituted and congested 1-chloro-2-nitrobenzene and phenylboronic acid.

For the same reasons to those that have already been discussed, analogous reactions involving metal-catalyzed C–F activation are rarer still due to the very strong carbon-fluorine bond. Nevertheless, different nickel catalyzed SMC [25] and some successful reactions involving cobalt [26], platinum [27] and titanium [28] species have been reported. There remain only a few other examples of analogous Pd-catalyzed syntheses [29]. In fact, Cargill et al. have reported [30] the first example of a Pd-catalyzed SMC reaction of perfluoroaromatic systems under MW irradiaion. They described how highly fluorinated nitrobenzene derivatives undergo regioselective SMC reactions via the Pd(0) catalyst's insertion into a C–F bond located *ortho* to the nitro group. The arylation of pentafluoronitrobenzene is the synthetic strategy used to obtain the previously unreported 2,3,4,5-tetrafluoro-6-nitrobiphenyl derivatives. Tetrafluoronitrobenzene and trifluoronitrobenzene systems are less reactive than pentafluoronitrobenzene, by virtue of the corresponding decrease in the aromatic ring electrophilicity. The MW irradiation for this SMC process was applied across a rapid and reproducible heating profile. C–F activation was ensured by a catalytic cycle in which the nitro group directs the nucleophilic Pd center to the contiguous C–F bond which also explains the *ortho* regiospecific arylation. Of the various catalytic systems tested, Pd(0) catalysts were noted to be far more efficent than Pd(II) analogues, especially in polar solvents. Pentafluoronitrobenzene was effectively coupled to different boronic acids and esters bearing both electron-withdrawing and -donating groups and giving yields of 53%–80% under MW irradiation (15 min) in the presence of Pd(PPh)$_4$ (5 mol %) (Scheme 2).

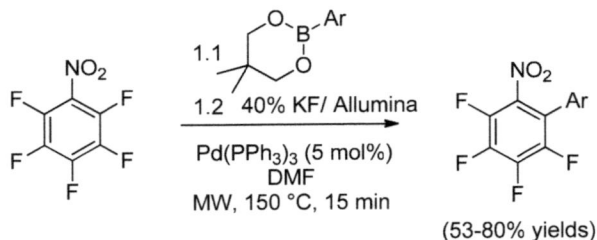

Scheme 2. Pd catalyzed MW promoted SMC of pentafluoronitrobenzene.

Arylation is an approach that affords structural modifications to quinones and naphthoquinones as well as being able to synthesize new substances with biological activity. Besides a few reports related to the arylation of unprotected 2-hydroxy-1,4-naphthoquinones, an interesting green approach has been published by Louvis et al. [31]. A set of 3-aryl-2-hydroxy-1,4-naphthoquinone analogues of atovaquone were synthesized from 3-iodo-2-hydroxy-1,4-naphthoquinone under aqueous conditions either using conventional heating or MW irradiation and phosphine-free sources of Pd (Scheme 3). MW irradiation improved the reaction rate (10 min versus 6 h) and good results were obtained using a lower loading of palladium (1% with irradiation versus 5% with conventional, $P = 300$ W, $T = 120$ °C, 10 min). In fact, intensive deiodination to obtain the Lawsone derivative occurred when 5% Pd(OAc)$_2$ was used, but side product production was surprisingly limited when using MW heating and a lower amount of catalyst. Furthermore, the authors also obtained the greatest conversion from 3-bromo-2-hydroxy-1,4-naphthoquinone with 5% of the Pd(OAc)$_2$ after 25 min at 120 °C under MW irradiation; 40% yield was obtained using phenylboronic acid.

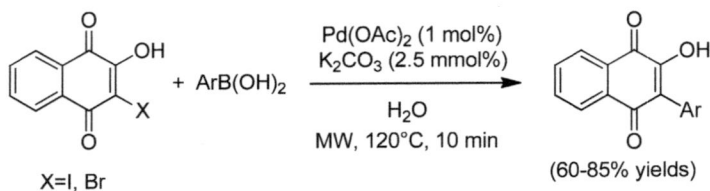

Scheme 3. Synthesis of 3-aryl-2-hydroxy-1,4-naphthoquinone analogues.

Biaryl chromone can be obtained in two steps route by performing a MW promoted SMC from 5-bromo-2-hydroxyacetophenone followed by condensation reaction with ethyl oxalate and intramolecular cyclization in HCl in acetic acid (Scheme 4).The Suzuki reaction gave high yields (73%–82%) when performed with a series of boronic acids in water with TBAB, K$_2$CO$_3$ and Pd(OAc)$_2$ at 150 °C for 15 min [32].

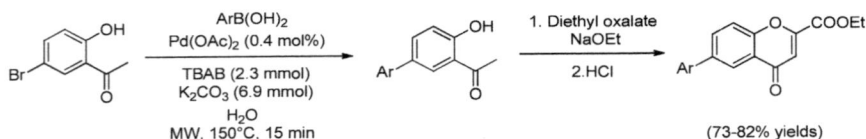

Scheme 4. Synthetic strategy used for the obtention of ethyl-6-phenyl-4-oxo-4*H*-chromene-2-carboxylate.

3-bromo-4*H*-chromen-4-one derivatives were also successfully arylated under MW irradiation at 60 °C in ethanol using K_2CO_3 and TBAB with 0.3 mol % of the Pd-complex derived from the chromen-4-one and naphthalene-1,2-diamine, (Scheme 5). The ligand was synthesized by the authors and a series of 24 different chromen-4-ones were obtained via SMC with excellent yields (81%–95%) [33].

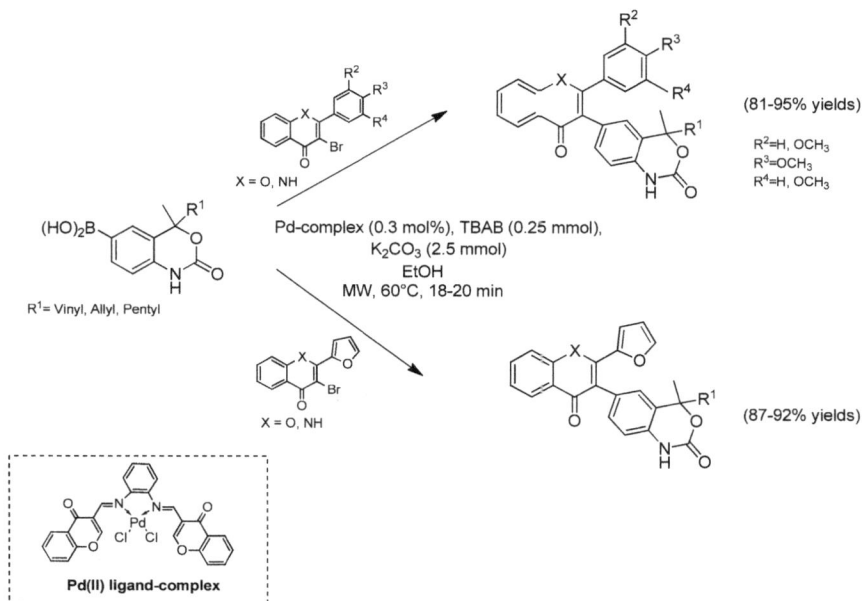

Scheme 5. Synthesis of 3-aryl-4*H*-chromen-4-one derivatives.

Pseudo halide such as sulfonate can be efficiently employed in the MW promoted SMC and recently 6-aryl salicilates were obtained from the triflates with 1% of Pd(PPh$_3$)$_4$ and 2 equiv. of NaHCO$_3$ in dimethyl ether (DME) at 110 °C in 4–10 min. The reaction tolerated significant structural modification of the aryl boronic acid and both electron rich, electron poor aryl as well as alkenyl boronic acid afforded good yields [34].

Furans are a key class of compounds that exist in many natural and pharmaceutical products. In fact, Kadam et al. [35] envisioned that MW irradiation may be used for the fast preparation of tetrasubstituted furanes. They described a fast three-step synthesis for tetrasubstituted furanes which included the condensation of a fluorous benzaldehyde with acetophenone followed by a Michael-type [3 + 2] cycloaddition with an 1,3-diketone and a Pd-catalyzed coupling reaction for fluorous linker cleavage. A fluorous sulfonate linker facilitated furane intermediate purification via fluorous solid-phase extraction and proved to be an effective triflate alternative for Pd-catalyzed SMC. Tetra-substituted furan products were produced from MW assisted SMC reactions between arylboronic acids and fluorous furan intermediates. Good yields (76%–85%) were described in 30 min at 130 °C when using Pd(dppf)Cl$_2$ (8 mol %) as a catalyst and Cs$_2$CO$_3$ as a base in acetone–H$_2$O–HFE7200 (hydrofluoroether 7200) (Scheme 6).

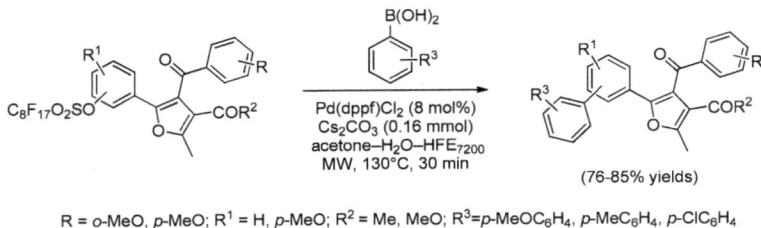

$$R = o\text{-MeO}, p\text{-MeO}; R^1 = H, p\text{-MeO}; R^2 = Me, MeO; R^3 = p\text{-MeOC}_6H_4, p\text{-MeC}_6H_4, p\text{-ClC}_6H_4$$

Scheme 6. Synthesis of tetra-substituted furans by SMC.

The considerable pharmacological activity of 7-substituted coumarins has made them the focus of great interest as the scientific community searches for new synthetic protocols for the preparation of biologically active molecules. SMC under MW irradiation has been studied by Joy et al. to modify coumarine 7-nonaflate, which shows higher stability and reactivity over the corresponding triflate [36]. In the optimization of the reaction conditions, the authors thought that the nature of the ligand would have a critical influence on achieving the desired product. 1,3-bis(diphenylphosphino) propane (dppp) gave the best results, as ligands with a moderate bite angle can form a divalent square planar complex which, in turn, enhances reductive elimination despite the presence of competitive side reactions, such as β-hydride elimination and substitution reactions. As depicted in Scheme 7, the reaction afforded the preparation of a large series of compounds in moderate to excellent yields using $Pd(OAc)_2$, dppp and tetrabutylammonium fluoride (TBAF) in DME-MeOH at 80 °C for 30 min. Electron poor boronic acid was the most critical substrate and provided lower yields even when reaction time was increased to 1 h. In an attempt to explain the mechanism, the authors suggested that TBAF was able to influence the stabilization of the oxidative adduct, which facilitates the transmetallation.

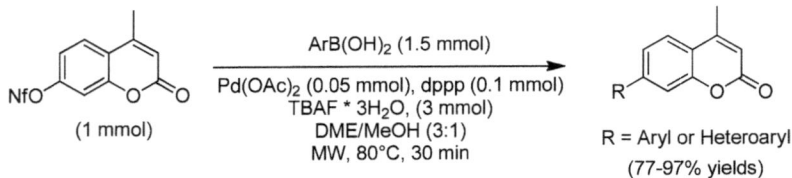

Scheme 7. SMC applied to 7-substituted coumarins preparation.

As part of the search for new procedures for the production of original small-molecules, a MW promoted SMC for the preparation of a wide range of 4(5)-arylated imidazoles has been introduced by Pochet et al. (Scheme 8) [37]. The C-4 arylation of (NH)-imidazole generally requires long reaction times (up to 48 h) and is limited to stable arylboronic acid. 4-iodo and 4-bromo-1H-imidazole were reacted with 2-naphthylboronic acid in the presence of 2 equiv. of CsF, 5 mol % of $PdCl_2$(dppf) and 5 mol % of $BnEt_3NCl$ in a 1/1 mixture of toluene/water under MW irradiation at 110 °C. After 2 h, the bromide derivative was totally converted and the expected compound was isolated in a 90% yield (92% of yield after 72 h under classical thermal conditions); total conversion of the iodine was obtained in 2 h. Furthermore, other boronic acids, bearing nitrogen (pyridinyl, isoquinoline and pyrimidinylboronic acids) and sensitive functional groups (aldehyde), were reacted with the 4-iodo-1H-imidazole with $Pd(PPh_3)_4$ as the catalyst and Na_2CO_3 in a mixture of dimethylformamide (DMF)/H_2O at 110 °C under MW irradiation, furnishing the desired products in moderate to excellent yields (35%–95%). Analogously, the imidazo[1,5-a]pyridine ring was used as a core structure and was decorated by SMC, in position 1, in the presence of $Pd(PPh_3)_4$ and K_2CO_3 in 1,4-dioxane at reflux temperature for 24 h. MW-irradiation (110 °C) was found to accelerate the reaction. In a separate work, a starting aryl iodide was fully converted to afford the desired products in a comparable yield in 1 h [38].

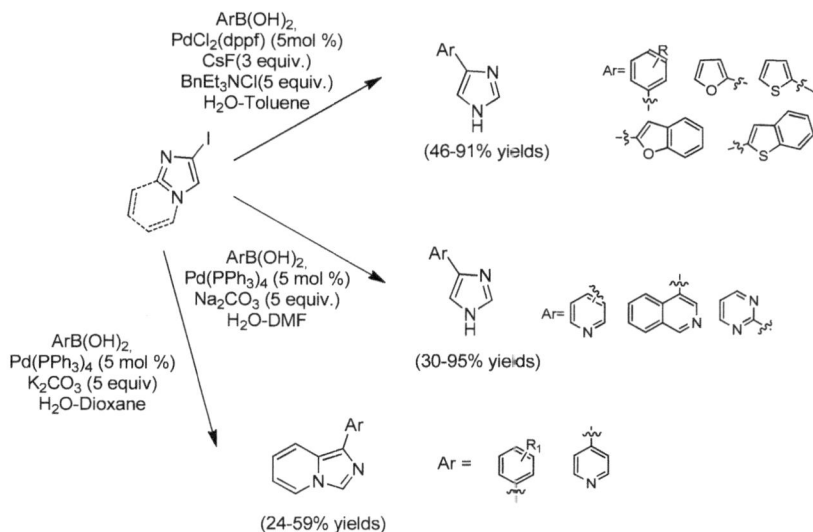

Scheme 8. Synthesis of arylated imidazoles and arylated imidazo[1,5-*a*]pyridine.

Sandtorv and Bjørsvik [39] have reported the first three-way switchable Pd-catalyzed protocol for the selective arylation and hydrodehalogenation of imidazole backbones. Using these MW promoted strategies, they prepared a wide range of 4,5-diaryl- and 4(5)-iodo-1*H*-imidazoles via the selective arylation and the hydrodehalogenation of the imidazole backbone, respectively. When the two strategies were combined in a sequential tandem reaction, they were able to successfully produce 4(5)-aryl-1*H*-imidazoles in excellent yields. Pd(OAc)₂/Xphos was used as the catalytic system in the presence of K₂HPO₄ for the SMC reaction, but two different products were achieved; a cross-coupling product and the unexpected hydrodehalogenation ones (Scheme 9). A series of control experiments was performed in order to shed light on this issue. The tuning of reaction conditions highlighted the fact that only a base was essential in the reaction mixture for hydrodehalogenation. However, they observed that it was possible to perform dehalogenation in the absence of PhB(OH)₂ and base by increasing the reaction temperature and time slightly. The author realized that hydrodehalogenation and cross-coupling processes could be coupled into a "dose-promoted" MW assisted tandem reaction in which the "trigger" for the subsequent catalytic cycle was a rise in the base and catalyst loadings to achieve a mono-arylated imidazole in excellent yields (83%).

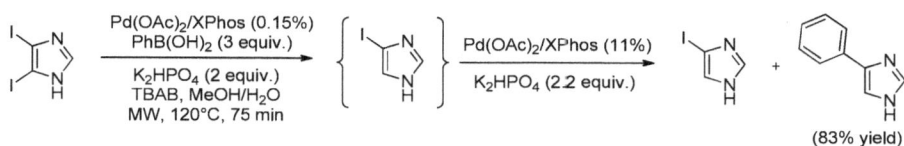

Scheme 9. "Dose promoted" assisted tandem reaction involving a hydrodehalogenation and a SMC to perform synthetic transformations on the imidazole.

Furthermore, (NH) free 3-bromo-indazol-5-amine can react with arylboronic acid in the presence of 5 mol % of $Pd(OAc)_2$, 10 mol % of RuPhos and K_3PO_4 in dioxane/H_2O = 1/1 at 140 °C under MW-assisted conditions for 30 min. In fact, a series of 19 azaindole derivatives have been synthesized, by Wu et al., with excellent yields (Scheme 10) using this method [40].Following the same procedure the authors reacted hetereoaryl boronic acid and moderate to good yields were obtained. Only thiophen-2-ylboronic acid and pyridin-3-ylboronic acid requested more excess and only pyridin-4-ylboronic acid provided a poor yield in this system even though the loading of the boronic acid was increased and the reaction time was prolonged.

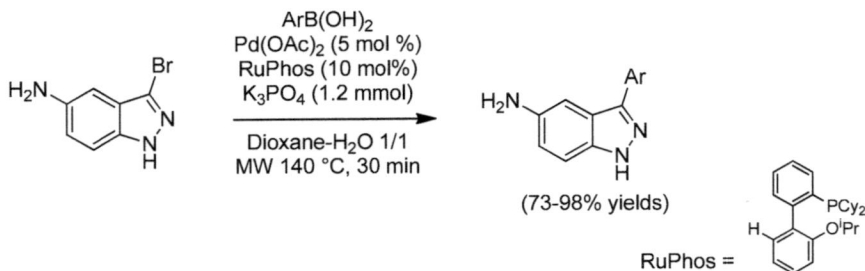

Scheme 10. SMC applied to derivatization of (NH) free 3-bromo-indazol-5-amine.

Another important class of bioactive heterocycle is isatin derivatives (indoline-2,3-diones). A limited number of reports have dealt with the the synthesis of 4-substituted-arylisatins. MW irradiation has been used in the synthesis of bulky 4-substituted-arylisatins via SMC, by Liu et al., using a wide range of substrates [41] (Scheme 11). All of the Pd catalyzed reactions were carried out using 5 mol % of $Pd(PPh_3)_4$ in the presence of 2 equiv. of $NaHCO_3$ and afforded very good yields (77%–92%) in DME/H_2O (5:1). This result indicates that electronic effects and steric modification have little influence on the reaction.

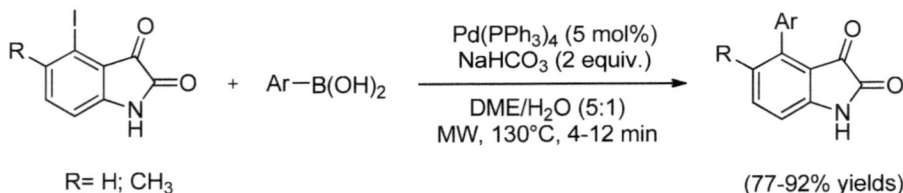

Scheme 11. MW-assisted reaction of various boronic acids and 4-iodo isatin.

El Akkaoui et al. (2010) [42] have demonstrated the flexibility of a MW-assisted, one-pot, two-step SMC/Pd-catalysed arylation process for the fast preparation of a library of different polysubstituted imidazo[1,2-*b*]pyridazine products in good yields (69%–78%) (Scheme 12). The first arylation step was performed in toluene/EtOH from 6-chloroimidazo[1,2-*b*]pyridazine in the presence of Pd acetate (0.1 equiv.), triphenylphosphine (0.2 equiv.), K_2CO_3 (2 equiv.) and various boronic acids (1.1 equiv.). The reaction was irradiated first for 15 min at 140 °C (SMC) and subsequently for 2 h at the same temperature for the final arylation, after aryl bromide was added (1.5 equiv.).

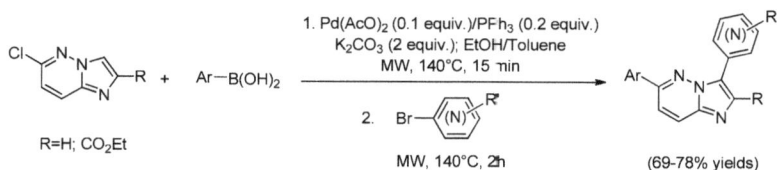

Scheme 12. Pd-catalysed one pot arylation of imidazo[1,2-*b*]pyridazines under MW irradiation.

Imidazo[2,1-*b*][1,3,4]thiadiazole scaffolds are another useful building block in pharmaceutical chemistry. Hence, there has been massive interest in establishing efficient synthetic methodologies for the regioselective preparation of polysubstituted imidazo[2,1-*b*][1,3,4]thiadiazoles. Copin et al. (2012) [43] has carried out some extensive research into developing preparative methods for these molecules, starting from a number of 2-bromo-imidazo[2,1-*b*][1,3,4]thiadiazole derivatives and using a MW assisted SMC strategy (Scheme 13). The Pd catalyzed coupling reaction was optimized under MW conditions using Pd(AcO)$_2$ (0.1 equiv.) as the catalyst in the presence of Xantphos (0.2 equiv.) and K$_2$CO$_3$ (2 equiv.) in 1,4-dioxane at 150 °C. The final compounds, which contain the rare imidazo[2,1-*b*][1,3,4]thiadiazole central skeleton, were obtained after 30 min MW irradiation in noticeable yields (73%–94%) in the presence of acid-sensitive- and both electron-withdrawing and electron-donating groups.

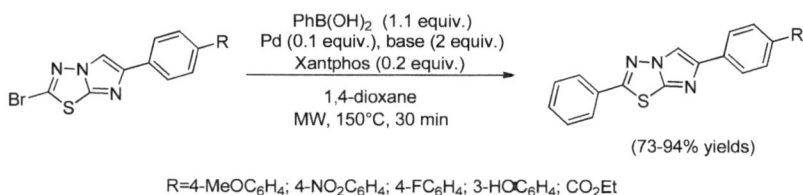

R=4-MeOC$_6$H$_4$; 4-NO$_2$C$_6$H$_4$; 4-FC$_6$H$_4$; 3-HOC$_6$H$_4$; CO$_2$Et

Scheme 13. Decoration of imidazo[2,1-*b*][1,3,4]thiadiazole derivatives through MW promoted SMC reaction.

The 3,4-dihydro-2*H*-1,4-benzoxazine and the 2*H*-1,4-benzoxazine-3-(4*H*)-one structures are wildely applied in the synthesis of a pletora of biologically active molecules. No reports on the MW-assisted synthesis of 6-aryl-1,4-benzoxazines/ones, with the most sterically hindered 5,7,8-trimethyl-1,4-benzoxazine/one core, had, until quite recently, appeared in the literature. It is in this context that Koini et al. (2012) [44] reported on an effective MW-assisted SMC strategy for the preparation of 6-substituted-5,7,8-trimethyl-1,4-benzoxazines(ones), using Pd(PPh$_3$)$_4$ as the catalyst (Scheme 14). This approach enabled the rapid incorporation of alkyl, aryl, and heteroaryl functionalities into benzoxazine moieties. Under MW heating, the SMC provided good yields at 160 °C in only 5 min using 5 mol % Pd catalyst, Na$_2$CO$_3$ and nearly equimolar boronic acid stoichiometry.

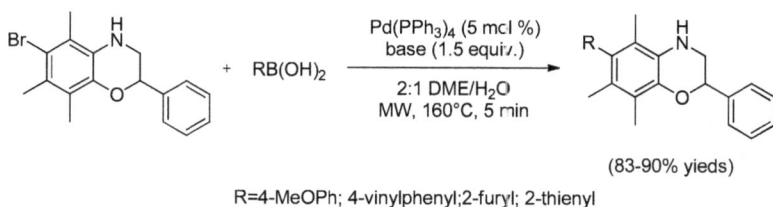

R=4-MeOPh; 4-vinylphenyl;2-furyl; 2-thienyl

Scheme 14. MW-promoted SMC of 2-phenyl-6-bromo-5,7,8- trimethyl-1,4-benzoxazine with various boronic acids.

Building blocks that bear a benzo[e]-annelated indoline moiety exhibit strong fluorescence and can be used as fluorescence emitting dyes. The synthesis of arylated imidazo- and pyrimido-[1,2-a]indolone compounds has been approached by Zukauskaite et al. SMC was performed in the presence of a series of boronic acids that possess a π-conjugated biaryl-based structural unit as a fluorofore (phenylboronic, naphthalen-2-ylboronic, pyren-1-ylboronic and dibenzo[b,d]thiophen-4-ylboronic acids). The synthetic protocol was performed in water with ligand-free Pd(OAc)$_2$ and MW irradiation (50 W) was chosen as the power source (Scheme 15) [45].

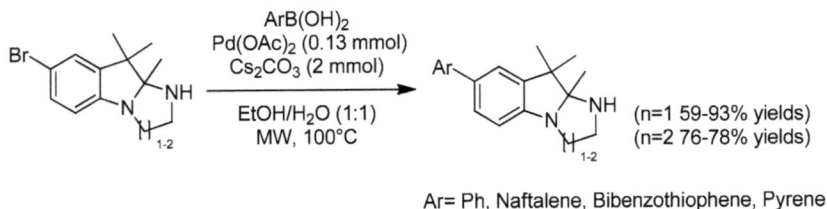

Ar= Ph, Naftalene, Bibenzothiophene, Pyrene

Scheme 15. Synthesis of arylated imidazo- and pyrimido-[1,2-a]indolone derivatives.

Arylation of purines and pyrrolopyrimidine at the C-4 and C-2 positions is an interesting task. Conventional condition approaches are known, however, SMC of 4-chloropyrrolopyrimidine derivatives has been performed by Prieur et al. under MW irradiation (Scheme 16) [46]. The synthesis of a series of compounds proved the versatility of the protocol and neither the position nor the nature of the aryl substituent on the boronic acid affected the reaction outcome (yield from 86% to 98%).

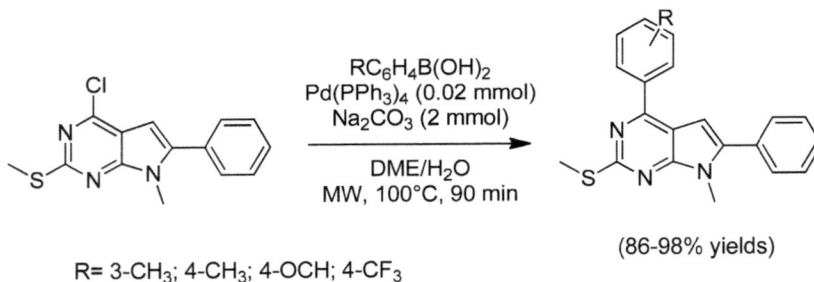

R= 3-CH$_3$; 4-CH$_3$; 4-OCH; 4-CF$_3$

(86-98% yields)

Scheme 16. Derivatization of 4-chloropyrrolopyrimidines derivatives by MW promoted SMC.

The synthesis of highly substituted spirooxindoles remains a great challenge for organic chemists due to their broad biolaogical activities mainly related to voltage-gated ion channels (Na$^+$, K$^+$, Ca^{2+}) regulation related to nociception. Guillaumet et al. (2015) [47] have recently developed a new original method for the preparation of multi-substituted spirooxindoles via SMC strategy (Scheme 17). Through a fast MW assisted palladium catalyzed protocol, the autors prepared a plethora of new C-5'- or C-5-monosubstitued and C-5,C-5'-disubstitued 1'-pentyl-2H-spiro[furo[2,3-b]pyridine-3,3'-indolin]-2'-one derivatives in only 10 min under MW irradiation at 150 °C by using Pd$_2$(PPh$_3$)$_4$ (0.1 equiv.) in presence of K$_2$CO$_3$. This SMC protocol was focused on the functionalization of the phenyl part of the oxindole core or the pyridine nucleus of the starting regioselective brominated or chlorinated spirooxindole moieties to give mono- or disubstituted- derivatives in good to excellent yields (70%–98%).

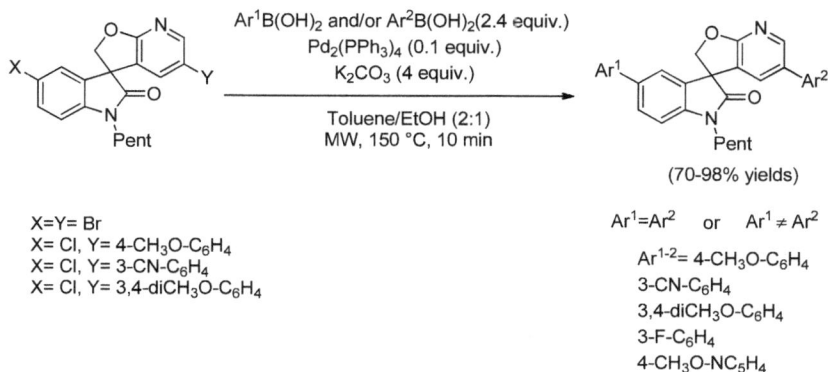

Scheme 17. MW-promoted C–C coupling leading to disubstituted 5,5′-di(het)aryl-1′-pentyl-2*H*-spiro[furo[2,3-*b*]pyridine-3,3′-indolin]-2′-ones.

Qu et al. [48] have described a fast and sustainable process for the synthesis of 6-arylpurines from 6-chloropurines and sodium tetraarylborates via a MW assisted SMC reaction in water at 100 °C (Scheme 18). Furthermore, most of the reactions involved are efficient when using Pd(PPh₃)₂Cl₂, 5 mol % catalyst in the presence of Na₂CO₃ (2 equiv.) and furnish the desired products in high yields (73%–98%) in short reaction times (30 min). This eco-friendly MW-assisted method is a promising, green strategy for the preparation of these main nucleoside compounds.

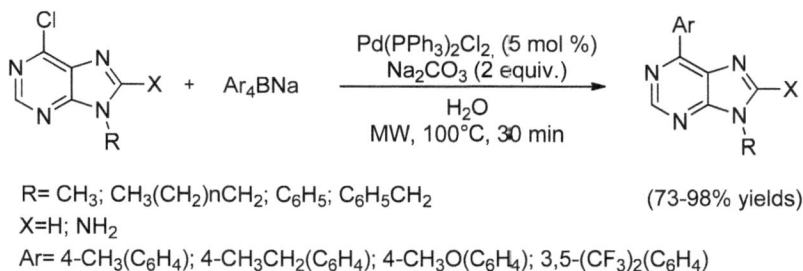

R= CH₃; CH₃(CH₂)nCH₂; C₆H₅; C₆H₅CH₂   (73-98% yields)
X=H; NH₂
Ar= 4-CH₃(C₆H₄); 4-CH₃CH₂(C₆H₄); 4-CH₃O(C₆H₄); 3,5-(CF₃)₂(C₆H₄)

Scheme 18. MW-assisted SMC of Ph₄BNa with various 6-chloropurines.

The 4*H*-pyrido[1,2-*a*]pyrimidin-4-one core has shown itself to be a useful scaffold for the disclosure of new bioactive compounds. In this context, Kabri et al. [49] have reported the first example of a MW assisted Pd catalyzed SMC reaction for the synthesis of a series of 3-aryl, 3-heteroaryl and 3-styryl-4*H*-pyrido[1,2-*a*]pyrimidin-4-ones. The coupling process was proven to be extremely tolerant to electron-poor, electron-rich and bulky boronic acid derivatives, giving the target products in good yields (70%–90%) in 2 h MW irradiation at 100 °C. The best conditions for the MW-assisted, Pd-mediated coupling reactions of aryl, heteroaryl and styrylboronic acid compounds with 7-chloro-3-iodo-4-oxo-4*H*-pyrido[1,2-*a*]pyrimidin-2-yl)methyl acetate were described in DME–EtOH (9:1), using PdCl₂(PPh₃)₂ (0.05 equiv.) as the catalyst in the presence of the base, K₂CO₃ (3 equiv.).

Another study [50] adopted the same efficient Pd-catalyzed procedure to design new and unique asymmetrical 3,9-bis-[(hetero)aryl]-4*H*-pyrido[1,2-*a*]pyrimidin-4-one compounds. They developed a MW one-pot chemoselective bis-SMC protocol that almost gave good 4*H*-pyrido[1,2-*a*]pyrimidin-4-ones yields (Scheme 19). Stepwise substitution at the 3- and then at the 7- and/or 9-position of 7,9-dichloro-

3-iodo-4-oxo-4*H*-pyrido[1,2-*a*]pyrimidin-2-yl)methyl acetate, using a range of boronic acids in a one-pot process, enabled the efficient and fast synthesis of chemical libraries of bioactive derivatives to be carried out. The same approach was more recently pursued by the same authors to obtain 2,6,8-trisubstituted 4-aminoquinazolines through MW-assisted consecutive one-pot chemoselective tris-SMC or SNAr/bis-SMC reactions in water [51].

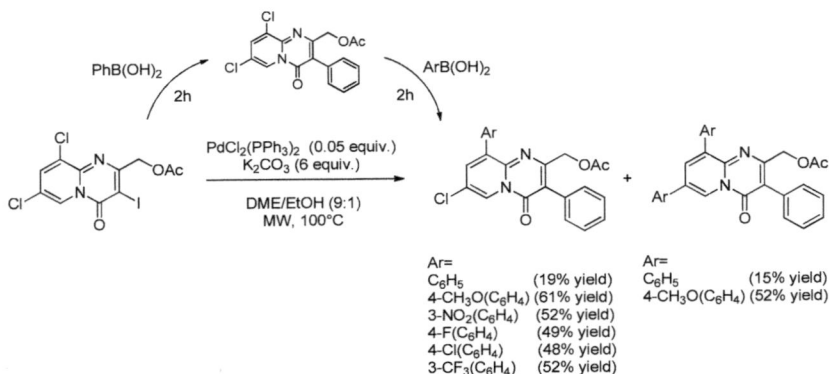

Scheme 19. One-Pot chemoselective bis-SMC gives simple access to 3,9-biaryl and 3,7,9-triaryl-4*H*-pyrido[1,2-*a*]pyrimidin-4-ones.

Several nucleoside analogues have recently attracted considerable attention due to their potential biological properties. Analogues that bear a C-aryl group on the glycone or aglycone section have been more intensively studied. In fact, an easy and efficient procedure for the direct synthesis of 5-aryl-2′-deoxyuridines has been reported, by Gallagher-Duval et al. [52], via a ligand-free SMC strategy which starts with totally deprotected 5-iodo-2′-deoxyuridine and different boronic acids and is carried out in pure water. The desired 5-arylated uridine derivatives were synthesized in satisfactory (57%–85%) yields in short reaction times (5–30 min) in the presence of very low Na_2PdCl_4 Pd catalyst loading amounts (0.05–0.1 mol %) and KOH (0.56 mmol) as the base. One protocol was carried out under classical thermal heating, another was performed under MW irradiation at 100 °C (Scheme 20). All reported results show that the desired cross-coupling products were obtained up to three times faster under MW irradiation than classic heating, even in the presence of electron-withdrawing groups in the boronic acid *para* position. Sterically demanding boronic acids proved to be most troublesome substrates for this process.

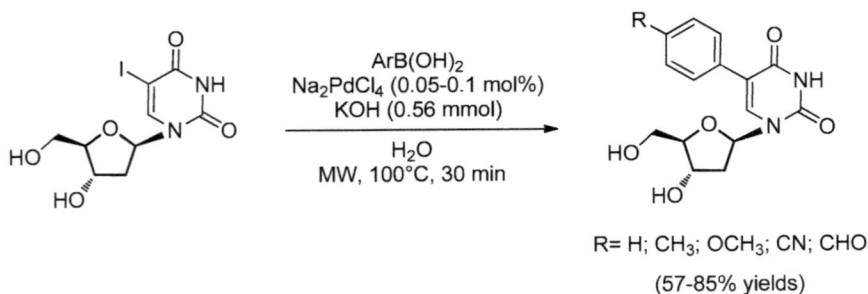

R= H; CH_3; OCH_3; CN; CHO

(57-85% yields)

Scheme 20. Ligandless MW-promoted SMC starting from 5-iodo-2′-deoxyuridine.

2.1.2. Organotrifluoroborates in MW-Assisted SMC

Potassium organotrifluoroborates have recently become powerful synthetic building blocks for the formation of new carbon–carbon bonds via SMC, as these compounds show greater nucleophilicity than their corresponding organoboranes or boronic acid derivatives and are simple to synthesize and purify. Moreover, as crystalline solids, they are air- and moisture-stable. The inertness of the trifluoroborate (-BF$_3$) group under common reaction conditions for direct transformation, using Pd or copper catalysts, enables the preparation of highly functionalized organotrifluoroborates to be performed [53,54]. In this context, Kim et al. [55] have recently described a sustainable MW assisted SMC process which starts from various triazole-containing trifluoroborates and which was successfully prepared via a regioselective, one-pot Cu-catalyzed azide/alkyne cycloaddition (CuAAC) reaction from ethynyltrifluoroborate. Potassium (1-organo-1H-1,2,3-triazol-4-yl)trifluoroborates were successfully cross-coupled with various functionalized aryl and alkenyl bromides under MW conditions. Good yields (88%–90%) were described in aqueous methanol for reactions performed at 150 °C for 40 min in the presence of PdCl$_2$(dppf)·CH$_2$Cl$_2$ (10 mol %), TBAB (1 equiv.) and K$_2$CO$_3$ under MW irradiation at 80 W (Scheme 21). C–C coupling products were achieved in reasonable yields (57%) even when using sterically bulky coupling alkenes, such as 1,2,2-triphenylvinyl bromide.

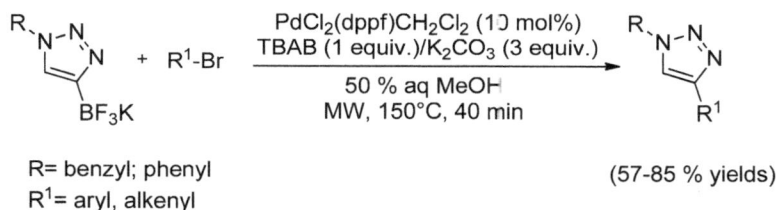

Scheme 21. MW-assisted C–C cross-coupling starting from (1-organo-1H-1,2,3-triazol-4-yl) trifluoroborates K salt.

The benzimidazole skeleton is an important heterocycle because of its wide range of pharmacological activities. The formation of a C–C bond at the 6-position of the electron-rich 1-,4-,6-trisubstituted benzimidazole nucleus is challenging and was not obtainable via Kumada, Negishi, Stille or Heck coupling strategies. As a result, Jain et al. [48] have also focused their attention on the advantages of organotrifluoroborate salts as coupling partners for SMC with 4-nitro-6-triflyl benzimidazoles under MW activation in aqueous tetrahydrofuran (THF) [56]. Novel functionalization of 1-,4-,6-trisubstituted benzimidazoles at the 6-position was presented and 37%–70% yields were reported under 1–2 h of MW irradiation at 100 °C in the presence of PdCl$_2$(dppf)·CH$_2$Cl$_2$ (1 equiv.) and Cs$_2$CO$_3$ (1.5 equiv.) (Scheme 22).

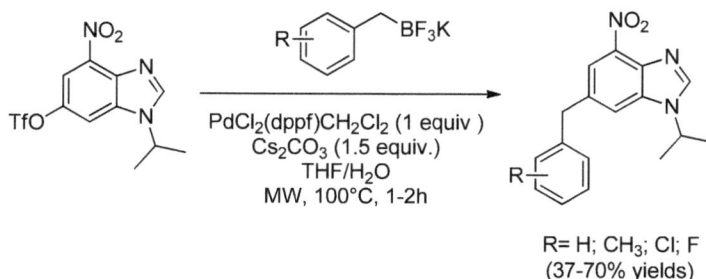

Scheme 22. MW-assisted SMC of potassium oraganoborates with 6-sulfonate bzenzimidazoles.

Several examples of SMC were perrormed in presence of vinyl boronic acid or vinyltrifluoroborate. Henderson et al. [57] have reported an optimized MW protocol for SMC between *N*-tosyl-2-bromo-benzylamines and -phenethylamines with vinylboronic acids. The corresponding 2-tosylaminomethyl- and 2-tosylaminoethyl-styrenes were achieved in very good yields under MW irradiation at 100 °C in 30 min using a catalyst, ligand and base pre-mixed system. The significance of this study was demonstrated when an attempted SMC between *N*-(2-bromobenzyl)-4-methylbenzenesulfonamide and (*E*)-2-(4-substituted phenyl)-vinylboronic acid, using relatively routine MW-assisted conditions, led to none of the expected (*E*)-4-substituted-*N*-(2-(4-methylstyryl)benzyl)benzenesulfonamide products in DMF, in presence of Pd(PPh$_3$)$_4$. The highly hindered phosphine ligand dtbpf (1,10-bis(di-*tert*-butylphosphino)ferrocene), combined with Pd(AcO)$_2$, was effective in this cross-coupling as was the use of potassium phosphate and the change of DMF to either aqueous DMF or aqueous EtOH (Table 1).

Table 1. Optimization of coupling reaction conditions. dtbpf, (1,10-bis(di-*tert*-butylphosphino)ferrocene).

Cat/Ligand/Base	Solvent	Yields (%)
Pd(PPh$_3$)$_4$–NaHCO$_3$	DMF	0
Pd(PPh$_3$)$_4$–NaOH	1:1 DMF–H$_2$O	47
Pd(dba)$_2$–dtbpf–K$_3$PO$_4$	1:1 DMF–H$_2$O	43
Pd(OAc)$_2$–dtbpf–K$_3$PO$_4$	1:1 DMF–H$_2$O	63
Pd(OAc)$_2$–dtbpf–K$_3$PO$_4$	1:1 EtOH–H$_2$O	90
Pd(OAc)$_2$–dtbpf–K$_3$PO$_4$	H$_2$O	72
Pd(OAc)$_2$–K$_2$CO$_3$	H$_2$O	37

In general, yields were excellent (81%–99%), when using a pre-mixed Pd(OAc)$_2$–dtbpf–K$_3$PO$_4$ system, and were not remarkably affected by the nature of the boronic acid, whether this was carrying an aromatic or an alkenyl substituent (Scheme 23). Not surprisingly, there was also a tiny change in the styrene yields when boronates, specifically 4,4,5,5,-tetramethyl-1,3,2-dioxaborolanes, were employed; these would be quickly converted into the corresponding boronic acids under the aqueous conditions used in the optimized protocol.

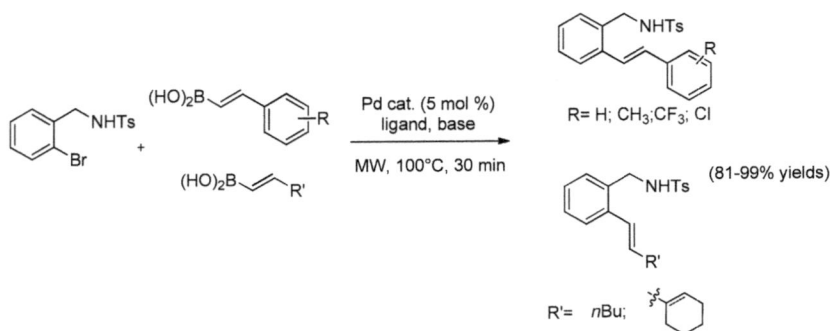

Scheme 23. MW-assisted SMCs of sulfonamides and vinylboronic acids using Pd(OAc)$_2$–dtbpf–K$_3$PO$_4$. dtbpf, (1,10-bis(di-*tert*-butylphosphino)ferrocene).

Brooker et al. [58] have reported the MW assisted SMC of sterically hindered and electron-rich ortho and ortho'-substituted aryl halides with potassium vinyltrifluoroborate. Good yields were described in short reaction times when the coupling reaction was performed in THF–H$_2$O (9:1 mixture) under MW irradiation (from 3 to 4 days under conventional conditions to 20 min) at 150 °C using Cs$_2$CO$_3$ as the base (Scheme 24). This MW vinylation methodology was also proven to be more tolerant to solvent ratios across the whole range of substrates (optimum concentration 0.051 up to

0.091 M with a slight drop in yields). Moreover, the PdCl$_2$(dppf)CH$_2$Cl$_2$ catalyst loading was reduced by over three-fold under MW (from 18 to 5 mol %). It maintained good conversions to styrene derivatives, while avoiding side product formation. In addition, it has been observed that vinylation with potassium vinyltrifluoroborate is better fulfilled using ortho and ortho'-substituted aryl halides with at least one electron-withdrawing group present, thus reducing the electron-rich nature of the aromatic halide (as is usual for SMC).

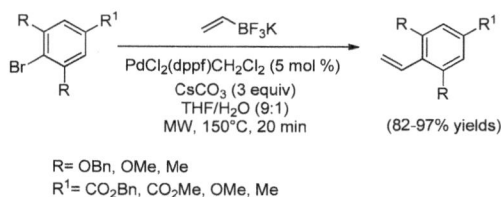

R= OBn, OMe, Me
R^1= CO$_2$Bn, CO$_2$Me, OMe, Me

Scheme 24. MW-promoted SMC vinylation of electron-rich, sterically hindered substrates using potassium vinyltrifluoroborate.

2.1.3. MW-Assisted One-Pot Protocols

An efficient synthesis of *N*-quinoline 3'/4'-biaryl carboxamides, which avoids the protection/deprotection of the amide function, has recently been described as being based around a one-pot, MW-assisted SMC and *N*-Boc-deprotection sequence [59]. The mixture with 2.0 mol % of Pd(PPh$_3$)$_4$, phenylboronic acid (1 equiv.) and NaOAc (2 equiv.) in dioxane:water was irradiated at 80 °C for 8–30 min. The optimal conditions involved the reaction temperature being increased for the first step and then increased further to 120 °C for 8 min. The reaction proved itself to be well tolerant of valuable, but unstable groups, such as hydroxyl and carbonyl groups, and excellent yields were obtained from the benzamido, naftalen, furane, thiophen and thiazole carboxamidoamido derivatives.

Boc protection of the sulfonamino group also greatly promoted the SMC and, as already described for the Boc carboxamides, MW irradiation in water enabled deprotection to be carried out quickly and efficiently. As depicted in Scheme 25, the one-pot procedure, which combines SMC and Boc-deprotection, can be performed starting with 4-bromo *N*-Boc-benzenesulfonamide via the addition of P(PPh$_3$)$_4$ and arylboronic at 85 °C (MW irradiation) for 8–10 min in the first step, followed by a second step at 130 °C for 8 min [60].

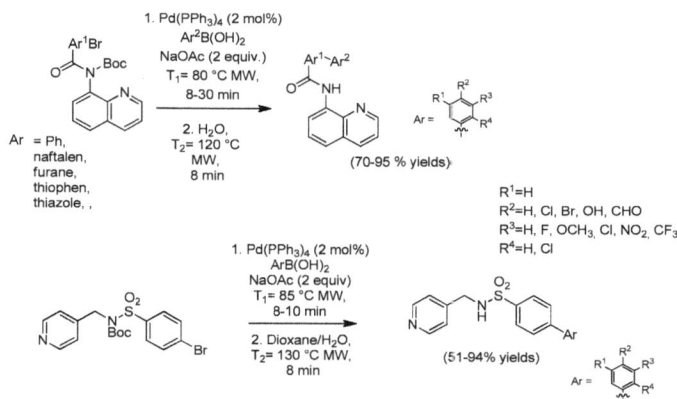

Scheme 25. Pd(PPh$_3$)$_4$-Catalyzed SMC and *N*-Boc cleavage, one-pot reaction.

The development of synthetic procedures which can perform the parallel synthesis of analogues, where diversity elements are introduced into a multistep sequence, has been the focus of significant interest over the last two decades. This investigation mainly makes use of reactions that are commonly used in the medicinal chemistry, including SMC, one of the most prevalent transformations in drug discovery parallel synthesis. As an application of this approach, a one-pot reductive amination-SMC sequence was proposed by Grob et al. in 2011 (Scheme 26) [61]. Boron-functionalized aryl aldehydes were first subjected to reductive amination and the C–C coupling then gave the diaryl derivative. MIDA (*N*-methyliminodiacetic acid) ligand boronates were selected on the basis of a preliminary screening because of their high stability when undergoing reductive amination. Furthermore, the proposed MW-promoted protocol includes both the deprotection of the MIDA boronate and the subsequent cross-coupling reaction. The authors observed that good yields were achieved with both electron-rich and electron-deficient haloarenes. Interestingly, heteroaryl halides react successfully with examples spanning a wide variation of ring size electronics and substitutions types; the only class that failed in the SMC was the five-membered NH free heterocycles. Relative success was obtained in the assessment of the versatility of reductive amination with amines containing amino-acid derivatives, ethers, esters and heterocycles.

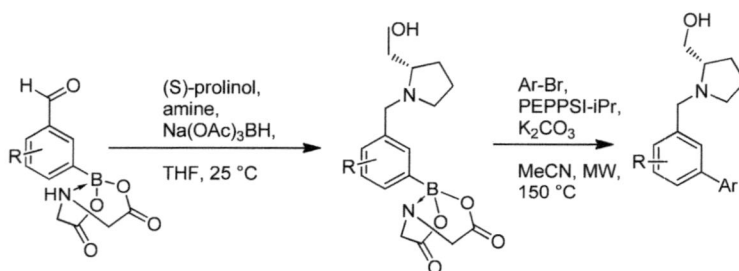

Scheme 26. Preparation of boron-functionalized aryl aldehydes and one-pot synthesis of biphenyl amine.

Park et al. [62] have developed a one-pot method that proceeds via SMC and aldol condensation reactions to obtain naphthoxindoles in good to excellent yields (63%–94%). This reaction afforded an oxindole moiety from 4-bromooxindoles and 2-formylarylboronic acids in 5 min under MW irradiation at 150 °C. Pd(PPh₃)₄ (5 mol %) was used as a catalyst in the presence of K₂CO₃ (1.5 mmol) as the one pot reaction was performed in toluene/EtOH (2:1) (Scheme 27). The described synthetic method was tolerant of various substituents and functional groups, while MW irradiation was used to improve the reaction rate of a wide set of naphthoxindole libraries.

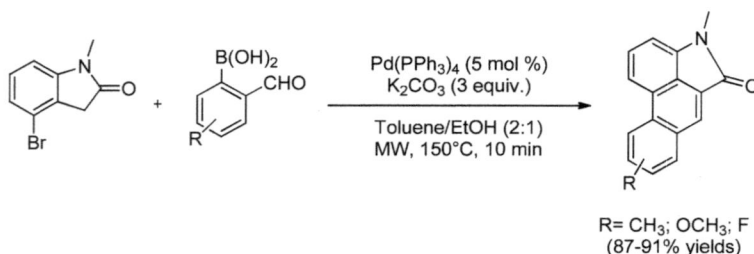

R= CH₃; OCH₃; F
(87-91% yields)

Scheme 27. One-pot synthesis of naphthoxindoles from 4-bromooxindoles by SMC and aldol condensation reactions.

MW irradiation has been applied to a versatile and efficient one-pot borylation/SMC in an attempt to overcome one of the main limitations of the SMC reaction; the lack of availability and/or the instability of certain boronic species [63]. The authors optimized the one-pot model reaction which starts from 5-bromoindanone and 3-bromopyridine to give 3-pyridinylindenone. The transformation is mediated by the formation of the intermediate bis(pinacolato)diboronic (Scheme 28). A change in the base used was crucial to this study. KOAc was used at first to activate the halide within the catalytic complex for the transmetallation step with bis(pinocolato)diboron. Na_2CO_3 was then introduced as a second base, after the formation of pinocalate, to form the biaryl-substituted palladium species. The optimized protocol was performed via the irradiation of a dioxane solution of arylbromide, Pd(PPh$_3$)$_4$ and KOAc in a MW oven at 120 °C for 45 min, followed by the addition of a second arylbromide, Na_2CO_3 and further irradiation at 120 °C for 30 min. The author synthesized a large series of diaryl and keto, Boc-protected aniline, halo, aryl, indanone, pyridyl, pyrazole, azaindole and quinoline functional groups to provide a set of hinge-binding fragments. Furthermore, heteroaromatic rings, which either contained a hydrogen bond acceptor, a hydrogen bond donor or both, were coupled in combination with four phenyl halides.

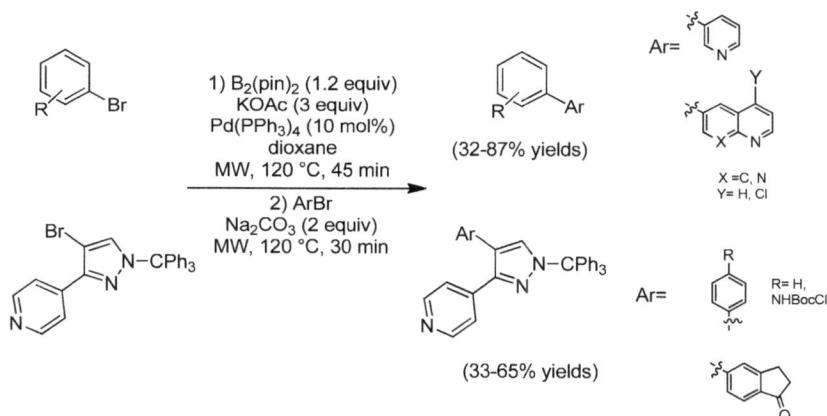

Scheme 28. One-pot borylation/SMC.

2.1.4. Ligands in MW-Assisted SMC

It is well known in literature that palladacycles are efficient pre-catalysts for the SMC reaction. Oxime-derived palladacycles are often chosen because they are known to be (1) stable at high temperature; (2) inert with air and moisture; (3) a pre-catalyst that releases Pd(0) and avoids the formation of large Pd metal particles. In this context, Cívicos et al. [64] have recently demonstrated the high catalytic activity of a number of different oxime palladacycles (Scheme 29) in the SMC of biaryls in aqueous solvents, using both conventional and MW irradiation conditions. The same authors presented a new and simple protocol for the Pd catalyzed SMC alkenylation of deactivated organic chlorides under MW irradiation conditions.

Alkenylboronic acids and potassium alkenyltrifluoroborates are effectively cross-coupled with these aryl and heteroaryl chlorides using the 4,4'-dichlorobenzophenone oxime-sourced palladacycles (**PcPd$_{1-2}$**) as pre-catalysts at 0.1 to 0.5 mol % Pd loading, tri(*t*-butyl)phosphonium tetrafluoroborate {[HP(*t*-Bu)$_3$]BF$_4$} as a ligand, tetra-*n*-butylammonium hydroxide as the co-catalyst and K$_2$CO$_3$ in DMF at 130 °C under MW heating (Scheme 29). Alkenylarenes, stilbenes and styrenes were obtained in 60%–95% yields, with high β/α regioselectivity and diastereoselectivity in only 20 min, under these conditions. The reported procedure is also very useful for the regioselective alkenylation of benzyl and allyl chlorides to obtain allylarenes and 1,4-dienes.

Oxime palladacycles

PcPd$_1$: R= CH$_3$; R^1= OH
PcPd$_2$: R= 4-ClC$_6$H$_4$;R^1= Cl

X=(OH)$_2$, F$_3$K

(60-95% yields)

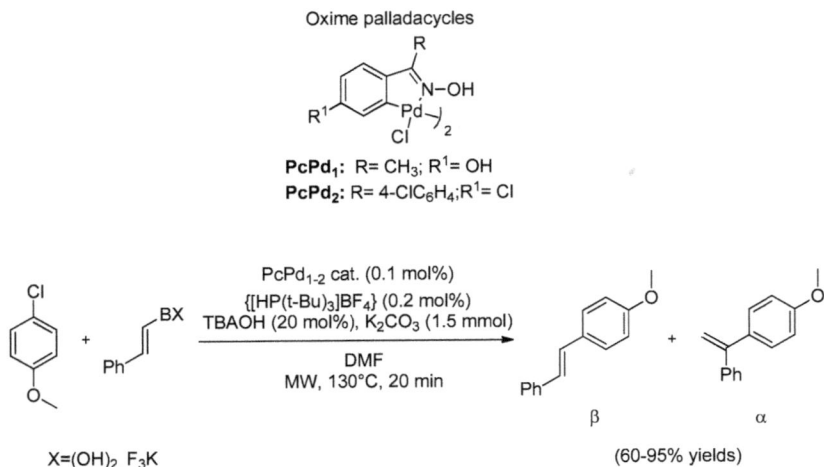

Scheme 29. SMC alkenylation of aryl chlorides.

The same authors proved that the oxime-palladacycle was efficient in arylation and alkenylation when using neutral, electron-rich and electron-poor phenyl imidazolesulfonates (Scheme 30) and also with sterically hindered electrophiles [65]. High isolated yields were obtained. 2-phenylpyridine was synthesized in a high isolated yield (72%) in aqueous conditions from phenylboronic acid and pyridin-2-yl 1*H*-imidazole-1-sulfonate.

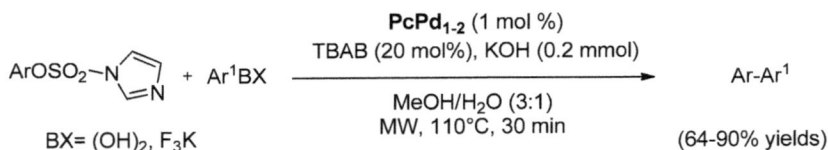

PcPd$_{1-2}$ (1 mol %)
TBAB (20 mol%), KOH (0.2 mmol)

ArOSO$_2$-N⟩⟩ + Ar^{1}BX → Ar-Ar^1

MeOH/H$_2$O (3:1)
MW, 110°C, 30 min

BX= (OH)$_2$, F$_3$K

(64-90% yields)

Scheme 30. SMC reaction of aryl imidazolesulfonates with arylboronic acids and potassium aryltrifluoroborates.

The authors underlined the efficiency of MW irradiation in all examples, but in particular with sterically challenging imidazolesulfonate. Remarkable, isolated yields (62%–93%) were also commonly described for the regio- and stereo-selective syntheses of stilbene and styrene compounds, regardless of which nucleophile was used. MW-assisted cross-coupling reactions were performed in the presence of different β-aryl- and β-alkyl-substituted alkenylboronic acids and potassium trifluoroborates. The improved reaction conditions have also proven to be effective in the coupling of electron-deficient electrophiles and heterocycles, such as pyridine-2-yl. Within a bi-functional starting material, the reactivity gap between the C–Br bonds and the C–O imidazolesulfonate bond was exploited by the authors to demonstrate that MW-promoted orthogonal cross-couplings with arylboronic acids primarily afforded the biphenyl-1*H*-imidazole-1-sulfonate via C–Br activation. This product can be subsequently submitted to SMC with phenylboronic acid under aqueous conditions to obtain the desired diarylated derivative (Scheme 31).

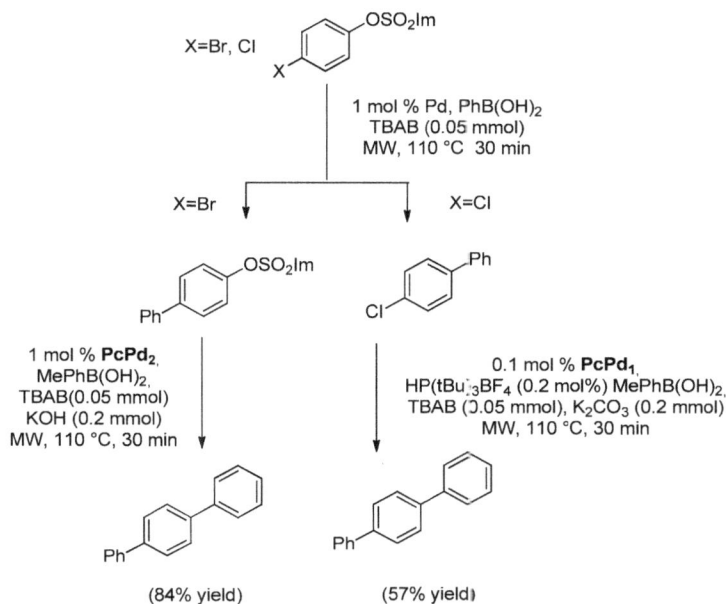

Scheme 31. Use of the oxime-palladacycle (**PcPd$_{1-2}$**) in MW promoted arylation.

Susanto et al. [66] have developed a fluorous, thermally stable oxime-derived palladacycle for MW-enhanced carbon–carbon SMC reactions in aqueous media. The palladacycle gave extremely low levels of Pd leaching (0.023–0.033 ppm over 5 cycles) and was reused five times with no significant loss in activity. The catalytic activity of the reported fluorous, oxime-based palladacycle **PcPd$_3$** (0.05 mol % Pd) (Scheme 32) was used in the SMC between phenylboronic acid and 4-bromobenzotrifluoride in the presence of K$_2$CO$_3$ as the base and TBAB as the phase transfer additive. The reaction proceeded efficiently under MW irradiation at 140 °C in water to give the biphenyl product in a 98% yield in 2 min.

Excellent yields (90%–98% corresponding to 2 × 10^4 TON (turnover number)) were obtained when Pd pre-catalyst loading was dropped down to 0.005 mol %, but the reaction time was longer at 1 h. It is worth noting that the author, after a mercury drop test, hypothesized that the oxime-based palladacycle formed Pd nanoparticles and that they were, in fact, the catalytic species in the reaction.

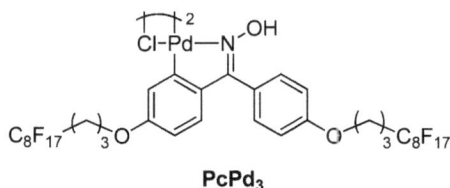

PcPd$_3$

Scheme 32. Fluorous, oxime-based palladacycle structure.

A fast MW assisted SMC reactions between either activating (2-acetyl) or deactivating (2-thiazol-4-yl) bromothiophenes moieties with a pletora of arylboronic acids were reported by Dawood et al. (2015) [67]. A benzothiazole-oxime Pd(II)-complex was successfully applied for this proposal either in water or DMF enabling good SMC product yields in 10–30 min under MW irradiation at 100–160 °C (Scheme 33).

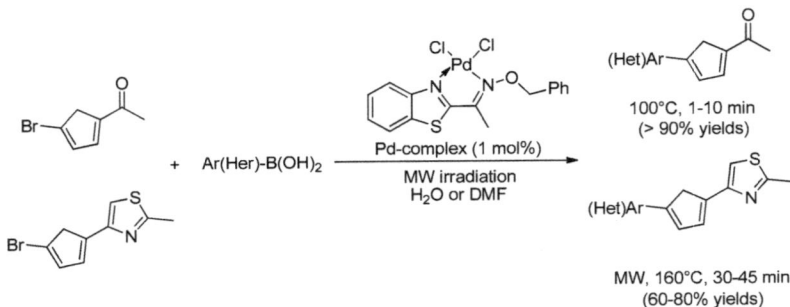

Scheme 33. MW-assisted synthesis of 2-acetyl-5-arylthiophenes and 4-(5-arylthiophen-2-yl)thiazoles via SMC.

Attempts to exploit aryl chlorides as SMC reaction substrates have led to several catalysts being developed. Of these new catalysts, the use of ferrocene derivatives (**Fc**) as ligands has attracted remarkable attention due to their large size and electron-rich nature. Ferrocene-containing phosphine ligands have been applied, however, the use of other ligands, such as diimines, has gained some limited attention. The use of diimine or pyridylimine ligands, in lieu of phosphine ligands, has major benefits, such as simple synthetic procedures, easy management and easy tuning of the electronic and steric properties of the final catalysts. In this context, Hanhan et al. (2012) [68] have reported the MW assisted SMC, in aqueous media, of different boronic acids with aryl chlorides using a ferrocene-containing Pd(II)–diimine complex as the catalyst. The use of the air-stable diimine ligand, which bears two ferrocene units, was observed to be powerful for SMC reactions due to the sterically demanding character of the ligand and the electron-rich properties of the ferrocene fraction. These features allow the active Pd(0) species to be stabilized in the catalytic cycle and support the reaction. Small amounts of **FcPd$_1$** (0.1%) (Scheme 34) were found to be powerful in the coupling of different boronic acids with non-activated aryl chlorides to supply sterically hindered ortho-substituted biaryls (yields > 90%) in aqueous medium in the presence of K_2CO_3 (1.5 equiv.) in only 15 min at 800 W. Rather, the use of very low quantities of catalyst (0.0001%) allowed the coupling of aryl iodides and bromides with boronic acids to be carried out in quantitative yields. When **FcPd$_1$** was compared with other catalysts, that have already been described as used in SMCs, **FcPd$_1$** did not required any additives, such as TBAB, or solvents other than H_2O.

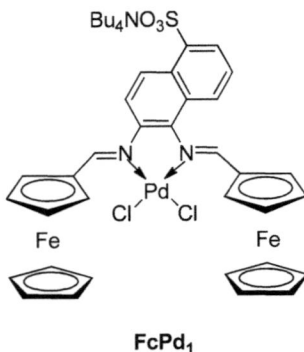

FcPd$_1$

Scheme 34. Ferrocene-based Pd(II)–diimine catalyst.

Palladacycles can possess widespread structural arrangements and synthetic accessibility and, as such, have gained attention as catalytic precursors. In this context, organosulfur ligands are usually precursors in the synthesis of palladacycles and can be successfully used in SMC reactions that are promoted by various heating sources, such as MW and infrared. In a comparison of the conversions obtained by coupling 4-iodotoluene and phenylboronic acid, Balam-Villarreal et al. used 0.1% mol of palladacyle FcC(S)OEtPdClZR$_3$ (ZR$_3$ = PPh$_3$, P(*o*-Tol)$_3$, and PMe$_3$) and K$_2$CO$_3$ in methanol [69]. The same reaction, conducted at 65 °C, produced a 99% yield in 240 min, while the same yield was obtained in 120 min (40 °C) under US irradiation and in 25 min using IR. The effectiveness of MW irradiation was proven as excellent coupling product yields were obtained in only 6 min (90 °C).

Yılmaz et al. [70] have described the synthesis of new *bis*-benzimidazole salts (**Bim$_{1-2}$**), which contain furfuryl and thenyl moieties (Scheme 35), and their further applications in MW-assisted SMC reactions in the presence of Pd(OAc)$_2$. In particular, it was reported that the SMC reaction, catalyzed by Pd(OAc)$_2$ (1 mol %) in the presence of bis-benzimidazolium salts (1 mol %), gave excellent yields (86%–99%)(Scheme 36) in only 5 min when using a DMF–H$_2$O (1:1) mixture as the solvent and either Cs$_2$CO$_3$ or K$_2$CO$_3$ as a base (2 mol %) at 145 °C/400 W MW heating.

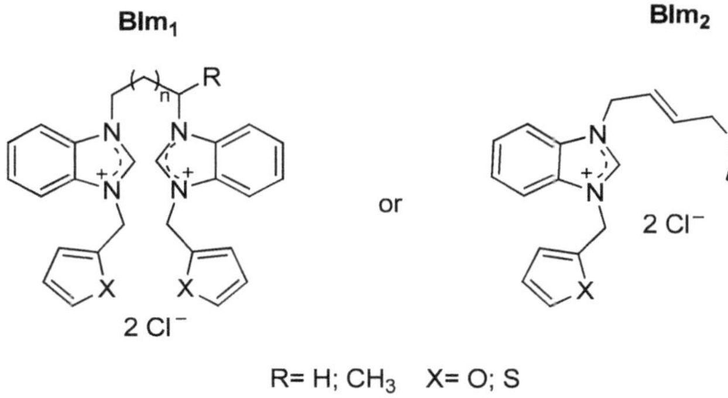

Scheme 35. Novel bis-benzimidazole salts bearing furfuryl and thenyl moieties.

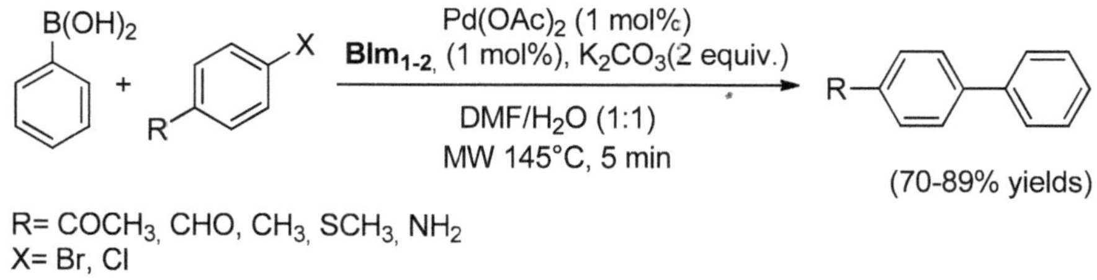

Scheme 36. MW-assisted SMC in presence of **Bim$_{1-2}$**.

No considerable enhancement in reaction yield was observed upon increasing the irradiation time to 60 min. MW irradiation accelerated SMC rate, even when aryl chlorides and especially when bearing electron-withdrawing substituents were used, resulting in 70%–89% yields. The same reaction yields were only achieved when performing the same SMC reactions in an oil bath (145 °C) in 90 min instead of MW heating. Furthermore, control experiments showed that the coupling reaction did not occur in 5 min in the absence of bis-benzimidazole salts.

Another set of benzimidazole salts, containing a trimethylsilylmethyl substituent (**BIm$_3$**), were then synthetized by the same authors (Scheme 37) [71]. The reported SMC reactions were performed using 300 W power MW irradiation at 120 °C in 10 min with a mixture of benzimidazole salts (2 mol %), Pd(OAc)$_2$ (1 mol %) and K$_2$CO$_3$ in DMF–H$_2$O (1:1). The use of the Pd catalyst system, including these benzimidazolium salts, gives better SMC reaction yields (87% vs. 13%) under MW-assisted conditions and lower reaction times than under conventional heating. Good reaction yields were also described under optimized MW conditions even when aryl chlorides were used and particularly so with electron withdrawing substituents (94% for $-NO_2$) (Scheme 38). On the other hand, electron-donating alkyl groups on benzimidazole salts led to better catalytic activity than electron withdrawing groups.

Blm$_3$

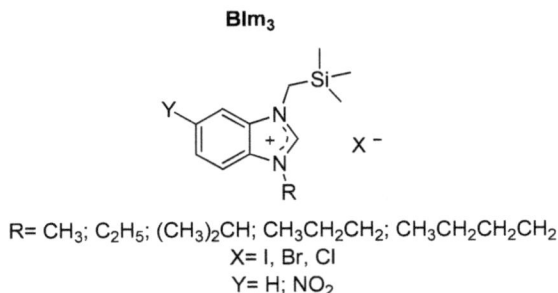

R= CH_3; C_2H_5; $(CH_3)_2CH$; $CH_3CH_2CH_2$; $CH_3CH_2CH_2CH_2$
X= I, Br, Cl
Y= H; NO_2

Scheme 37. General structure of benzimidazole salts which contain trimethylsilylmethyl substituent (**Bim$_3$**).

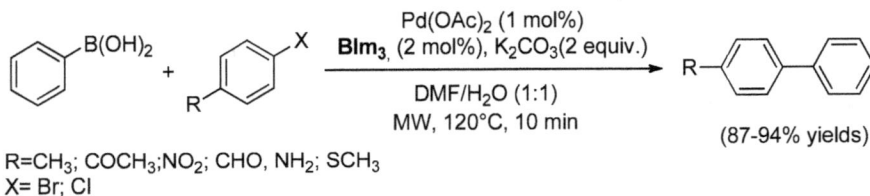

R=CH_3; $COCH_3$;NO_2; CHO, NH_2; SCH_3
X= Br; Cl

Scheme 38. SMC using **Bim$_3$** under MW irradiation.

Of the phosphine-free (ligand-free) Pd catalysts available for use, anionic Pd complexes with imidazolium based ionic liquids have recently attracted increased attention. In this context, Pd complexes of the $[IL]_2[PdCl_4]$ (IL = imidazolium cation) type have been shown to be effective catalysts when used in the SMC of phenylboronic acid with 2-bromotoluene, in the presence of aqueous or pure 2-propanol at 40 °C under MW. In 2-propanol, the maximum yields (89% and 85%) have been described by Silarska et al. [72] for $[dmiop]_2[PdCl_4]$ and $[dmdim][PdCl_4]$ (dmiop = 1,2-dimethyl-3-propoxymethylimidazolium cation, dmdim = 3,3'-[1,7-(2,6-dioxaheptane)]bis(1,2-dimethylimidazolium) cation) which contain cations with a methyl group at the C2 position. When water was used, all $[IL]_2[PdCl_4]$ complexes produced ca. 90% of the 2-methylbiphenyl product. Excellent results were also reported in the C–C cross coupling reaction of various aryl bromides and chlorides (81%–96% yields) (Scheme 39). For instance, the conversion of 2-chlorotoluene was 71% at 70 °C. The formation of Pd(0) nanoparticles during the catalytic reaction was identified by transmission electron microscopy (TEM) and confirmed by mechanistic studies, also involving Hg(0) tests, showing that active soluble Pd species could be generated from Pd(0) nanoparticles in catalytically amounts.

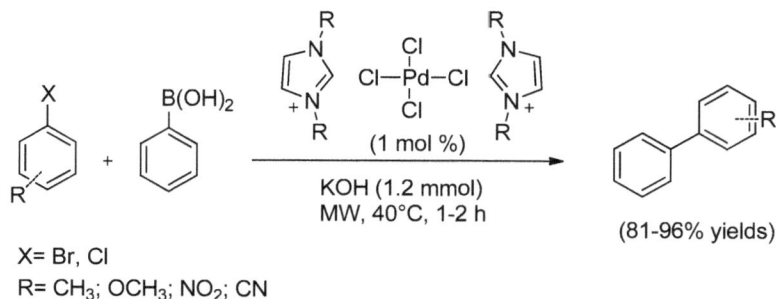

Scheme 39. SMC in the presence of [IL]₂[PdCl₄] complexes, used as efficient catalysts.

Pyridyl–imine and diimine type ligands are very helpful for designing homogeneous catalysts due to their easily switchable properties, but they suffer from rapid decomposition in water. In this context, Hanhan and Senemoglu [73] have prepared a series of unsymmetrical, ionic sulfonated pyridyl imine ligands and their Pd(II) complexes which are suitable for MW-assisted SMC in water (**PyIPd**) (Scheme 40). These ionic Pd(II) complexes proved to be effective when used as 0.1 mol % catalysts in water for SMC reactions in the presence of K₂CO₃ (2 mmol) and terabutylammonium bromide (TBAB) (0.5 mmol), used as the phase transfer reagents. Ionic Pd(II) complexes tolerate a broad range of functional groups on the substrate phenyl rings and good yields (71%–98%) were described in all cases, in only 5 min of MW irradiation (850 Watt), and bulky groups on the ligand increased the efficacy of these catalysts. Furthermore, it was possible to use Pd complexes for up to four cycles before decomposition under MW irradiation. The hydrolysis of the imine C=N bonds in water to give aldehydes and Pd-black is probably more easily prevented by the faster MW assisted SMC.

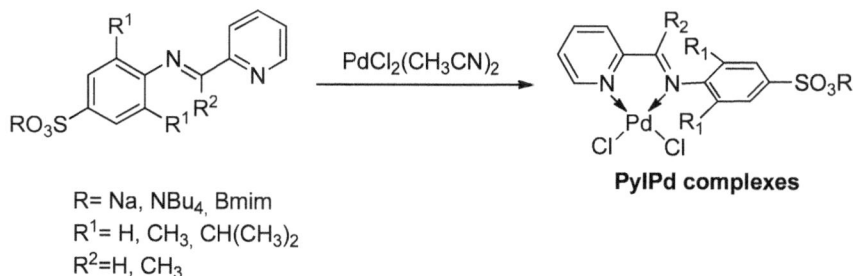

Scheme 40. Water soluble and unsymmetrical sulfonated Pd(II)-pyridyl imine complexes.

A new and efficient pyridine-pyrazole/Pd(II)catalyst system (**PP-Pd**) (Scheme 41) for MW-mediated SMC reactions has been developed by Shen et al. [74]. To the best of our knowledge, this is the first report of such complexes for the catalysis of SMC. The reactions were carried out under MW irradiation in a water/EtOH mixture and were applied to the synthesis of different biaryls. The MW protocol has the advantage of quick reaction (2 min, 60 W), while also avoiding anaerobic conditions or the use of a nontoxic solvent. The best results (75%–90%) were obtained using KOH as the base, in the presence of 0.1 mol % complex **PP-Pd**. The catalysts were applied for up to five cycles and the methodology is an effective synthetic route to biaryl synthesis.

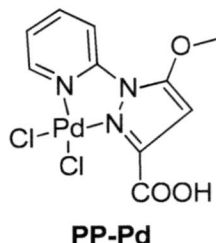

PP-Pd

Scheme 41. New pyridine-pyrazole/Pd(II) species as a SMC catalyst in aqueous media.

In the past decade, the synthesis and design of polyphosphine ligands has been a hot research topic due to the wide structural diversity of their metal complexes.

This type of ligand can easily form multinuclear complexes and also lead two metals into close proximity if the bite distances are small. In this context, the bis-amino(diphosphonite) ligand, p-$C_6H_4\{N\{P(OC_6H_4C_3H_5\text{-}o)_2\}_2\}_2$ (**PL**), has recently been prepared, by Naik et al. [75], by reacting p-C_6H_4-$\{N(PCl_2)_2\}_2$ with o-allylphenol (4 equiv.) in an 85% yield (Scheme 42). When the **PL** ligand was reacted with [Pd(COD)Cl$_2$](2 equiv.) a di-nuclear complex [$\{PdCl_2\}_2\{p$-$C_6H_4\{N(P(OC_6H_4C_3H_5\text{-}o)_2)_2\}_2\}$] (**PL-Pd**) was reached which was proven to be suitable for SMC under MW. This was the first example where a dipalladium(II) complex, containing a bis(diphosphonite) ligand, has been used in a MW–assisted SMC reaction. Of the various solvents and bases tested, the best results (>90%) were obtained under MW irradiation in methanol and K_2CO_3 (1.5 equiv.). An excellent turnover frequency (TOF) of 22,800 was obtained with bromotoluene in 5 min at 60 W and at a catalyst loading of 0.05 mol %. Even aryl halides, in particular sterically hindered bromides with *ortho* substituents, reacted with phenylboronic acid in excellent conversions and yields. An enhanced TOF of 24,000 was obtained for acyl and formyl substituents.

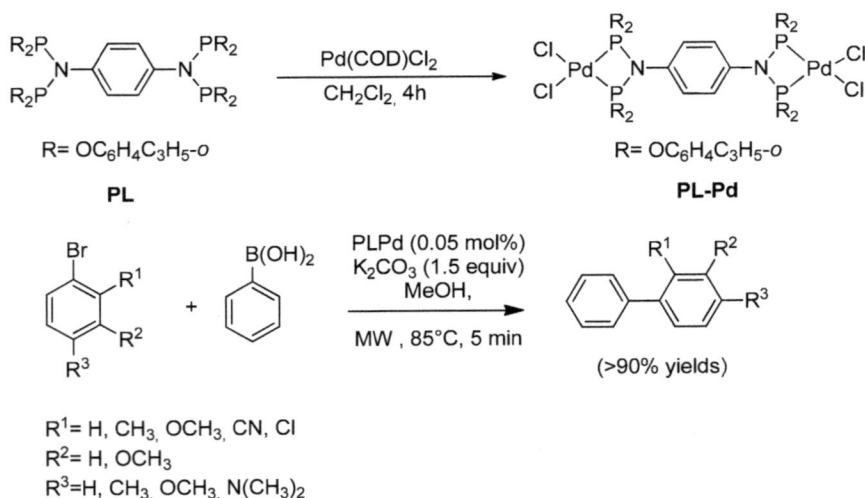

PL

PL-Pd

R= $OC_6H_4C_3H_5$-o

R= $OC_6H_4C_3H_5$-o

PLPd (0.05 mol%)
K_2CO_3 (1.5 equiv)
MeOH,

MW , 85°C, 5 min

(>90% yields)

R^1= H, CH$_3$, OCH$_3$, CN, Cl
R^2= H, OCH$_3$
R^3=H, CH$_3$, OCH$_3$, N(CH$_3$)$_2$

Scheme 42. Pd complex synthesis from bis-amino(diphosphonite) ligand.

Morales-Morales [76] have reported the preparation of a family of SCS pincer compounds, of the [PdCl{C_6H_3-2,6-(CH$_2$SR)$_2$}] {R = tBu, sBu, iBu} type, which are suitable for SMC reaction, under both conventional and MW heating. These series of alkyl-substituted SCS-Pd^{II} pincer complexes

(Scheme 43) were prepared via the direct C–H activation of a series of α,α-*bis*(butylthio)metaxylene proligands (C$_6$H$_4$-1,3-(CH$_2$SR)$_2$, for R = *t*Bu, *s*Bu, *i*Bu substituents) starting from a PdCl$_2$ suspension in toluene. The C–C coupling reaction between the various aryl halides and phenylboronic acid were then carried out in DMF using a catalyst loading of 0.2 mol % in the presence of Na$_2$CO$_3$, used as a base. Quantitative conversions were described for all ligands tested under dielectric heating (75 W) in 10 min irradiation while 10 h were required with conventional heating. In addition, a correlation has been established between the Hammet parameter (σ) of the *para* electron-withdrawing substituents on aryl halide substrates used and the SMC conversion.

R= *t*-Bu; *s*Bu; *i*Bu
SCS Pd^{II} pincer complexes

Scheme 43. General structure of SCS-Pd^{II} pincer complexes.

Palladacycles have been reported to exhibit weak water solubility. However, an example of a comprehensive study into the use of Pd–PTA–imidate complexes in palladium-catalyzed SMC in water has nevertheless been published by Gayakhe et al. [77]. Four new, water-soluble palladacyclic derivatives [Pd(C^N) (imidate) (PTA)] were prepared via bridge-splitting reactions of the corresponding di-μ-imidate complexes with PTA (1,3,5-Triaza-7-phosphaadamantane). The palladacycles were then tested in SMC with four nucleosides and a wide variety of aryl and heteroarylboronic acids (Scheme 44). Good to excellent yields (89%–94%) were obtained in 5 min under MW irradiation for the pyrimidine nucleosides, while an appreciable increase in product formation was observed for cytidine. Similarly, an improvement in reactivity was observed in the coupling reaction with purine nucleosides (adenosine and guanosine) under MW irradiation, in comparison to conventional heating, albeit in longer reaction times than their pyrimidine analogues (10 min).

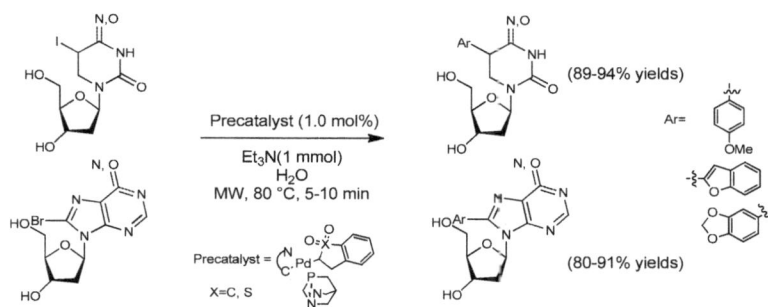

Scheme 44. Pd–PTA (1,3,5-Triaza-7-phosphaadamantane)–imidate complexes in palladium-catalyzed SMC.

2.1.5. MW Promoted Ni Catalyzed SMC

The clear advantages such as, cheapness and earth abundance, of the first row transition metals compared to precious metals have moved the scientific community to study Ni catalyzed SMC [78]. A few papers have studied the effect that MW irradiation has on SMC in the absence of Pd [79,80]. The replacement of traditionally used Pd with less expensive Ni-based catalyst systems was approached by Kappe et al. in 2011 [81]. The Ni-catalyzed SMC reactions were performed using aryl carbamates and/or sulfamates as the electrophilic coupling partners for reactions, because they are readily available and more stable under a variety of reaction conditions than the more reactive triflates. The commercially available and air/moisture stable $Ni(PCy_3)_2Cl_2$ catalyst was employed and the coupling was performed in toluene at 150–180 °C for 10 min in a single-mode MW reactor with accurate internal temperature monitoring. 87% of isolated yield was obtained from the *N,N*-diethyl naftalene carbamate with 5 mol % of the Ni catalyst and boronic acid (2.5 equiv.) at 180 °C for 10 min, whereas the reaction required 20–24 h under conventional heating. The rate enhancement was demonstrated by the authors to be the result of a purely thermal effect as highlighted by control experiments which made use of a reaction vessel made out of strongly MW absorbing silicon carbide (SiC).

2.2. Moving from Homogeneous to Heterogeneous Catalysis in MW Promoted SMC

Supported Pd catalysts (1.0 wt %) and Pd_2dba_3 (1.0 mmol) have been found, by Martins et al. [82], to be active catalysts in the SMC of aryl chlorides with arylboronic acids. Excellent yields, in terms of aromatic ketones, were described (79%–99%) under MW irradiation without the addition of phosphine ligands in 5 min at 120 °C (250 Watt) (Scheme 45).

Selectivity toward benzophenone was not good under optimized conditions in MW when Pd/C was used as the catalyst. Pd/C is as strong MW absorber. The high superficial temperatures reached on the Pd/C surface under dielectric heating probably lead to rapid biphenyl by-product formation. Selectivity toward benzophenone dramatically improved upon reducing the reaction temperature from 120 to 80 °C. Moreover, Pd/C was proven to be more effective when acetonitrile was exploited as reaction solvent under MW irradiation instead of toluene. This improved result can be ascribed to the optimized solubilization of the base ($Na_3PO_4 \cdot 12H_2O$) in a more polar solvent. This catalytic system gave excellent selectivity, when acetonitrile was used as the solvent and Pd_2dba_3 as the Pd source, but the reaction was slower. In fact, the improved base solubility in acetonitrile was found during the hydrolysis of the benzoyl chloride and the discovery of large amounts of benzoic acid. Better selectivity was finally observed using K_2CO_3 in toluene.

The effect of phase transfer catalysts in the Pd_2dba_3 catalyzed coupling reaction was also investigated. Of the phase transfer catalysts studied, polyethylene glycol 200 (PEG-200) was found to be a good choice probably because it avoids the acyl chloride hydrolysis issue as well as promoting better base solubilization.

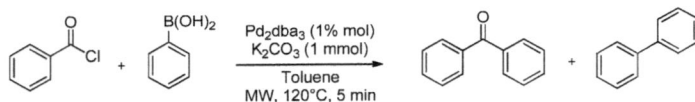

Scheme 45. Synthesis of aromatic ketones without the addition of phosphine ligands under MW irradiation.

2.3. Solid Supported Pd Catalyzed MW Promoted SMC

The combination of MW heating, supported metal (mainly Pd) nanoparticles and ligand-free C-C coupling reactions give very high reaction rates using very small catalyst amounts, therefore increasing the sustainability of the processes. However, separating the homogeneous catalyst (generally palladium metal ions) is a challenging requisite, especially from a pharmaceutical point of view. A first example of maintaining high Pd dispersion in the presence of a support can be found in $Pd(OAc)_2$ species immobilised on mesoporous γ-Al_2O_3 which displayed excellent catalytic activity in the coupling of a number of aryl halides with phenylboronic acid [83]. A DMF/water mixture was employed and it was necessary to heat the reaction via MW irradiation, at 150 °C for 10 min, despite the low catalyst amount (0.2 mol % Pd). It was proposed that the highly dispersed nature of $Pd(OAc)_2$ coupled with its high affinity for γ-Al_2O_3 render the catalyst both active and very stable. XPS measurements revealed that the original Pd(II) was basically transformed to Pd(0) after reaction. The use of metallic nanoparticles, instead of isolated Pd(II) species, can therefore be advantageous. Indeed, the increase in surface area that accompanied the decrease in the nanoparticle size implies that more active sites are available for catalysis. Pd is the most investigated metal, however, the introduction of more than one metal allows the tuning of the electronic properties of the material to be carried out by opportunely choosing the metal or alloy.

Novel CuPd bimetallic nanoparticles have recently been synthesized in oleylamine which acted as both surfactant and solvent, therefore avoiding the need for additional surfactants, ligands or reducing agents [84]. Moreover, the introduction of Cu did not only lower costs because of the need for less Pd in preparation, but catalyst recovery was also more efficient as there were no phosphine ligands, which can interact with the reaction product. The alloyed nanoparticles (0.01 mol %) showed high catalytic activity in the SMC reaction of aryl halide and phenylboronic acid with a number of functionalized substrates in the presence of K_2CO_3 using H_2O; ethanol was used as an environmentally benign solvent system under MW irradiation at 120 °C for 10 min. The high 92% biphenyl product yield, an excellent TON (6000) and good TOF of 72,000 h^{-1} were achieved. The outstanding activity was explained by the presence of electron transfer from Pd to Cu (as revealed by the co-presence of both Pd and Cu metals and oxides) that allows the conversion of Pd^{2+} to Pd^0 to catalyze the reaction. Furthermore, the CuPd nanoparticles were successfully recycled, giving 95%, 90% and 78% conversions after 3 consecutive runs. The PdCu nanoparticles were easily recovered by centrifugation and washed with ethanol after which the solution was decanted and reused in the subsequent run. Both electron donor and electron withdrawing functional groups, such as methoxy, ethoxy, cyano, aldehyde, thiomethyl and dimethylamino, in both aryl halides and arylboronic acids were used and gave excellent yields.

In order to promote separation, the nanoparticles can be strategically supported on insoluble supports which can stabilize the nanoparticles without decreasing the accessibility of the catalytic sites. In addition, the manipulation of supported nanoparticles further improves the sustainability of the material by facilitating both recovery and reuse. The huge effort made to obtain green protocols for reaction has given rise to the synthesis of water soluble ligands on one hand, and polymer supported ligands or Pd-based catalysts on the other [85]. The research was focused on reactions carried out in mild and ligandless conditions, using water as a solvent and at moderate temperatures, including room temperature. The use of heterogeneous catalysts, instead of homogeneous catalysts, therefore has several advantages, caused by their facility of reusability and good compatibility with flow reactors, which facilitate the effective production of materials using continuous processes [86–88]. Several strategies have been designed to effectively immobilize catalysts on solid supports, which include inorganic oxides, polymers and organic–inorganic hybrid materials, in order to obtain, on one hand, a highly dispersed and stable metal phase and, on the other, an active phase which is able to maintain its catalytic activity under MW irradiation. For example, Soni et al. (2014) [89] have described a suitable heterogeneous catalyst for SMC with the synthesis of a new palladium doped silica (Pd/SiO_2) mesoporous catalyst, which proceeds via the sol–gel route, using the P123 surfactant as a structure directing agent. Of the wide plethora of supports available, ordered mesoporous silica materials are very attractive for immobilizing/doping Pd catalysts, due to their large pore size, high surface area and tunable pore structure. The Pd/SiO_2 catalyst (0.5 wt %) exhibited high activity in the coupling of various aryl bromides and a number of arylboronic acids at 85 °C in water with KOH after 7 min of MW irradiation at 245 W power. The catalyst was recycled 9 times without a significant loss in activity. It is worth noting that the reactions were completed in 2 h at 60 °C under conventional heating conditions. In a previous paper, it was shown that SMC was mediated by a 0.5–1 mol % Pd-organosilica heterogeneous catalyst (SiliaCat Pd(0)) at room temperature, with complete conversion of the substrate after 6.5 h [90]. MW heating was needed to convert readily available chloroarenes. Under 200 W power irradiation, aryl chlorides reacted with arylboronic acid in good yields, whilst aryl bromides were coupled in remarkably short reaction times (from 20-to-100 times faster than under reflux).

Verho et al. have observed a 15-fold increase in reaction rate, compared to conventional heating in an oil bath, in the SMC which gave 87%–90% yields in the presence of Pd nanoparticles immobilized on aminopropyl (AmP)-functionalized siliceous mesocellular foam (MCF) (Pd^0-AmPMCF) under MW irradiation [91]. The corresponding reaction under conventional heating, employing an oil bath, gave rise to a four-fold decrease in yield. The authors explained such trends in terms of the existence of more proficient heat transfer to the Pd nanoparticles which was achieved by MW irradiation, giving rise to so-called "hot spots" [92] which possess a temperature higher than the neighboring solvent. Additionally, two blank experiments were carried out; the first reaction was carried out in the absence of Pd^0-AmPMCF and the second was performed with MCF. As expected, no product formation was detected in the absence of Pd.

A MW-assisted SMC was also performed which tested two catalysts, Pd/MCM-41 and Pd/SBA-15, under solvent-free conditions and heating at 120 °C for 10 min in the presence of a base (K$_2$CO$_3$, Cs$_2$CO$_3$ or CsF) [93]. Phenylboronic acid and phenyl iodide produced biphenyl in a 97.4% yield using Pd/MCM-41, whereas phenylboronic acid and phenyl bromide gave biphenyl in excellent yields with both Pd/MCM-4 and Pd/SBA-15. Nevertheless, the reaction with phenyl chloride gave poor yields in the same experimental conditions.

Zheng et al. [94] have reported the MW-assisted SMC of a series of Pd and Pd/Au alloy nanoparticles (with uniform size 1–3 nm) that had been either encapsulated or stabilized by G4-poly (amido-amine) (G4-PAMAM, G stands for the generation) planted in SBA-15. A water/ethanol (3:2) solvent was heated with iodobenzene (1 mmol), benzeneboronic acid (1.2 mmol) and 0.5% mol of catalyst in the presence of K$_3$PO$_4$ (3 equiv.) at 100 °C. In conventional conditions, a 97% yield was observed after the first cycle of 8 h, whereas the yield dropped to 31% after the 2nd cycle. However, MW irradiation granted a 97% yield after only 10 min, a 48-fold rate increase, and the reactions were completed after 30 min. In order to clarify the influence of the catalyst in the MW-assisted reaction, the authors performed SMC with and without a catalyst under MW. Almost no product was obtained, which demonstrates that the effect of MW irradiation was limited to the promotion of the catalyzed reactions. Moreover, it was demonstrated that air retained its influence on the reaction as the procedure that was carried out without N$_2$ bubbling gave a 96% yield. It is worth noting that the presence of a mild base is required to avoid the dissolution of the SBA-15 holding dendrimers in solvent. Interestingly, they found that Pd/Au alloy nanoparticles display higher activity than monometallic catalysts, at similar metal loading values, in catalyzing the coupling between aryl bromide/aryl chloride and arylboronic acid. Although, pure gold gave no catalytic activity, the presence of Au^{3+} in the dendrimer before reduction may be able to decrease Pd^{2+} complexation by internal amine groups or other groups, resulting in the complete reduction of Pd^{2+} to Pd^0. This feature may justify the higher catalytic activity shown by the Pd/Au alloy catalyst over the monometallic Pd catalyst.

An effective methodology has been developed for the coupling of aryl halides (including aryl chloride) and phenylboronic acid under MW irradiation for 30 min in the presence of Pd/Fe$_3$O$_4$@SiO$_2$ and K$_2$CO$_3$ [95]. The catalyst was prepared by simply supporting PdCl$_2$ on Fe$_3$O$_4$@SiO$_2$ in ethylene glycol and it was easily recovered using an external magnet after the reaction and reused without further treatment. The MW-assisted reaction time was shortened and increased yields (99.4% and 99.8%) were achieved, as compared to those obtained for traditional heating (90.6%) in 12 h of reaction time. Interestingly, Pd/Fe$_3$O$_4$@SiO$_2$ exhibited considerable catalytic activity in the SMC involving aryl chlorides and a 68.2% yield was observed in the coupling of chlorobenzene with PhB(OH)$_2$ under MW irradiation for 90 min. In addition, no deactivation was observed even after 6 runs. At the same time, Nehlig et al. reported the exceptional catalytic performance displayed by palladium (0.01 mol % Pd), immobilised on a maghemite nanoparticle core bearing proline at the surface, in the coupling of 4-tolylboronic acid (0.22 mM) to 4-iodonitrobenzene (0.20 mM) in water–ethanol (1:1), under aerobic conditions and using a base [96]. The TOF values were 1940 and 18,000 mol p-I-C$_6$H$_4$NO$_2$ (mol Pd)$^{-1}$·h^{-1}. MW irradiation was used to heat the reaction mixture in accordance with the green chemistry approach and to obtain careful control of the reaction conditions (temperature, stirring, cooling). It was reported that the test reactions performed under thermal heating gave exactly the same results. Moreover, γ-Fe$_2$O$_3$@Cat-Pro(Pd) was reused for seven runs providing total conversion and contained Pd leaching. The catalyst was also extremely stable; it was still active with the same efficiency after being stored in aqueous solution, under aerobic conditions at room temperature for more than 6 months after its synthesis.

Cravotto et al. have observed the advantages on offer when working under MW irradiation and/or ultrasound irradiation [97]. To our knowledge, this is the sole example found in the literature in which the coupled techniques have been successfully employed. They have performed a comparison of a number of different activation methodologies, such as conductive heating in an oil bath (OB), MW irradiation (MW), US horn irradiation coupled with conductive heating in a thermostated oil bath (US/OB) and contemporary US/MW irradiation (US/MW) (as reported in Table 2) in the study of a series of metal-catalyzed C–C couplings in glycerol, along with the SMC (Scheme 46).

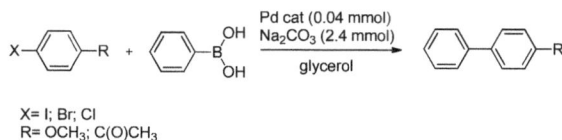

X= I; Br; Cl
R= OCH$_3$; C(O)CH$_3$

Scheme 46. SMC reaction in glycerol (for specific reaction conditions see the Table 2).

Table 2. Cross-coupling yields in glycerol of 4-iodobenzene and phenylboronic acid via a variety of techniques.

Method [1]	T (min)	Yield (%)		
		Pd(OAc)$_2$	PdCl$_2$	Pd/C
OB [2]	60	85	75	44
US/OB	60	99	98	86
MW	15	57	70	60
MW	60	74	92	67
US/MW	60	100	98	94

[1] Reaction conditions: 4-iodomethoxbenzene (2 mmol), phenylboronic acid (2.4 mmol), Na$_2$CO$_3$ (2.4 mmol), ligand-free catalyst (0.04 mmol), glycerol (42 mmol, 4.0 g), 80 °C; [2] OB = oil bath.

In particular, they investigated the coupling between 4-iodoanisole and phenylboronic acid employing ligand-free palladium salts or palladium on charcoal as a model reaction. All the reactions were performed at 80 °C, because glycerol guarantees excellent acoustic cavitation even at high temperatures. PdCl$_2$ and Pd(OAc)$_2$ were more efficient than Pd/C (see Table 2) and it was found that US/OB, MW and simultaneous US/MW irradiation significantly enhanced the reaction rate, with the US/MW and US/OB methods giving the best results due to enhanced heat and mass transfer. In addition to classic palladium salts, a solid ligand free catalyst, a Pd loaded cross-linked chitosan [98], gave outstanding catalytic activity. Enhanced reaction rates, in the order MW/US > US > MW, were observed in a number of Pd catalyzed SMC in glycerol. It is worth noting that ultrasound and MWs significantly improved the reaction of halobenzenes, such as chloroacetophenone, which are poorly reactive toward C–C coupling. Further modification of the Schiff base chitosan by reaction with 2-pyridinecarboxaldehyde and the subsequent palladium deposition resulted in catalysts that are highly active in the coupling of phenylboronic acid with *p*-bromophenol and *p*-bromoacetophenone giving high product selectivity and yields (>99%) in water under MW irradiation without the need for a phase-transfer catalyst [99].

In order to make any catalytic process "greener", the catalyst itself should be "green" in nature and the process should not demand harmful solvents. With such an approach in mind, Baran et al. [100] have prepared a green cross-linked-chitosan-cellulose composite microbead-based catalyst support for Pd ions and tested it in the synthesis of biaryl compounds, via an environmentally safe MW irradiation technique in a solvent free medium. The cellulose particles were incorporated into the chitosan matrix to improve mechanical strength on one hand and the interaction with Pd(II) on the other (see Figure 1).

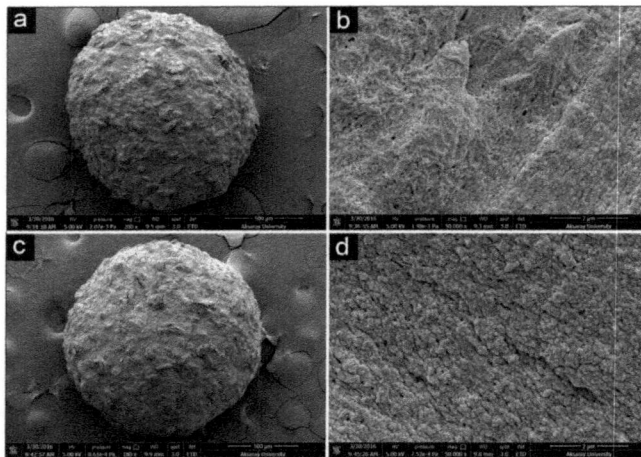

Figure 1. Scanning electron microscopy (SEM) images of (**a,b**) cross linked-chitosan-cellulose composite micro beads; (**c,d**) green chitosan/cellulose-Pd(II) catalyst. Reprinted from [100]. Copyright (2016), with permission from Elsevier.

The reactions between phenylboronic acid and 16 different arylhalides were also examined. The 5 min MW-assisted synthesis of biaryl compounds was carried out in a solvent free medium adopting a small amount of catalyst (0.015 mol %) at 50 °C. The activity of the catalyst was compared with that of commercially available Pd salts $PdCl_2$, $PdCl_2(CH_3CN)_2$ and Na_2PdCl_4, under the same reaction conditions. In addition, the authors tested the performance of the activation method against a conventional heating reflux system (90 °C, 24 h reaction time). They found TON and TOF values of 6600 and 82,500. Na_2PdCl_4 (the best performing commercial product) gave a 47% yield, whilst a much higher product yield (99%) was obtained using the green chitosan/cellulose-Pd(II) catalyst.

In order to reveal the efficiency of the MW irradiation method, 4-methoxybiphenyl was synthesized in the presence of the chitosan/cellulose-Pd(II) catalyst in a conventional heating-reflux system (24 h, 90 °C) using toluene as the solvent, obtaining a 46% product yield vs. the 99% reaction yield produced by green method (5 min, 50 °C) in solvent-free media. The catalyst worked efficiently for up to nine cycles, giving high TONs (2867) and TOFs (35,837). It was concluded that this environmentally harmless catalyst can be transferred and scaled-up into industrial operations.

A new series of solid cross-linked cyclodextrin (α-, β-, and γ-cyclodextrin) based catalysts that are obtained via reticulation with hexamethylene diisocyanate in solutions containing either Pd(II) or Cu(I) cations was investigated by the same group [101]. Diisocyanates are efficient cross-linking agents for cyclodextrins, due to their high reactivity towards hydroxyl groups. The Pd(II) based catalysts have been successfully tested in C–C couplings (Heck and Suzuki reactions). The catalyst proved to be extremely versatile because of its polar structure and was found to be particularly suitable for MW-assisted reactions. Indeed, both native cyclodextrins and cross-linked derivatives are very sensitive to dielectric heating due to their polar structure. Such effect was further enhanced by the embedded cations.

One of the strategies exploited for the preparation of active catalysts that contain solid supported nanoparticles is the functionalization of the surface with organic ligands. Martina et al. have reported [102] the synthesis of a novel cyclodextrin/silica support to host Pd nanoparticles with enhanced stability and reactivity, according to Scheme 47. The authors ascribed the formation of homogeneously dispersed

Pd nanoparticles of small size to the presence of a coordinating group. Indeed, the amino alcohol groups and triazole on the spacer are also able to coordinate Pd species and influence metal nanoparticle content, size and distribution on the silica surface.

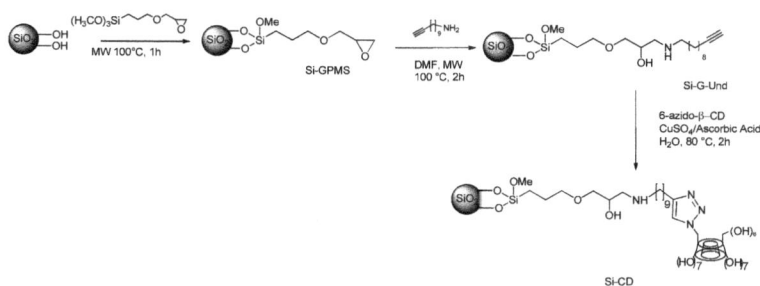

Scheme 47. Synthetic procedure for the preparation of Si–cyclodextrin [102].

An extensive study of the catalytic performance of a Pd nanoparticle supported hybrid cyclodextrin derivative in ligand-free C–C SMC with a large number of aryl iodides and bromides was also carried out. The catalyst exhibited excellent results and MW irradiation abated reaction times. The catalyst underwent 5 cycles and no appreciable loss in activity was detected. Inductively coupled plasma (ICP) analyses of the catalyst after recycling showed negligible Pd leakage, whereas XRD measurements indicated a slight increase in Pd nanoparticle size after usage.

Isfahani et al. [103] have described, for the first time, the synthesis of a new catalyst which is based on Pd nanoparticle immobilized on a nano-silica triazine dendritic polymer (Pdnp-nSTDP) and its subsequent application in C–C coupling reactions. This catalytic system showed high activity in the SMC of aryl iodides, bromides and chlorides with arylboronic acids. These reactions were best performed under MW irradiation (200 W, 70 °C) in a dimethylformamide (DMF)/water mixture (1:3) in the presence of K_2CO_3 (1.5 equiv.) and of only 0.006 mol % of Pdnp-nSTDP (Pd size 3.1 ± 0.5 nm) and gave (2–10 min) the desired coupling products in high yields (>90%). Pdnp-nSTDP was also applied as an effective catalyst for the MW preparation of a series of star- and banana-shaped compounds, using benzene, pyridine, pyrimidine or 1,3,5-triazine units as the central core, via a SMC reaction (see Scheme 48). 2,6-dibromopyridine (for banana-shaped) or 1,3,5-tribromobenzene, 2,4,6-trichloropyrimidine and 2,4,6-trichlorotriazine (for star-shaped) were used as the starting materials. The Pdnp-nSTDP catalyst was easily recovered and reused without significant loss in catalytic activity making this a sustainable process. The analysis of palladium leaching from the Pdnp-nSTDP catalyst by ICP pointed out that only a trace amount of palladium leached in the first two runs.

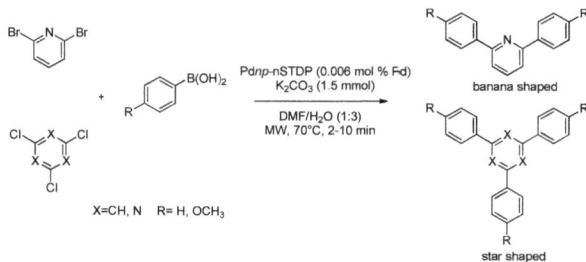

Scheme 48. Star- and banana-shaped compounds with benzene, pyridine, pyrimidine or 1,3,5-triazine units as the central core.

Following a similar approach, Borkowski et al. [104] achieved very good results using Pd supported on a siloxane polymer functionalized with imidazole groups (Scheme 49) in a SMC model reaction between 2-bromotoluene and phenylboronic using conventional heating and MW energy. Environmentally friendly solvents, such as H_2O and a 2-propanol/H_2O mixture, were employed.

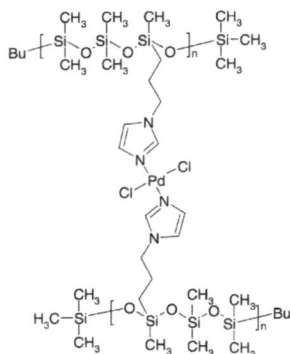

Scheme 49. Proposed structure of Pd catalyst supported on a siloxane polymer functionalized with an imidazole group. Reprinted from [104], Copyright (2016), with permission from Elsevier.

The Pd supported catalyst showed very good recyclability (eight runs) and high activity at 40 or 60 °C. However, even better results and yields of 90%–100% were obtained after 1 h in three sequential runs with the same catalyst when MW heating was used instead of conventional heating. Interestingly, the application of MW conditions enabled the less reactive 4-chlorotoluene and 1-chloro-4-nitrobenzene to react.

Ceramic materials, based on aluminosilicates, react quickly under MW irradiation with consequent rapidity and uniformity of heating. MW irradiation was very recently employed during a SMC reaction between phenylboronic acid and a broad range of aryl halides in the presence of K_2CO_3 (1.12 equiv.), TBAB as an additive (0.05 mmol) and 0.016 mol % (0.7 mg) of Pd nanoparticles immobilised into Poly(N-isopropylacrylamide) (PNIPAAM), which had previously been grafted onto halloysite nanotubes (HNT) using water as the solvent [105] (see Scheme 50).

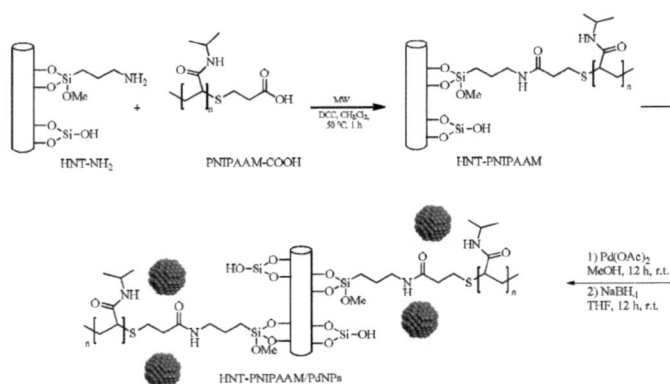

Scheme 50. Representation of the synthesis of Pd nanoparticles immobilised into Poly(N-isopropylacrylamide) (PNIPAAM) previously grafted onto halloysite nanotubes (HNT). Reprinted from [105]. Copyright (2016), with permission from Royal Society of Chemistry.

The reactions were performed at varying temperatures (between 25 and 120 °C) under MW irradiation. The catalyst was recuperated via centrifugation and used again for five cycles (Figure 2). Yields ranged from 73% to 99% and no by-products were formed. Nevertheless, aryl chlorides gave almost no conversions under the same reaction conditions.

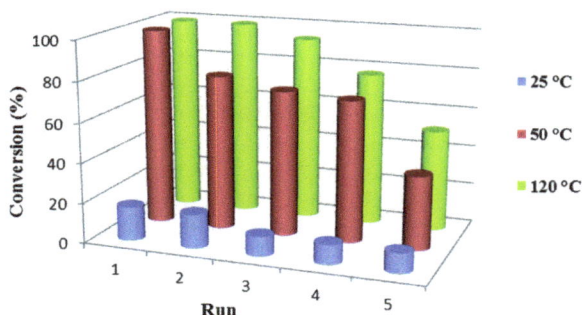

Figure 2. Effect of temperature on the SMC between 4-bromoacetophenone and phenylboronic acid (reaction conditions: phenylboronic acid (0.547 mmol), 4-bromoacetophenone (0.55 mmol), K_2CO_3 (0.615 mmol), EtOH/H_2O 1:1 (1.2 mL), catalyst (0.16 mol %, 7 mg) under MW irradiation). Reprinted from [105]. Copyright (2016), with permission from Royal Society of Chemistry.

The good results may well be due to the thermo-responsive behavior of PNIPAAM. When recycling tests were carried out at temperatures above the lower critical solution temperature (LCST > 32 °C) of PNIPAAM, the system catalyzed the formation of the desired biphenyl-4-acetophenone product in good yields, even after 4 runs, ranging from 99% to 77%. On the other hand, very low conversions (about 10%) were obtained at 25 °C as expected for the thermo-responsive behavior of the polymer. It was proposed that, the PNIPAAM chains collapse, at above the LCST, to form a dense hydrophobic layer covering the surface of HNTs hence promoting the mass-transfer of the hydrophobic reagents (Figure 3). However, it was observed that prolonged MW irradiation induced a decrease in both conversion and yield, due to homocoupling, dehalogenation products and the degradation of both reactants and products.

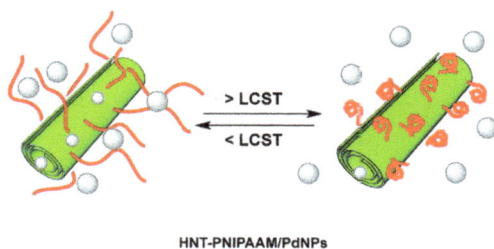

HNT-PNIPAAM/PdNPs

Figure 3. Sketch representation of the swelling behavior of HNT–PNIPAAM/Pd nanoparticles at the critical solution temperature (32 °C). Reprinted from [105]. Copyright (2016), with permission from Royal Society of Chemistry.

Environmentally friendly halloysite-dicationic triazolium salts (HNT-IL) were used as supports for palladium catalysts and used in the SMC between phenylboronic acid and a set of aryl halides. The reaction was carried out in water at 120 °C and using K_2CO_3 [106]. All the reactions were run for 10 min. HNT-IL/Pd was used at 0.1 mol % loading. Conversions were high with yields up to 99% in the case of anisole derivative substrates. On the other hand, the conversion of other substrates increased when the H_2O/EtOH (1:1) mixture was used, due to the good solubility of the organic reactants and of the inorganic base in the co-solvent. It was proposed that the impressive conversions for anisole derivatives may be justified by a synergic effect caused by the ethereal alkyl chains, or the aromatic rings, of the support and by MW irradiation on the cross-coupling of anisole in water. The catalyst was recovered via centrifugation and reused for 5 runs, affording biphenyl-3-anisole in 99%–90%. The same authors [107] found that the use of MW irradiation cut down the reaction time and enhanced the conversion, with respect to traditional heating, when using the same catalyst, for which the modification of the external halloysite nanotube surface with octylimidazolium moieties (HNT-IL) was performed using MW irradiation under solvent-free conditions. The HNT-IL/Pd catalyst (1 mol %) was tested in the SMC of a number of different kinds of aryl bromides, iodides and less reactive aryl chlorides with phenylboronic acid in under 10 min MW irradiation at 120 °C in the presence of K_2CO_3. No by-products were detected in all cases and the conversions reported in Table 3 correspond to yields. Full conversions were obtained when aryl bromides and iodides were used, whilst less reactive aryl chlorides gave lower yields. The authors pointed out that no product was detected when traditional heating was used on aryl chlorides. Moreover, comparable results were achieved with both electron-donating and electron-withdrawing substituents on aryl halides.

Table 3. SMC reaction of phenylboronic acid with a number of different halides under optimized reaction conditions under MW irradiation [1].

Ar-X	Yield (%) [2]
4-Bromoacetophenone	>99
3-Bromoacetophenone	80
4-Bromobenzaldehyde	>99
3-Bromobenzaldehyde	>99
4-Bromoanisole	>99
3-Bromoanisole	90
4-Bromotoluene	>99
2-Bromotoluene	>99
4-Bromoaniline	78
2-Bromobenzonitrile	>99
3,5-Bis(trifluoromethyl)bromobenzene	>99
1-Bromo-2,4,6-triisopropylbenzene	<5
1-Bromo 4-nitrobenzene	>99
4-Iodoacetophenone	>99
4-Iodotoluene	>99
Methyl 4-iodobenzoate	>99
2-Chlorobenzaldehyde	14
4-Chlorobenzaldehyde	33

[1] Reaction conditions: aryl halide (1.01 mmol), phenylboronic acid (1 mmol), K_2CO_3 (1.12 mmol), solvent (1.2 mL), HNT-IL/Pd (1 mol %), 10 min MW irradiation at 120 °C; [2] Determined by ^{1}H NMR.

It was therefore shown that MW irradiation can decrease the reaction time and improve conversion when compared to traditional heating (50 °C for 19 h). The outstanding reactivity was explained by the introduction of an ionic liquid onto the external surface of the halloysite nanotubes. Such a small amount was able to induce a dramatic change in the heating profiles by changing the overall dielectric properties of the reaction mixture, which consequently led to improved yields.

The biogenic synthesis of cellulose supported Pd(0) nanoparticles (NPs) employing the hearth wood extract of *Artocarpus lakoocha* Roxb and their use as a flexible and effective catalyst in Suzuki and Heck couplings under MW heating has been reported [108]. Catalyst synthesis was followed by the observation of color changes (Figure 4); the aqueous solution C, containing oxyresveratrol, $PdCl_2$ and cellulose initially displayed a light brown color and then gradually turn to black (Figure 4D). The black solid deposited on the bottom of the test tube was solid Pd(0) NPs@cellulose (Figure 4E). In a 10 mL MW glass vial, 0.5 mol % (0.028 g) Pd(0) NPs@cellulose, 0.5 mmol 2-bromobenzaldehyde,

0.75 mmol phenylboronic acid and 1.5 mmol K_2CO_3 were mixed in 5 mL H_2O. The mixture was stirred at 80 °C for the stipulated time in a MW oven. It was found that the $PdCl_2$ salt gave disappointing results (yield < 30%) as a catalyst, compared to Pd(0) NPs@cellulose (product yield > 90%), under the same reaction conditions. In addition, the catalyst was filtered off and simply washed with acetone and stored for further reactions. Its reusability was proven for up-to 10 runs without measurable Pd leaching.

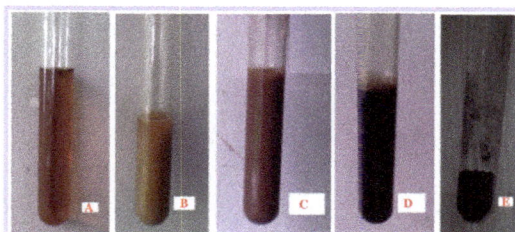

Figure 4. Photograph of (**A**) hearth wood extract of A. lakoocha Roxb in water; (**B**) solution A with cellulose; (**C**) Solution B with $PdCl_2$; (**D**) Solution C after 15 min warming in heating bath at 60 °C; (**E**) solid recyclable Pd(0) NPs@cellulose. Reprinted from [108]. Copyright (2015), with permission from Elsevier.

Furthermore, two comparative cross coupling reactions were carried out using the same substrates, the former was performed in MW heating and the latter using an oil bath at 100 °C. It was shown that the MW-assisted reaction required shorter times and gave higher yields than conventional heating. The SMC was then carried out on a large number of substituted arylbromides (0.5 mmol) and arylboronic acids (0.75 mmol) as substrates using Pd(0) NPs@cellulose (0.5 mol %) in the presence of K_2CO_3(1.5 mmol), H_2O (5 mL) at 80 °C under MW irradiation. The results are reported in Scheme 51.

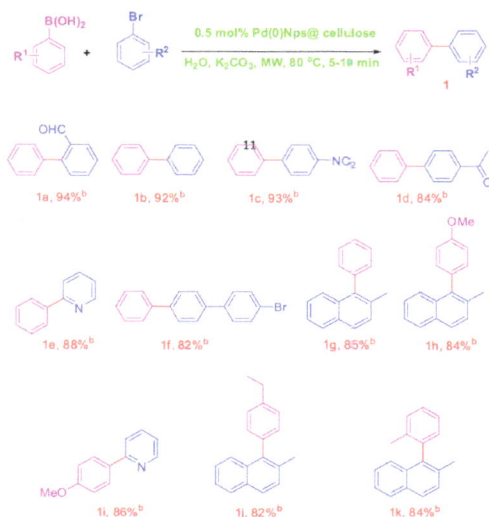

Scheme 51. SMC reaction between phenylboronic acid and substituted arylbromides under MW heating. Reaction conditions: arylbromides (0.5 mmol), phenylboronic acid (0.75 mmol), Pd(0) NPs@cellulose (0.5 mol %), K_2CO_3(1.5 mmol), H_2O (5 mL), 80 °C, under MW irradiation (isolated yield). [b] Isolated yields. Reprinted from [108], Copyright (2015), with permission from Elsevier.

Generally, all reactions provided excellent coupling product yields under MW heating (yields in the 82%–94% range). However, arylbromides, containing electron withdrawing groups (**1a, 1c**), gave yields that were higher (93%–94%) that those obtained with their electron donating counterparts. It was found that the coupling reaction occurred also in the case of hetero aryl bromides (**1e, 1i**) and ortho substituted aryl bromide (**1a, 1g, 1h, 1j, 1k**) as substrates. In conclusion, substrate steric and electronic factors did not markedly influence the product yield, due to the high activity of the cellulose supported Pd(0) catalyst.

Baran et al. [109] have very recently reported the design of highly thermally stable silane and Schiff base-modified sporopollenin microcapsules as a support for Pd(II) catalyst (see Scheme 52). This catalyst displayed high selectivity towards SMC reactions by producing TONs and TOFs as high as 40,000 and 400,000. In order to prove the efficiency of MW heating, the authors tested the behavior of the catalyst on the reaction in a reflux heating setup at 100 °C for 48 h in toluene and found that MW irradiation proved to be more efficient than reflux-heating.

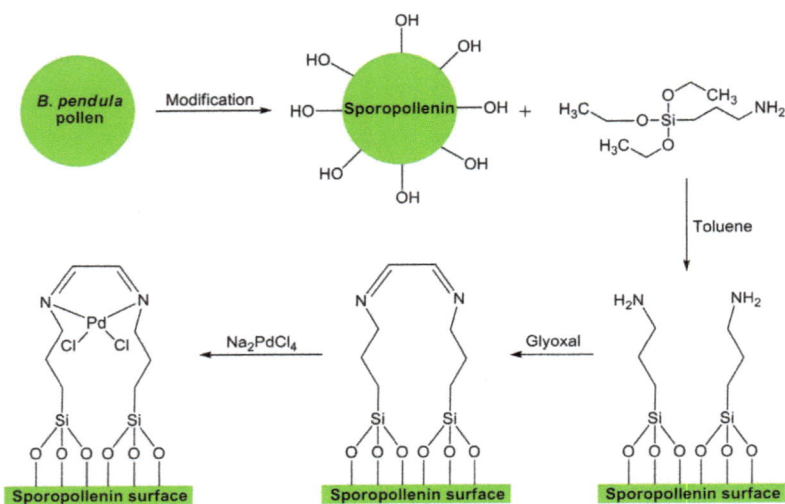

Scheme 52. Design of sporopollenin microcapsule supported Pd(II) catalyst. Reprinted from [109]. Copyright (2016), with permission from Elsevier.

In a related paper, where a chitosan-pyridil-based Pd(II) catalyst was investigated in the MW-assisted synthesis of biaryls in SMC reactions [110], outstanding TON and TOF values were obtained with very low catalyst loading values (5×10^{-3} mol %) in 5 min and catalytic activity was retained up to 7 cycles. In addition, the authors demonstrated that the catalyst was more suitable for MW irradiation than for a conventional reflux-heating system, for which low reaction yields, TON and TOF values were found. In another very recent paper [111], *Ulva* sp. (a fast growing macroalga) particles were incorporated into a chitosan matrix to favor interactions with Pd ions. The presence of functional groups, such as thiol, hydroxyl, carboxyl, amino and imidazole moieties, on the cell surface, mean that the biomass derived from such macroalga can be used to uptake metal ions. Moreover, Pd ions can coordinate with the imine groups belonging to the glutaraldehyde cross-linked chitosan Schiff base. The catalytic performance of chitosan-*Ulva* supported Pd(II) green catalysts was investigated in the biaryl synthesis via the SMC reaction without using any solvent under MW irradiation. High selectivity and efficiency were found in the reactions of phenylboronic acid with a number of aryl halides under MW irradiation in only 4 min at 50 °C (TON and TOF values: 4950 and 75,000) without showing any activity drop for 8 cycles.

Shah et al. [112] have compared the time and yield benefits of MW heating reactions with those carried out under conventional heating. The authors prepared and employed Pd nanoparticles supported on cross-linked polystyrene resins to successfully catalyze the one pot synthesis of asymmetric terphenyls, a useful industrial commodity, starting from 1-bromo-4-iodobenzene via the sequential addition of various boronic acids, according to Scheme 53:

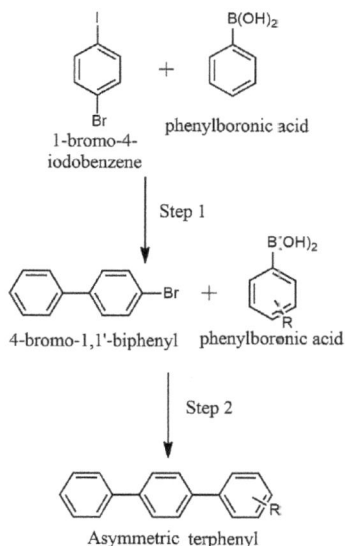

Scheme 53. Synthesis of asymmetric terphenyls. Reprinted from [112]. Copyright (2016), with permission from Elsevier.

The reaction was carried out according to a two-step procedure. Step 1: 4-iodo bromobenzene (1.0 mmol) phenyl-boronic acid (1.0 mmol) sodium carbonate (2.0 mmol), Pd catalyst (200 mg), ethanol (1.5 mL) and water (1.0 mL) were added into a 10 mL vial. Then the mixture was heated at 140 °C under microwave irradiation (100 W). Step 2: 1.0 mmol of substituted phenylboronic acid (1.0 mmol) was added after 10 min and the mixture was further heated under the same conditions for 10 min. The same reaction was also performed under conventional heating and the authors contrasted reaction times and yields with those found upon MW heating. Heating with MW was not only very effective, but is also green. Indeed, the reaction time was reduced (8 min) compared to conventional heating (12 h). The yields obtained were 75% and 82%, respectively. It is worth noting that the continuous purging of inert gas was not required during MW heating. The authors did not observe any side products. The excellent reactivity was explained by the choice of non-functional resin, which probably rendered the reaction sites more accessible and less than 0.1% palladium was sufficient to carry out the C–C coupling reactions. The presence of permanent pores and the hydrophobic nature of the resin can favor the mass transfer of the materials inside the resin matrix. Effective contact among the reactants and the catalyst leads to high reaction rates (TOFs in the order of >103) for bromo derivatives.

Kaur et al. [113] have optimized a very simple procedure for a simple, robust, efficient and recyclable material that is based on Pd nanoparticles encapsulated inside the matrix of Amberlite XAD-4 (a commercial polystyrene–divinylbenzene cross-linked macroporous resin). XAD-4 can be employed in a variety of solvents, due to the low degree of cross-linking, it guarantees facile mass transfer and, above all, is chemically and mechanically stable under MW heating. Encapsulated Pd nanoparticles were successfully tested in the ligand-less SMC reactions of a variety of aryl bromides under MW heating conditions giving high yields. The reaction conditions were reported as the following; phenylboronic acid (1.8 mmol), aryl bromide (1.5 mmol), sodium carbonate (2.0 mmol), ethanol (1.5 mL), water (1.0 mL) and catalyst (200 mg wet resin) were heated in a 10 mL vial using a CEM MW (140 °C, 100 W) for 8 min. Interestingly, catalytic activity was maintained even in the presence of a thio-containing substrate and a 54% yield was obtained with chlorobenzene.

There are few examples in the literature in which a supported Pd catalyst shows higher activity than a homogeneous system [114,115]. An explanation for this behavior can be found in the improved stability of the active Pd species in the heterogeneous catalyst. Giacalone et al. [116] have adopted a synthetic strategy to achieve the efficient stabilization of very small Pd nanoparticles and therefore grant enhanced catalytic activity. They prepared a C_{60}-ionic liquid hybrid covalently linked to three different solid supports (amorphous silica, SBA-15 and $Fe_2O_3@SiO_2$) and the resulting materials were used to immobilize and stabilize the metal active phase. The synthetic approach was successful as outstanding catalytic activity was achieved in SMC both under classical heating and under MW irradiation (TOFs up to 3,640,000 h^{-1}) and the silica-based catalyst showed full recyclability even after 10 cycles. The modular synthesis of these supported catalysts, depicted in Figure 5, started from the reaction between triethoxy-3-(2-imidazolin-1-yl)propylsilane and the C_{60} hexakis-adduct **1** to form compound **2**. **2** was then grafted onto varying supports to obtain compounds **3a–c**, which reacted with 1-methylimidazole and gave the corresponding **4a–c**. The obtained C60-IL hybrids **4a–c** were used for the immobilization of palladium nanoparticles via anion metathesis with tetrachloropalladate ions. Further reduction with $NaBH_4$ supplied catalysts **5a–c**.

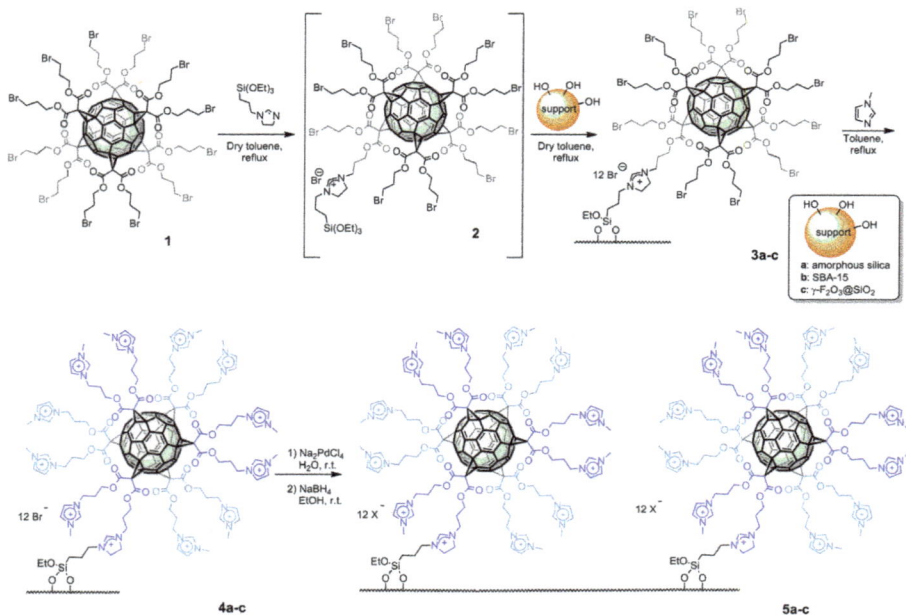

Figure 5. Synthesis of catalysts **5a–c**. Reprinted from [116]. Copyright (2016), with permission from the Royal Society of Chemistry.

Catalysts **5a–c** were also tested at just 0.01 mol % loading in a SMC carried out between various aryl bromides and phenylboronic acid at 50 °C in a mixture of ethanol and water (1:1) in the presence of K_2CO_3. High yields were obtained in short times (1–2 h) in almost all cases. The silica-containing catalyst **5a** was less active than the corresponding SBA-15 and g-$Fe_2O_3@SiO_2$ based catalysts **5b** and **5c**.

Interestingly, catalysts **5a** and **5b** were chosen to carry out experiments at higher temperatures (120 °C) both under MW irradiation and conventional heating, according to Scheme 54, in order to investigate the occurrence of any benign effect that may be due to the use of MWs during the reactions.

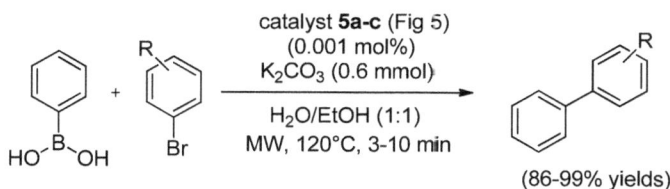

Scheme 54. Reaction conditions used for the SMC reaction performed in the presence of the **5a–c** catalysts.

The results of the comparative catalytic tests are summarized in Table 4. In more detail, the reactions between 4-bromoacetophenone or 3-bromoanisole and phenylboronic acid were performed in the presence of catalyst **5b** (0.001 mol %), which was the most active at 120 °C. The authors found no specific effect on the formation of the 4-phenylacetophenone product when the reaction was carried out for 10 min using MWs, since both reactions reached 100% conversion within this time (Table 4, entries 1 and 2).

Table 4. SMC catalysed by materials **5a–b** at 120 °C [1].

Entry	Catalyst	R	T (°C)	T (min)	Conv. [2] (%)	TON [3]	TOF [4] (h^{-1})
1	5b	4-COCH$_3$	120 Δ [5]	10	>99	100,000	600,000
2	5b	4-COCH$_3$	120 MW	10	>99	100,000	600,000
3	5b	4-COCH$_3$	120 Δ	3	88	88,000	1,760,000
4	5b	4-COCH$_3$	120 MW	3	>99	100,000	2,000,000
5	5a	4-COCH$_3$	120 MW	3	>99	100,000	2,000,000
6	5b	3-OCH$_3$	120 Δ	10	56	56,000	336,000
7	5b	3-OCH$_3$	120 MW	3	73	73,000	1,460,000
8 [6]	5b	4-COCH$_3$	120 MW	3	91	182,000	3,640,000

[1] Reaction conditions: phenylboronic acid (2.2 mmol), 4-bromobenzaldehyde or 3-bromoanisole (2 mmol), K$_2$CO$_3$ (2.4 mmol), EtOH (0.8 mL), H$_2$O (0.8 mL), and catalyst (0.001 mol %) were stirred under MW irradiation (11 W) or under conventional heating for the indicated time; [2] Determined by ^{1}H NMR; [3] TON defined as mol of product/mol of catalyst; [4] TOF = TON/h; [5] Δ, conventional heating; [6] Phenylboronic acid (4.4 mmol), 4-bromobenzaldehyde (4 mmol), K$_2$CO$_3$ (4.8 mmol), EtOH (1.6 mL), H$_2$O (1.6 mL), and catalyst (0.0005 mol %).

However, a small difference between the two activation methods was observed when the heating time was decreased to 3 min (see entries 3 and 4). This was probably due to the faster and uniform heating of the reaction mixture during MW irradiation, as compared to conventional heating. It was proposed that the higher heating ability of the MW irradiation is the discriminating factor that led to a quantitative yield in such a short time. On the other hand, 10 min was enough time to mitigate any difference between the two methods and a quantitative yield was also obtained when conventional heating was adopted (Table 4, entries 1 and 2). In addition, the less active catalyst SiO$_2$-C60-IL-Pd (**5a**) also efficiently catalyzed the reaction under MW irradiation giving the coupling product in a quantitative yield in 3 min (entry 5).

A TON of 100,000 and a TOF of 2,000,000 h^{-1} were reached in **5b** (entry 4). Nevertheless, the MW-assisted reaction of the less reactive 3-bromoanisole substrate, for 3 min at 120 °C, led to improved conversions as compared to the same reaction carried out under conventional heating for 10 min, proving that there was a beneficial effect (entries 6 and 7). Finally, a further decrease in catalyst **5b** loading, to 0.0005%, gave the outstanding TON and TOF values of 182,000 and 3,640,000 h^{-1}, respectively (entry 8).

A novel Fe$_3$O$_4$@SiO$_2$/*N*-[3-(trimethoxysilyl)propyl] ethylenediamine-Pd(II) catalyst was successfully synthesised and tested in the Suzuki cross-coupling reaction between phenyl boronic acid and

4-iodoanisol [117]. The authors proposed an elegant multivariate approach to optimise the reaction by varying different parameters. In the first stage, the effects of time and temperature, as well as of the interaction between them were evaluated by using factorial design. In the second step, the Doehlert matrix was used to find the optimal reaction conditions, as the choice of the solvent, the nature of the base and the catalyst loading for the Suzuki cross-coupling. The authors claimed that the optimized protocol can be applied to many substrates, with yields ranging from 71% to 96% after 6 min under microwave irradiation. In addition, the catalyst was recovered and reused for 3 runs with restrained activity loss [117]. Moussa et al. [118] have reported the optimization of an easy laser reduction method to synthesise Pd nanoparticles immobilised on partially reduced graphene oxide (PRGO) and have demonstrated their high activity in Suzuki, Heck and Sonogashira cross-coupling reactions. The strategy adopted in improving this procedure makes use of the photogenerated electrons in GO that can reduce the Pd ions and, at the same time, can partially reduce GO, thus forming PRGO-supported Pd nanoparticles. It was shown that it is possible to modify the reducing environment by tuning the composition of the solvent, thus controlling the growth kinetics of the Pd nanoparticles. The authors compared the catalytic activity of the Pd-PRGO nanocatalysts A (20 µL of $Pd(NO_3)_2$ added to 6 mL of GO solution in 2 mg of GO/10 mL of deionized water), B (in 50 vol % ethanol–water), and C (in 50% methanol–water mixtures) in the SMC reaction of bromobenzene and phenylboronic acid in a mixture of H_2O/EtOH (1:1) at room temperature. In more detail, A and B show comparable reactivity and result in high conversions of 100% and 95%, respectively, after 45 min at room temperature. Catalyst C displays only 88% conversion in 45 min. The same trend was noted when the catalyst loading was decreased to 0.008 mol % after 8 h at room temperature (A, B, and C result in 95, 92, and 80% conversions, respectively).

Interestingly, the reaction was surprisingly fast for catalyst A in the presence of the same Pd loading at 120 °C under MW irradiation, converting 62% of the bromobenzene after 2 min. The complete formation (100%) of the biphenyl product was observed after 5 min under the same reaction conditions. The authors also claimed that MW heating at 120 °C was needed to reach full conversion with this very low catalyst loading. An outstanding turn over number (TON) of 7800 and a turnover frequency (TOF) of 230,000 h^{-1} were obtained for catalyst A.

It is worth noting that a small conversion (5%) was observed after 6 h in refluxing conditions at 120 °C when a similar SMC reaction was carried out under conventional thermal heating, using 0.008 mol % of catalyst A. This finding definitely proves the beneficial effect of MW irradiation. Indeed MW, as a direct and rapid heating source, can increase the cross coupling reaction rates. This is in agreement with recent studies on the non-uniform heating at the surface of heterogeneous catalysts and on the production of hot spots by MW irradiation, resulting in non-equilibrium local heating localized at the surface of the metal nanoparticles present on the catalysts. This phenomenon was observed as occurring to the dimethylsulfoxide (DMSO) molecules in the proximity of Co nanoparticles under MW irradiation by real-time in situ Raman spectroscopy [119]. Interestingly, non-equilibrium local heating was only induced under MW irradiation, not under conventional heating. The Raman spectra collected for dimethylsulfoxide (DMSO) under MW heating at 0.25 s intervals, in the temperature range of 300–500 K, are shown in Figure 6a. The T_r (temperatures measured by the intensity ratios of Stokes to anti-Stokes lines employing in situ Raman scattering measurements) and T_f (temperatures determined by the fiber-optical thermometer) plotted vs. t (time) are reported in Figure 6b,c. Some spikes were detected in the measured temperatures, as defined by Raman, demonstrating that abnormally high temperatures occurred at 3.4 min (433 K), 5.8 min (473 K), and 7.3 min (423 K).

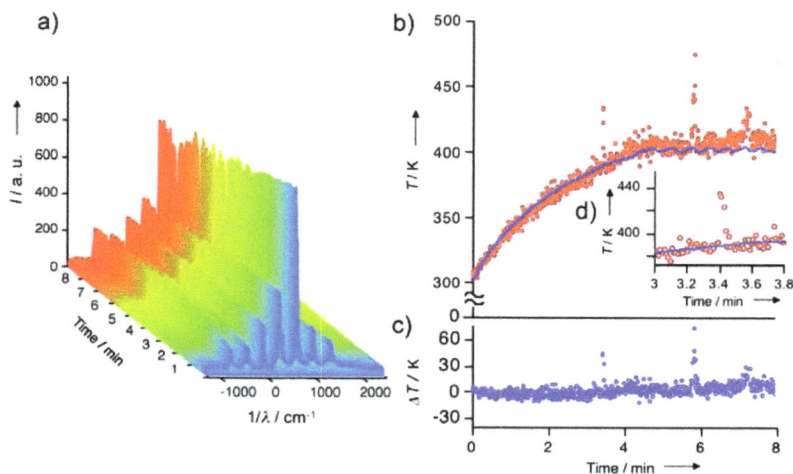

Figure 6. Non-equilibrium local heating induced by MW irradiation. (**a**) Time-dependent Raman spectra of DMSO heated by MW irradiation at 0.25 s intervals for 8 min in the temperature range 300–500 K; (**b**) T_r (temperatures determined by the intensity ratios of Stokes to anti-Stokes lines using in situ Raman scattering measurements) (red circles) and T_f (temperatures monitored by the fiber-optical thermometer) (blue line) vs. t (time) plot; (**c**) ΔT ($\Delta T = T_r - T_f$) vs. t (time) plot; (**d**) Enlarged view of Figure 6b in the range 3.0–3.8 min. Reprinted from [119]. Copyright (2010), with permission from American Chemical Society.

Horikoshi et al. [120] investigated the hot-spot generation on Pd nanoparticles supported on activated carbon during the synthesis of 4-methylbiphenyl by a Suzuki coupling reaction occurring in a nonpolar solvent by recording the events with a high-speed camera. The authors found that the hot-spots are produced during reaction under exposure to the microwave electric field, whilst they are scarcely formed under magnetic field conditions. It was demonstrated that the efficiency in hot-spot generation is related to the type of microwave generator (semiconductor or magnetron) that is employed. Interestingly, the microwave H-field originated by a semiconductor generator has a positive effect on the reaction, resulting in enhanced yields. On the contrary, the authors claimed that the formation of hot-spots on the activated carbon under E-field conditions negatively influences the product yields, due to the Pd nanoparticle aggregation. Therefore, hot-spots can have both positive and negative effect and it is of pivotal importance to control the formation of hot-spots under E-field conditions whn the reaction is performed in the presence of a heterogeneous catalyst in a nonpolar solvent. Nevertheless, the reaction can be carried out with microwaves under H-field conditions. It is worth noting that hot-spots also play an important role in the synthesis of heterogeneous catalysts, as will be discussed in detail in the following. Pd nanoparticles supported on graphene platelets have been successfully employed in a SMC between 4-bromoanisole and potassium phenyltrifluoroborate using 1 mol % of Pd and K_2CO_3, used as a base, in $MeOH/H_2O$ as the solvent at 80 °C under conventional and MW heating [121]. The separation of the catalyst from the reaction mixture after each run was easily achieved by washing with a solvent mixture ($EtOAc/MeOH/H_2O$) and further centrifugation. Interestingly, Pd/graphene was reused at least 8 times without losing activity when MW irradiation (MW) conditions were adopted (first row of Table 5). However, the activity significantly dropped after 5 cycles under conventional heating (Δ) reaction conditions (second row of Table 5).

Table 5. Pd/graphene recyclability and yields in the SMC across different cycles.

Cycle		1	2	3	4	5	6	7	8
Yield (%)	MW [1]	>99	>99	>99	>99	91	95	97	>99
	Δ [1]	>99	>99	>99	90	83	58	44	40

[1] MW irradiation for 2 h or Δ for 20 h.

The authors performed an inductively coupled plasma mass spectrometry (ICP-MS) analysis of the washings after the 1st cycle under both reaction conditions (MWs and conventional heating), and found varying Pd leaching levels. In particular, 1103 ppb of leached Pd were detected under MW irradiation conditions, whilst 15 ppb of Pd were found after the 1st run using conventional heating. Pd leaching was therefore not the cause of the activity decrease observed after 5 cycles under conventional thermal conditions. Transmission electron microscopy (TEM) measurements highlighted that the average diameter of the nanoparticles after 8 cycles (originally 4.5 nm) increased significantly under conventional heating compared to MW heating (13.4 nm vs. 6.6 nm), explaining the lower reactivity of the recycled catalyst after being tested under conventional heating reaction conditions. It was elegantly demonstrated that the Pd leached for the duration of the reaction is largely re-deposited on the solid support after the reaction and that centrifugation facilitated this recovery process.

García-Suárez et al. have compared conventional and MW heating in the use of Pd nanoparticles that are immobilised on mesoporous carbon bead catalysts were evaluated in the aqueous SMC reactions between phenylboronic acid and arylhalides, which contained a number of different electron withdrawing and donor substituents [122]. Excellent activity was obtained at short reaction times (5–10 min) while TOF values of up to 3000 h^{-1} were reached under MW heating.

It was found that it was possible to recycle catalysts up to 10 (conventional heating) and to 5 (MW irradiation) reaction runs without any loss in activity. The reaction media was found to have a strong influence on the recyclability of the catalysts, which could be improved in the presence of polyethylene glycol. The conditions of a typical experiment, under an open atmosphere, are reported as follows; catalyst (1.5 or 0.375 mol % of Pd), arylhalide (0.25 mmol), phenylboronic acid (1.4 equiv.), either K_2CO_3, NaOH or KOAc (2 equiv.), internal standard (*n*-decane; 25 μL) and solvent (3 mL), namely distilled H_2O or a PEG in H_2O mixture (50 wt %). The reaction mixture was stirred at 50 °C for 5–30 min. As for the reactions performed under MW conditions, the reaction medium was heated in a CEM-DISCOVER reactor for 5 min at a power output of 100 W, reaching a maximum temperature of 150 °C. Very good conversion values, >90%, were achieved in polyethylene glycol/H_2O mixtures after 10 min of reaction with all catalysts. It was found that the catalytic activity is influenced by the chemical surface and porosity of the carbon supports. SMC reactions were also attempted under MW irradiation using para-arylbromides and phenylboronic acid in the polyethylene glycol/H_2O mixture solvent in the presence of a reduced amount of catalyst (from 1.5 to 0.375 mol %). A comparison of the results obtained under MW irradiation and conventional heating lead to the conclusion that that MW heating considerably enhanced the reaction rate for all the catalysts, particularly when using para-bromobenzaldehyde as the substrate. Moreover, the trend for the activity of the catalysts is the same irrespective of the activation method used. Almost complete conversion (96%) was achieved after 5 min of reaction time using Pd supported on mesoporous carbon beads previously heated at 2000 °C for 1 h, while a a TOF of 3236 h^{-1} was found, which is higher than the 650 h^{-1} obtained under conventional heating. Interestingly, MW heating provided beneficial effects in the cross coupling of para-bromoanisole, with significantly improved conversions as compared with bromobenzene. A similar trend was also reported by Gomez et al. and explained by the interaction between the methoxy group and the metallic surface which induces higher C–Br bond activation than expected for the intrinsically electronic factors behind para-bromoanisole [123]. Finally, coupling reactions were also attempted using the more challenging *p*-chlorobenzaldehyde, but, unfortunately, very low conversions were achieved after 5 min of reaction.

Schmidt et al. [124] have investigated the protecting group-free synthesis of 2,2′-biphenols via heterogeneously catalyzed SMC under MW irradiation (see Scheme 55). Palladium on charcoal was adopted as the simple and conveniently available heterogeneous catalyst. Because of its nature, this catalyst can be recycled and reused and, at the same time, the contamination of the reaction products by metal residues can be diminished. A plethora of o-halophenols and o-boronophenols underwent SMC in high yields (85%–98%) and selectivity, in terms of 2,2-biphenols. The reactions proceeded in H_2O in the presence of simple additives, such as K_2CO_3, KOH, KF and TBAF and 2 mol % Pd catalysts. The described protocol was simple and extremely sustainable, because no elaborate pre-catalysts and ligands were needed to achieve synthetically useful yields and short reaction times. While iodophenols successfully reacted in 2.5 h in excellent yields even under thermal conditions, the coupling of the analogous bromophenols only proceeded effectively under MW irradiation in 0.5 h.

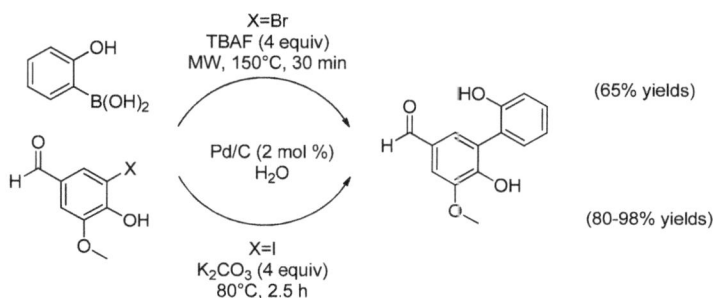

Scheme 55. Synthesis of 2,2′-Biphenols via protecting group-free MW-promoted SMC in water.

Continuing investigations into catalyst immobilization onto varying supports have led Al-Amin et al. [125] to obtain a low-leaching Pd-catalyst supported on a sulfur-modified Au supported material (SAPd). This immobilized Pd-catalyst has proven to be very feasible in both C–C and C–N bond-forming reactions. SAPd was able to release a trace of highly active Pd into the reaction mixture, which enabled it to be used in the MW-assisted SMC reactions.

In this context, the reported SAPd-mediated reactions proceeded under ligand-free conditions under the coupled use of two different MW irradiation chambers; a leaching chamber, in which traces of the active Pd phase would leach from SAPd via the effects of arylhalides under single-mode MW irradiation and a catalytic reaction chamber, in which the catalytic reaction would occur towards the desired transformations under multi-mode MW irradiation. The combined use of two MW reactors employing differing irradiation methods made the protocol very effective, due to the easy control of the leaching throughout the 2 step process.

The considerable accomplishment of this MW protocol was that less reactive arylbromides (i.e., bromobenzene, 4-bromoanisole, p-bromotoluene, 4-bromonitrobenzene and 4-bromobenzonitrile) were effectively coupled to the desired products in ethanol, a feat which has not yet been reached under conventional heating conditions, with such catalytic system, while the reaction time for aryliodide coupling was decreased from 12 to 2 h in a toluene/water solvent mixture. In addition, the low leaching characteristics of SAPd make it recyclable for more than 10 runs meaning that the reaction can be simply scaled up.

Use of MWs for the Synthesis of the Catalysts for the SMC Reaction

MW-assisted reactions are considered to be a very interesting approach and one which rapidly prepares uniform and phase pure materials. MW irradiation has been proven to be effective in the synthesis of a variety of nanomaterials, which are prepared with controlled size and shape without the need for high temperatures or high pressures. Heinrich et al. [126] have synthesized two new square-planar *trans* Pd complexes, [Pd(MEA)$_2$Cl$_2$] and [Pd(MEA)$_2$Br$_2$] [MEA = (2-methoxyethyl)amine], by reacting 2 equiv. of MEA with either PdCl$_2$ or [(cod)PdBr$_2$] (cod = cycloocta-1,5-diene). The complexes were then used as precursors to the preparation of Pd nanoparticles by MW-assisted synthesis.

The authors investigated the effect of the reaction temperature, irradiation time and surfactant (polyvinylpyrrolidone) amount on the size of the produced particles (5–40 nm). It was found that the growth mechanism of the nanoparticles depended on the type of halide ligand used. The Pd particles were embedded in carbonized wood and tested as a catalyst in C–C cross-coupling Heck, Suzuki and Sonogashira reactions, resulting in TON values of 4321 (Heck), 6173 (Sonogashira) and 8223 (Suzuki).

Pd nanoparticles supported on graphene (G) and graphene oxide (GO) have been prepared using the MW irradiation method [127]. The uniform and fast temperature rise produced by MWs in the presence of hydrazine hydrate, used as a reducing agent, offered a facile and efficient procedure to effectively reduce Pd(II) and GO into a dispersion of metallic nanoparticles supported on graphene sheets with large surface area. Power and time irradiation can be optimized to yield the nearly complete reduction of both GO and the Pd salt precursor. Therefore, it was claimed that, unlike conventional thermal heating, activation by MW provides better control of the extent of graphene oxide reduction by hydrazine hydrates. On the contrary, Pd/GO was synthesized by the MW-assisted deposition of Pd(NO$_3$)$_2$ in a GO dispersion and no reducing agent was added.

The SMC of variously substituted aryl bromide and phenylboronic acid reagents was performed in the presence of 0.3 mol % Pd/G and K$_2$CO$_3$ (3 equiv.) in H$_2$O/EtOH (1:1), used as environmentally benign solvents. The reactions were carried out at room temperature or heated under MW irradiation at 80 °C for 10 min. It was observed that the rapid temperature rise during MW heating in H$_2$O induced quick Pd nucleation and consequently produced small uniform nanoparticles with sizes of 7–9 nm. As revealed by TEM measurements, the nanoparticles produced were well dispersed on the graphene sheets, explaining the great reactivity displayed by the Pd/C catalyst.

Elazab et al. have reported a simple MW-assisted (1000 W, 2.45 MHz for 120 s) one-step synthesis of Pd/Fe$_3$O$_4$ nanoparticles supported on graphene nanosheets (Pd/Fe$_3$O$_4$/G) that display extraordinary catalytic activity in Suzuki and Heck coupling reactions [128]. The SMC reaction is reported together with the reaction conditions in Scheme 56.

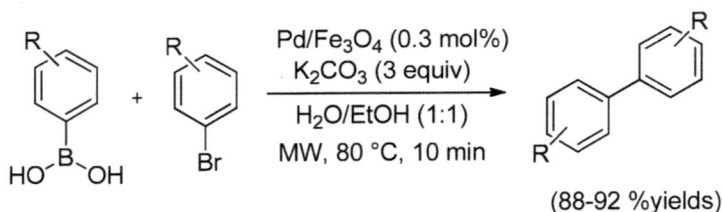

Scheme 56. MW-assisted SMC with Pd/Fe$_3$O$_4$/C nanoparticles catalyst.

The catalyst was magnetically separated from the reaction mixture and recycled many times without losing activity. The synthetic procedure envisaged the reduction of palladium and ferric nitrates in the presence of graphene oxide (GO) nanosheets under MW irradiation. Hydrazine hydrate was employed as the reducing agent. It was found that the most active and recyclable catalyst (7.6 wt % Pd loading) was made up of Pd(0) nanoparticles with sizes in the 4–6 nm diameter range which strongly interacted with the 30 wt % Fe$_3$O$_4$ nanoparticles, with 12–16 nm diameters, on highly reduced GO containing a C/O ratio of 8.1. Such a combination was proven to have a synergic effect on the catalytic activity of SMC reactions under MW irradiation. The reactions tested in the presence of the 7.6 wt % Pd/Fe$_3$O$_4$ catalyst are summarized in Table 6.

Table 6. SMC reactions using 7.6 wt % Pd/Fe$_3$O$_4$ [1].

Aryl-Halide	Boronic Acid	Isolated Yields
		92%
		88%
		94%
		70%
		92%
		90%

[1] Aryl halide (0.32 mmol), boronic acid (0.382 mmol), potassium carbonate (0.96 mmol) and Pd/Fe$_3$O$_4$/G (0.3 mol %) in 4 mL (H$_2$O/EtOH) (1:1) was heated at 80 °C under MW for 10 min. 1st and 2nd reactions were completed at r.t. after 30 min.

Indeed, extremely high TON (9250) and TOF (111,000 h^{-1}) values were attained at 80 °C. The magnetic properties caused by the Fe$_3$O$_4$ component guaranteed easy separation, as illustrated in Figure 7, thus greatly simplifying the purification of the reaction products and increasing the economic value of the catalyst.

Figure 7. (**a**) Image of the reaction mixture using 0.3 mol % Pd/Fe$_3$O$_4$/G catalyst after reaction; (**b**) Separation of spent catalyst from reaction mixture using a simple magnet. Reprinted from [128]. Copyright (2015), with permission from Elsevier.

Indeed, the synthesis of materials under MW irradiation is a relatively novel method that can be used in many applications, such as the preparation of nanoporous structures. The synthesis of metal-organic frameworks (MOFs) under MW irradiation not only extensively reduces reaction time, but also can grant control of the size and shape of the crystals [129]. Li et al. [130] have developed an efficient MW assisted procedure to synthesize UiO-66 in high yields in the presence of benzoic and acetic acids. They found that MW irradiation not only shortened the reaction time, but also improved the surface area of UiO-66. Recently, Dong et al. [131] have successfully synthesized UiO-66 under MW irradiation and then used the MOF as a support for Pd nanoparticles. A 0.075 mol % amount of the obtained catalyst was able to efficiently catalyze the SMC reaction between bromobenzene and phenylboronic acid in ethanol–water mixture (1:1) at 30 °C. Moreover, the complete reaction was achieved in 0.5–1.5 h with moderate or excellent yields, in the presence of substrates that bore electron-donating groups and electron-withdrawing groups. Pd@UiO-66 proved to be very stable since it was separated via filtration and recycled 5 times without significant loss in catalytic activity. Unfortunately, the authors only observed a 29% yield using aryl chlorides.

A novel heterobimetallic Pd/Y-MOF catalyst was synthesized using MW irradiation as reported by Huang et al. [132]. This catalyst was proven to be very active in the water-medium SMC reactions of a number of aryl halides with 4-methylphenylboronic acid and in the Sonogashira reaction. It was found that the catalytic activity of Pd/Y-MOF is as good as that of the traditional Pd(OAc)$_2$ homogeneous catalyst. This was due to, on one hand, the cooperative effect of Y(III) coordination and, on the other hand, to the highly dispersed and accessible Pd(II) active sites present in the layered structure. More interestingly, besides high activity, the Pd/Y-MOF catalyst was also able to select the size of the reactant molecules because of the diffusion limit within the micropores. The coordination of Pd(II) and N atoms with 2,2'-bipyridine-5,5-dicarboxylate acid blocked the leaching of Pd(II) active species under the reaction conditions. Moreover, it can be successfully recycled and reused repetitively thanks to excellent MOF framework stability, which is derived from the strong interaction between Y(III) and carboxyl, and therefore shows great potential in practical applications. This study demonstrated that the MW-assisted synthetic strategy can be considered a general procedure for developing MOF-based heterogeneous catalysts by building homogeneous counterparts inside MOFs frameworks for practical catalytic applications.

A simple MW assisted technique was recently used to prepare Cu$_2$O microcubes decorated with nano Au and Pd [133]. The SEM and TEM images, together with Energy-dispersive X-ray Spectroscopy (EDXS) analyses, are shown in Figure 8.

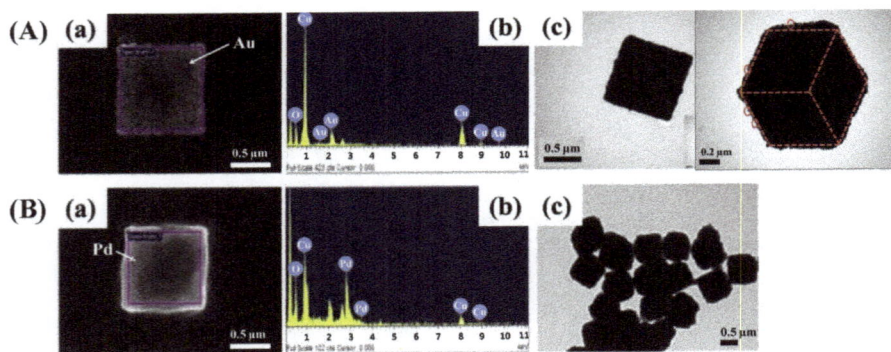

Figure 8. (a) SEM; (b) EDXS; and (c) TEM images of (**A**) Au and (**B**) Pd nanoparticle decorated Cu$_2$O microcubes. Reprinted from [133]. Copyright (2014), with permission from Elsevier.

More specifically, Cu$_2$O microcrystals were synthesized by adding 0.03 g of polyvinylpyrrolidone to 5.0 mL of a 10.0 mM CuCl$_2$ aqueous solution, followed by the addition of 0.18 mL 1 N NaOH and 0.38 mL deuterium-depleted H$_2$O under stirring. Glucose (0.0072 g) was subsequently added as a reducing agent and the solution was placed into a MW oven for 3 min at 100 W. 1.0 mL of

10.0 mM $HAuCl_4$ was subsequently added to the Cu_2O colloidal solution, which then had a brick red color, and underwent 100 W MW irradiation for 1 min, forming a blackish, winered colloidal solution of Cu_2O/Au. A Cu_2O/Pd catalyst was synthesized by following the same procedure, but using 1.0 mL of 10.0 mM H_2PdCl_4. A dark greenish brown color Cu_2O/Pd colloid was obtained and was tested in the coupling reaction synthesis of phenylboronic acid and 4-iodobenzonitrile (see Figure 9 for all the synthetic steps). The SMC was performed overnight, using water as the solvent in the presence of 8.0 mg Cu_2O/Pd colloid and K_2CO_3 at 80 °C. 4-cyanobiphenyl was produced with good selectivity and a satisfactory yield (over 85%). Moreover, the authors verified the versatile nature of Cu_2O/Pd catalyst in the coupling reactions of iodoarenes by performing reactions with electron donating and withdrawing groups. The results of this paper demonstrate that the MW-assisted technique successfully simplifies the synthesis process for metal/metal oxide catalysts.

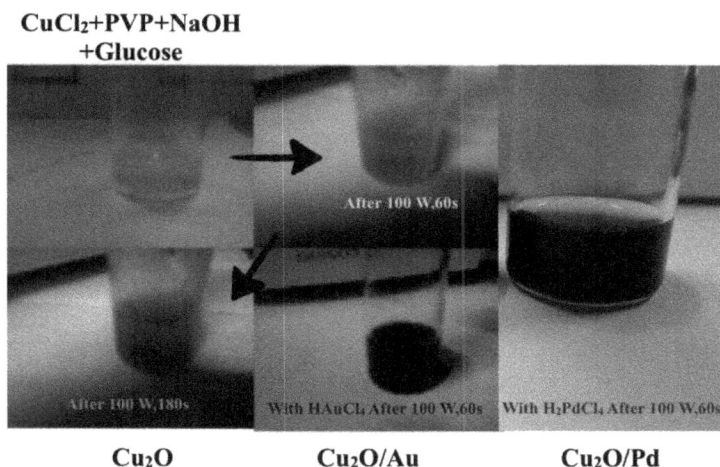

Figure 9. Digital photographs of reagents involved during the preparation of Cu_2O based catalyst. Reprinted from [133]. Copyright (2014), with permission from Elsevier.

Veerakumar et al. [134] have reported the preparation of highly dispersed palladium nanoparticles (average size of ca. 5 nm, PdNPs) immobilized on carbon porous materials (CPMs) via a MW assisted procedure, during which the Pd^{2+} ions were successfully reduced to Pd^0 and extremely well dispersed onto the carbonaceous support. These heterogeneous Pd/CPM catalysts (5.0 mg) exhibit high activity for C–C coupling reactions of various aryl halides with phenylboronic acid in aqueous dimethylformamide (DMF/H_2O = 1:1) with desirable product yields of >88% in the presence of K_2CO_3 (2 mmol) when irradiated for 10–15 min in MW at 100 °C. The reported catalyst can be easily recovered from the reaction mixture by centrifugation and reused in a few reaction cycles.

Along with supplying robust, stoichiometric and complex compounds, MW preparation routes have also proven to be suitable for the synthesis of complex substituted oxides, where a dopant ion resides on a metal cation site. Misch et al. [135] have recently reported a rapid MW-assisted combustion/sol–gel preparation as a suitable procedure for the preparation of noble metal-substituted perovskites. The authors performed a detailed investigation of the phase formation process thanks to the very brief and controllable heating times enabled by MWs. It was also demonstrated that substituted perovskites were able to provide a Pd source for SMC catalysis. The La containing perovskites were more active than the Y containing perovskites (the structures are represented in Figure 10), and the authors proposed that a varying inductive effect on the perovskite can occur depending on the differing A-site cation, resulting in differing Pd ion stability in the material.

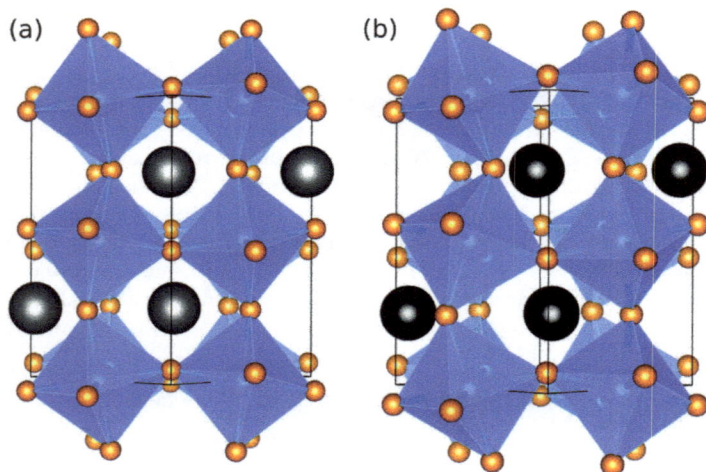

Figure 10. Crystal structures for (**a**) LaFeO$_3$; and (**b**) YFeO$_3$. The A site cations are displayed as grey spheres and FeO$_6$ octahedra in blue. Reprinted from [135]. Copyright (2013), with permission from Royal Society of Chemistry.

Furthermore, it was unambiguously established, using the 2-dicyclohexylphosphine-2′,6′-dimethoxybiphenyl ligand, that catalytic activity does not originate from the perovskite itself, but from reduced Pd^0 that is liberated and then bound by the ligand. Interestingly, these materials were proven to be suitable for aryl chloride coupling under more mild conditions than had previously been described. The reported method may furthermore allow the synthesis of materials with even higher surface areas to be carried out, thereby regulating their properties.

3. Ultrasound-Assisted SMC Reactions

3.1. Ultrasound-Assisted SMC Reactions

Among the wide range of US application in organic chemistry, a great nuber of US promoted SMC reaction have been reported in literature in the last decade. [136] An US-induced continuous flow technique has been developed by Shil et al. [137] for the gram scale SMC reaction of haloarenes (chloro, bromo and iodo) with phenylboronic acid catalyzed by solid supported Pd(0) nano/microparticle (SS-Pd). They previously reported [138] a facile process for the in situ generation of Pd(0) nano- and microparticles and their stable deposition over the amberlite IRA 900 resin. The hydrophobic and hydrophilic combination of the borohydride exchanged polymer used for SS-Pd preparation played a crucial role in reducing the Pd(II) ion to Pd(0), and impregnating it into the hydrophobic pockets of the polymer matrix as nano- and microparticles. The obtained SS-Pd catalyst was noted as being extremely stable in aqueous media, easily separable and recyclable for up to five runs in SMC without relevant loss in activity under mild basic conditions (Figure 11). Furthermore, ultrasonication provided enough excitation energy to perform the reaction in methanol-water and increased the solubility of K$_2$CO$_3$ to supply the extraordinary basic conditions, which drove the reaction into products at ambient temperature with satisfactory yields (~80% for 3 mol % of Pd). Furthermore, the simple isolation of the crude products with negligible metal contamination and recyclability of the SS-Pd catalyst can furnish huge green impact and lead to industrial interest.

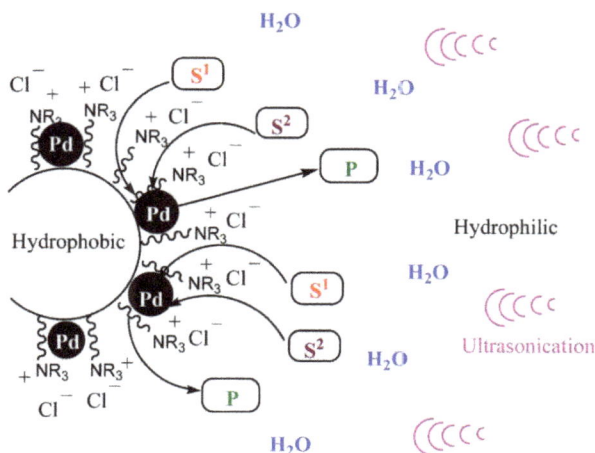

Figure 11. A schematic diagram of the reaction outline on the catalyst surface; S^1 = aryl halide, S^2 = phenylboronic acid and P = product (Reproduced from [137]. Copyright (2013), with permission from Royal Society of Chemistry).

Azua et al. [139] have reported the synthesis of novel Pd *N*-heterocyclic-carbene (NHC)-based complexes with 3,4,5-trimethoxybenzyl, alkyl and sulfonate *N*-substituents, suitable for SMC (Figure 12). The new complexes were exploited as pre-catalysts (1 mol % of Pd) in the cross-coupling reactions of various aryl halides/boron moieties in glycerol under pulsed-US activation. US activation allows scientists to overcome one of the major drawbacks of the application of glycerol; its high viscosity, which limits mass transport and the efficient mixing of the reaction mixtures. High yields (70%–85%) were obtained in 30 min under pulsed US activation, without the production of undesired by-products. Catalyst fate was investigated by the author using TEM and X-ray Photoelectron Spectroscopy (XPS) analyses. Not quite rigidly classifying the transformation as a homogenous or heterogeneous process, they preferred to postulate that different active catalytic species (cocktail of compounds) were present in solution.

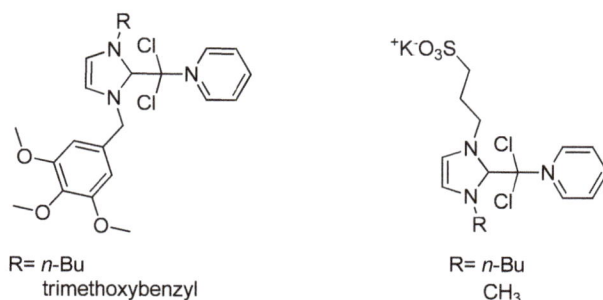

R= *n*-Bu
trimethoxybenzyl

R= *n*-Bu
CH$_3$

Figure 12. Pd(II)-NHC (*N*-Heterocyclic-Carbene) based complexes.

A series of 1-(α-aminobenzyl)-2-naphthols have been prepared, by Chaudhary and Bedekar, [140] from 2-naphthols using an aromatic Mannich reaction [141] and screened as air-stable phosphine-free ligands for Pd-catalyzed US assisted SMC. The amino phenol structure, as reported in Figure 13, has a suitable arrangement of heteroatoms to form a six-membered stable chelate, a prerequisite for use as a ligand in metal-catalyzed reactions. The results of the Pd(AcO)$_2$ catalyzed SMC, in the presence of

1-(α-aminobenzyl)-2-naphthols, showed very efficient conversion with aryl iodides and aryl bromides when using a small quantity of TBAB and aqueous dioxane as the solvent. Chlorobenzene was considerably less reactive after a long reaction time but iodobenzene gave biphenyls with good conversion amounts even at very low catalyst quantities. Varying the amount of catalyst, to as low as 0.01 mol % of Pd, resulted in the US assisted formation of biphenyl in high yields and TON values.

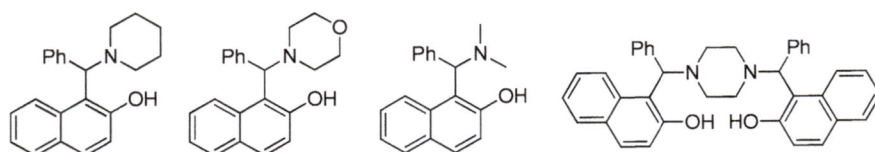

Figure 13. Air-stable phosphine-free ligands used for SMC based on 1-(α-aminobenzyl)-2-naphthol structure.

A new and extremely stable Pd(EDTA)$^{2-}$-salt has been prepared as a catalyst by Ghotbinejad et al. [142] employing a counter-cation of *N*-methylimidazolium bonded to 1,3,5-triazine-tethered superparamagnetic iron oxide nanoparticles (SPIONs) (Figure 14). This complex efficiently catalyzed SMC reactions between various arylbromides and arylboronic acids in the presence of K_2CO_3, used as a base, in an aqueous solution of DMF. The cross-coupled products were synthesized under conventional heating (70 °C) and US irradiation (160 W, 30 °C) using a very low catalyst loading (as low as 0.032 mol % Pd). Results indicated that conventional preparation took longer (4 h) and gave moderate yields, while the reaction occurred quickly (10 min) in the presence of US irradiation and gave high to excellent yields (95%). The SPION-A-Pd(EDTA) catalyst was quickly recovered using an applied external magnetic field and was reused for several cycles without any loss in its high catalytic performance (TOF 1.1×10^5 h^{-1}).

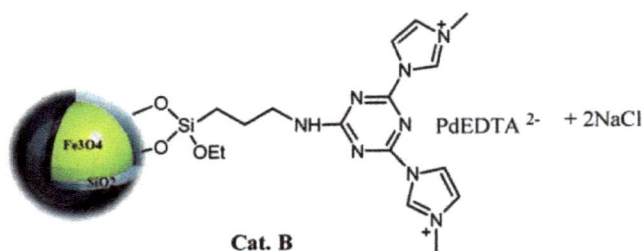

Cat. B

Figure 14. General structure of SPION-A-Pd(EDTA) nanocatalyst. (Reproduced from [142]. Copyright (2014), with permission from Royal Society of Chemistry).

A heterogeneous catalyst, involving Pd embedded within porous carbon (CMK-3), has been prepared by Wang et al. (2014) [143] via a three-step method: immersion, ammoniahydrolysis and heating. Accurate characterization of the Pd@CMK-3 catalyst showed that Pd(0) and Pd(II) species co-exist and were embedded inside the matrix of the porous carbon CMK-3 (Figure 15). This catalyst has shown high activity toward standard SMC in the presence of K_2CO_3. Resistance to mass transfer in the pore channels was significantly reduced when the reaction mixture was homogenized by two minutes of ultrasonication rather than magnetic stirring before heating. As a result, the reactions proceeded quickly and a four-fold increase in the turnover frequency (~2800 h^{-1}) was described. When ultrasonication (40 kHz, 150 W, 30 °C) was used throughout the total reaction process, conversion exceeded 90% even without the protection of an inert gas.

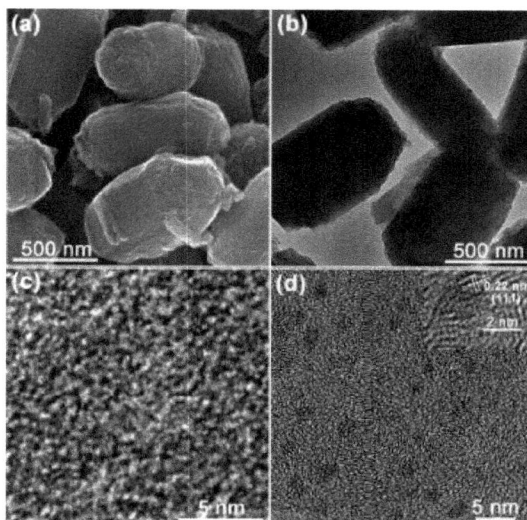

Figure 15. Electron microscope images of Pd@CMK-3. (**a**) SEM image; (**b**) TEM image; (**c**) typical HRTEM image; (**d**) HRTEM image of the sample after 30 min irradiation (300 keV), and the inset showing the lattice fringe of a typical Pd NP (Reproduced from [143]. Copyright (2014), with permission of Springer).

Pd/C has been identified, by Suresh et al. [144], as an efficient catalyst for the US-assisted SMC coupling of 2-aryl-3-bromoflavones with 4,4-disubstituted (1,4-dihydro-2-oxo-2*H*-3,1-benzoxazin-6-yl)boronic acids. The coupling reactions were performed in aqueous DMF, without the need for expensive phosphine ligands in the presence of 10% Pd/C and TBAB-K_2CO_3, and afforded the desired 6-flavonyl substituted 1,4-dihydro-benzo[*d*][1,3]oxazin-2-ones in good yields (82%–91%) and short reaction times (20 min) (Scheme 57). Studies have shown that both catalyst and US played a key role in the faster coupling reaction. The recyclability of Pd/C was documented at four cycles when the reactions were executed under 35 kHz at 55–60 °C. A reaction mechanism describing the generation of actual catalytic species is presented followed by the catalytic cycle. Overall, the methodology may find wide applications in constructing a diverse library of small molecules containing the 6-flavonyl substituted 1,4-dihydro-benzo[d][1,3]oxazin-2-one framework, which would be of potential pharmacological interest.

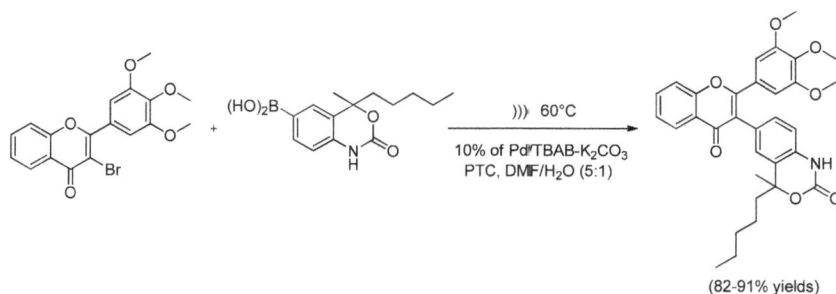

(82-91% yields)

Scheme 57. US assisted 6-flavonyl substituted 1,4-dihydro-benzo[*d*][1,3]oxazin-2-ones synthesis via Pd/C catalyzed SMC strategy.

Kulkarni et al. [145] have described the application of an energy-efficient surface acoustic wave (SAW) device for driving SAW-assisted, ligand-free SMCs in aqueous media. SAWs of well-defined wavelength and frequency (19.50 MHz) were generated on a piezoelectric substrate (LiNbO$_3$). The reactions were performed in a closed vessel using a high boiling point, low viscosity fluid as the liquid couplant. Under these conditions, the SAWs are transmitted into the paraffin oil as a bulk sound wave at a Rayleigh angle of 22.2 degrees. As the bulk wave approaches the glass vessel, a Lamb wave is formed in the base of the vessel which transmits the acoustic energy into the reaction solution within. The reactions were performed on a mmolar scale using low to ultra-low Pd(AcO)$_2$ catalyst loadings (0.08%–0.0008%). The reactions were boosted by the heating which resulted from the acoustic energy penetration, derived from the radio frequency (RF) Raleigh waves generated by the piezoelectric chip via a renewable fluid coupling layer. The yields were uniformly high (>90%) when starting from a variety of aryl bromides and phenylboronic acid, while the reactions were executed without the addition of ligands in water. In terms of energy density, this new technology was labeled as being roughly as effective as MW and superior to US (Table 7).

Table 7. Surface acoustic wave (SAW), ultrasound (US) and MW energy density comparison for SMC reactions between bromoanisole and phenylboronic acid.

Energy Source	Reaction Time (min)	Power (W)	Energy Density (kJ/mmol)	Catalyst Loading (mol %)	Yields (%)
SWA	25	8	12	0.08	89
MW	17	5	5.1	0.8	72
US	10	120	144	0.9	85

Cravotto et al. [97] have compared results obtained from metal-catalyzed SMC reactions preformed under MW, US and conventional thermal conditions. The coupling between 4-iodoanisole and phenylboronic acid, using either ligand-free Pd salts or Pd on charcoal, was used as a model reaction in glycerol. Both US and MW irradiation strongly improved the reaction rate in glycerol. Only by combining these two enabling technologies was it possible to achieve quantitative conversions in 60 min at 80 °C in the presence of 0.04 mmol Pd catalyst. Under these conditions, glycerol showed excellent acoustic cavitation even at high temperatures. In all cases, the Pd salts used were more efficient than Pd on charcoal, but less effective than previously reported Pd-loaded cross-linked chitosan for the SMC of various aryl halides (86% reaction yields).

3.2. Ultrasound Assisted Heterogeneous Preparation of Catalysts Suitable for SMC Reactions

The US assisted synthesis of new palladium nanoparticles (Pd NP) that are supported on cobalt ferrite magnetic nanoparticles (Pd–CoFe$_2$O$_4$ MNPs) has been introduced by Senapati et al. [146] (Figure 16). The CoFe$_2$O$_4$ MNPs were prepared in aqueous medium (under alkaline pH) without using surfactant or organic capping agents via a combined sonochemical and co-precipitation technique. The catalytic activity of the Pd incorporated cobalt ferrite nanoparticles was evaluated for standard SMC reaction in ethanol at reflux in the presence of Na$_2$CO$_3$ under ligand free and aerobic conditions. The cross coupling was performed with low catalyst loading (1.6 mol %) and gave high reaction yields (81%–92%) (TON value = 6250). The recyclability of this magnetic catalyst was also assessed in the SMC reaction of phenylboronic acid and 4-methyl iodobenzene. After each cycle, Pd–CoFe$_2$O$_4$ MNPs were magnetically recovered and reused with no significant loss in activity. Moreover, the morphology of the recovered MNPs (40–50 nm) was unaltered, even after four consecutive cycles.

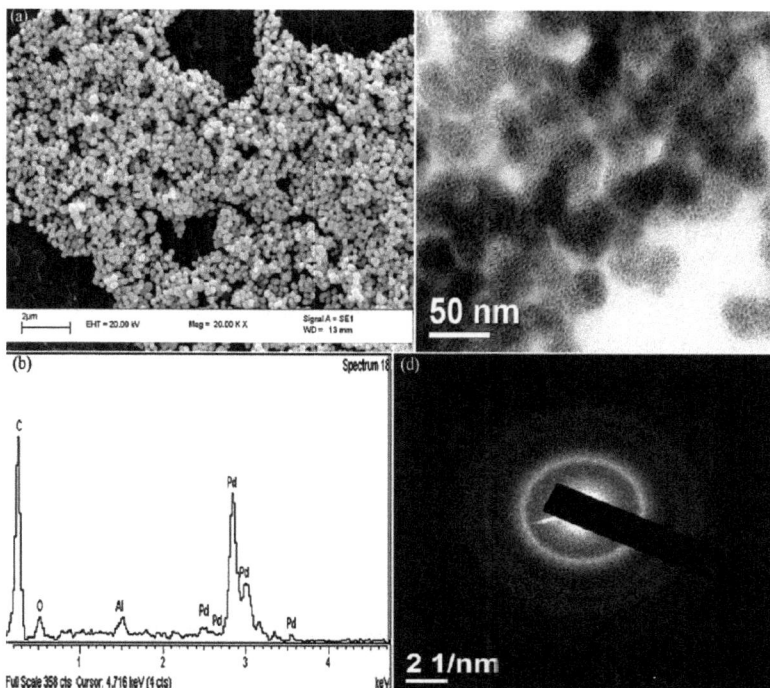

Figure 16. SEM with EDX (**a,b**) and TEM image with Selected Area Electron Diffraction (SAED); (**c,d**) of ferrite magnetic nanoparticles (Pd–CoFe$_2$O$_4$ MNPs) (Reproduced from [146]. Copyright (2012), with permission from Elsevier).

Magnetic nanoparticles (MNPs) have recently been used as catalyst supports for different organic transformations due to their high surface area and their easy recovery. In this context, an efficient super-paramagnetic solid catalyst, which is suitable for SMC reaction, has been prepared by Singh et al. [147]. Pd(0) species were loaded onto the paramagnetic catalyst support during the synthesis of zinc ferrite nanoparticles using a co-precipitation technique in aqueous medium under ultrasound irradiation, without the need for additional surfactants or organic capping agents. The crystallite size for the Pd–ZnFe$_2$O$_4$ nanoparticles was determined by X-ray diffraction and was found to be 10 ± 5 nm, whereas the iron/zinc ratio, measured by EDAX analysis, was found to be 2.05. The activity and stability of this Pd–ZnFe$_2$O$_4$ nanocatalyst was proven in model SMC reactions. Good yields (>85%), in terms of the C–C coupling product, were reported for reactions performed under reflux in pure ethanol in the presence of K$_2$CO$_3$ and 4.62 mol % of Pd–ZnFe$_2$O$_4$–MNPs, for a broad range of aryl halides and phenylboronic acid. Moreover, the super-paramagnetic nature of the catalyst meant that it was recovered using an external magnet and reused without any appreciable loss in activity for five cycles.

Of all the supports available for metal nanocatalyst preparation, widespread attention has recently been paid to so-called g-C$_3$N$_4$ which is the most stable allotrope of carbon nitride. In this context, highly uniform Pd NPs, with an average size of 4 nm, were supported on the g-C$_3$N$_4$ surface by means of an ultrasound assisted solution-reduction method by Su et al. (2015) [148]. This supported catalyst (Figure 17) was very efficient in SMC and provided isolated yields of up to 99% in aqueous ethanol at 80 °C when used with a 1 mol % Pd content combined with K$_2$CO$_3$. Good results were reported for aryl bromides bearing both electron-withdrawing and donating groups, with a TOF value of 378 h^{-1}, whereas no C–C couplings were detected using aryl chlorides. The results denoted that the electronic transfer from N atoms to the γ-C$_3$N$_4$ framework and finally to Pd NPs improved the interaction between the Pd NPs and the support. It was established that the interaction is useful for improving the Pd NPs dispersion and stability on the g-C$_3$N$_4$. The catalyst may be easily recycled without any leak in activity and selectivity at least 3 times showing negligible metal leaching (<0.5%). The outstanding performance of the catalyst, in terms of activity and recyclability, may be ascribed to the considerable interaction between g-C$_3$N$_4$ and Pd NPs and the robustness of the CN framework.

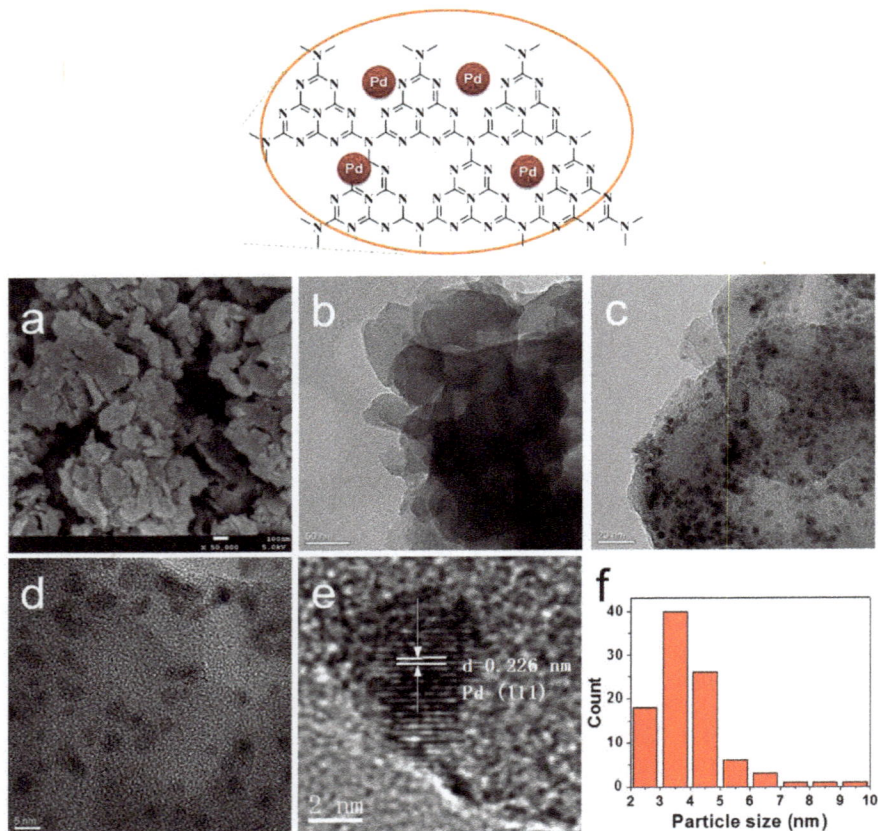

Figure 17. Top: Pd NPs, supported on the g-C$_3$N$_4$ surface. Bottom: SEM image of g-C$_3$N$_4$ (**a**); TEM images of g-C$_3$N$_4$ (**b**); and Pd/g-C$_3$N$_4$ (**c**); HRTEM images of Pd/g-C$_3$N$_4$ (**d,e**); and size distribution of Pd NPs (**f**) (Reproduced from [148]. Copyright (2015), with permission of Springer).

Supported Pd NPs play a crucial role in SMC reactions. Nevertheless, some issues, such as aggregation and the leaching of Pd NPs onto the support's surface, are still great challenges to be addressed. The uniform dispersion of Pd NPs may well be the first step towards a solution to these problems. Layered double hydroxides (LDHs), also famous as hydrotalcites, have recently fostered increasing attention as they can play a role as a new support for the most profitable layered crystals applied in the synthesis of nanocomposites; this interest is due to some specific features of LDHs, such as their capability to exchange anions and undergo reconstruction ("structure memory effect"). Focusing on this topic, Li and Bai [149] have synthesized, using a one-step ultrasonic method, new Pd

NPs supported on the surface of sodium dodecylsulfonate (SDS)-intercalated LDH nanocomposites that are suitable for SMC. The Pd NPs described were uniformly dispersed on the SDS–LDH surface and showed an average size of 3.56 nm (Figure 18).

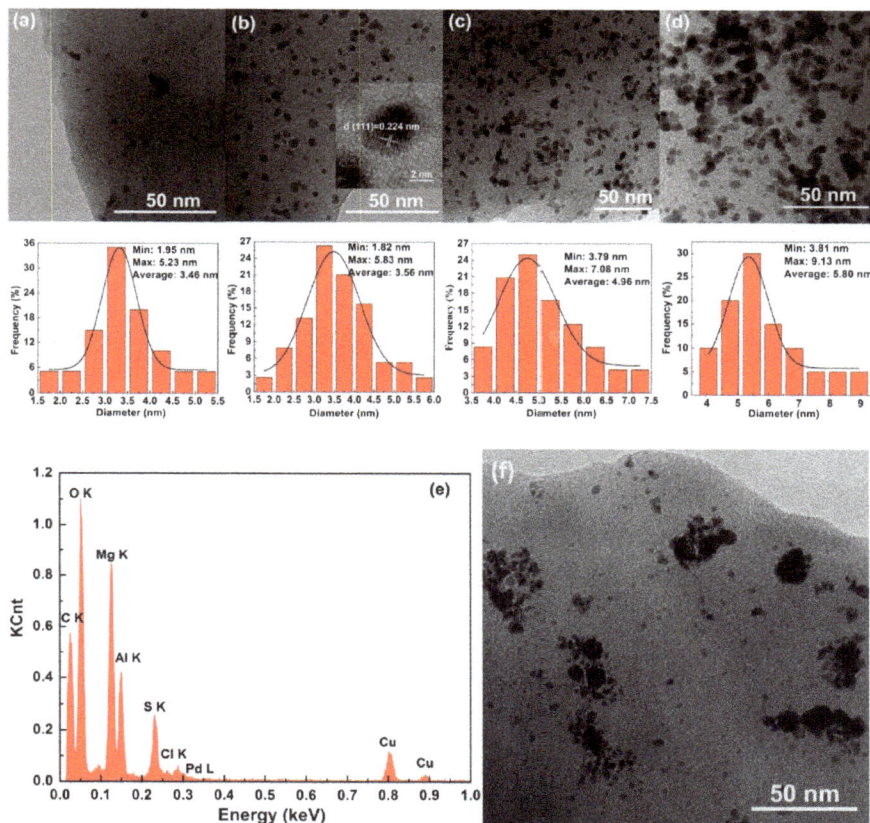

Figure 18. TEM images and corresponding size distributions of the $Pd_{0.005}$/SDS–LDHs (Sodium Dodecylsulfonate-Layered double hydroxides) (**a**); $Pd_{0.02}$/SDS–LDHs (**b**) (HRTEM images (inset)); $Pd_{0.05}$/SDS–LDHs (**c**); and $Pd_{0.10}$/SDS–LDHs (**d**); EDS profile of the $Pd_{0.02}$/SDS–LDHs (**e**); $Pd_{0.02}$/SDS–LDHs prepared in the absence of US (**f**). (Reproduced from [149]. Copyright (2016), with permission from Springer).

The conversion of 4-bromotoluene was catalyzed by Pd/SDS–LDHs in the presence of boronic acid, reaching 98.16% yield with 0.1 mmol % catalyst in $EtOH/H_2O$ at room temperature without any phase transfer agents (Figure 19). Moreover, the conversion was much higher than that of the Pd/SDS–LDHs that were synthesized without US. This enhancement was ascribed to Pd/SDS–LDH's size uniformity and high dispersion. In particular, Pd/SDS–LDHs showed greater catalytic activity than Pd/C catalysts with the equivalent Pd content due to the considerable interaction between the Pd species and SDS–LDHs within the Pd/SDS–LDH nanocomposites. These catalysts were easily separated and recycled five times without any significant leak in activity.

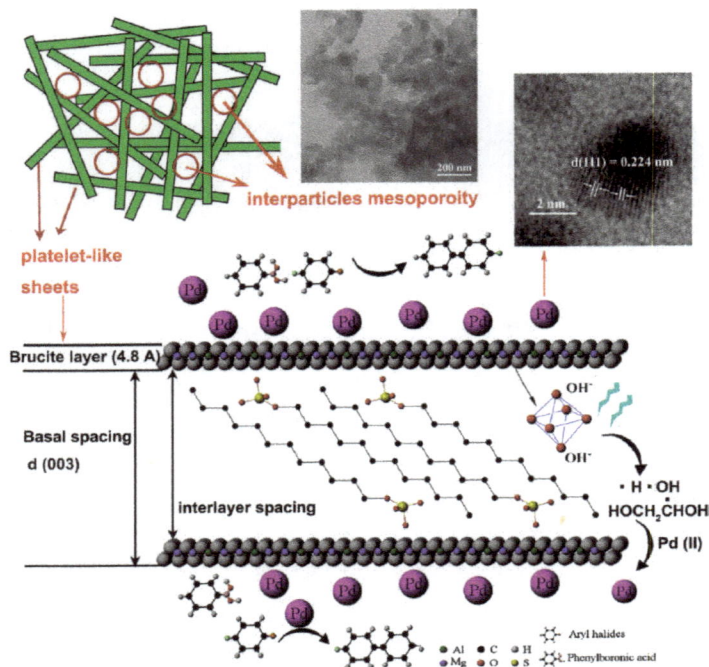

Figure 19. Schematic representation of the Pd/SDS–LDH nanocomposites. (Reproduced from [149]. Copyright (2016), with permission from Springer).

3.3. Ultrasound Assisted Heterogeneous Catalyst Preparation for MW Promoted SMC Reactions

Martina et al. [98] have reported a simple US-assisted procedure to prepare solid supported Pd(II) catalysts on chitosane (CS). They synthesized cross-linked chitosan derivatives (CS–Pd) by reacting chitosan in the presence of hexamethylene diisocyanate and the corresponding metal salt. The in situ polymerization, carried out in water under sonochemical conditions (90 min under 19.5 kHz irradiation at 30 W), was extremely fast and efficient. The activity of these cross-linked CS–Pd(II) catalysts (1 or 0.5 mol %) was tested on a model reaction between 4-bromoacetophenone and phenylboronic acid in a H_2O/dioxane 9:1 mixture. Satisfactory biaryl product yields (90%) were reported to have occurred under MW irradiation in 1 h at 90 °C without the addition of a phase transfer catalyst. The recovered CS–Pd(II) catalyst was reused several times in SMC reactions with a minimal loss in activity.

Cravotto et al. [101] have described the use of a new series of solid cross-linked cyclodextrin (α-, β- and γ-CD) based catalysts that were obtained via US-assisted reticulation with hexamethylene diisocyanate (HDI) in solutions containing Pd(II). Their polar structure makes both native CDs and their cross-linked derivatives very sensitive to dielectric heating. Moreover, this property is strongly enhanced in the Pd(II)-CD cross-linked catalyst by the embedded Pd(II) cations, rendering them extremely active in MW assisted SMC reactions. Best yields (75%–99%) were described as occurring under MW irradiation in water for 45 min using aryl bromides as the C–C coupling partner at 65 °C in the presence of Na_2CO_3. Metal leaching from the Pd(II)-CD catalyst is negligible (0.5%–1.5% less), which allows it to be recycled.

Silica has always found widespread applicationin the synthesis of solid-supported PdNPs because of its remarkable chemical and thermal stability, mechanical robustness and high accessibility. Silica is a remarkably versatile support and one that is able to host metal nanoparticles (NPs) and improve their reactivity and stability. Martina et al. [102], have recently prepared a novel cyclodextrin/silica support for Pd NPs (Pd/Si-CD) (Scheme 45). Cyclodextrin (CD) grafting into an inorganic silica framework provided a hybrid organic/inorganic material suitable for further metal impregnation. It displayed modified surface reactivity, with respect to material bulk properties, due to its new hydrophilic/hydrophobic profile. The extremely efficient and homogeneous impregnation of small Pd NPs into this support was successfully carried out under US irradiation at 20.4 kHz (90–100 W) for 1 h. The Pd/Si-CD catalyst exhibited excellent activity in ligand-free C–C SMC with different aryl iodides and bromides (both electron rich and electron deficient).

4. Mechanochemical Activation of SMC

Several publications have recently reported upon the solid-state grinding route for the preparation of new types of supported Pd catalyst. An example of environmentally friendly preparation of $PdCl_2$ with a bidentate 1,5-bis(diphenylphosphino)-pentane has been described in the Journal of Chemical Education; this procedure consists of grinding a mixed powder against the side of a 50 mL round-bottomed flask so to achieve high metal loading and high particle dispersion capacities [150]. Many other supports with high pore volume have been used, such as mesoporous silica, carbon and one reported example of Pd dispersed in ascorbic acid. Based on our knowledge, some examples of the preparation of Pd supported catalysts by mechanochemical activation are reported in Table 8.

Table 8. Preparation of supported Pd catalyst by mechanochemical activation under solvent-free conditions.

Catalyst	Loading Conditions	Technology	Pd Loading (wt %)	SMC Synthetic Protocol	Reference
Magnetically separable mesoporous SBA-15 nanocomposites	1. SBA-15 silica support, Fe(NO$_3$)$_3$·9H$_2$O 10 min at 350 rpm. 2. propionic acid at 85 °C for 3 h	Retsch PM-100 planetary ball mill 18 stainless steel balls (10 mm)	19.2%	Bromobenzene, phenylboronic acid, Pd catalyst K$_2$CO$_3$, H$_2$O, MW 150 °C, 20 min (59% yield)	[151]
Palladium nanoparticles supported on carbon nanotubes	MWCNT powder/Pd(OAc)$_2$ (10:1) mechanical shaking for 30 min	Ball-mill mixer (SPEX CertiPrep 8000D), two ceramic balls (d $\frac{1}{4}$ 1.3 cm), 1060 lateral cycles per minute.	9%	Bromobenzene, phenylboronic, Pd/MWCNT (Multiwall Carbon Nanotubes) (0.5 mol %) K$_2$CO$_3$, H$_2$O–EtOH (1:1), MW 80 °C for 10 min (100% conv.)	[152]
Monodispersed Pd in ascorbic acid	Pd(NO$_3$)$_2$·2H$_2$O, and Ascorbic Acid (1.0568 g, 6 mmol) 10 min	Mortar	-	Phenyl iodide, phenylboronic acid, Na$_2$CO$_3$, TBAB, Pd (10 mol %), H$_2$O, 80 °C, 24 h (Yield 97%)	[153]

Examples in the literature also address the topic of using mechanochemical activation in solvent free SMC [154,155]. Cravotto et al. have reported the use of a new catalyst, based on chitosan that is cross-linked with hexamethylene diisocyanate using Pd(OAc)$_2$, in solid-state SMC with aryl chloride. The reactions were carried out in a planetary ball mill for 120 min at 600 min^{-1} and high yields were obtained with unsubstituted and *p*-nitro chloro benzene [11].

The advantage of the strongly basic nature of KF–Al$_2$O$_3$, which can replace organic bases in a number of reactions, was explored by Braga et al. in 2004 in solvent free SMC under mechanochemical activation [156]. Stolle et al. have reported the same approach and reactions were performed in a mechanical manner via the co-grinding of the reactants with agate milling balls using a planetary ball mill as the source for alternative energy input. A model reaction was performed with bromo acetophenone and phenylboronic acid and the inherent basicity of the aluminas proved to be beneficial for the reaction, as compared to MgO, SiO$_2$, TiO$_2$, CeO$_2$ and Fe$_2$O$_3$. In contrast to the results obtained by Saha et al., who studied a MW assisted Pd(PPh$_3$)$_4$-catalyzed solid-state SMC [157], neutral γ-Al$_2$O$_3$, and not the basic α-Al$_2$O$_3$, yielded the best results with the tested aryl bromides. The reaction was effective in the presence of Pd-loadings higher than 1 mol % and high yields were obtained when KF loading was higher than 20 wt % on aluminas. In addition, KF–Al$_2$O$_3$ activity strongly depends on the residual water content, while the extent of water influence on the performance of the coupling reaction is related to the polarity of the aryl bromides [158].

The effect of liquid-assisted grinding has recently been studied using mechanical SMC for the coupling of aryl chlorides as the model reaction [159]. Catalytic systems that used Davephos and PCy3 were tested and the author demonstrated the strong influences that various liquids had, added at the amount of 0.045 μL/mg. Alcohols produce unexpected improvements in yield, when used as additives. This is perhaps due to the in situ formation of alkoxides and their participation in oxidative addition, while the reactions performed neat gave the worst results. Based on this evidence, the authors proposed a mechanism and the reaction was optimized over a series of chloro benzene reactions to reduce the amount of catalyst to 2% (chloro benzene derivative, phenylboronic, Pd(OAc)$_2$ (2 mol %), PCy$_3$·HBF$_4$ (4 mol %), K$_2$CO$_3$ (5.0 equiv.) and MeOH (η = 0.045 μL/mg, two stainless steel balls (ø = 1.4 cm), 30 Hz, 99 min).

5. When the SMC Reaction Is Driven by Light

Solar light is a copious and safe source of energy that is attracting huge interest in efforts to figure out the matter of growing energy demands and environmental concerns. The ultimate aim for us here is to carry out visible light driven photocatalytic reactions. Indeed, solar light absorption is a convenient and sustainable means for generating electronically excited states in photocatalysts. In fact, plasmonic metal nanoparticles are able to transform solar energy into chemical energy via localized surface plasmon resonance (LSPR) [160]. In particular, many advantages, in terms of light harvesting and energy of the electrons, can be achieved when metal NPs are immobilized on insulating solids [161]. Indeed, the photo-excited electrons of NPs attain energy and gather on the surface, therefore aiding the molecule activation for chemical reactions. Higher photon efficiency can be achieved since the light harvesting and reaction occur at a single site [162]. Pd is judged as one of the most active metals for many reactions, among which the SMC is counted. However, its inability to show plasmonic absorption inhibits the utilization of visible light energy, whereas other metals, such as Ag nanoparticles, display a marked light harvesting ability due to their size-dependent plasmonic absorption [163]. Plasmonic Pd/Ag bimetallic catalysts supported on mesoporous silica SBA-15 have recently been used in the SMC coupling of iodobenzene and phenylboronic acid in the presence of K$_2$CO$_3$, used as a base, with ethanol as the solvent for 6 h (Figure 20) [164]. Pd nanoparticles were synthesized by LSPR-assisted deposition on the highly dispersed Ag nanoparticles under visible light irradiation. It was found that the integrated system, made up of Pd and plasmonic Ag nanoparticles, favored the formation of an efficient light harvesting system with unique photocatalytic reactions. It is worth noting that no catalytic activity (yield < 1%) was observed under dark conditions at room temperature (at 25 °C) due to the lack of the LSPR effect induced by silver and to the very low amount of Pd NPs. The authors observed that the yield of biphenyl was determined by the color of the Ag catalysts, according to the following order; Pd/Ag/SBA-15 (yellow) (2.16) < Pd/Ag/SBA-15 (red) (3.00) < Pd/Ag/SBA-15 (blue) (3.33), the values in brackets are the rates of activity enhancement. It was supposed that the occurrence of the reaction under visible light irradiation was related to the heating effect of the infrared component of light. Indeed, a 10 °C temperature increase in the reaction mixture was observed under light irradiation. The reaction was therefore performed at 35 °C and the contribution of the conversion efficiency under light irradiation was evaluated.

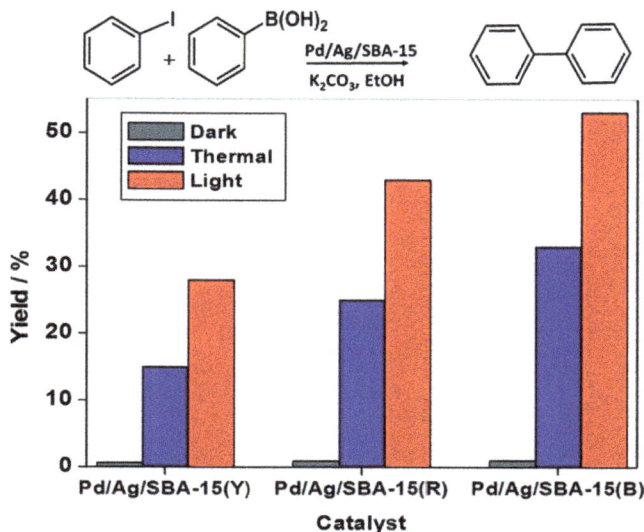

Figure 20. SMC for Pd/Ag/SBA-15 catalysts under dark, thermal (35 °C) and light irradiation conditions. Reprinted from [164]. Copyright (2015), with permission from Royal Society of Chemistry.

Superior performance was observed under light irradiation, followed by thermal conditions and then the reactions in the dark. This clearly indicates that the enhanced activity was strictly connected to the synergistic effect between Pd and Ag nanoparticles which mutually interact with the assistance of the LSPR effect. It was proposed that, due to the charge heterogeneity at the interface between the two metals, the energetic electrons generated on the silver transfer to the Pd nanoparticles upon irradiation with visible light. This can be understood if we consider that the work function of Pd metal is 5.00 eV in a vacuum, which is larger than the work function of Ag metal, which is 4.30 eV in a vacuum. The amount of energy required to eject the electron determines the position of the Fermi level, suggesting that the position of the Fermi level of Pd is lower to that of Ag, making the electron transfer process achievable [165,166]. Activated Pd species can therefore accelerate the oxidative addition step, resulting in an increase in the intrinsic catalytic activity of palladium. The energetic electrons also release energy into the surrounding environment, leading to an increase in temperature, which also contributes to the improvement in activity. All Pd/Ag bimetallic catalysts were active in the coupling reaction under visible light irradiation which was emitted at $\lambda > 420$ nm.

In a very recent paper by Yamashita et al. [167], bimetallic Pd/M (M = Ag, Au) nanoparticles embedded into a SBA-15 (20 mg) matrix have been tested in the coupling of aryl halides and boronic acid in ethanol and in the presence of K_2CO_3 (41.5 mg) and phenylboronic acid (36.6 mg) at room temperature for 2 h under a Xe lamp irradiation with a glass filter cutting out the ultraviolet portion of light. In these systems, Pd (0.5 wt %) was deposited onto M/SBA-15 (M = Au, Ag) using an LSPR assisted deposition method under visible light irradiation. Interestingly, both Au and Ag (1 wt %) were incorporated into the mesoporous silica by the MW (500 W, 2450 ± 30 MHz, 3 min) polyol method using either ethylene glycol or 1-hexanol as the solvent and reducing agent.

PdAg/SBA-15 and PdAu/SBA-15 were proven to be more active than the monometallic Pd-containing catalyst, while the beneficial effects of photoactivity are higher in the presence of gold than in the presence of Ag, according to the results of the photocatalytic tests reported in Figure 21.

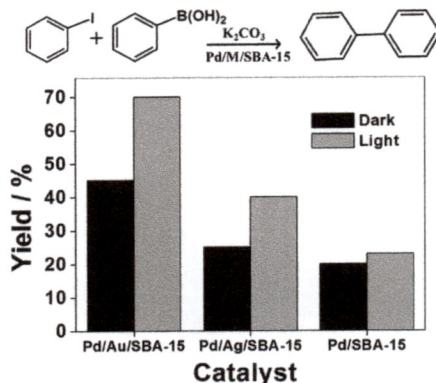

Figure 21. Results for Pd/M/SBA-15 in the SMC reaction under dark and visible light irradiation conditions. Reprinted from [167]. Copyright (2016), with permission from Royal Society of Chemistry.

A supposable mechanism has been proposed to explain the tentative reaction pathway that may occur under visible light irradiation. The resonant excitation of plasmonic nanoparticles, which is accomplished by matching frequencies with those of the incident electromagnetic radiation, undergoes relaxation within tens of femtoseconds [168–170]. Relaxation can occur via either radiative or non-radiative emission. The possible relaxation mode for metal nanoparticles with a diameter < 30 nm is the non-radiative Landau damping process, which causes the formation of electron–hole pairs [171,172]. The electron–hole pairs are generated by the elevated electric fields caused by the resonance condition between the plasmonic metal nanoparticles and the electromagnetic radiation [173]. These energetic electrons, or so-called hot electrons, are available at the surface of the Pd sites in the bimetallic catalysts due, as explained before, to the considerably higher work function values of Pd (5.0 eV) than Ag (4.3 eV) and Au (4.7 eV), making the electron transfer process quite feasible. These energetic electrons can be accumulated in the lowest unoccupied molecular orbital (LUMO) of the reactant arylhalide, through the Pd nanoparticles, which is a transformed ionic/transient species and leads to bond weakening, as illustrated in Figure 22.

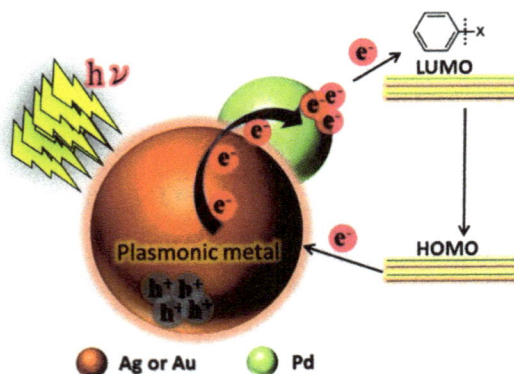

Figure 22. Proposed mechanism for enhanced catalytic activity under visible light irradiation in the presence of plasmonic metal NPs. Adapted from [167]. Copyright (2016), with permission from Royal Society of Chemistry.

The electrons return to the HOMO and ultimately back to the plasmonic metal nanoparticles after the activation of the reactants. The differing behavior of Ag and Au was explained on the basis of their tunability sensitivity upon electron injection. Ag is much more sensitive than Au towards the plasmonic tunability of wavelength, however both showed significant damping of intensity.

Sarina et al. [174] have observed that in reactions, such as the SMC (see Scheme 58), where the catalytic activity is driven by Pd, the intrinsic catalytic activity of palladium in PdAu alloy nanoparticles is significantly enhanced by light irradiation, even at room temperature. On the other hand, AuPd alloy nanoparticles did not perform better than mono metallic Au in other reactions in which the activity is dominated by gold.

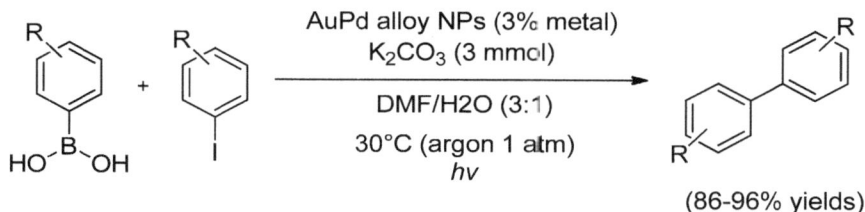

AuPd alloy NPs (3% metal)
K_2CO_3 (3 mmol)

DMF/H2O (3:1)
30°C (argon 1 atm)
hv

(86–96% yields)

Scheme 58. SMC under light irradiation.

Additionally, catalytic performance was found to be strongly dependent on Au/Pd molar ratio in both the presence and the absence of light, as demonstrated by the data reported in Table 9.

Table 9. SMC catalyzed by Au–Pd@ZrO_2 with different Au/Pd molar ratios under visible light irradiation and in the dark [1].

R_1	R_2	Au/Pd [1]	Yield (%)		Selectivity (%)		TON		TOF (h^{-1})		Q.Y. [7] (%)
			Light	Dark	Light	Dark	Light	Dark	Light	Dark	
		1:1.86	96 [2]	37	99	99	87	34	14.5	5.7	2.7
		1:1.00	55 [2]	17	98	99	56	17	9.3	2.8	1.7
3-CH$_3$	4-H	1:5.58	40 [2]	10	99	99	34	8	5.7	1.3	1.4
		1:0.62	28 [2]	6	99	100	33	6	5.5	1.0	1.0
		1:0	2 [2]	0	100	-	3	0	0.5	0	0.1
		0:1	26 [2]	11	98	99	18	8	3.0	1.3	0.7
4-CH$_3$	4-H	1:1.86	94 [2]	46	98	97	95	46	15.8	7.7	2.2
2-CH$_3$	4-H	1:1.86	86 [2]	30	98	98	87	30	14.5	5.0	2.6
4-OCH$_3$	4-H	1:1.86	96 [3]	41	98	99	97	41	19.4	8.2	2.5
4-H	4-OCH$_3$	1:1.86	99 [4]	58	99	99	100	99	50.0	0.0	1.9
4-H	4-CHO	1:1.86	98 [5]	65	99	96	99	67	24.8	0.0	1.5
4-H	4-N(CH$_3$)$_2$	1:1.86	80 [6]	55	68	60	81	56	3.0	0.0	0.3

[1] Molar ratio; [2] Reaction time 6 h; [3] Reaction time 5 h; [4] Reaction time 2 h; [5] Reaction time 4 h; [6] Reaction time 22 h. [7] Q.Y., quantum yield. Reaction conditions: 1 mmol of aryl iodide, 1.5 mmol of arylboronic acid, 50 mg (containing 3% of metals) of catalyst and 3 mmol of base K_2CO_3 in DMF/H_2O = 3:1 (solvent) at 30 °C and 1 atm of argon. TON and TOF values were calculated based the total amount of metal(s).

AuPd nanoparticles are able to absorb light energy and photo-excited conduction electrons are generated at the surface of Pd sites where the reactant molecules are activated. Moreover, the increase in light intensity produced a linear increase in conversion, while the reaction temperature was carefully checked and kept at 30 ± 1 °C in order to avoid the influence of thermal effects on performance. It was observed that the largest contribution (>64%) to the catalytic activity came from radiation with wavelengths in the 490–600 nm range, which is where the LSPR peak of gold nanoparticles is usually observed. This suggests that Au atoms present in the alloy nanoparticles work as an antenna for visible light absorption. Furthermore, the apparent activation energy of the SMC was calculated both in the dark (~49.2 kJ/mol) and under visible light illumination (~33.7 kJ/mol). It was found that the activation energy is reduced by 15.5 kJ/mol under visible light irradiation, which represents a 31% decrease in activation energy.

It was proposed that the intrinsic catalytic activity of these alloy AuPd nanoparticles is related to charge heterogeneity at the surface of the alloy nanoparticles, hence to the Au/Pd ratio of the alloy, resulting in an enhanced interaction between the alloy nanoparticles and the reactant molecules. The catalytic activity's observed dependence on alloy composition was believed to be related to electron redistribution between the two metals. Using a free electron-gas model, the same authors found that the number of transferred electrons (ΔN) is at its highest when the ratio of the electrons of the two metals in the alloy NPs is approximately equal [175]. In more detail, the electron transfer (ΔN) predicted by the model is a function of the Au electron concentration (%) in the AuPd alloy nanoparticles and/or of the Au/Pd molar ratio. A strong correlation between ΔN and the conversion efficiency of reactants was observed, as illustrated in Figure 23. The alloy nanoparticles which possess an Au/Pd electron ratio near to 1:1 (corresponding to the Au/Pd molar ratio 1:1.62) have the largest electron transfer number and displayed significant Au–Pd ionic bond character because of the Pd rich surface of the NP.

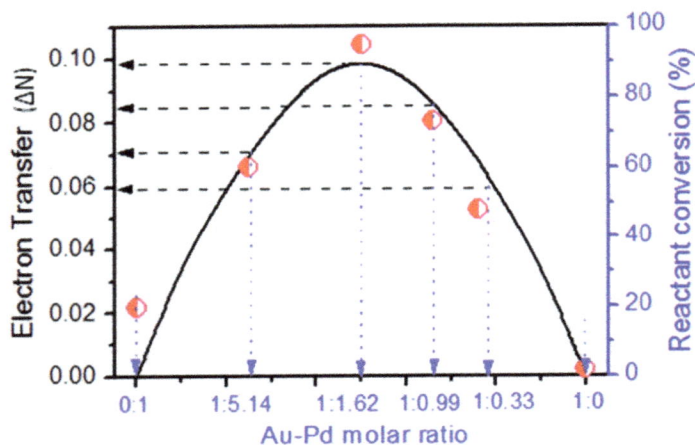

Figure 23. Electron transfer from gold to palladium in the alloy NPs, expressed as ΔN, varies with the composition of the alloy NPs (the curve). ΔN reaches a maximum when the Au/Pd molar ratio is 1:1.62. The Au/Pd molar ratio in the alloy NPs (horizontal axis) and photocatalytic conversions (red symbols) of the SMC in the present study (vertical axis on the right) are given respectively, reaction conversions are based on the average values of three runs for each experiment. Reprinted from [175]. Copyright (2014), with permission from Royal Society of Chemistry.

These papers highlight the fact that the intimate interaction between light-absorbing plasmonic and catalytically active components plays a key role in achieving efficient light harvesting as well as catalytic activity. Following the same idea, Wang et al. [176] has proposed different Au–Pd plasmonic metal nanostructures with synthetically tunable absorption wavelengths to efficiently harvest light. According to Figure 24, the most promising new nanostructures were made up of gold nanorods decorated with small Pd nanoparticles that had previously been heteroepitaxially grown on the nanorods [177,178].

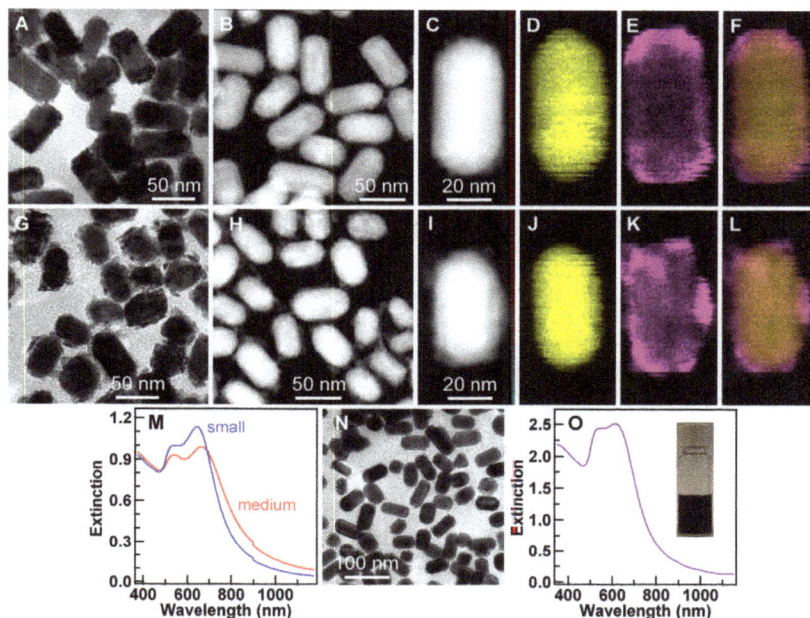

Figure 24. Nanostructures for catalytic reactions under solar radiation. (**A–F**) Medium Au–Pd nanostructures: (**A**) TEM image; (**B**) high-angle annular dark-field scanning transmission electron microscopy (HAADF-STEM) image; (**C**) HAADF-STEM image of a single nanostructure; (**D**) elemental map of gold for the nanostructure shown in (**C**); (**E**) corresponding elemental map of palladium; (**F**) merged elemental map; (**G–L**) Corresponding imaging results for the small Au–Pd nanostructures; (**M**) Extinction spectra recorded in 0.5 cm cuvettes for the medium and small Au–Pd nanostructure samples; (**N**) TEM image of the mixture of the medium, small and spherical nanostructure samples; (**O**) Extinction spectrum of the mixture at 1 cm path length. The inset shows the digital photo of the reaction solution containing the nanostructure mixture in a glass vessel for experiments under solar radiation. Reprinted with permission from [176]. Copyright (2013), with permission from American Chemical Society.

The authors demonstrated that the use of a mixture of these nanostructures, of varying size and shape (medium, small and spherical), guaranteed the maximum exploitation of sunlight and the best photoactivity. In particular, it was proposed that the Au nanorods can function as the light-harvesting component, because of their synthetically tunable longitudinal plasmonic absorption. At the same time, the Pd nanoparticles are active catalysts for the C–C cross coupling reaction. The strong interaction between the two metals means that the chemical potentials of the electrons in the Pd nanoparticles and in the Au nanorods are in equilibrium [179]. The electrons belonging to the Pd nanoparticles are therefore involved in the plasmon resonance occurring across the entire nanostructure upon light irradiation. As a consequence, plasmon excitation promotes the Pd-catalyzed SMC reaction. The authors also synthesized a similar nanostructure with a 25-nm-thick TiO_x shell that plays the role of separating the Au nanorod from the Pd nanoparticles. A comparison of the photocatalytic performances of these two nanostructures, i.e., Au–Pd and Au–TiO_x–Pd, further confirmed the key role played by the longitudinal plasmon of the Au nanorods. Several reactions (involving the coupling of mainly arylbromides with different arylboronic acids) performed under solar radiation gave quite satisfying yields, of above 50% even >90% with nine of them. The catalyst used in these catalytic tests contains 326 µg of gold and 16.6 µg of palladium.

More recently, a very interesting paper [180] reported the intriguing possibility of extending the plasmonic absorption of Pd nanoparticles into the visible-NIR (Near-Infrared) region of the spectrum. This was exploited by synthesizing Pd hexagonal nanoplates with well-defined and tunable longitudinal localized surface plasmon resonance (LSPR). As shown in Figure 25, each Pd nanoplate is indeed a twin crystal, with its top and bottom faces enclosed by {111} facets with stacking faults and with side surfaces surrounded by a mixture of six {111} and six {100} facets. These new nanostructures efficiently catalyze the SMC reaction of iodobenzene and phenylboronic acid under light irradiation. Experimental and theoretical results proved that the increase in catalytic activity granted by the Pd hexagonal nanoplates was due to the photocatalytic effect of plasmon induced hot electrons.

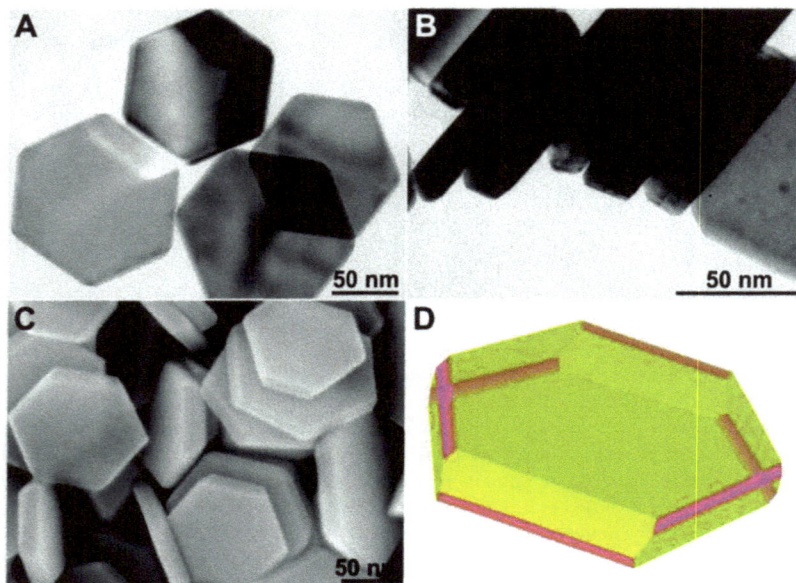

Figure 25. (**A**) Top-view; and (**B**) side-view TEM; (**C**) SEM images; and (**D**) sketch of the Pd hexagonal nanoplates. Yellow and pink represent the {111} and {100} planes, respectively. Reprinted from [180]. Copyright (2015), with permission from Royal Society of Chemistry.

As shown in Figure 26, the obtained TOF values were 2.5 and 2.7 times higher than those related to non-plasmonic {111}-enclosed Pd nano-octahedra and to {100}-enclosed Pd nanocubes, respectively, under illumination with a Xe lamp (λ = 300–1000 nm, 176 mW·cm^{-2}) and 1.7 times higher than the value obtained when the reaction was thermally heated to the same temperature.

Basing on the mechanism of activation of homogeneous Pd-containing catalysts which depend on the electron-enrichment caused by the introduction of special ligands, Li et al. [181] were interested in increasing the activity of heterogeneous Pd nanoparticle-based catalysts by increasing their electron density via support effects. The authors employed the stimulated electron transfer at the metal-semiconductor interface from optically active mesoporous carbon nitride (g-C$_3$N$_4$) nanorods to Pd nanoparticles. It was proposed that the presence of noble metal nanoparticles on the surface or inside of the pores of mesoporous g-C$_3$N$_4$ would give rise to metal-semiconductor contact (Mott-Schottky heterojunction). Charge transfer can be envisioned at the interface and would result in a positively charged region (the depletion region with a thickness of a few nanometers measured from the interface) and a nanoparticle with negative charge, due to the Schottky effect [182]. Theoretically, the energetic electrons located at the noble metal NPs and the holes placed at the g-C$_3$N$_4$ plane would be able to activate two substrates, towards electron-rich and electron-deficient intermediates respectively, to facilitate C±C coupling reactions. The Schottky effect can also be amplified by photo-excited electrons under irradiation, simultaneously amplifying the effective charge transfer at the interface towards the metal nanoparticle. It was found that mesoporous g-C_3N_4 nanorod supported Pd nanoparticles are highly efficient in the Mott-Schottky accelerated SMC of aryl halides with various coupling partners under photo irradiation and very mild conditions. The wavelength-dependence of the activity, which corresponded to the optical absorption spectrum of carbon nitride, was elegantly demonstrated, as shown in Figure 27.

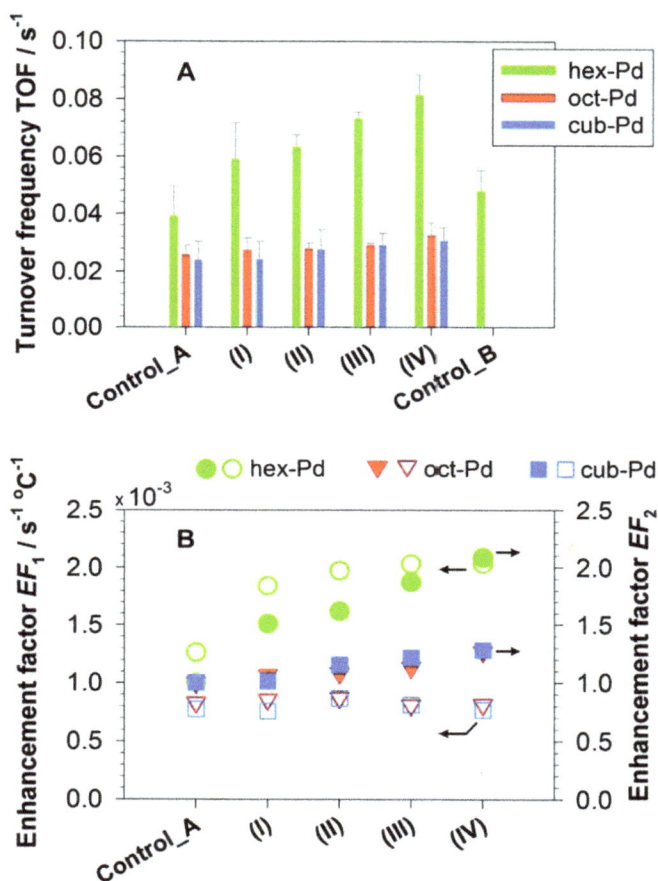

Figure 26. (A) Turnover frequencies (TOFs) per surface Pd atom of hexagonal-Pd were compared with octahedral-Pd and cubic-Pd in SMC reactions carried out using different light sources: 350 nm, 2.6 mW·cm^{-2} (I); 380 nm, 3.4 mW·cm^{-2} (II); 420–1000 nm, 84.0 mW·cm^{-2} (III); and 300–1000 nm, 176.0 mW·cm^{-2} (IV); (B) The enhancement factors EF_1 (open symbols) and EF_2 (solid symbols) for hexagonal-Pd were estimated from the data in (A); Control_A was performed without light illumination in a dark room and Control_B was performed under isothermal heating at 40 °C. All reactions were performed in an isothermal environment (25 °C) for 3 h. Reprinted from [180]. Copyright (2015), with permission from Royal Society of Chemistry.

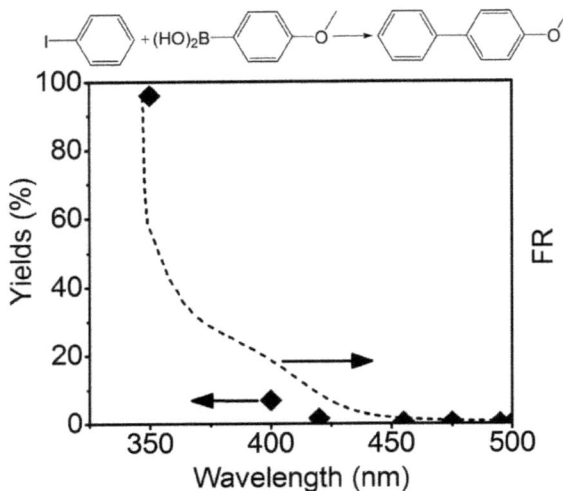

Figure 27. Wavelength dependent activity of Mott-Schottky photocatalyst and corresponding UV-vis absorbance spectra (FR) of mesoporous g-C$_3$N$_4$ nanorods. The coupling reaction of iodobenzene and 4-methoxybenzeneboronic acid was conducted under the following conditions: water (2.5 mL), EtOH (2.5 mL), 138 g of K$_2$CO$_3$ (1 mmol), 73 mg of 4-methoxybenzeneboronic acid (0.6 mmol), 0.017 mL of iodobenzene (0.15 mmol) and 10 mg of m-CNR-Pd (3 wt %) as the catalyst, 150 W Xe lamp, room temperature (25 ± 5 °C), 1 h. The nominal spectral output of the Xe lamp is located between 350 and 800 nm. The wavelength of the incident light was controlled using an appropriate cut-off filter. Reprinted from [181]. Copyright (2013), with permission from Macmillan Publishers Ltd.

Decreased product yield (49% after 12 h) was observed by filtering off the UV part of the Xe lamp (≤420 nm), however no product yield was observed when wavelengths of ≤460 nm were filtered off. Interestingly, the catalyst gave high conversion and selectivity in the coupling of iodobenzene and 4-methoxybenzeneboronic acid at elevated temperatures, even in the dark, but no conversion was observed at temperatures lower than 50 °C. In the same reaction system, light irradiation enabled the coupling reaction to occur, even at room temperature, which gave excellent conversion and selectivity. Both reactions could not occur without the presence of K$_2$CO$_3$, as it is essential to activate boronic acid. No inert atmosphere was required.

Inspired by recent research efforts into the design of electron-rich catalytic sites which could carry out the oxidative addition of Pd^0 with arylhalide and facilitate the coupling reactions, Zhang et al. have successfully performed the activation of aryl chlorides, using heterogeneous multifunctional Pd/Au/porous nanorods of CeO$_2$ (PN-CeO$_2$) which are catalysts with a well-defined spatial configuration, under the irradiation of visible light (>400 nm) at room temperature [183]. In this heterogeneous catalyst, Au nanoparticles were deposited on ceria using the classical deposition–precipitation method. Pd was then selectively photo-deposited onto the surface of Au nanoparticles. It is worth noting that PN-CeO$_2$ have a band gap of 2.63 eV, which is lower than the value of an ideal CeO$_2$ crystal (3.2 eV), and can absorb part of visible light, generating electron/hole pairs. These photogenerated electrons can be injected into Au nanoparticles [184]. Gold not only absorbs visible light via LSPR excitation, but also serves as a charge mediator for the transfer of electrons to Pd nanocatalysts. When Pd/Au/PNCeO$_2$ catalysts is under the illumination of visible light, the hot electrons generated by Au, which display an excited hot state for up to 0.5–1 ps, will flow across the Au and Pd interface immediately, which enriches the electron density of Pd and allows the SMC to occur. This possible electron transfer has also been proven both in previous experiments and in theoretical calculation results [174,176]. Along with the consumption of electrons on Pd during reaction, more and more hot electrons should be provided. Au^+ therefore needs to be supplemented with electrons for its recovery to the Au^0 state by the photogenerated electrons from PN-CeO$_2$ to mediate the electron transfer. The spatially selective deposition of Pd nanocatalysts onto the surface of Au nanoparticles favors the possibility of electron transfer from PN-CeO$_2$ to Au nanoparticles. Under such synergistic action, the multifunctional Pd/Au/PN-CeO$_2$ catalyst is able to activate different aryl chlorides at room temperature, as seen in Figure 28. The electron-rich Pd nanoparticles can activate aryl chlorides and facilitate the first step of the oxidative addition reaction of SMC by accelerating the formation of active radical ligand ArPd^{II}Cl. At the same time, electron/hole pairs created in PN-CeO$_2$, upon the absorption of the incident visible light, activate various arylboronic acids by cleaving the C–B bonds. The oxidized arylboronic acids close to Pd can react with the activated aryl chlorides, giving the final products.

Reaction conditions were as follows; water (1 mL), DMF (1 mL), K$_2$CO$_3$ (0.6 mmol), chlorobenzene (0.2 mmol), phenylboronic acid (0.24 mmol) and catalyst (15 mg). 91.6% conversion and 89.6% yield were achieved in the presence of the strong electron-withdrawing –NO$_2$ group (*p*-substituted with respect to Cl) after 6 h. When electron-donating substituents, *p*-substituted with respect to the halogen position were introduced to phenyl chloride, the yields of the cross-coupling products were reduced to 80.6% (–OCH$_3$), 56.8% (–CH$_2$OH) and 15.3% (conjugated alkynyl) for the cross-coupling product after 6 h of visible light irradiation. Furthermore, high catalytic activities were observed for arylboronic acids with electron-donating groups, again in the *p*-position with respect to Cl, such as –CH$_3$, aryl and tertiary butyl. On the other hand, the insertion in the same position of the electron-withdrawing group, –CHO, into arylboronic acid caused a drop in the yield of the cross-coupling product.

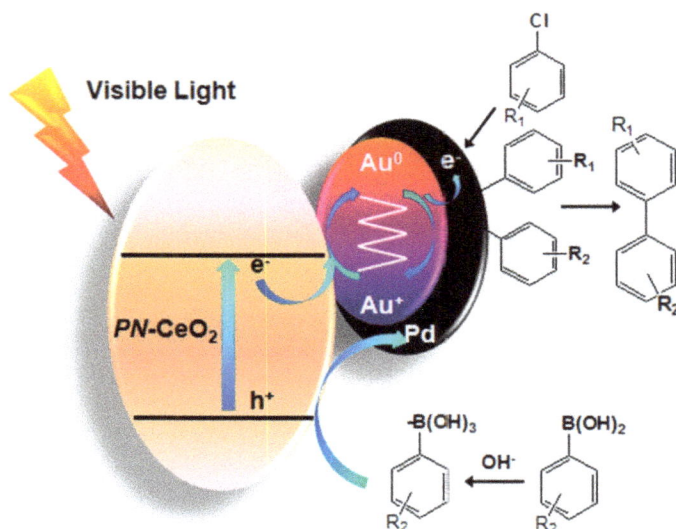

Figure 28. Schematic view of the proposed photocatalytic reaction mechanism over Pd/Au/PN-CeO$_2$ catalysts under visible light irradiation. Reprinted with permission from [183]. Copyright (2015), with permission from American Chemical Society.

A new Cu$_7$S$_4$@Pd heteronanostructured photocatalyst, in which Pd nanoparticles and Cu$_7$S$_4$ domains are in intimate interfacial contact, were found to be excellent catalysts for solar-driven organic synthesis reactions [185]. Both simulation studies and experimental investigations demonstrated that a combination of the NIR LSPR light-harvesting property of Cu$_7$S$_4$ and the catalytic features of Pd resulted in the enhanced photocatalytic activity that these systems show towards the SMC reaction.

In particular, Pd displays no LSPR absorption above 1000 nm, whereas Cu_7S_4 or a physical mixture of Pd and Cu_7S_4 gave an absorption peak at about 1500 nm (this absorption was also observed by simulation, compare Figure 29a,b). Interestingly, $Cu_7S_4@Pd$ showed a red shift of the LSPR peak of about 500 nm compared to bare Cu_7S_4. An explanation for such a spectroscopic feature is that the Pd on the Cu_7S_4 surface is possibly bound to S atoms, hence lowering its exposure to oxygen. This might, in turn, annihilate vacancies in Cu_7S_4, and result in a red shift of the LSPR peak. The red shift also correlated to a refractive index change which was caused by the incorporation of palladium into the Cu_7S_4 surface.

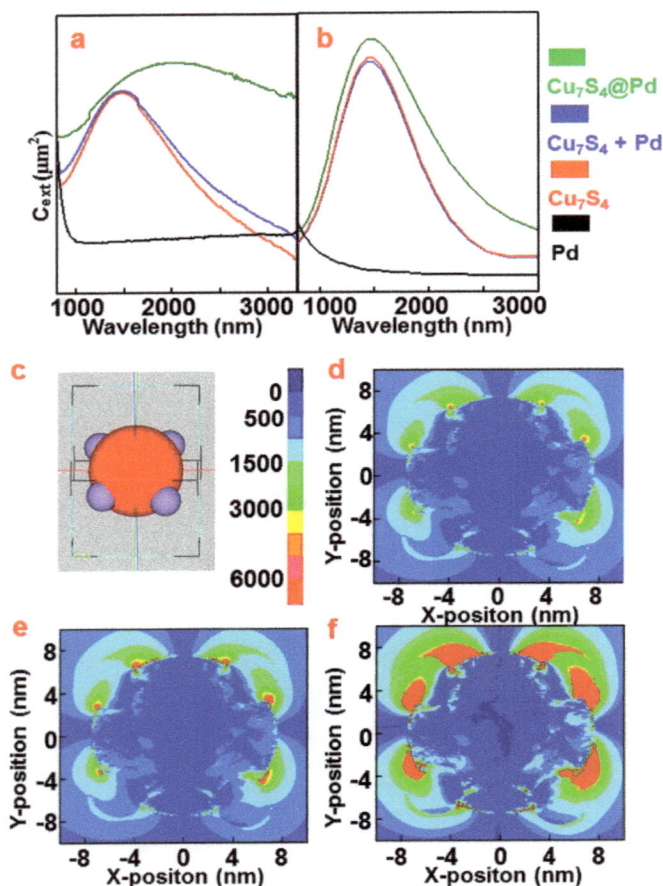

Figure 29. Localized surface plasmon resonance (LSPR) absorption spectra of different nanostructures obtained from experiment (**a**); and finite difference time domain (FDTD) simulation (**b**); Green curves: $Cu_7S_4@Pd$ heteronanostructures. Blue curves: physical mixture of Cu_7S_4 and Pd nanoparticles. Red curves: Cu_7S_4 nanoparticles. Black curves: Pd nanoparticles; (**c**) FDTD simulation setup for $Cu_7S_4@Pd$ and electrical field intensity scale; (**d–f**) The 2D contour of the electric field intensities around the $Cu_7S_4@Pd$ heteronanostructures under illumination of 808 nm (**d**); 980 nm (**e**); and 1500 nm (**f**); respectively. Reprinted from [185]. Copyright (2015), with permission from American Chemical Society.

The authors demonstrated the presence of a local electrical field enhancement, due to the LSPR effect, and studied how this enhancement is correlated with various wavelengths using a FDTD simulation (the Cu_7S_4@Pd nanocrystal was located in the center of a simulated box, as shown in Figure 29c) of the 2D contour of the electron field intensity (XOY-plane, $z = 0$ nm) under different illumination sources (808, 980 and 1500 nm, Figure 29d–f). An electrical field enhancement was observed under irradiation at all examined wavelengths. In particular, the electrical field intensity for 1500 nm irradiation appears to be far stronger than that for 808 and 980 nm, and the same trend was obtained when evaluating the catalytic activity of Cu_7S_4@Pd in the SMC reaction between iodobenzene and phenylboronic acid at different reaction time intervals (10, 20, and 30 min). As an example, 97% conversion was reached under 1500 nm irradiation in 30 min, whereas 45% and 50% conversions were obtained using the 808 and 980 nm irradiation wavelengths, respectively. Similar results were collected when phenylboronic acid was replaced by *p*-tolylboronic acid.

In this case, the use of cetyltrimethylammonium bromide (CTAB) was necessary to help the hydrophobic Cu_7S_4@Pd nanoparticles disperse into water. It was proposed that the LSRP absorption of Cu_7S_4 gives rise to "hot holes" under 1500 nm illumination. These hot holes have enough energy to overcome the small Schottky barrier and inject themselves into the Pd domain, possibly rendering the Pd surface electron-deficient. In terms of the SMC reaction, theoretical studies indicated that the rate-limiting step is the oxidative addition of aryl halides onto Pd(0) [186]. Photoactivation favors the positively charged Pd surface, analogously to the action of electron-withdrawing ligands, and can promote the activation of aryl halides. Finally, differently to the majority of the traditional semiconductor–metal photocatalysts in which the LSPR effect of the noble metal nanoparticles serves the visible range absorption of the semiconductor, here Cu_7S_4 nanoparticles take advantage of NIR range photons via LSPR and grants the benefits to the metal.

6. Conclusions and Future Perspectives

Within the exploitation of SMC reactions, the use of unconventional technologies is allowing the advancement of a new chemistry with enormous potential, because it can promote processes and products that are more sustainable from both environmental and economic points of view. In particular, it has been shown that SMC is strongly facilitated by physical activation with MW, US, ball milling and light in a greener approach to process intensification. Despite the large fraction of papers here cited is dealing with the MW activation, the development of environmentally benign and often cost-effective procedures for SMC reactions is attracting a lot of interest. It was found that mainly either isolated Pd(II) species or supported Pd nanoparticles display a key role in the field of SMC catalyzed reaction in which the effects due to the combined use of the new methodologies can be investigated. Indeed, a great number of novel palladium-based catalysts has been reported in recent years for SMC. A careful examination of the catalysts reveals that it is important not only to regulate the Pd dispersion in terms of Pd(II) immobilization and of particle size, but also to consider the nature of the support as well as to optimise the synthetic procedure to control the morphology and the electronic properties of the obtained catalyst. It is worth noting that the optimised catalyst is required to be stable, recyclable and robust under the experimental conditions achieved by non conventional activation.

All these advances are well documented in recent literature, however, there are still a lot of problems that urgently require a solution: among these, to avoid the use of ligands, which can have environmental impact, is one of the main goals. Moreover, the requirement of ever-smaller amounts of metal catalyst to abate costs, as well as the need to produce recyclable catalysts are also main issues. The possibility to employ less reactive, but more attractive chloroarenes remains a challenge. In many cases, the selectivity is unsatisfactory when the reaction is driven by light, due to the involvement of free radical intermediates in UV-induced chemical reactions, giving for example, undesired autooxidation products, therefore leading to low selectivity of the desired product.

The current efforts in catalyst design can allow a more efficient exploitation of unconventional techniques. Recent discoveries here reported have demonstrated that highly selective reactions are feasible in the presence of an opportunely optimised catalyst and the right reaction conditions.

The results here described represent a helpful starting point to new green processes for SMC in the presence of innovative activation methodologies and open the door for significant improvement connected with the use of heterogeneous catalysts that permit product isolation by simple separation. In the near future, the green methodologies here examined combined with the catalyst design will be strategic to improve atom and energy efficiency, especially for industrial applications. In this frame, more intense efforts are required in the area of scaling up. In a future scenario, the most challenging problems that have to be overcome are: (i) to develop new active catalysts able to work in combination with unconventional activation methods; (ii) to provide cost analysis and (iii) to achieve progress in reactor engineering with new reliable flow systems that are well suited for industrial applications.

Acknowledgments: This work was supported by the University of Turin (Progetti Ricerca Locale 2015).

Conflicts of Interest: The authors declare no conflict of interest.

References

1. Phan, N.T.S.; Van Der Sluys, M.; Jones, C.W. On the nature of the active species in palladium catalyzed Mizoroki–Heck and Suzuki–Miyaura couplings—Homogeneous or heterogeneous catalysis, A critical review. *Adv. Synth. Catal.* **2006**, *348*, 609–679. [CrossRef]
2. Kappe, O.K.; Pieber, B.; Dallinger, D. Microwave effects in organic synthesis: Myth or reality? *Angew. Chem. Int. Ed.* **2013**, *52*, 1088–1094. [CrossRef] [PubMed]
3. Cravotto, G.; Cintas, P. Power ultrasound in organic synthesis: Moving cavitational chemistry from academia to innovative and large-scale applications. *Chem. Soc. Rev.* **2006**, *35*, 180–196. [CrossRef] [PubMed]
4. Lang, X.; Chen, X.; Zhao, J. Heterogeneous visible light photocatalysis for selective organic transformations. *Chem. Soc. Rev.* **2014**, *43*, 473–486. [CrossRef] [PubMed]
5. Suzuki, A. Kreuzkupplungen von Organoboranen: Ein einfacher Weg zum Aufbau von C-C-*Bindungen* (Nobel-Aufsatz). *Angew. Chem.* **2011**, *123*, 6854–6869. [CrossRef]
6. Miyaura, N.; Yamada, K.; Suzuki, A. A new stereospecific cross-coupling by the palladium-catalyzed reaction of 1-alkenylboranes with 1-alkenyl or 1-alkynyl halides. *Tetrahedron Lett.* **1979**, *20*, 3437–3440. [CrossRef]
7. Lennox, A.J.J.; Lloyd-Jones, G.C. Selection of boron reagents for Suzuki–Miyaura coupling. *Chem. Soc. Rev.* **2014**, *43*, 412–443. [CrossRef] [PubMed]
8. Suzuki, A. Cross-coupling reactions of organoboranes: An easy way to construct C–C bonds. *Angew. Chem. Int. Ed.* **2011**, *50*, 6722–6737. [CrossRef] [PubMed]
9. Hassan, J.; Sévignon, M.; Gozzi, C.; Schulz, E.; Lemaire, M. Aryl–Aryl bond formation one century after the discovery of the Ullmann reaction. *Chem. Rev.* **2002**, *102*, 1359–1470. [CrossRef] [PubMed]
10. Han, F.S. Transition-metal-catalyzed Suzuki–Miyaura cross-coupling reactions: A remarkable advance from palladium to nickel catalysts. *Chem. Soc. Rev.* **2013**, *42*, 5270–5298. [CrossRef] [PubMed]
11. Suzuki, A. Cross-coupling reactions via organoboranes. *J. Organomet. Chem.* **2002**, *653*, 83–90. [CrossRef]
12. Lipshutz, B.H.; Abela, A.R.; Boskovic, Z.V.; Nishikata, T.; Duplais, C.; Krasovskiy, A. "Greening up" cross-coupling chemistry. *Top. Catal.* **2010**, *53*, 985–990. [CrossRef]
13. Paul, S.; Islam, M.M.; Islam, S.M. Suzuki–Miyaura reaction by heterogeneously supported Pd in water: Recent studies. *RSC Adv.* **2015**, *5*, 42193–42221. [CrossRef]
14. Polshettiwar, V.; Decottignies, A.; Len, C.; Fihri, A. Suzuki–Miyaura cross-coupling coupling reactions with low catalyst loading: A green and sustainable protocol in pure water. *ChemSusChem* **2010**, *3*, 502–522. [CrossRef] [PubMed]
15. Chatterjee, A.; Ward, T.R. Recent advances in the palladium catalyzed Suzuki–Miyaura cross-coupling reaction in water. *Catal. Lett.* **2016**, *146*, 820–840. [CrossRef]
16. Cravotto, G.; Garella, D.; Tagliapietra, S.; Stolle, A.; Schuessler, S.; Leonhardt, S.E.S.; Ondruschka, B. Suzuki cross-couplings of (hetero)aryl chlorides in the solid-state. *New J. Chem.* **2012**, *36*, 1304–1307. [CrossRef]
17. Larhed, M.; Moberg, C.; Hallberg, A. Microwave-accelerated homogeneous catalysis in organic chemistry. *Acc. Chem. Res.* **2002**, *35*, 717–727. [CrossRef] [PubMed]
18. Mehta, V.P.; Van der Eycken, E.V. Microwave-assisted C–C bond forming cross-coupling reactions: An overview. *Chem. Soc. Rev.* **2011**, *40*, 4925–4936. [CrossRef] [PubMed]

19. Bai, L.; Wang, J.X. Environmentally friendly Suzuki aryl–aryl cross-coupling reaction. *Curr. Org. Chem.* **2005**, *9*, 535–553. [CrossRef]
20. Hoogenboom, R.; Meier, M.A.R.; Schubert, U.S. The introduction of high-throughput experimentation methods for Suzuki–Miyaura coupling reactions in university education. *J. Chem. Educ.* **2005**, *82*, 1693–1696. [CrossRef]
21. Costa, N.E.; Pelotte, A.L.; Simard, J.M.; Syvinski, C.A.; Deveau, A.M. Discovering green, aqueous Suzuki coupling reactions: Synthesis of ethyl (4-phenylphenyl)acetate, a biaryl with anti-arthritic potential. *J. Chem. Educ.* **2012**, *89*, 1064–1067. [CrossRef]
22. Hill, N.J.; Bowman, M.D.; Esselman, B.J.; Byron, S.D.; Kreitinger, J.; Leadbeater, N.E. Ligand-free Suzuki–Miyaura coupling reactions using an inexpensive aqueous palladium source: A synthetic and computational exercise for the undergraduate organic chemistry laboratory. *J. Chem. Educ.* **2014**, *91*, 1054–1057. [CrossRef]
23. Soares, P.; Fernandes, C.; Chavarria, D.; Borges, F. Microwave-assisted synthesis of 5-phenyl-2-hydroxyacetophenone derivatives by a green Suzuki coupling reaction. *J. Chem. Educ.* **2015**, *92*, 575–578. [CrossRef]
24. Elumalai, V.; Sandtorv, A.H.; Bjorsvik, H.-R. A highly efficient Pd(PPh$_3$)$_4$-catalyzed suzuki cross-coupling method for the preparation of 2-nitrobiphenyls from 1-chloro-2-nitrobenzenes and phenylboronic acids. *Eur. J. Org. Chem.* **2016**, *2016*, 1344–1354. [CrossRef]
25. Liu, J.; Robins, M.J. Fluoro, alkylsulfanyl, and alkylsulfonyl leaving groups in Suzuki cross-coupling reactions of purine 2′-deoxynucleosides and nucleosides. *Org. Lett.* **2005**, *7*, 1149–1151. [CrossRef] [PubMed]
26. Korn, T.J.; Schade, M.A.; Wirth, S.; Knochel, P. Cobalt(II)-catalyzed cross-coupling between polyfunctional arylcopper reagents and aryl fluorides or tosylates. *Org. Lett.* **2006**, *8*, 725–728. [CrossRef] [PubMed]
27. Wang, T.; Alfonso, B.J.; Love, J.A. Platinum(II)-catalyzed cross-coupling of polyfluoroaryl imines. *Org. Lett.* **2007**, *9*, 5629–5631. [CrossRef] [PubMed]
28. Guo, H.; Kong, F.; Kanno, K.-I.; He, J.; Nakajima, K.; Takahashi, T. Early transition metal-catalyzed cross-coupling reaction of aryl fluorides with a phenethyl Grignard reagent accompanied by rearrangement of the phenethyl group. *Organometallics* **2006**, *25*, 2045–2048. [CrossRef]
29. Ruiz, J.R.; Jimenez-Sanchidrian, C.; Mora, M. Suzuki cross-coupling reaction of fluorobenzene with heterogeneous Palladium catalysts. *J. Fluorine Chem.* **2006**, *127*, 443–445. [CrossRef]
30. Cargill, M.R.; Sandford, G.; Tadeusiak, A.J.; Yufit, D.S.; Howard, J.A.K.; Kilickiran, P.; Nelles, G. Palladium-catalyzed C–F activation of polyfluoronitrobenzene derivatives in Suzuki–Miyaura coupling reactions. *J. Org. Chem.* **2010**, *75*, 5860–5866.
31. Louvis, A.D.; Silva, N.A.; Semaan, F.S.; Da Silva, F.D.; Saramago, G.; de Souza, L.C.; Ferreira, B.L.; Castro, H.C.; Salles, J.P.; Souza, A.L.; et al. Synthesis, characterization and biological activities of 3-aryl-1,4-naphthoquinones—Green palladium-catalysed suzuki cross coupling. *New J. Chem.* **2016**, *40*, 7643–7656. [CrossRef]
32. Fernandes, C.; Soares, P.; Gaspar, A.; Martins, D.; Gomes, L.R.; Low, J.N.; Borges, F. Synthesis of 6-aryl/heteroaryl-4-oxo-4*H*-chromene-2-carboxylic ethyl ester derivatives. *Tetrahedr. Lett.* **2016**, *57*, 3006–3010. [CrossRef]
33. Kumar, S.; Ahmed, N. A facile approach for the synthesis of novel 1-oxa- and 1-aza-flavonyl-4-methyl-1*H*-benzo[*d*][1,3]oxazin-2(4*H*)-ones by microwave enhanced Suzuki–Miyaura coupling using bidentate chromen-4-one-based Pd(II)–diimine complex as catalyst. *RSC Adv.* **2015**, *5*, 77075–77087. [CrossRef]
34. Qu, R.Y.; Liu, Y.C.; Wu, Q.Y.; Chen, Q.; Yang, G.F. An efficinet method for the syntheses of funtionalized 6-bulkysubstituted salicilates under microwave irradiation. *Tetrahedron* **2015**, *71*, 8123–8130. [CrossRef]
35. Kadam, J.; Buckley, S.B.; Dinh, T.; Fitzgerald, R.; Zhang, W. Convertible fluorous linker assisted synthesis of tetrasubstituted furans. *Synlett* **2011**, *11*, 1608–1612.
36. Joy, M.N.; Bodke, Y.D.; Khader, K.K.A.; Sajith, A.M.; Venkatesh, T.; Kumar, A.K.A. Simultaneous exploration of TBAF·3H$_2$O as a base as well as a solvating agent for the palladium catalyzed Suzuki cross-coupling of 4-methyl-7-nonafluorobutylsulfonyloxy coumarins under microwave irradiation. *J. Fluorine Chem.* **2016**, *182*, 109–120. [CrossRef]
37. Vichier-Guerre, S.; Dugue, L.; Pochet, S. A convenient synthesis of 4(5)-(hetero)aryl-1*H*-imidazoles via microwave-assisted Suzuki–Miyaura cross-coupling reaction. *Tetrahedron Lett.* **2014**, *55*, 6347–6350. [CrossRef]

38. Fuse, S.; Ohuchi, T.; Asawa, Y.; Sato, S.; Nakamura, H. Development of 1-aryl-3-furanyl/thienyl-imidazopyridine templates for inhibitors against hypoxia inducible factor (HIF)-1 transcriptional activity. *Bioorg. Med. Chem. Lett.* **2016**, *26*, 5887–5890. [CrossRef] [PubMed]

39. Sandtorv, A.H.; Bjørsvika, H.-R. A three-way switchable process for Suzuki cross-coupling, hydrodehalogenation, or an assisted tandem hydrodehalogenation and Suzuki cross-coupling sequence. *Adv. Synth. Catal.* **2013**, *355*, 3231–3243. [CrossRef]

40. Wang, S.; Guo, R.; Li, J.; Zou, D.; Wu, Y.; Wu, Y. Efficient synthesis of 3-aryl-1*H*-indazol-5-amine by Pd-catalyzed Suzuki–Miyaura cross-coupling reaction under microwave-assisted conditions. *Tetrahedron Lett.* **2015**, *56*, 3750–3753. [CrossRef]

41. Liu, Y.-C.; Ye, C.-J.; Chen, Q.; Yang, G.-F. Efficient synthesis of bulky 4-substituted-isatins via microwave-promoted Suzuki cross-coupling reaction. *Tetrahedron Lett.* **2013**, *54*, 949–955. [CrossRef]

42. El Akkaoui, A.; Berteina-Raboin, S.; Mouaddib, A.; Guillaumet, G. Direct arylation of imidazo[1,2-*b*]pyridazines: Microwave-assisted one-pot suzuki coupling/Pd-catalysed arylation. *Eur. J. Org. Chem.* **2010**, 862–871. [CrossRef]

43. Copin, C.; Henry, N.; Buron, F.; Routier, S. Synthesis of 2,6-disubstituted imidazo[2,1-*b*][1,3,4]thiadiazoles through cyclization and Suzuki–Miyaura cross-coupling reactions. *Eur. J. Org. Chem.* **2012**, *2012*, 3079–3083. [CrossRef]

44. Koini, E.N.; Avlonitis, N.; Martins-Duarte, E.S.; de Souza, W.; Vommaro, R.C.; Calogeropoulou, T. Divergent synthesis of 2,6-diaryl-substituted 5,7,8-trimethyl-1,4-benzoxazines via microwave-promoted palladium-catalyzed Suzuki–Miyaura cross coupling and biological evaluation. *Tetrahedron* **2012**, *68*, 10302–10309. [CrossRef]

45. Zukauskaite, Z.; Buinauskaite, V.; Solovjova, J.; Malinauskaite, L.; Kveselyte, A.; Bieliauskas, A.; Ragaite, G.; Sackus, A. Microwave-assisted synthesis of new fluorescent indoline-based building blocks by ligand free Suzuki–Miyaura cross-coupling reaction in aqueous media. *Tetrahedron* **2016**, *72*, 2955–2963. [CrossRef]

46. Prieur, V.; Pujol, M.D.; Guillaumet, G. A strategy for the triarylation of pyrrolopyrimidines by using microwave-promoted cross-coupling reactions. *Eur. J. Org. Chem.* **2015**, *2015*, 6547–6556. [CrossRef]

47. El Bouakher, A.; Allouchi, H.; Abrunhosa-Thomas, I.; Troin, Y.; Guillaumet, G. Suzuki–Miyaura reactions of halospirooxindole derivatives. *Eur. J. Org. Chem.* **2015**, 3450–3461. [CrossRef]

48. Qu, G.-R.; Xin, P.-Y.; Niu, H.-Y.; Jin, X.; Guo, X.-T.; Yang, X.-N.; Guo, H.-M. Microwave promoted palladium-catalyzed Suzuki–Miyaura cross-coupling reactions of 6-chloropurines with sodium tetraarylborate in water. *Tetrahedron* **2011**, *67*, 9099–9103. [CrossRef]

49. Kabri, Y.; Crozet, M.D.; Szabo, R.; Vanelle, P. Efficient and original microwave-assisted Suzuki–Miyaura cross-coupling reaction in the 4*H*-pyrido[1,2-*a*]pyrimidin-4-one series. *Synthesis* **2011**, *19*, 3115–3122. [CrossRef]

50. Kabri, Y.; Crozet, M.D.; Primas, N.; Vanelle, P. One-pot chemoselective bis(Suzuki–Miyaura cross-coupling): Efficient access to 3,9-bis[(hetero)aryl]-4*H*-pyrido[1,2-*a*]pyrimidin-4-one derivatives under microwave irradiation. *Eur. J. Org. Chem.* **2012**, *28*, 5595–5604. [CrossRef]

51. Kabri, Y.; Crozet, M.D.; Terme, T.; Vanelle, P. Efficient access to 2,6,8-trisubstituted 4-aminoquinazolines through microwave-assisted one-pot chemoselective tris-Suzuki–Miyaura or SNAr/bis-Suzuki–Miyaura reactions in water. *Eur. J. Org. Chem.* **2015**. [CrossRef]

52. Gallagher-Duval, S.; Hervé, G.; Sartori, G.; Enderlin, G.; Len, C. Improved microwave-assisted ligand-free Suzuki–Miyaura cross-coupling of 5-iodo-2′-deoxyuridine in pure water. *New J. Chem.* **2013**, *37*, 1989–1995. [CrossRef]

53. Darses, S.; Genet, J.-P. Potassium organotrifluoroborates: New perspectives in organic synthesis. *Chem. Rev.* **2008**, *108*, 288–325. [CrossRef] [PubMed]

54. Gary, A. Molander organotrifluoroborates: Another branch of the mighty oak. *J. Org. Chem.* **2015**, *80*, 7837–7848.

55. Kim, T.; Song, J.H.; Jeong, K.H.; Lee, S.; Ham, J. Potassium (1-organo-1*H*-1,2,3-triazol-4-yl)trifluoroborates from ethynyltrifluoroborate through a regioselective one-pot Cu-catalyzed azide–alkyne cycloaddition reaction. *Eur. J. Org. Chem.* **2013**. [CrossRef]

56. Jain, P.; Yi, S.; Flaherty, P.T. Suzuki–Miyaura cross-coupling of potassium organoborates with 6-sulfonate benzimidazoles using microwave irradiation. *J. Heterocycl. Chem.* **2013**, *50*, E166–E173. [CrossRef]

57. Henderson, L.; Knight, D.W.; Rutkowski, P.; Williams, A.C. Optimised conditions for styrene syntheses using Suzuki–Miyaura couplings and catalyst-ligand-base pre-mixes. *Tetrahedron Lett.* **2012**, *53*, 4654–4656. [CrossRef]

58. Brooker, M.D.; Cooper, S.M., Jr.; Hodges, D.R.; Carter, R.R.; Wyatt, J.K. Studies of microwave-enhanced Suzuki–Miyaura vinylation of electron-rich sterically hindered substrates utilizing potassium vinyltrifluoroborate. *Tetrahedron Lett.* **2010**, *51*, 6748–6752. [CrossRef] [PubMed]

59. Huang, Z.-Y.; Yang, J.-F.; Song, K.; Chen, Q.; Zhou, S.-L.; Hao, G.-F.; Yang, G.-F. One-pot approach to *N*-quinolyl 3′/4′-biaryl carboxamides by microwave-assisted Suzuki–Miyaura coupling and *N*-boc deprotection. *J. Org. Chem.* **2016**, *81*, 9647–9657. [CrossRef] [PubMed]

60. Huang, Z.-Y.; Yang, J.-F.; Chen, Q.; Cao, R.-J.; Huang, W.; Hao, G.-F.; Yang, G.-F. An efficient one-pot access to *N*-(pyridin-2-ylmethyl) substituent biphenyl-4-sulfonamides through water-promoted, palladium-catalyzed, microwave-assisted reactions. *RSC Adv.* **2015**, *5*, 75182–75186. [CrossRef]

61. Grob, J.E.; Nunez, J.; Dechantsreiter, M.A.; Hamann, L.G. One-pot reductive amination and Suzuki–Miyaura cross-coupling of formyl aryl and heteroaryl mida boronates in array format. *J. Org. Chem.* **2011**, *76*, 4930–4940. [CrossRef] [PubMed]

62. Park, K.-Y.; Kim, B.T.; Heo, J.-N. Direct one-pot synthesis of naphthoxindoles from 4-bromooxindoles by Suzuki–Miyaura coupling and aldol condensation reactions. *Eur. J. Org. Chem.* **2014**, *1*, 164–170. [CrossRef]

63. Hooper, A.; Zambon, A.; Springer, C.J. A novel protocol for the one-pot borylation/suzuki reaction provides easy access to hinge-binding groups for kinase inhibitors. *Org. Biomol. Chem.* **2016**, *14*, 963–969. [CrossRef] [PubMed]

64. Cívicos, J.F.; Alonso, D.A.; Nájera, C. Oxime palladacycle-catalyzed Suzuki–Miyaura alkenylation of aryl, heteroaryl, benzyl, and allyl chlorides under microwave irradiation conditions. *Adv. Synth. Catal.* **2011**, *353*, 1683–1687. [CrossRef]

65. Cívicos, J.F.; Alonso, D.A.; Najera, C. Oxime-palladacycle-catalyzed Suzuki–Miyaura arylation and alkenylation of aryl imidazolesulfonates under aqueous and phosphane-free conditions. *Eur. J. Org. Chem.* **2012**, *2012*, 3670–3676. [CrossRef]

66. Susanto, W.; Chu, C.-Y.; Ang, W.J.; Chou, T.-C.; Lo, L.-C.; Lam, Y. Development of a fluorous, oxime-based palladacycle for microwave-promoted carbon–carbon coupling reactions in aqueous media. *Green Chem.* **2012**, *14*, 77–80. [CrossRef]

67. Dawood, K.M.; Elamin, M.B.; Farag, A.M. Microwave-assisted synthesis of 2-acetyl-5-arylthiophenes and 4-(5-arylthiophen-2-yl)thiazoles via Suzuki coupling in water. *ARKIVOC* **2015**, *7*, 50–62.

68. Hanhan, M.E.; Martínez-Máñez, R.; Ros-Lis, J.V. Highly effective activation of aryl chlorides for Suzuki coupling in aqueous media using a ferrocene-based Pd(II)–diimine catalyst. *Tetrahedron Lett.* **2012**, *53*, 2388–2391. [CrossRef]

69. Balam-Villarreal, J.A.; Sandoval-Chavez, C.I.; Ortega-Jimenez, F.; Toscano, R.A.; Carreon-Castro, M.P.; Lopez-Cortes, J.G.; Ortega-Alfaro, M.C. Infrared irradiation or microwave assisted cross-coupling reactions using sulfur-containing ferrocenyl-palladacycles. *J. Organomet. Chem.* **2016**, *818*, 7–14. [CrossRef]

70. Yılmaz, Ü.; Şireci, N.; Deniz, S.; Küçükbay, H. Synthesis and microwave-assisted catalytic activity of novel bis-benzimidazole salts bearing furfuryl and thenylmoieties in Heck and Suzuki cross-coupling reactions. *Appl. Organomet. Chem.* **2010**, *24*, 414–420.

71. Yılmaz, Ü.; Küçükbay, H.; Şireci, N.; Akkurt, M.; Günald, S.; Durmaz, R.; Tahir, M.N. Synthesis, microwave-promoted catalytic activity in Suzuki–Miyaura cross-coupling reactions and antimicrobial properties of novel benzimidazole salts bearing trimethylsilyl group. *Appl. Organomet. Chem.* **2011**, *25*, 366–373. [CrossRef]

72. Silarska, E.; Trzeciak, A.M.; Pernak, J.; Skrzypczak, A. [IL]₂[PdCl₄] complexes (IL = imidazolium cation) as efficient catalysts for Suzuki–Miyaura cross-coupling of aryl bromides and aryl chlorides. *Appl. Catal. A* **2013**, *466*, 216–223. [CrossRef]

73. Hanhan, M.E.; Senemoglu, Y. Microwave-assisted aqueous Suzuki coupling reactions catalyzed by ionic palladium(II) complexes. *Transit. Met. Chem.* **2012**, *37*, 109–116. [CrossRef]

74. Shen, L.; Huang, S.; Nie, Y.; Lei, F. An efficient microwave-assisted Suzuki reaction using a new pyridine-pyrazole/Pd(II) species as catalyst in aqueous media. *Molecules* **2013**, *18*, 1602–1612. [CrossRef] [PubMed]

75. Naik, S.; Kumaravel, M.; Mague, J.T.; Balakrishna, M.S. Bisamino(diphosphonite) with dangling olefin functionalities: Synthesis, metal chemistry and catalytic utility of Rh^I and Pd^{II} complexes in hydroformylation and Suzuki–Miyaura reactions. *Dalton Trans.* **2014**, *43*, 1082–1095. [CrossRef] [PubMed]

76. Basauri-Molina, M.; Hernández-Ortega, S.; Morales-Morales, D. Microwave-assisted C–C and C–S couplings catalysed by organometallic Pd-SCS or coordination Ni-SNS pincer complexes. *Eur. J. Inorg. Chem.* **2014**, *27*, 4619–4625. [CrossRef]

77. Gayakhe, V.; Ardhapure, A.; Kapdi, A.R.; Sanghvi, Y.S.; Serrano, J.L.; Garcia, L.; Perez, J.; Garcia, J.; Sanchez, G.; Fischer, C.; et al. Water-soluble Pd-imidate complexes: Broadly applicable catalysts for the synthesis of chemically modified nucleosides via Pd-catalyzed cross-coupling. *J. Org. Chem.* **2016**, *81*, 2713–2729. [CrossRef] [PubMed]

78. Glorius, F. Asymmetric cross-coupling of non-activated secondary alkyl halides. *Angew. Chem. Int. Ed.* **2008**, *47*, 8347–8349. [CrossRef] [PubMed]

79. Glasson, C.R.K.; Meehan, G.V.; Motti, C.A.; Clegg, J.K.; Turner, P.; Jensen, P.; Lindoy, L.F. New nickel(II) and iron(II) helicates and tetrahedra derived from expanded quaterpyridines. *Dalton Trans.* **2011**, *40*, 10481–10490. [CrossRef] [PubMed]

80. Lee, G.M.; Loechtefeld, R.; Menssen, R.; Bierer, D.E.; Riedl, B.; Baker, R.T. Synthesis of bromodifluoromethyl (arylsulfonyl) compounds and microwave-assisted nickel catalyzed cross coupling with arylboronic acids. *Tetrahedron Lett.* **2016**, *57*, 5464–5468. [CrossRef]

81. Baghbanzadeh, M.; Pilger, C.; Kappe, C.O. Rapid nickel-catalyzed Suzuki–Miyaura cross-couplings of aryl carbamates and sulfamates utilizing microwave heating. *J. Org. Chem.* **2011**, *76*, 1507–1510. [CrossRef] [PubMed]

82. De Luna Martins, D.; Aguiar, L.C.S.; Antunes, O.A.C. Microwave promoted Suzuki reactions between aroyl chlorides and boronic acids catalyzed by heterogeneous and homogeneous phosphine-free palladium catalysts. *J. Organomet. Chem.* **2011**, *696*, 2845–2849. [CrossRef]

83. Feng, G.; Liu, F.; Lin, C.; Li, W.; Wang, S.; Qi, C. Crystalline mesoporous γ-Al$_2$O$_3$ supported palladium: Novel and efficient catalyst for Suzuki–Miyaura reaction under controlled microwave heating. *Catal. Commun.* **2013**, *37*, 27–31. [CrossRef]

84. Smith, S.E.; Siamaki, A.R.; Gupton, B.F.; Carpenter, E.E. CuPd nanoparticles as a catalyst in carbon–carbon cross-coupling reactions by a facile oleylamine synthesis. *RSC Adv.* **2016**, *6*, 91541–91545. [CrossRef]

85. Molnár, Á. Efficient, selective, and recyclable palladium catalysts in carbon–carbon coupling reactions. *Chem. Rev.* **2011**, *111*, 2251–2320. [CrossRef] [PubMed]

86. Kim, S.W.; Kim, M.; Lee, W.Y.; Hyeon, T. Fabrication of hollow palladium spheres and their successful application to the recyclable heterogeneous catalyst for Suzuki coupling reactions. *J. Am. Chem. Soc.* **2002**, *124*, 7642–7643. [CrossRef] [PubMed]

87. Fihri, A.; Bouhrara, M.; Nekoueishahraki, B.; Basset, J.-M.; Polshettiwar, V. Nanocatalysts for Suzuki cross-coupling reactions. *Chem. Soc. Rev.* **2011**, *40*, 5181–5203. [CrossRef] [PubMed]

88. Scheuermann, G.M.; Rumi, L.; Steurer, P.; Bannwarth, W.; Mulhaupt, R. Palladium nanoparticles on graphite oxide and its functionalized graphene derivatives as highly active catalysts for the Suzuki–Miyaura coupling reaction. *J. Am. Chem. Soc.* **2009**, *131*, 8262–8270. [CrossRef] [PubMed]

89. Kotadia, D.A.; Patel, U.H.; Gandhi, S.; Soni, S.S. Pd doped SiO$_2$ nanoparticles: An efficient recyclable catalyst for Suzuki, Heck and Sonogashira reactions. *RSC Adv.* **2014**, *4*, 32826–32833. [CrossRef]

90. Ciriminna, R.; Pandarus, V.; Gingras, G.; Beland, F.; Demma Cara, P.; Pagliaro, M. Heterogeneously catalyzed Suzuki–Miyaura conversion of broad scope. *RSC Adv.* **2012**, *2*, 10798–10804. [CrossRef]

91. Verho, O.; Nagendiran, A.; Johnston, E.V.; Tai, C.; Baeckvall, J.-E. Nanopalladium on amino-functionalized mesocellular foam. An efficient catalyst for Suzuki reactions and transfer hydrogenations. *ChemCatChem* **2013**, *5*, 612–618. [CrossRef]

92. Parvulescu, A.N.; Van der Eycken, E.; Jacobs, P.A.; De Vos, D.E. Microwave-promoted racemization and dynamic kinetic resolution of chiral amines over Pd on alkaline earth supports and lipases. *J. Catal.* **2008**, *255*, 206–212. [CrossRef]

93. Chang, W.; Chae, G.H.; Jang, S.R.; Shin, J.; Ahn, B.J. An efficient microwave-assisted Suzuki reaction using Pd/MCM-41 and Pd/SBA-15 as catalysts in solvent-free condition. *J. Ind. Eng. Chem.* **2012**, *18*, 581–585. [CrossRef]

94. Zheng, Z.; Li, H.; Liu, T.; Cao, R. Monodisperse noble metal nanoparticles stabilized in SBA-15: Synthesis, characterization and application in microwave-assisted Suzuki–Miyaura coupling reaction. *J. Catal.* **2010**, *270*, 268–274. [CrossRef]

95. Luo, M.; Dai, C.; Han, Q.; Fan, G.; Song, G. Preparation and characterization of palladium immobilized on silica-coated Fe$_3$O$_4$ and its catalytic performance for Suzuki reaction under microwave irradiation. *Surf. Interface Anal.* **2016**, *48*, 1066–1071. [CrossRef]

96. Nehlig, E.; Waggeh, B.; Millot, N.; Lalatonne, Y.; Motte, L.; Guenin, E. Immobilized Pd on magnetic nanoparticles bearing proline as a highly efficient and retrievable Suzuki–Miyaura catalyst in aqueous media. *Dalton Trans.* **2015**, *44*, 501–505. [CrossRef] [PubMed]

97. Cravotto, G.; Orio, L.; Calcio Gaudino, E.; Martina, K.; Tavor, D.; Wolfson, A. Efficient synthetic protocols in glycerol under heterogeneous catalysis. *ChemSusChem* **2011**, *4*, 1130–1134. [CrossRef]

98. Martina, K.; Leonhardt, S.E.; Ondruschka, B.; Curini, M.; Binello, A.; Cravotto, G. In situ cross-linked chitosan Cu(I) or Pd(II) complexes as a versatile, eco-friendly recyclable solid catalyst. *J. Mol. Catal. A* **2011**, *334*, 60–64. [CrossRef]

99. Leonhardt, S.E.S.; Stolle, A.; Ondruschka, B.; Cravotto, G.; De Leo, C.; Jandt, K.D.; Keller, T.F. Chitosan as a support for heterogeneous Pd catalysts in liquid phase catalysis. *Appl. Catal. A* **2010**, *379*, 30–37. [CrossRef]

100. Baran, T.; Sargin, I.; Kaya, M.; Mentes, A. Green heterogeneous Pd(II) catalyst produced from chitosan-cellulose micro beads for green synthesis of biaryls. *Carbohydr. Polym.* **2016**, *152*, 181–188. [CrossRef] [PubMed]

101. Cravotto, G.; Calcio Gaudino, E.; Tagliapietra, S.; Carnaroglio, D.; Procopio, A. A green approach to heterogeneous catalysis using ligand-free, metal-loaded cross-linked cyclodextrins. *Green Proc. Synth.* **2012**, *1*, 269–273. [CrossRef]

102. Martina, K.; Baricco, F.; Caporaso, M.; Berlier, G.; Cravotto, G. Cyclodextrin-grafted silica-supported Pd nanoparticles: An efficient and versatile catalyst for ligand-free C–C coupling and hydrogenation. *ChemCatChem* **2016**, *8*, 1176–1184. [CrossRef]

103. Isfahani, A.L.; Mohammadpoor-Baltork, I.; Mirkhani, V.; Khosropour, A.R.; Moghadam, M.; Tangestaninejad, S.; Kia, R. Palladium nanoparticles immobilized on nano-silica triazine dendritic polymer (Pdnp-nSTDP): An efficient and reusable catalyst for Suzuki–Miyaura cross-coupling and Heck reactions. *Adv. Synth. Catal.* **2013**, *355*, 957–972. [CrossRef]

104. Borkowski, T.; Zawartka, W.; Pospiech, P.; Mizerska, U.; Trzeciak, A.M.; Cypryk, M.; Tylus, W. Reusable functionalized polysiloxane-supported palladium catalyst for Suzuki–Miyaura cross-coupling. *J. Catal.* **2011**, *282*, 270–277. [CrossRef]

105. Massaro, M.; Schembri, V.; Campisciano, V.; Cavallaro, G.; Lazzara, G.; Milioto, S.; Noto, R.; Parisi, F.; Riela, S. Design of PNIPAAM covalently grafted on halloysite nanotubes as a support for metal-based catalysts. *RSC Adv.* **2016**, *6*, 55312–55318. [CrossRef]

106. Massaro, M.; Riela, S.; Cavallaro, G.; Colletti, C.G.; Milioto, S.; Noto, R.; Parisi, F.; Lazzara, G. Palladium supported on Halloysite-triazolium salts as catalyst for ligand free Suzuki cross-coupling in water under microwave irradiation. *J. Mol. Catal. A* **2015**, *408*, 12–19. [CrossRef]

107. Massaro, M.; Riela, S.; Lazzara, G.; Gruttadauria, M.; Milioto, S.; Noto, R. Green conditions for the Suzuki reaction using microwave irradiation and a new HNT-supported ionic liquid-like phase (HNT-SILLP) catalyst. *Appl. Organomet. Chem.* **2014**, *28*, 234–238. [CrossRef]

108. Baruah, D.; Das, R.N.; Hazarika, S.; Konwar, D. Biogenic synthesis of cellulose supported Pd(0) nanoparticles using hearth wood extract of Artocarpus lakoocha Roxb—A green, efficient and versatile catalyst for Suzuki and Heck coupling in water under microwave heating. *Catal. Commun.* **2015**, *72*, 73–80. [CrossRef]

109. Baran, T.; Sargin, I.; Kaya, M.; Mentes, A.; Ceter, T. Design and application of sporopollenin microcapsule supported palladium catalyst: Remarkably high turnover frequency and reusability in catalysis of biaryls. *J. Coll. Interface Sci.* **2017**, *486*, 194–203. [CrossRef] [PubMed]

110. Baran, T.; Sargin, I.; Mentes, A.; Kaya, M. Exceptionally high turnover frequencies recorded for a new chitosan-based palladium(II) catalyst. *Appl. Catal. A* **2016**, *523*, 12–20. [CrossRef]

111. Baran, T.; Sargin, I.; Kaya, M.; Mentes, A. An environmental catalyst derived from biological waste materials for green synthesis of biaryls via Suzuki coupling reactions. *J. Mol. Catal. A* **2016**, *420*, 216–221. [CrossRef]

112. Shah, D.; Kaur, H. Supported palladium nanoparticles: A general sustainable catalyst for microwave enhanced carbon-carbon coupling reactions. *J. Mol. Catal. A* **2016**, *424*, 171–180. [CrossRef]

113. Kaur, H.; Shah, D.; Pal, U. Resin encapsulated palladium nanoparticles: An efficient and robust catalyst for microwave enhanced Suzuki–Miyaura coupling. *Catal. Commun.* **2011**, *12*, 1384–1388. [CrossRef]

114. Djakovitch, L.; Dufaud, V.; Zaidi, R. Heterogeneous palladium catalysts applied to the synthesis of 2- and 2,3-functionalised indoles. *Adv. Synth. Catal.* **2006**, *348*, 715–724. [CrossRef]

115. Djakovitch, L.; Koehler, K. Heck reaction catalyzed by Pd-modified zeolites. *J. Am. Chem. Soc.* **2001**, *123*, 5990–5999. [CrossRef] [PubMed]

116. Giacalone, F.; Campisciano, V.; Calabrese, C.; La Parola, V.; Liotta, L.F.; Aprile, C.; Gruttadauria, M. Supported C_{60}-IL-PdNPs as extremely active nanocatalysts for C–C cross-coupling reactions. *J. Mat. Chem. A* **2016**, *4*, 17193–17206. [CrossRef]

117. Guerra, R.R.G.; Martins, F.C.P.; Lima, C.G.S.; Gonçalves, R.H.; Leite, E.R.; Pereira-Filho, E.R.; Schwab, R.S. Factorial design evaluation of the Suzuki cross-coupling reaction using a magnetically recoverable palladium catalyst. *Tetrahedron Lett.* **2017**, *58*, 903–908. [CrossRef]

118. Moussa, S.; Siamaki, A.R.; Gupton, B.F.; El-Shall, M.S. Pd-partially reduced graphene oxide catalysts (Pd/PRGO): Laser synthesis of Pd nanoparticles supported on PRGO nanosheets for carbon-carbon cross coupling reactions. *ACS Catal.* **2012**, *2*, 145–154. [CrossRef]

119. Tsukahara, Y.; Higashi, A.; Yamauchi, T.; Nakamura, T.; Yasuda, M.; Baba, A.; Wada, Y. In situ observation of nonequilibrium local heating as an origin of special effect of microwave on chemistry. *J. Phys. Chem. C* **2010**, *114*, 8965–8970. [CrossRef]

120. Horikoshi, S.; Osawa, A.; Abe, M.; Serpone, N. On the generation of hot-spots by microwave electric and magnetic fields and their impact on a microwave-assisted reaction in the presence of metallic Pd nanoparticles on an activated carbon support. *J. Phys. Chem. C* **2011**, *115*, 23030–23035. [CrossRef]

121. Gomez-Martinez, M.; Buxaderas, E.; Pastor, I.M.; Alonso, D.A. Palladium nanoparticles supported on graphene and reduced graphene oxide as efficient recyclable catalyst for the Suzuki–Miyaura reaction of potassium aryltrifluoroborates. *J. Mol. Catal. A* **2015**. [CrossRef]

122. Garcia-Suarez, E.J.; Lara, P.; Garcia, A.B.; Ojeda, M.; Luque, R.; Philippot, K. Efficient and recyclable carbon-supported Pd nanocatalysts for the Suzuki–Miyaura reaction in aqueous-based media: Microwave vs. conventional heating. *Appl. Catal. A* **2013**, *468*, 59–67. [CrossRef]

123. Sanhes, D.; Raluy, E.; Rétory, S.; Saffon, N.; Teuma, E.; Gomez, M. Unexpected activation of carbon–bromide bond promoted by palladium nanoparticles in Suzuki C–C couplings. *Dalton Trans.* **2010**, *39*, 9719–9726. [CrossRef] [PubMed]

124. Schmidt, B.; Riemer, M.; Karras, M. 2,2'-biphenols via protecting group-free thermal or microwave-accelerated Suzuki–Miyaura coupling in water. *J. Org. Chem.* **2013**, *78*, 8680–8688. [CrossRef] [PubMed]

125. Al-Amin, M.; Akimoto, M.; Tameno, T.; Ohki, Y.; Takahashi, N.; Hoshiya, N.; Shuto, S.; Arisawa, M. Suzuki–Miyaura cross-coupling reactions using a low-leaching and highly recyclable gold-supported palladium material and two types of microwave equipments. *Green Chem.* **2013**, *15*, 1142–1145. [CrossRef]

126. Heinrich, F.; Kessler, M.T.; Dohmen, S.; Singh, M.; Prechtl, M.H.G.; Mathur, S. Molecular palladium precursors for Pd^0 nanoparticle preparation by microwave irradiation: Synthesis, structural characterization and catalytic activity. *Eur. J. Inorg. Chem.* **2012**, *2012*, 6027–6033. [CrossRef]

127. Siamaki, A.R.; Abd El Rahman, S.K.; Abdelsayed, V.; El-Shall, M.S.; Gupton, B.F. Microwave-assisted synthesis of palladium nanoparticles supported on graphene: A highly active and recyclable catalyst for carbon-carbon cross-coupling reactions. *J. Catal.* **2011**. [CrossRef]

128. Elazab, H.A.; Siamaki, A.R.; Moussa, S.; Gupton, B.F.; El-Shall, M.S. Highly efficient and magnetically recyclable graphene-supported Pd/Fe_3O_4 nanoparticle catalysts for Suzuki and Heck cross-coupling reactions. *Appl. Catal. A* **2015**, *491*, 58–69. [CrossRef]

129. Ni, Z.; Masel, R.I. Rapid production of metal-organic frameworks via microwave-assisted solvothermal synthesis. *J. Am. Chem. Soc.* **2006**, *128*, 12394–12395. [CrossRef]

130. Li, Y.; Liu, Y.; Gao, W.; Zhang, L.; Liu, W.; Lu, J.; Wang, Z.; Deng, Y.J. Microwave-assisted synthesis of UIO-66 and its adsorption performance towards dyes. *Cryst. Eng. Commun.* **2014**, *16*, 7037–7042. [CrossRef]

131. Dong, W.; Feng, C.; Zhang, L.; Shang, N.; Gao, S.; Wang, C.; Wang, Z. Pd@UiO-66: An efficient catalyst for Suzuki–Miyaura coupling reaction at mild condition. *Catal. Lett.* **2016**, *146*, 117–125. [CrossRef]

132. Huang, J.; Wang, W.; Li, H. Water-medium organic reactions catalyzed by active and reusable Pd/Y heterobimetal-organic framework. *ACS Catal.* **2013**, *3*, 1526–1536. [CrossRef]

133. Prasad, K.S.; Noh, H.-B.; Reddy, S.S.; Reddy, A.E.; Shim, Y.-B. Catalytic properties of Au and Pd nanoparticles decorated on Cu_2O microcubes for aerobic benzyl alcohol oxidation and Suzuki–Miyaura coupling reactions in water. *Appl. Catal. A* **2014**, *476*, 72–77. [CrossRef]

134. Veerakumar, P.; Madhu, R.; Chen, S.-M.; Veeramani, V.; Hung, C.-T.; Tang, P.-H.; Wang, C.-B.; Liu, S.-B. Highly stable and active palladium nanoparticles supported on porous carbon for practical catalytic applications. *J. Mater. Chem. A* **2014**, *2*, 16015–16022. [CrossRef]

135. Misch, L.M.; Birkel, A.; Figg, C.A.; Fors, B.P.; Hawker, C.J.; Stucky, G.D.; Seshadri, R. Rapid microwave-assisted sol-gel preparation of Pd-substituted $LnFeO_3$ (Ln = Y, La): Phase formation and catalytic activity. *Dalton Trans.* **2014**, *43*, 2079–2087. [CrossRef] [PubMed]

136. Lennox, A.J.J.; Lloyd-Jones, G.C. Organotrifluoroborate hydrolysis: Boronic acid release mechanism and an acid–base paradox in cross-coupling. *J. Am. Chem. Soc.* **2012**, *134*, 7431–7441. [CrossRef] [PubMed]

137. Shil, A.K.; Guha, N.R.; Sharma, D.; Das, P. A solid supported palladium(0) nano/microparticle catalyzed ultrasound induced continuous flow technique for large scale Suzuki reactions. *RSC Adv.* **2013**, *3*, 13671–13676. [CrossRef]

138. Das, P.; Sharma, D.; Shil, A.K.; Kumari, A. Solid-supported Pd nano and microparticles: An efficient heterogeneous catalyst for ligand-free Suzuki–Miyaura cross coupling reaction. *Tetrahedron Lett.* **2011**, *52*, 1176–1178. [CrossRef]

139. Azua, A.; Mata, J.A.; Heymes, P.; Peris, E.; Lamaty, F.; Martinez, J.; Colacino, E. Palladium *N*-heterocyclic carbene catalysts for the ultrasound-promoted Suzuki–Miyaura reaction in glycerol. *Adv. Synth. Catal.* **2013**, *355*, 1107–1116. [CrossRef]

140. Chaudhary, A.R.; Bedekar, A.V. 1-(α-Aminobenzyl)-2-naphthol as phosphinefree ligand for Pd-catalyzed Suzuki and one-pot Wittig–Suzuki reaction. *Appl. Organomet. Chem.* **2012**, *26*, 430–437. [CrossRef]

141. Chaudhary, A.R.; Bedekar, A.V. Application of 1-(α-aminobenzyl)-2-naphthols as air-stable ligands for Pd-catalyzed Mizoroki–Heck coupling reaction. *Synth. Commun.* **2012**, *42*, 1778–1785.

142. Ghotbinejad, M.; Khosropour, A.R.; Mohammadpoor-Baltork, I.; Moghadam, M.; Tangestaninejad, S.; Mirkhani, V. Ultrasound-assisted C–C coupling reactions catalyzed by unique SPION-A-Pd(EDTA) as a robust nanocatalyst. *RSC Adv.* **2014**, *4*, 8590–8596. [CrossRef]

143. Wang, Z.; Chen, W.; Han, Z.; Zhu, J.; Lu, N.; Yang, Y.; Ma, D.; Chen, Y.; Huang, S. Pd embedded in porous carbon (Pd@CMK-3) as an active catalyst for Suzuki reactions: Accelerating mass transfer to enhance the reaction rate. *Nano Res.* **2014**, *7*, 1254–1262. [CrossRef]

144. Suresh, N.; Prasanna, G.L.; Rao, M.V.B.; Pal, M. Ultrasound assisted synthesis of 6-flavonyl substituted 1,4-dihydro-benzo[*d*][1,3]oxazin-2-ones via Suzuki–Miyaura coupling under Pd/C catalysis. *Arab. J. Chem.* **2016**. [CrossRef]

145. Kulkarni, K.; Friend, J.; Yeo, L.; Perlmutter, P. An emerging reactor technology for chemical synthesis: Surface acoustic wave-assisted closed-vessel Suzuki coupling reactions. *Ultrason. Sonochem.* **2014**, *21*, 1305–1309. [CrossRef] [PubMed]

146. Senapati, K.K.; Roy, S.; Borgohain, C.; Phukan, P. Palladium nanoparticle supported on cobalt ferrite: An efficient magnetically separable catalyst for ligand free Suzuki coupling. *J. Mol. Catal. A* **2012**, *352*, 128–134. [CrossRef]

147. Singh, A.S.; Patil, U.B.; Nagarkar, J.M. Palladium supported on zinc ferrite: A highly active, magnetically separable catalyst for ligand free Suzuki and Heck coupling. *Catal. Commun.* **2013**, *35*, 11–16. [CrossRef]

148. Su, X.; Vinu, A.; Aldeyab, S.S.; Zhong, L. Highly uniform Pd nanoparticles supported on g-C_3N_4 for efficiently catalytic Suzuki–Miyaura reactions. *Catal. Lett.* **2015**, *145*, 1388–1395. [CrossRef]

149. Li, J.; Bai, X. Ultrasonic synthesis of supported palladium nanoparticles for room-temperature Suzuki–Miyaura coupling. *J. Mater. Sci.* **2016**, *51*, 9108–9122. [CrossRef]

150. Berry, D.E.; Carrie, P.; Fawkes, K.L.; Rebner, B.; Xing, Y. The mechanochemical reaction of palladium(II) chloride with a bidentate phosphine. *J. Chem. Educ.* **2010**, *87*, 533–534. [CrossRef]

151. Ojeda, M.; Balu, A.M.; Barron, V.; Pineda, A.; Coleto, A.G.; Romero, A.A.; Luque, R. Solventless mechanochemical synthesis of magnetic functionalized catalytically active mesoporous SBA-15 nanocomposites. *J. Mater. Chem. A* **2014**, *2*, 387–393. [CrossRef]

152. Siamaki, A.R.; Lin, Y.; Woodberry, K.; Connell, J.W.; Gupton, B.F. Palladium nanoparticles supported on carbon nanotubes from solventless preparations: Versatile catalysts for ligand-free suzuki cross coupling reactions. *J. Mater. Chem. A* **2013**, *1*, 12909–12918. [CrossRef]

153. Zhang, A.; Liu, M.; Liu, M.; Xiao, Y.; Li, Z.; Chen, J.; Sun, Y.; Zhao, J.; Fang, S.; Jia, D.; et al. Homogeneous Pd nanoparticles produced in direct reactions: Green synthesis, formation mechanism and catalysis properties. *J. Mater. Chem. A* **2014**, *2*, 1369–1374. [CrossRef]

154. Schneider, F.; Ondruschka, B. Mechanochemical solid-state Suzuki reactions using an in situ generated base. *ChemSusChem* **2008**, *1*, 622–625. [CrossRef] [PubMed]

155. Schneider, F.; Stolle, A.; Ondruschka, B.; Hopf, H. The Suzuki–Miyaura reaction under mechanochemical conditions. *Org. Process Res. Dev.* **2009**, *13*, 44–48. [CrossRef]

156. Braga, D.; D'Addario, D.; Polito, M.; Grepioni, F. Mechanically induced expeditious and selective preparation of disubstituted pyridine/pyrimidine ferrocenyl complexes. *Organometallics* **2004**, *23*, 2810–2812. [CrossRef]

157. Saha, P.; Naskar, S.; Paira, P.; Hazra, A.; Sahu, K.B.; Paira, R.; Banerjee, S.; Mondal, N.B. Basic alumina-supported highly effective Suzuki–Miyaura cross-coupling reaction under microwave irradiation: Application to fused tricyclic oxa-aza-quinolones. *Green Chem.* **2009**, *11*, 931–934. [CrossRef]

158. Bernhardt, F.; Trotzki, R.; Szuppa, T.; Stolle, A.; Ondruschka, B. Solvent-free and time-efficient Suzuki–Miyaura reaction in a ball mill: The solid reagent system KF-Al$_2$O$_3$ under inspection. *Beilstein J. Org. Chem.* **2010**. [CrossRef]

159. Jiang, Z.-J.; Li, Z.-H.; Yu, J.-B.; Su, W.-K. Liquid-assisted grinding accelerating: Suzuki–Miyaura reaction of aryl chlorides under high-speed ball-milling conditions. *J. Org. Chem.* **2016**, *81*, 10049–10055. [CrossRef] [PubMed]

160. Linic, S.; Christopher, P.; Ingram, D.B. Plasmonic-metal nanostructures for efficient conversion of solar to chemical energy. *Nat. Mater.* **2011**, *10*, 911–921. [CrossRef] [PubMed]

161. Christopher, P.; Xin, H.; Marimuthu, A.; Linic, S. Singular characteristics and unique chemical bond activation mechanisms of photocatalytic reactions on plasmonic nanostructures. *Nat. Mater.* **2012**, *11*, 1044–1050. [CrossRef] [PubMed]

162. Sarina, S.; Bai, S.; Huang, Y.; Chen, C.; Jia, J.; Jaatinen, E.; Ayoko, G.A.; Bao, Z.; Zhu, H. Visible light enhanced oxidant free dehydrogenation of aromatic alcohols using Au–Pd alloy nanoparticle catalysts. *Green Chem.* **2014**, *16*, 331–341. [CrossRef]

163. Li, G.; Wang, Y.; Mao, L. Recent progress in highly efficient Ag-based visible-light photocatalysts. *RSC Adv.* **2014**, *4*, 53649–53661. [CrossRef]

164. Verma, P.; Kuwahara, Y.; Moriab, K.; Yamashita, H. Synthesis and characterization of a Pd/Ag bimetallic nanocatalyst on SBA-15 mesoporous silica as a plasmonic catalyst. *J. Mater. Chem. A* **2015**, *3*, 18889–18897. [CrossRef]

165. Tanaka, A.; Fuku, K.; Nishi, T.; Hashimoto, K.; Kominami, H. Functionalization of Au/TiO$_2$ plasmonic photocatalysts with Pd by formation of a core–shell structure for effective dechlorination of chlorobenzene under irradiation of visible light. *J. Phys. Chem. C* **2013**, *117*, 16983–16989. [CrossRef]

166. Tanaka, A.; Nakanishi, K.; Hamada, R.; Hashimoto, K.; Kominami, H. Simultaneous and stoichiometric water oxidation and Cr(VI) reduction in aqueous suspensions of functionalized plasmonic photocatalyst Au/TiO$_2$–Pt under irradiation of green light. *ACS Catal.* **2013**, *3*, 1886–1891. [CrossRef]

167. Verma, P.; Kuwahara, Y.; Moriab, K.; Yamashita, H. Pd/Ag and Pd/Au bimetallic nanocatalysts on mesoporous silica for plasmon-mediated enhanced catalytic activity under visible light irradiation. *J. Mater. Chem. A* **2016**, *4*, 10142–10150. [CrossRef]

168. Smith, J.G.; Faucheaux, J.A.; Jain, P.K. Plasmon resonances for solar energy harvesting: A mechanistic outlook. *Nano Today* **2015**, *10*, 67–80. [CrossRef]

169. Del Fatti, N.; Voisin, C.; Achermann, M.; Tzortzakis, S.; Christofilos, D.; Vallée, F. Nonequilibrium electron dynamics in noble metals. *Phys. Rev. B* **2000**, *61*, 16956–16966. [CrossRef]

170. Link, S.; Link, S.; El-Sayed, M.A.; El-Sayed, M. Spectral properties and relaxation dynamics of surface plasmon electronic oscillations in gold and silver nanodots and nanorods. *J. Phys. Chem. B* **1999**, *103*, 8410–8426. [CrossRef]

171. Burda, C.; Chen, X.; Narayanan, R.; El-Sayed, M.A. Chemistry and properties of nanocrystals of different shapes. *Chem. Rev.* **2005**, *105*, 1025–1102. [CrossRef] [PubMed]

172. Evanoff, D.D.; Chumanov, G. Synthesis and optical properties of silver nanoparticles and arrays. *ChemPhysChem* **2005**, *6*, 1221–1231. [CrossRef] [PubMed]

173. Tanaka, A.; Fuku, K.; Nishi, T.; Hashimoto, K.; Kominami, H. Gold–titanium(IV) oxide plasmonic photocatalysts prepared by a colloid-photodeposition method: Correlation between physical properties and photocatalytic activities. *J. Phys. Chem. C* **2013**, *117*, 16983–16989. [CrossRef]

174. Sarina, S.; Zhu, H.; Jaatinen, E.; Xiao, Q.; Liu, H.; Jia, J.; Chen, C.; Zhao, J. Enhancing catalytic performance of palladium in gold and palladium alloy nanoparticles for organic synthesis reactions through visible light irradiation at ambient temperatures. *J. Am. Chem. Soc.* **2013**, *135*, 5793–5801. [CrossRef] [PubMed]

175. Xiao, Q.; Sarina, S.; Jaatinen, E.; Jia, J.; Arnold, D.P.; Liu, H.; Zhu, H. Efficient photocatalytic Suzuki cross-coupling reactions on Au–Pd alloy nanoparticles under visible light irradiation. *Green Chem.* **2014**, *16*, 4272–4285. [CrossRef]

176. Wang, F.; Li, C.; Chen, H.; Jiang, R.; Sun, L.-D.; Li, Q.; Wang, J.; Yu, J.C.; Yan, C.-H. Plasmonic harvesting of light energy for Suzuki coupling reactions. *J. Am. Chem. Soc.* **2013**, *135*, 5588–5601. [CrossRef]

177. Wang, F.; Sun, L.-D.; Feng, W.; Chen, H.J.; Yeung, M.H.; Wang, J.F.; Yan, C.-H. Heteroepitaxial growth of core–shell and core–multishell nanocrystals composed of palladium and gold. *Small* **2010**, *6*, 2566–2575. [CrossRef]

178. Wang, F.; Li, C.H.; Sun, L.-D.; Wu, H.S.; Ming, T.; Wang, J.F.; Yu, J.C.; Yan, C.-H. Heteroepitaxial growth of high-index-faceted palladium nanoshells and their catalytic performance. *J. Am. Chem. Soc.* **2011**, *133*, 1106–1111. [CrossRef]

179. Hu, J.-W.; Li, J.-F.; Ren, B.; Wu, D.-Y.; Sun, S.-G.; Tian, Z.-Q. Palladium-coated gold nanoparticles with a controlled shell thickness used as surface-enhanced Raman scattering substrate. *J. Phys. Chem. C* **2007**, *111*, 1105–1112. [CrossRef]

180. Trinh, T.T.; Sato, R.; Sakamoto, M.; Fujiyoshi, Y.; Haruta, M.; Kurata, H.; Teranishi, T. Visible to near-infrared plasmon-enhanced catalytic activity of Pd hexagonal nanoplates for the Suzuki coupling reaction. *Nanoscale* **2015**, *7*, 12435–12444. [CrossRef] [PubMed]

181. Li, X.-H.; Baar, M.; Blechert, S.; Antonietti, M. Facilitating room-temperature Suzuki coupling reaction with light: Mott-Schottky photocatalyst for C–C-coupling. *Sci. Rep.* **2013**, *3*, 1743. [CrossRef]

182. Möhlmann, L.; Baar, M.; Rieß, J.; Antonietti, M.; Wang, W.; Blecher, S. Carbon nitride-catalyzed photoredox C–C bond formation with *N*-aryltetrahydroisoquinolines. *Adv. Synth. Catal.* **2012**, *354*, 1909–1913. [CrossRef]

183. Zhang, S.; Chang, C.; Huang, Z.; Ma, Y.; Gao, W.; Li, J.; Qu, Y. Visible-light-activated Suzuki–Miyaura coupling reactions of aryl chlorides over the multifunctional Pd/Au/porous nanorods of CeO$_2$. *ACS Catal.* **2015**, *5*, 6481–6488. [CrossRef]

184. Zhang, S.; Li, J.; Gao, W.; Qu, Y. Insights into the effects of surface properties of oxides on the catalytic activity of Pd for C–C coupling reactions. *Nanoscale* **2015**, *7*, 3016–3021 [CrossRef] [PubMed]

185. Cui, J.; Li, Y.; Liu, L.; Chen, L.; Xu, J.; Ma, J.; Fang, G.; Zhu, E.; Wu, H.; Zhao, L.; Wang, L.; Huang, Y. Near-infrared plasmonic-enhanced solar energy harvest for highly efficient photocatalytic reactions. *Nano Lett.* **2015**, *15*, 6295–6301. [CrossRef] [PubMed]

186. Miyaura, N.; Suzuki, A. Palladium-catalyzed cross-coupling reactions of organoboron compounds. *Chem. Rev.* **1995**, *95*, 2457–2483. [CrossRef]

Review

The Suzuki–Miyaura Cross-Coupling as a Versatile Tool for Peptide Diversification and Cyclization

Tom Willemse [1,2], Wim Schepens [3], Herman W. T. van Vlijmen [3], Bert U. W. Maes [2] and Steven Ballet [1,*]

[1] Research Group of Organic Chemistry, Departments of Bioengineering Sciences and Chemistry, Vrije Universiteit Brussel, 1050 Brussels, Belgium; tomwille@vub.ac.be
[2] Organic Synthesis Division, University of Antwerp, 2020 Antwerp, Belgium; bert.maes@uantwerpen.be
[3] Discovery Sciences, Janssen Research and Development, 2340 Beerse, Belgium; wschepen@its.jnj.com (W.S.); hvvlijme@its.jnj.com (H.W.T.V.V.)
* Correspondence: sballet@vub.ac.be; Tel.: +32-629-3292

Academic Editor: Ioannis D. Kostas
Received: 25 January 2017; Accepted: 21 February 2017; Published: 25 February 2017

Abstract: The (site-selective) derivatization of amino acids and peptides represents an attractive field with potential applications in the establishment of structure–activity relationships and labeling of bioactive compounds. In this respect, bioorthogonal cross-coupling reactions provide valuable means for ready access to peptide analogues with diversified structure and function. Due to the complex and chiral nature of peptides, mild reaction conditions are preferred; hence, a suitable cross-coupling reaction is required for the chemical modification of these challenging substrates. The Suzuki reaction, involving organoboron species, is appropriate given the stability and environmentally benign nature of these reactants and their amenability to be applied in (partial) aqueous reaction conditions, an expected requirement upon the derivatization of peptides. Concerning the halogenated reaction partner, residues bearing halogen moieties can either be introduced directly as halogenated amino acids during solid-phase peptide synthesis (SPPS) or genetically encoded into larger proteins. A reversed approach building in boron in the peptidic backbone is also possible. Furthermore, based on this complementarity, cyclic peptides can be prepared by halogenation, and borylation of two amino acid side chains present within the same peptidic substrate. Here, the Suzuki–Miyaura reaction is a tool to induce the desired cyclization. In this review, we discuss diverse amino acid and peptide-based applications explored by means of this extremely versatile cross-coupling reaction. With the advent of peptide-based drugs, versatile bioorthogonal conversions on these substrates have become highly valuable.

Keywords: Suzuki–Miyaura reaction; peptide diversification; peptide cyclization

---

1. Introduction

The Pd-catalyzed Suzuki–Miyaura reaction has found widespread use in the synthesis of carbon–carbon bonds [1,2]. The success of this cross-coupling reaction can be attributed to the generally mild reaction conditions required and the versatility of the boron species as organometallic partners [3–9]. Compared to related C–C bond formations, the environmentally benign nature, high functional group tolerance, wide commercial availability of the reactants, and compatibility with aqueous conditions are advantageous [10–12]. As a result, various research groups have utilized and reviewed this reaction for the derivatization of complex molecules such as natural products [13–16] and nucleosides [17–23]. The mechanism of this reaction originally proposed by Miyaura and Suzuki has been revisited recently, which provides insights useful for rational reaction optimization [24–26]. Besides classical organoboronic acids, new boron-derived reagents have been developed with improved shelf life and controlled release of reactant and stability versus deborylation during cross-coupling [4–9,27,28]. In this review, we focus on the derivatization of amino acids up to protein substrates. In general, through bioorthogonal reactions, the site-selective derivatization of peptides and proteins is possible after the incorporation of non-proteinogenic amino acids and subsequent labeling of the (bioactive) substrate [29–33]. Additionally, this cross-coupling possesses a huge potential with regards to the improvement of peptide-based therapeutics and pharmacological probes. To achieve the envisaged site-selective modification, a single residue must undergo transformation in the presence of a significant amount of other side chains or functionalities In contrast, when multiple reactive sites are present, non-selective derivatization methods often yield less active conjugates that are difficult to separate [34]. To access the desired, pin-pointed modification, a wide array of halogenated and borylated residues (selected examples are shown in Figure 1) can for instance be incorporated in peptidic substrates, all of which allow diverse substitutions or cyclizations.

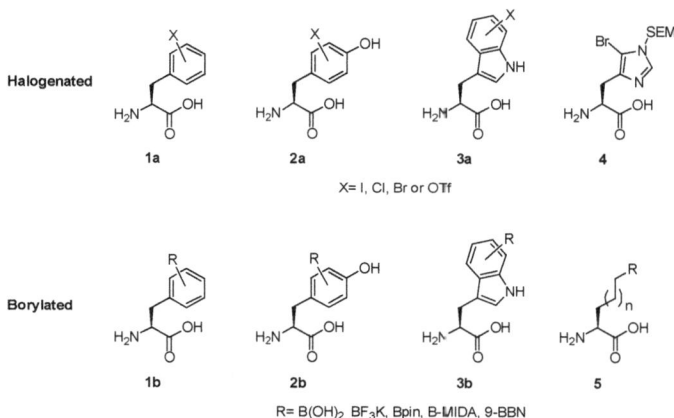

Figure 1. Representative examples of building blocks used in Suzuki–Miyaura-based derivatizations.

Halophenylalanines are generally commercially available, while (bio)synthetic methods have been developed for the preparation of halogenated tryptophans [35–38]. Boron groups are readily introduced in aromatic and aliphatic amino acids via borylation or hydroboration [39,40]. Insertion of these residues can be realized via solid-phase peptide synthesis (SPPS) methods utilizing the halogenated or borylated amino acid analogue. Next, peptide derivatization can be performed in solution-phase, after cleavage from the solid support. Alternatively, the direct Pd-catalyzed solid-phase derivatization has also been widely reported. The latter strategy offers the added value of a rapid purification by filtration, and hence removal of excess reactants, reagents, by- or side-products.

Due to the significant role that natural aromatic amino acids play in peptide-protein interactions [41], it can easily be perceived that additional interactions with the biological targets are within reach through usage of derivatized amino acids, prepared by the Suzuki–Miyaura reaction. The benefit of using non-natural amino acids has been well established [42–44], as the biological activity profile of the targeted peptide and its overall stability can be greatly enhanced through such modifications [45]. Ready access to a broad variety of unusual amino acids is therefore highly desirable for medicinal chemists, since they allow to perform structure–activity relationship studies. Hence, the Suzuki–Miyaura reaction serves as a powerful tool for peptide/medicinal chemists and it has certainly secured a place in their "toolbox" to improve peptide lead agents [45–47].

Furthermore, since both the halogen substituent and the boronate group are readily introduced, cyclization of the peptide can be achieved through Suzuki–Miyaura cross-coupling. Traditionally, cyclic peptides are often used as analogues of their linear counterparts to protect the peptide from a premature enzymatic degradation, to stabilize secondary motifs and lock the active conformation for an improved pharmacological profile [48]. Peptide macrocyclization has thus been developed to overcome the intrinsic drawbacks of peptide therapeutics [49–52]. In this regard, various strategies have emerged for the synthesis of constrained cyclic peptides with different structural motifs emerging at the site of macrocyclization. Commonly applied strategies include disulfide or lactam bridge formation [53], ring-closing metathesis (RCM) [54], azido-alkyne cycloaddition [55], and palladium-catalyzed reactions [15,56,57]. Of specific importance to the current review are the biaryl-bridges, which gained interest for the stabilization of α-helices and β-sheets [58,59]. Interestingly, biaryl-linkages and moieties are also found in a variety of naturally occurring peptides and even synthetic drugs (e.g., Figure 2) [60,61].

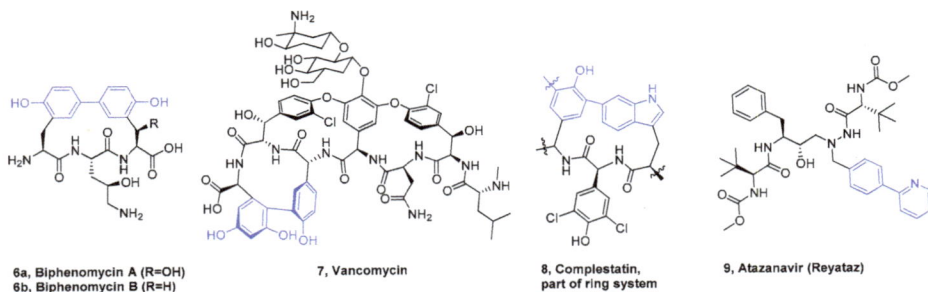

6a, Biphenomycin A (R=OH)
6b, Biphenomycin B (R=H)

7, Vancomycin

8, Complestatin, part of ring system

9, Atazanavir (Reyataz)

Figure 2. Naturally occurring peptides **6–8** and anti-retroviral drug Atazanavir **9** containing a biaryl moiety.

In light of the natural environment of peptide substrates, the importance of water as a (co-)solvent has been evaluated during Suzuki–Miyaura reactions applied on peptide substrates. The need for aqueous or mixed aqueous conditions is important for enabling the derivatization of unprotected amino acids or proteins, since such solvent systems ensure solubility of these complex products. The development of water soluble ligands and further improvement of catalysts for classical organic reactions [62] has encouraged peptide chemists to also explore mild conditions for amino acid/peptide derivatizations [63,64]. Such reaction conditions are preferred for amino acid(-based) substrates in order to avoid epimerization or other unwanted side reactions, while mild conditions are simply mandatory for protein substrates. Physiological conditions (37 °C, buffered aqueous solutions) are often considered to maintain the natural protein folding [63,65].

In this review, the aim is to present, in a comprehensive manner, the major advances of the Suzuki–Miyaura cross-coupling in the field of peptide chemistry. Ever since the first reports in 1992, replacement of the typical organic reaction conditions, along with developments in chemical biology, has opened a plethora of potential applications. The usefulness of this cross-coupling has been shown on simple amino acids, synthetically prepared peptides, and large protein complexes. Herein, we focus on methodology development for both solution- and solid-phase derivatizations of increasingly complex substrates. Efforts towards peptidic natural product synthesis and peptide cyclization are discussed. In the final part, several biological applications in different therapeutic areas are briefly highlighted in which the Suzuki–Miyaura reaction represented a key reaction.

2. Derivatization of Amino Acids and Peptides in Solution

2.1. Access to and Derivatization of (Pseudo)halogenated Aromatic Amino acid Substrates

Halogenated residues can be introduced in peptide sequences via replacement of aromatic amino acids by their halogenated (Cl, Br, I) or pseudohalogenated (OTf, OMs) analogues. In contrast to these aromatic amino acids bearing (pseudo)halogenated moieties, amino acid-based alkyl halides are not used directly as substrates in the Suzuki–Miyaura reaction. To date, their use is limited to the Negishi reaction of β-iodoalanine, β-iodohomoalanine, and iodobishomoalanine with various aryl iodides, to gain access to non-proteinogenic phenylalanine analogues [66–69]. For peptides that are prepared by chemical methods, SPPS easily enables the replacement of an aromatic amino acid with a non-natural (pseudo)halogenated analogue. Most of the (pseudo)halogenated phenylalanines are available from commercial sources. In contrast, the use of halotryptophans as substrates in Suzuki–Miyaura couplings was initially hampered by their limited commercial availability (NB: pseudohalogenated tryptophans and derivatization thereof have not been reported to date). Chemical synthesis of these building blocks often lack generality and require specific procedures for each regioisomer [70–73]. In view of this issue, a straightforward procedure for the enzymatic synthesis of halogenated tryptophans was described via use of tryptophan synthase on the corresponding readily available indole analogues [37,74]. This methodology affords Gram-scale quantities of enantiomerically pure chlorotryptophans, bromotryptophans, and more recently 7-iodotryptophan [75]. Site-specific halogenases of tryptophan have also been developed for the preparation of halotryptophans. [35–38,76,77] Complex protein substrates that contain halogenated residues and which are not accessible via chemical synthesis could also be specifically obtained by means of auxotrophic strains or reassignment of stop codons [78,79] Pioneering work by the group of P. G. Schultz expanded the genetic code via a unique tRNA/aminoacyl-tRNA synthetase pair, resulting in a solid protocol to insert 4-iodophenylalanine [80] and 4-boronophenylalanine [81]. The introduction of 4-iodophenylalanine, for example, opened the gateway for protein modification via Pd-catalyzed reactions [65,82,83].

While early cross-couplings on amino acid substrates include Sonogashira, Mizoroki–Heck, and Stille cross-couplings, the Suzuki–Miyaura reaction is most prominent in literature for the bioorthogonal derivatization of (pseudo)halogenated aromatic amino acid-containing substrates [21,84,85]. Initial work reported by Shieh et al. described the derivatization of protected L-tyrosine triflate **10** with different aryl boronic acids [86]. By combination of Pd(PPh₃)₄ and anhydrous K₂CO₃ in DMF (Figure 3A), racemization was observed (an ee as low as 66% was observed). However, by changing base to Et₃N, racemization could be avoided, but the reaction time was significantly increased to two days. Improvement of the reaction conditions was realized by use of toluene and anhydrous K₂CO₃, which combined shorter reaction times with excellent yields and high optical purity. Derivatization of halogenated substrates **12** (Figure 3B) was first described by Burk and coworkers, and their pioneering work involved biphasic conditions (DME/aqueous base), and di- and tripeptide substrates [87]. During their optimalization, it was found that using Pd(OAc)₂ in combination with P(o-tol)₃ yielded high conversion and no observable amounts of racemization. These conditions could be applied on protected 2-,3- and 4-bromophenylalanine **12**, whereas the successful derivatization of several dipeptides and a tripeptide (containing methionine) demonstrated the versatility of this reaction. Kotha et al. further explored the derivatization on Boc-Phe(4-I)-OMe and small peptides (up to five residues: Boc-Ala-Leu-Phe(4-I)-D-Val-Val-OMe, not shown) [88]. In their work, Pd(PPh₃)₄ in a mixed THF/toluene solvent system, combined with an aqueous solution of inorganic base (Na₂CO₃), afforded cross-coupled products starting from protected 4-iodophenylalanine with excellent yields and no reported racemization.

In following years, the substrate scope was further extended. Wang et al. showed that an asymmetric hydrogenation of **14**, (with (S/S)/(R/R)[Et-DuPHOS-Rh]OTf), followed by Suzuki–Miyaura cross-couplings on protected 5-bromotrypthophan **15**, afforded ready access to a series of novel tryptophan analogues **16** (Figure 4A) [89]. Intermediate **14** was prepared via Horner-Wadsworth-Emmons olefination of 5-bromo-3-formylindole and phosphonate (MeO)₂P(O)CH(NHCbz)-COOMe. The protected 5-bromotryptophan **15** was then submitted to similar conditions, as described by Burk [87], yielding bulky and hydrophobic 5-aryltryptophans (Figure 4A). In addition, Espuña published the first derivatization of an iodinated aspartame **17**, bearing no protecting group at the N-terminus and a side chain carboxylic acid (Figure 4B) [90].

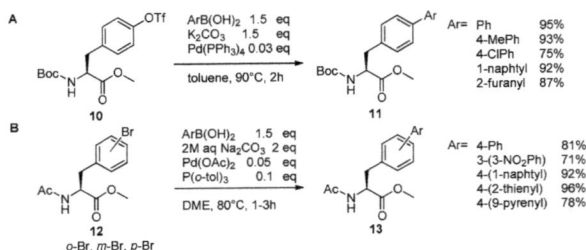

Figure 3. (**A**) First report of Suzuki–Miyaura reaction on 4-triflyloxyphenylalanine [86]. (**B**) Protected bromophenylalanines as substrates for cross-coupling.

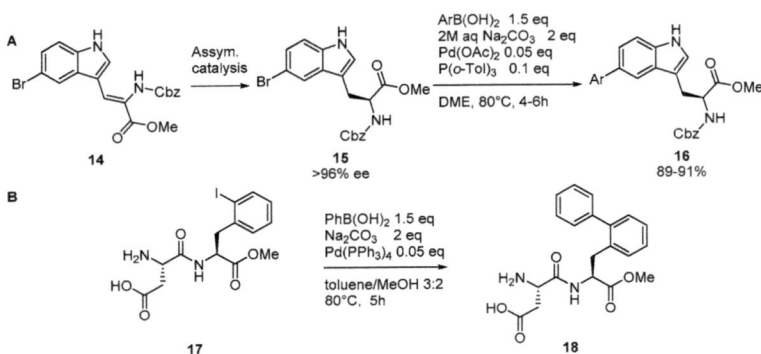

Figure 4. (**A**) Access to 5-aryltryptophans **16** via sequential asymmetric hydrogenation and Suzuki–Miyaura reaction. (**B**) Iodinated aspartame **17** transformed via cross-coupling.

Work published by Limbach et al. showed that protected 4-bromo-β^3-homophenylalanine **19** could be derivatized in mixed organic/aqueous (toluene/H_2O 8:1) conditions with a catalyst loading as low as 5 mol % (Figure 5A) [91]. The substrate scope was further expanded by Knör and coworkers via introduction of protected 3-iodo-L-tyrosine as a substrate [92]. In their work, biphasic conditions (DME/aq. Na_2CO_3) were applied in combination with $Pd(OAc)_2$ and P(o-tol)$_3$. Under these conditions, it was shown that protection of both the main chain termini and the phenol group was required. Cerezo reported the synthesis of 5-arylhistidines **22** under microwave conditions [93]. Ac–His–OMe was first treated with N-bromosuccinimide (NBS) and subsequently protected with a 2-(trimethylsilyl)ethoxymethyl (SEM) group, to yield **21**. Due to the coordination potential of the imidazole ring of histidine, elevated reaction temperatures, combined with relatively high catalyst loading (up to 40 mol %) and KF as a base, were required (Figure 5B). Under these optimized conditions, a series of novel 5-substituted histidines **22** were prepared, but only in moderate yields and facilitated under microwave (MW) conditions. In an attempt to suppress the racemization of protected 3-bromo-4-methoxyphenylglycine (not shown), a substrate sensitive for racemization, Prieto and colleagues studied the Suzuki–Miyaura cross-coupling of protected phenylglycine with various aryl boronic acids [94]. It was found that, by means of mild bases, such as sodium succinate and potassium acetate, racemization could be avoided. Recently, Dumas et al. proposed the use of Pd-nanoparticles (ca. 5.6 nm diameter) as catalyst for Suzuki–Miyaura reactions in aqueous media for the derivatizations of Boc-protected halophenylalanine as the substrate [95]. In their work, the nanoparticles were stabilized on the surface of larger poly(D,L-lactide–co–glycolide)–block–poly(ethylene glycol) copolymer nanoparticles (PLGA–PEG NPs, ca. 158 nm diameter). These nano-assemblies were found to be excellently suited to catalyze the Suzuki–Miyaura coupling (1 mol %) of brominated and halogenated amino acids with arylboronic acids at 37 °C for 18 h (98% conversion) (Figure 5C). Remarkably, when potassium phenyltrifluoroborate or phenylboronic acid MIDA-ester (*N*-methyliminodiacetic ester) were used, respectively 4% and no cross-coupled product was found. When cyclic triolborate derivatives (for example, potassium 4-methyl-1-phenyl-2,6,7-trioxa-1-borate-bicyclo-[2.2.2]-octane **24**) were used, the reaction rate was improved in comparison with boronic acids. Moreover, the reaction could be carried out in slightly acidic media (phosphate buffer, pH = 6.0), which can be beneficial for substrates containing base-sensitive groups.

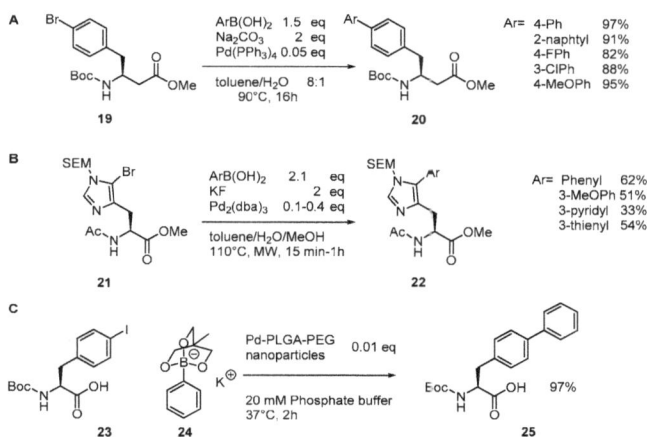

Figure 5. (A) Derivatization of β³-homophenylalanine **19** in mixed aqueous conditions. **(B)** Protected 5-bromohistidine **21** as a substrate for arylation. **(C)** Cross-coupling with palladium nanoparticles yielding **25**.

The derivatization of unprotected halotryptophans was first described by Deb Roy et al. under purely aqueous conditions, with the aid of the water-soluble trisulfonated triphenylphosphine (TPPTS) ligand, and Na₂PdCl₄ as a precatalytic system (Figure 6A), to generate tryptophan analogues **27** with fluorescent properties [64]. As expected, the obtained yields were highly dependent on the halogen atom present (Br > Cl), and derivatization on the 7-position afforded low yields. Further optimization indicated that a related ligand, bearing sulfonated *m*-xylene groups instead of phenyl (TXPTS), allowed for work at a lower temperature (40 °C) while still presenting excellent yields and significantly improved conversions when introducing a phenyl substituent on the 7-position. Alteration of the ligand to sulfonated 2-dicyclohexylphosphino-2′-6′-dimethoxybiphenyl (SSPhos) was necessary to maintain the applicability of the Suzuki–Miyaura reaction on 7-Cl-tryptophans, as beautifully illustrated by derivatization of the natural product chloropacidamycin in mixed aqueous conditions (water/acetonitrile (AcN) 5:1) [96]. The synthesis of 7-vinyltryptophan **29** with potassium vinyltrifluoroborate and PdCl₂(dppf), starting from 7-iodotryptophan **28** in mixed aqueous conditions (Figure 6B), broadened the scope of halotryptophan derivatization, along with the derivatization of unprotected bromotryptophan and bromotryptophan-containing dipeptides [97]. Recently, Frese et al. combined a biocatalytic and regioselective tryptophan halogenation and Suzuki–Miyaura cross-coupling in a multi-step one-pot reaction [98]. The benign halogenation is performed via immobilized tryptophan 5-, 6-, or 7-halogenases using NaBr as a halogen source, followed by biocatalyst removal and subsequent Suzuki–Miyaura reaction. The cross-couplings proceeded in high conversion with either SPhos in mixed aqueous conditions or SSPhos in purely aqueous conditions (Figure 6C).

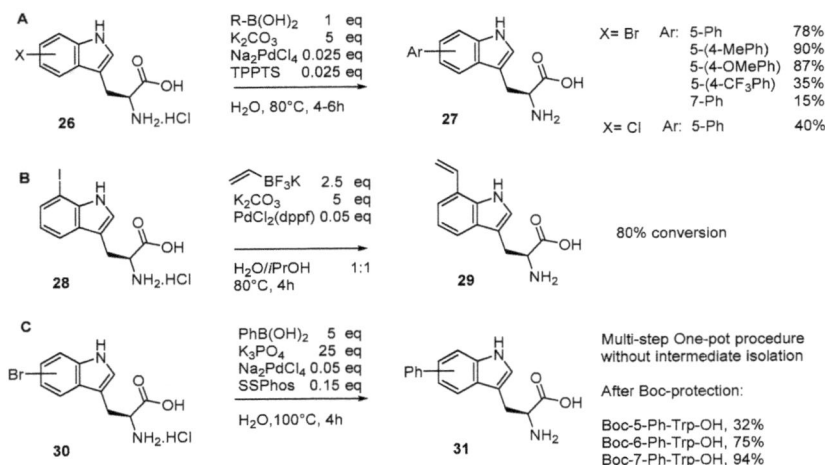

Figure 6. Derivatization of unprotected halotryptophans in aqueous conditions: (**A**) Arylation. (**B**) Vinylation. (**C**) Multistep one-pot biocatalytic halogenation presented 30, followed by Suzuki–Miyaura reaction.

As the Suzuki–Miyaura reaction is performed in basic conditions, the base labile 9-fluorenylmethyloxycarbonyl (Fmoc) protecting group is often not considered as part of a substrate. Nevertheless, Willemse reported Fmoc-iodophenylalanine as a substrate, and in their hands demonstrated that at 80 °C Fmoc-deprotection occurs when using PdCl$_2$(dppf) in mixed aqueous conditions (*i*PrOH/H$_2$O 1:1) with K$_2$CO$_3$ (not shown) [97]. However, by lowering the reaction temperature to 40°C, no deprotection occurred, but the reaction only reached a maximum conversion of 80% after 24 h. Similarly, Maity et al. investigated the fluorescent properties of an amino acid containing a 4'-acetamido[1,1'-biphenyl] moiety [99]. The synthesis involved the successful Suzuki–Miyaura cross-coupling reaction of 4-acetamidophenyl-1-pinacolatoboron ester with Fmoc-4-bromophenylalanine step in 81% yield (not shown). The cross-coupling proceeded in the presence of PdCl$_2$ and Na$_2$CO$_3$ in a mixture of THF and ethylene-glycol (10:1) at 66 °C. Recently, Qiao and coworkers reported the one-step synthesis of Fmoc-protected aryl substituted phenylalanines 33 (Figure 7) [100]. Formation of des-Fmoc product was minimized by using relatively mild conditions (<80 °C) and an organic solvent (THF or *t*-amylOH). Microscale high-throughput screening of diverse phosphine ligands indicated that 1,1'-bis(di-*tert*-butylphosphino)ferrocene (DTBPF) was optimal in combination with Pd(OAc)$_2$ as Pd-source and K$_3$PO$_4$ as a base. Under the optimized conditions, no ee erosion was observed. The developed methodology proved successful for generating Fmoc-Bip (biphenyl) derivatives with both aryl boronic acids containing e-withdrawing and e-donating groups. Preliminary experiments on a Fmoc-protected pentapeptide suggested that the reaction rate was significantly reduced and after 50 h only des-Fmoc products (3:1 starting bromide vs. product) were observed.

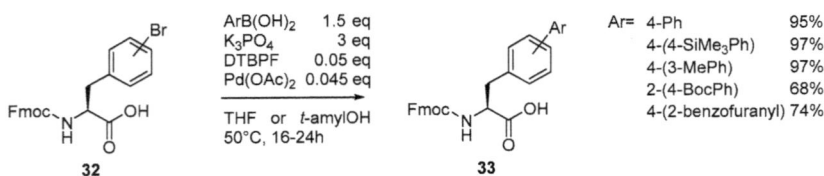

Figure 7. Synthesis of Fmoc-protected biphenylalanines 30.

2.2. Synthesis of Borylated Aromatic and Aliphatic Amino Acids and Subsequent Derivatization

Most studies of the Suzuki–Miyaura reaction on biomolecules, such as amino acids and peptides, are performed on halogenated Trp, Phe, or peptide sequences containing these residues as a consequence of their good accessibility and straightforward introduction into sequences by means of SPPS. The incorporation of borylated amino acids into peptide sequences equally allows cross-coupling with aryl or alkyl halides. Coupling with alkyl halides in particular has the added advantage of an excellent commercial availability of structurally diverse halogenated reactants and their improved stability in comparison to alkylboronic acids. The latter compounds possess low stability and limited commercial availability, and easily undergo deborylation or β-hydride elimination after transmetallation [101]. The biological stability and role of boronic acids in therapeutic peptides has already been studied in detail since this functional group has been used in protease inhibitors, boron neutron-capture therapy, and cancer therapy (e.g., Bortezomib) [102].

Multiple borylated amino acids have been synthesized, both aliphatic and aromatic residues [39,40]. Introduction of boron can be achieved through different strategies which depend on the envisaged application. Synthetic methods standardly introduce a protected analogue of a boronic acid. The preparation of pinacolyl arylboronate esters in particular has been well described, either starting from Seebach's imidazolidinone [103,104], via the Miyaura borylation of a (pseudo)halogenated phenylalanine [105–108], metal–halogen exchange [109], through direct Iridium-catalyzed borylation [110–112], or even Friedel–Crafts alkylation (selected examples are shown in Figure 8) [113]. Alternatively, direct synthesis of amino acids containing the MIDA-boronate ester group is also possible via Negishi coupling (not shown) [114].

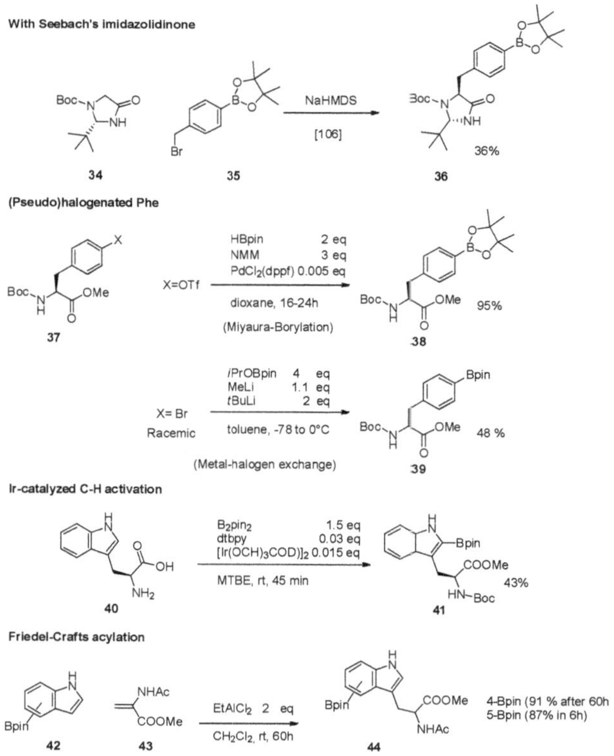

Figure 8. Representative examples of methods to prepare pinacolyl arylboronate esters.

The cross-coupling of 4-pinacolylborono-L-phenylalanine [L-Phe(4-Bpin)] **45** with aryl iodides was first reported by Satoh et al. When aryl bromides and triflates in combination with the pre-catalyst PdCl$_2$(dppf) were used, moderate yields resulted (Figure 9A) [103]. In the same year, an extension of this study was reported by Firooznia, describing the derivatization of racemic Boc–Phe(4-Bpin)–OEt with aryl chlorides, conversions which required the alkylphosphine-based precatalyst PdCl$_2$(PCy$_3$)$_2$ [104]. Their methodology was however limited to electron-donating aryl chlorides. Enantiomerically pure Boc-L-Phe(4-Bpin)–OMe was also derivatized by the same group through application of the previously published conditions of Satoh et al. [103], which used PdCl$_2$(dppf) in DME. It was also shown that racemic, unprotected 4-boronophenylalanine (4-BPA) could be derivatized under microwave conditions (150 °C, 5–10 min) [115]. Optimal conversion was achieved when PdCl$_2$(PPh$_3$)$_2$ was used in combination with Na$_2$CO$_3$ in water/acetonitrile (1:1) as a solvent. Čapek et al. investigated the cross-coupling of unprotected 4-boronophenylalanine with halogenated nucleosides [116,117]. Surprisingly, use of protected nucleosides in combination with Pd(OAc)$_2$ and JohnPhos in dioxane proved to be ineffective. However, when the water-soluble TPPTS ligand was employed with unprotected 8-bromoadenine nucleosides and 4-boronophenylalanine, efficient cross-couplings followed. These reactions were facilitated by microwave irradiation at 150 °C for 5 min, yielding the desired compounds in excellent optical purity, as determined by FDAA analysis [116]. Following these results, Čapek and coworkers also successfully coupled 6- or 8-halogenated purine nucleosides with 4-BPA in the presence of TPPTS and Pd(OAc)$_2$ [117]. In addition to pinacol esters, the MIDA boronate group was also introduced on aromatic amino acids such as **49** by Colgin et al. [114]. Starting from compounds **47** and **48**, a Negishi coupling and subsequent Suzuki–Miyaura reaction yielded **50** in mixed aqueous conditions (dioxane/H$_2$O 10:1) with Pd(OAc)$_2$ as the precatalyst and SPhos as ligand (Figure 9B). Very recently, Bartocinni showed that racemic (7-pinacolborono)tryptophan **51** can be transformed to biaryl products **52** via Suzuki–Miyaura reaction in mixed aqueous conditions (Figure 9C) [118]. Similarly, access to (7-pinacolborono)tryptophan via selective Ir-catalysis allowed Loach to prepare arylated tryptophan derivatives [119]. Meyer et al. reported the selective borylation of phenylalanine (in the presence of a directing group, for example Cl on the 3-position) and heteroaromatic amino acids, such as 2-thienylalanine, 3-pyridinylalanine, and tryptophan [111]. To show the applicability of their methodology, protected borylated thienylalanine **54** was converted to **55** (Figure 9D). When comparing the optimized reaction conditions found for the derivatization of arylboronate esters with conditions commonly applied for the Suzuki–Miyaura reaction of halogenated amino acids and peptides, one can conclude that in the former case higher reaction temperatures and/or prolonged reaction times are often required. In addition, the limited stability of pinacolyl boronate esters towards conditions commonly applied during solid-phase peptide synthesis requires careful design of the peptide assembly [120]. Although the use of other protected alkylboronic acid analogues may circumvent this problem, as exemplified via MIDA-boronates [114], the pinacol ester functionality is used in most cases.

In contrast to the absence of Suzuki–Miyaura reactions involving amino acid-based alkyl halides, the synthesis and derivatization of borylated aliphatic amino acids have been described extensively. These have been prepared via hydroboration of protected allylglycine [121–125], quenching of a lithiated serine analogue with triisopropylborate [126], a Cu-catalyzed cross-coupling of *gem*-diborylalkanes with primary alkyl halides [127], and very recently by Pd-catalyzed C(sp^3)-H bond activation (selected examples are shown in Figure 10, left side) [128].

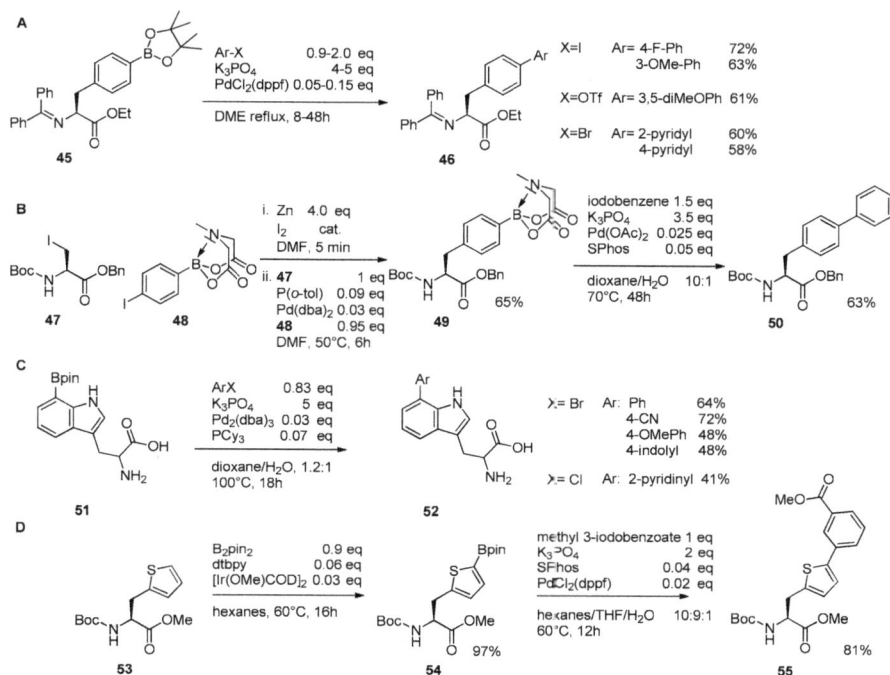

Figure 9. (A) Derivatization of *N*-benzophenone protected pinacolboronophenylalanine **45** with aryl halides. (B) Biphenylalanine **50** synthesis from the corresponding MIDA-boronate **49**. (C) Preparation of racemic 7-aryltryptophan **52** by Suzuki–Miyaura reaction of the unprotected pinacolborono-D/L-tryptophan **51**. (D) Protected thienylalanine **53** transformed to biaryl **55** via subsequent Ir-catalyzed borylation and Suzuki–Miyaura cross-coupling.

Campbell prepared bishomophenylalanine derivatives **58** by hydroboration of protected allylated oxazolidine **56** with 9-borabicyclo[3,3,1]nonane (9-BBN), followed by Suzuki–Miyaura reaction with aryl iodides and phenyl bromide in good yields (Figure 10A) [121]. The homologated phenylalanine analogues were obtained in mediocre yields (30%–50%) after oxidation with Jones' reagent (CrO_3/H_2SO_4) (not shown). Comparable work was also published by Sabat, which expanded the scope of the reaction to aryl triflates and heteroaryl bromides as substrates, giving way to oxazolidines like **58** in decent yields [122]. The oxidation protocol was further optimized to obtain the α-amino acid analogues in yields up to 82% with excellent optical purity. Collier et al. further improved the procedure by performing the hydroboration-Suzuki protocol directly on protected allylglycine **59**, yielding bishomophenylalanine derivatives **61** in two steps (Figure 10B) [123]. Extension to higher phenylalanine homologues was achieved by Rodríguez and coworkers [124]. In their work, but-3-enylglycine was prepared via Negishi coupling and after hydroboration, access to higher homologues was realized via Suzuki–Miyaura cross-couplings (not shown). Alternatively, a borylated serine analogue **63** was prepared by Harvey et al. by quenching an organolithium species (derived from **62**) with triisopropylborate. This building block was transformed to non-proteinogenic phenylalanine analogues **64** by Suzuki–Miyaura reaction in the presence of Ag_2O (Figure 10C). The corresponding amino acid was obtained after consecutive 2-(trimethylsilyl)ethoxymethyl (SEM) deprotection and a two-step oxidation with Dess–Martin periodinane and Pinnick oxidation (not shown).

Synthesis of alkylboronates

Hydroboration with 9-BBN

Suzuki-Miyaura coupling of alkylboronates

A

ArX	1.1 eq	X= I	Ar: Ph	79%
aq. K_3PO_4	2.2 eq		2-(*N*-Boc-aniline)	70%
$PdCl_2$(dppf)	0.05 eq		3-NO_2-Ph	69%
			4-anisol	72%
DMF, 16h, rt		X= Br Ar: Ph		72%

B

ArX	1.1 eq	X= I	Ar: Ph	62%
aq. K_3PO_4	2.2 eq		2-NO_2-Ph	53%
$PdCl_2$(dppf)	0.05 eq		4-anisol	60%
DMF, 16h, rt		X= Br	Ar: 4-CO_2Me-Ph	72%
			2-pyridinyl	58%

Quenching of lithiated serine analogue

i. *n*BuLi, 1.1 eq
Li-napthalenide 3 eq
THF, -78°C, 15 min

ii. Triisopropylborate 2.3 eq
THF, -78 to -40°C overnight
aq. sat. NH₄Cl, aq. 0.5 M HCl

62 → **63**

C

ArBr	2.2 eq		Ar: 4-anisol	65%
K_2CO_3	3 eq		3-NO_2-Ph	67%
Ag_2O	2.5 eq		1-napthyl	73%
$PdCl_2$(dppf)	0.09 eq		3-thienyl	19%
THF, 70°C, 21h		**64**		

Cu-catalyzed coupling with *gem*-diborylalkanes

*t*BuOLi 3 eq
CuI 0.2 eq

DMF, 60°C, 24h

65 **66** → **67**

Pd-catalyzed C-H bond activation

B₂pin₂ 2 eq
KHCO₃ 2 eq
HOAc 0.2 eq
2,4-diMeOquinoline 0.2 eq
Pd(OAc)₂ 0.1 eq

AcN, O₂, 80°C, 15h

68 → **69**

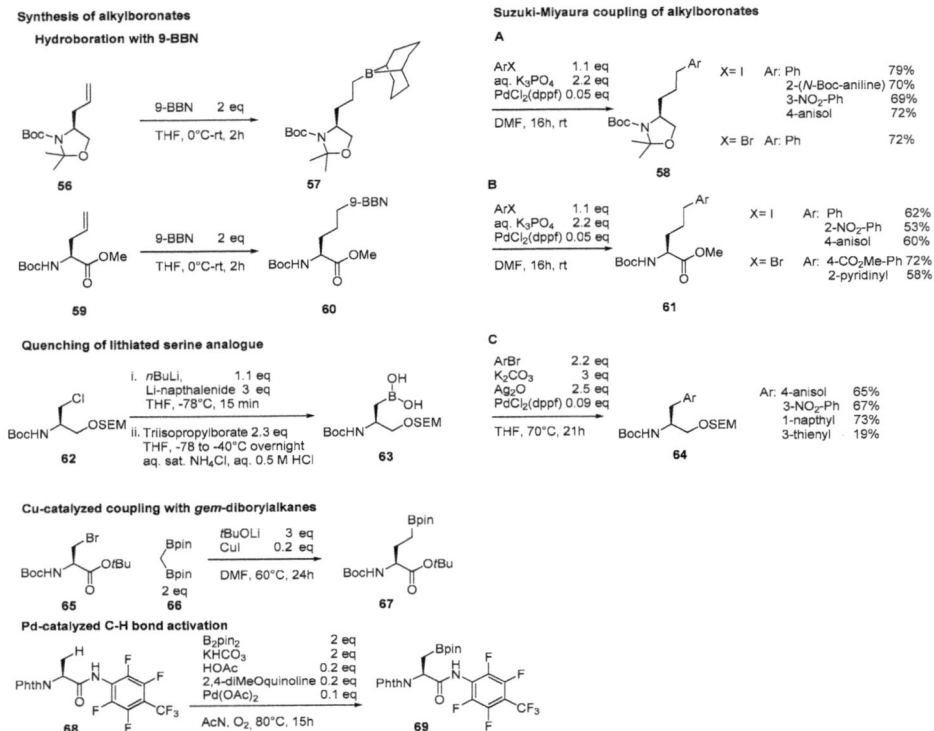

Figure 10. (**Left**) Representative examples of the preparation of amino acid alkylboronates. (**Right**) Synthesis of phenylalanine and analogues thereof prepared (**A**) via hydroboration/Suzuki–Miyaura protocol on protected allylated oxazolidine **56** (**B**) directly from protected allylglycine **59** or (**C**) from borylated serine analogue **63**.

2.3. Coupling of Amino Acids: Synthesis of Biaryl-Bridged Dipeptides

Yoburn investigated the synthesis of protected dityrosine substrate **70** (Figure 11) [129]. Initially, homocoupling of Ac–Tyr(3-I, 4-OBn)–OMe was performed in a one-pot borylation-Suzuki–Miyaura protocol with $PdCl_2$(dppf) and B₂pin₂. Here, K_2CO_3 was essential as a base in order to obtain efficient cross-coupling, since use of KOAc only led to borylation of the iodotyrosine educt. Reaction rates and conversions were however significantly lower when the substrate's complexity was increased, for example, to tripeptides. Next, isolation of the borylated peptide intermediate was envisaged and coupling proceeded smoothly when KOAc was used as a base. As such, Suzuki–Miyaura cross-coupling between peptide substrates bearing iodo- and boronotyrosine residues proceeded efficiently, and a viable strategy for homodimerizations (up to heptapeptides) was realized. These findings were simultaneously reported by Hutton et al. [130]. Similarly, Kotha synthesized dimers of protected phenylalanine **71** with racemic 4-boronophenylalanine and 4-iodophenylalanine (Figure 11), along with several unnatural bis-armed amino acids containing a variable number (one to five) of benzene rings [131]. In addition to their investigations on the racemization of phenylglycine under Suzuki–Miyaura conditions [94], Prieto further elaborated on the cross-coupling between 4-hydroxyphenylglycine and tyrosine or tryptophan derived building blocks [132]. Using standard reaction conditions, up to 32% of undesired diastereoisomer was found. Racemization could be suppressed (with up to 99% ee) via usage of the Buchwald ligand SPhos [9], eventually yielding biaryl building block **72** (Figure 11). Synthesis of these biaryl-bridged building blocks **71–72** represents a basis to obtain cyclic peptides when intramolecular couplings are performed (*vide infra*).

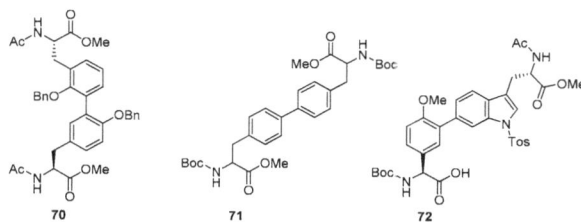

Figure 11. Amino acid homo- and heterodimers **45–47** by intermolecular Suzuki–Miyaura cross-coupling.

2.4. Suzuki–Miyaura Reaction on Peptidic Substrates

Transposition of the Suzuki–Miyaura reaction from amino acid to larger, more complex peptides substrates required adapted methodologies. While solubility of protected amino acid substrates in organic solvents is not an issue, peptide substrates, and by extension proteins, are preferentially treated in aqueous conditions. Circumvention of the traditional solvents (e.g., toluene, DMF, dimethoxyethane) and lower reaction temperatures complicated the envisaged transformations. Moreover, application of the reaction with water as (co-)solvent demands a catalytic system that achieves high conversions in this environment, whilst avoiding side product formation. One needed to move away from the use of $Pd(PPh_3)_4$, $Pd_2(dba)_3$ or $Pd(OAc)_2$, combined with classical (phosphine) ligands such as the ones used in sections above (Figure 12, compounds **73–76**), and out-of-the-box catalytic systems were required. One solution was found in designing more hydrophilic analogues of traditional ligands [11,62]. For example, 3,3′,3″-phosphanetriyltris(benzene sulfonic acid) trisodium salt (TPPTS **77a**, Figure 12) [18,64], its xylene analogue TXPTS **77b** [64,133], sodium 2-dicyclohexylphosphino-2′,6′-dimethoxybiphenyl-3′-sulfonate (SSPhos **78**) [10,96], and water-soluble NHC-ligands (*N*-heterocyclic carbene), such as **79** [134–137], have been proposed for Pd-catalysis in aqueous media (Figure 12). Suzuki–Miyaura reactions encompassing other water soluble ligands [11,12,138], or additives such as polyethyleneglycol and surfactants were also applied [11,34,139] Notably for Suzuki–Miyaura reaction in aqueous conditions, ligands containing the guanidine functionality, such as 2-aminopyrimidine-4,6-diol disodium salt (ADHP **80a**) [63,140], its dimethylated analogue (*N,N*-diMeADHP **80b**) [34], and (methyl)guanidines **81** and **82** [34,141,142], have been reported.

The development of peptide diversification through Suzuki–Miyaura reactions in a purely aqueous environment was first reported by Vilaró et al. [143]. By combination of $Pd(OAc)_2$ and the water-soluble sulfonated triphenylphosphine (TPPTS) ligand, couplings were realized in aqueous media on unprotected peptides **83** in very mild conditions (Figure 13A). Nearly quantitative yields were reported starting from the iodinated Leu-enkephalin analogue **83**, bearing neither main chain nor side chain protection. These transformations were realized upon application of potassium aryltrifluoroborates as reactants [7,9]. As expected, it was found that not only the length of the peptide chain, but also its respective amino acid composition were critical. For example, conversions in the presence of a histidine residue were greatly reduced. This can easily be rationalized by a potential complexation of unprotected side chains - but also the main chain termini- with the palladium catalyst, hence interfering with the envisaged derivatizations. During their exploration for phosphine-free catalyst systems, Chalker et al. observed that an aqueous solution of the sodium salt of ADHP **80a** and $Pd(OAc)_2$ could be used for cross-couplings of partially unprotected substrates like **85** in purely aqueous conditions (Figure 13B) [63]. In addition, mono- and bis-arylations were realized when, respectively, 3-iodotyrosine or 3,5-diiodotyrosine was introduced in the unprotected peptide sequence. Despite limited catalyst loadings (1–4 mol %), high conversions were obtained for halogenated phenylalanines, tyrosines, and *p*-iodobenzyl cysteine (Pic, such as in **85**). In contrast, in case of reactions on chlorophenylalanine- or substrates bearing Cys (unprotected in the side chain), these derivatization reactions were unsuccessful, presumably, due to the fact that the palladium catalyst is not electron-rich enough. Extension of their work showed that both the dimethyl analogue of ADHP (i.e., Me$_2$ADHP **80b**) and methylated guanidine **81** or **82** outperformed the ADHP catalyst system when a complex PEG-bound phenylboronic acid was utilized [34]. The latter consisted of monomethoxy-PEG (MW 2 k or 20 kDa) linked to 4-propylcarbamoylphenylboronic acid.

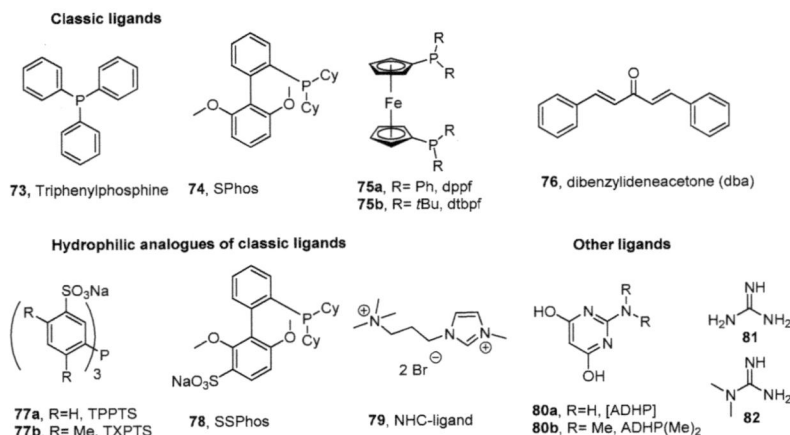

Classic ligands

73, Triphenylphosphine   **74**, SPhos   **75a**, R= Ph, dppf / **75b**, R= *t*Bu, dtbpf   **76**, dibenzylideneacetone (dba)

Hydrophilic analogues of classic ligands   **Other ligands**

77a, R=H, TPPTS / **77b**, R= Me, TXPTS   **78**, SSPhos   **79**, NHC-ligand   **80a**, R=H, [ADHP] / **80b**, R= Me, ADHP(Me)$_2$ **81**   **82**

Figure 12. Ligands used Suzuki–Miyaura cross-coupling reactions on peptides. Classical ligands (**73–76**) that have commonly been used in the early developments on protected peptide substrates. Hydrophilic analogues (**77–79**) and non-classical ligands (**80–82**) used in aqueous media.

In order to unravel side/main chain functionality that hampers the application of Suzuki–Miyaura conversions on peptides, Willemse et al. performed a systematic screening of unprotected peptide main and side chain functionalities with various aqueous soluble ligands on dipeptide substrates of type **87** (Figure 13C) [97]. It was shown that, in purely aqueous conditions, a precatalyst solution of ADHP and Pd(OAc)$_2$ led to efficient derivatizations of dipeptides containing 4-bromophenylalanine. Presence of histidine, bearing an imidazole moiety, and asparagine, bearing a primary amide, were revealed to be troublesome during the targeted cross-couplings. Alteration of the catalytic system to the ferrocene-containing catalyst PdCl$_2$(dppf) resulted in improved conversions in these cases. However, in order to achieve solubility of the catalyst and efficient cross-couplings, a water/*i*PrOH mixture was required. For the investigated histidine-containing peptides, maximum conversions of approximately 50% were reached, despite efforts to pre-saturate the imidazole moiety by the addition of sacrificial Lewis acids (e.g., MgCl$_2$) prior to the cross-coupling, as was previously successfully reported for ring-closing metatheses (RCM) and Heck reactions [82,144,145]. Derivatizations on a large peptide (**89**, 34 amino acid residues) in aqueous conditions were performed by Ojida et al. [146] (Figure 13D). Synthetic analogues of the WW domain (6–39) of PinI protein were prepared via SPPS assembly of a peptide containing an iodophenylalanine residue and subsequent Suzuki–Miyaura reactions. Experiments indicated that the addition of glycerol (10% v/v), known to stabilize proteins, to a Tris-HCl buffer (pH 8) was mandatory to reach high yields. Remarkably, the reaction temperature could be limited to 40 °C when using 1 equivalent of Na$_2$PdCl$_4$.

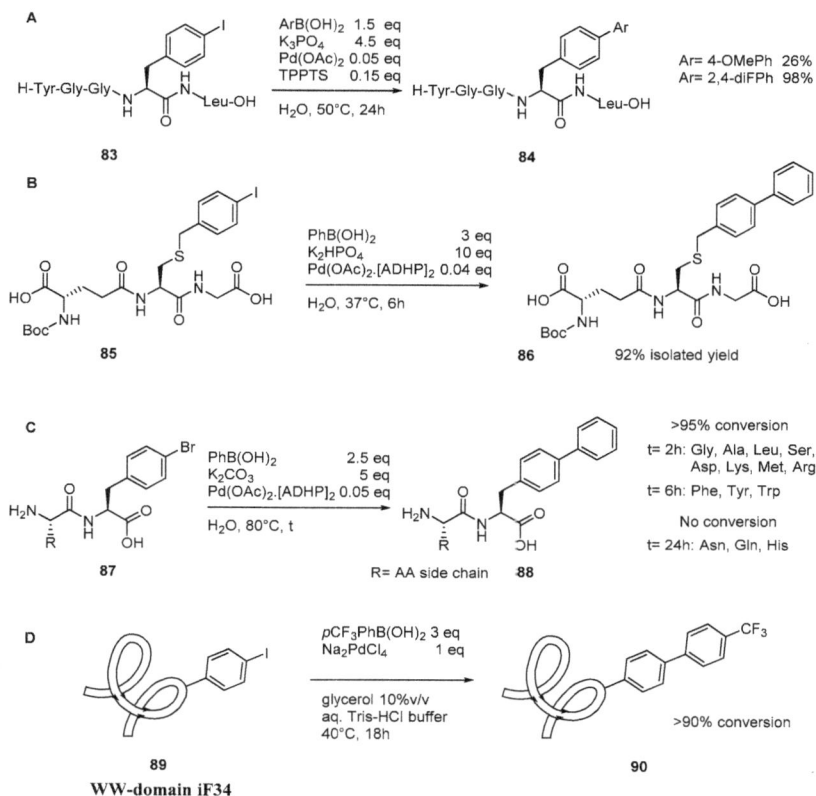

Figure 13. Suzuki–Miyaura reactions on peptides (**A**) Unprotected pentapeptide **83** as a substrate. (**B**) Internal *p*-iodobenzylcysteine residue transformed to biaryl tripeptide **86**. (**C**) Use of unprotected dipeptides **87**. (**D**) Derivatization of 34-mer **90**.

3. Solid-Phase Derivatizations of Peptide Substrates

Alternatively to Pd-catalyzed cross-couplings in solution, the derivatization of solid supported substrates offers advantages such as convenient washing steps and easy removal of any excess or unreacted reagents and reactants through filtrations [147,148]. For these cross-couplings on resin, the solvent is of high importance, as it is related to the swelling properties of the solid support. Typically, polystyrene-based resins are used for solid-phase peptide synthesis (SPPS) in organic media (e.g., DMF, CH_2Cl_2), whereas very limited swelling is observed in aqueous conditions or upon use of polar solvents. Alternatively, PS-PEG copolymer resins have been developed to improve swelling in mixed aqueous or polar conditions [149–151]. Of relevance to the current review, the preparation of a solid supported phenylalanine scaffold was first performed by Colombo [152]. Herein, it was shown that the Suzuki–Miyaura reaction of 4-iodophenylalanine anchored on Rink Amide resin was possible by means of biphasic conditions (EtOH/toluene 7:3), Pd(PPh₃)₄ precatalyst, and an aqueous solution of Na_2CO_3, albeit full conversion was not reached in these reactions. In their work, Wang resin was the first to be evaluated as the solid support; however, under the applied conditions, significant loss of product because of benzyl ester hydrolysis (i.e., linkage between amino acid and support), and low purity was observed. Performing the reaction on Rink Amide resin afforded desirable transformations. Limbach et al. reported Suzuki–Miyaura reactions on a Rink

Amide resin linked pentapeptide containing a brominated β^3-homophenylalanine [91]. In their hands, Pd(PPh$_3$)$_4$ and Na$_2$CO$_3$ in a mixture of EtOH/toluene/H$_2$O solvent system resulted in high conversion. The "on support" derivatization was further elaborated by Nielsen et al. on hydroxymethylbenzoic acid-polyethylene glycol (HMBA-PEG) resin (Figure 14A) [153]. Peptide substrates anchored on this base-labile resin and containing either a 3- or 4-iodophenylalanine residue were derivatized to yield Compound **92** after coupling with a variety of arylboronic acids, K$_3$PO$_4$ and PdCl$_2$(dppf) in a biphasic (*t*BuOH/toluene/H$_2$O) medium.

Similarly, Le Quement and co-workers reported the solid-phase derivatization of thienylalanine containing peptides **93** (Figure 14B) on HMBA-PEG$_{800}$ resin [154]. By solid-phase peptide synthesis, the 3-thienylalanine residue was introduced and subjected to Pictet–Spengler cyclization and bromination, to obtain **94** on support. From this compound, a set of cross-coupled products **95** could be obtained in good yields. Under analogous conditions, a derivatization strategy was employed to access biaryl pentapeptides with antagonistic effect on inhibitors of apoptotic proteins [155]. Additionally, a library of antimicrobial peptides was prepared by Haug et al. via the Suzuki–Miyaura reaction on tripeptides containing 4-iodophenylalanine [156]. Here, a Rink Amide linker on NovaGel resin (i.e., a PS-PEG resin) was employed as the solid support for the preparation of the Boc–Arg(Pbf)–Phe(4-I)–Arg(Pbf)–Rink amide educt. Successful cross-couplings with different bulky arylboronic acids were realized with Pd(OAc)$_2$ and P(*o*-tolyl)$_3$ as the catalyst system in mixed DME/H$_2$O (not shown).

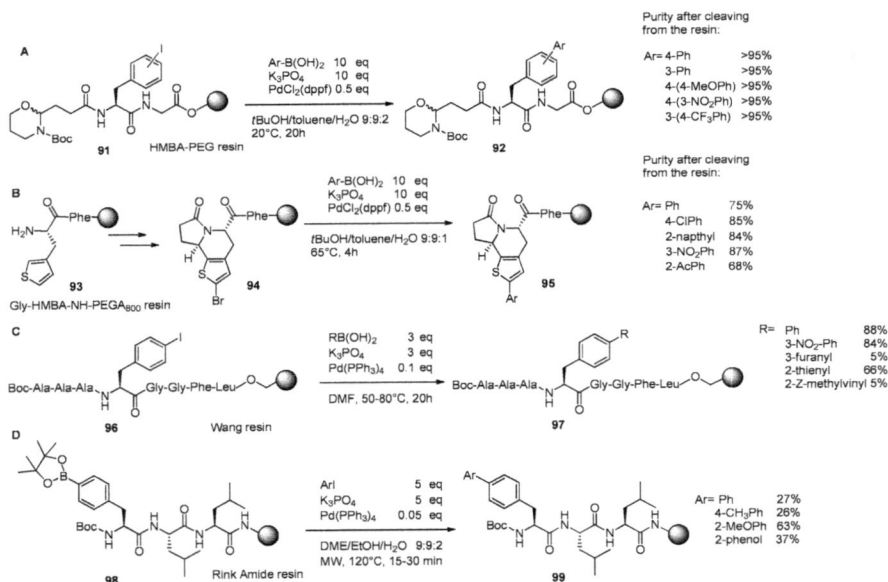

Figure 14. Solid-Phase derivatization via Suzuki–Miyaura cross-coupling (**A**) with iodophenylalanine containing **91**, (**B**) with bromothienylalanine containing **94**, (**C**) with 4-iodophenylalanine containing octapeptide **96**, and (**D**) with tripeptide **98** bearing pinacol protected 4-borono-L-phenylalanine.

Doan et al. reported the solid-phase derivatization of an enkephalin-like octapeptide sequence **96** (Figure 14C) [157]. The peptide, prepared on Wang resin, was modified using a variety of bases, solvents, catalysts and temperatures. Based on earlier reports on the instability of resin-bound peptides with unprotected N-termini [153], the final amino acid was introduced with a Boc protecting group. In their hands, aqueous solutions of weak bases (K$_3$PO$_4$ or Na$_2$CO$_3$) in combination with DMF as solvent proved to be optimal. It was also found that the use of Pd(PPh$_3$)$_4$ as catalyst at a maximum temperature of 80 °C for 20 h provided access to a range of biaryl containing peptides **97**. Not surprisingly, low yields were obtained upon insertion of several heteroaryl moieties, and significant side products were formed upon prolonged exposure to high temperature (>100 °C).

Interestingly, Afonso et al. disclosed for the first time the preparation of biaryl-containing peptides from 4-iodophenylalanine by sequential solid-phase borylation and Suzuki–Miyaura conversion [158]. Their approach relied on the introduction of the arylboron group by Miyaura borylation on Rink Amide resin. Treatment of polymer-bound halogenated peptide with the optimized catalytic conditions, being B_2pin_2/KOAc/$PdCl_2$(dppf) in DMSO, afforded the boronopeptidyl resin **98** (Figure 14D) with excellent conversion, as determined by HPLC after small scale cleavage. Next, the cross-coupling was investigated with iodobenzene in degassed DMF under microwave irradiation. Initially, only minor conversions were attained, and side products, including deborylated or homocoupled peptide, were formed. By switching the solvent system to DME/EtOH/H_2O 9:9:2, in combination with the use of an aqueous solution of K_3PO_4, improved reaction efficiency was obtained, and gratifyingly, target arylated peptides **99** were accessed in acceptable isolated yields. In an extension of their work in solution phase [93], Cerezo reported the first solid-phase derivatization of 5-bromohistidine containing tripeptides (not shown) [159]. The 5-bromo–histidine residue was introduced as the N-terminal amino acid via SPPS on Rink MBHA resin and subjected to Suzuki–Miyaura reactions. The imidazole moiety was protected with a SEM-group (as in Figure 5B). In comparison with their optimal solution-phase conditions, elevated temperatures (170 °C) and a solvent change to DME/H_2O/MeOH was required.

4. Formation of Biaryl-Bridges in Peptidic Natural Products and Peptide Macrocyclics

By introduction of halogen and boron groups onto amino acid side chains, the preparation of biaryl products starting from two amino acids can be envisaged (see Section 2.3). First reports of these biaryl linkages in cyclic structures via Suzuki–Miyaura reactions were inspired on peptidic natural molecules such as complestatin. Elder and coworkers synthesized a model ring system **105** based on complestatin [160]. In their work, two syntheses are reported, going either via intramolecular cross-coupling of **103**, containing brominated tryptophan and borylated tyrosine residues (Figure 15A) or via lactamization of a biaryl-bridged intermediate **104**. The D-tryptophan residue in **103** and **104** was prepared via a key enantioselective ene-reaction between **100** and **101** with (S)-(−)-2,2'-*p*-tolyl–phosphino)-1,1'-binaphthyl ((S)-Tol–BINAP). The unnatural protected amino acid **102** was obtained in excellent ee (94%). After peptide assembly in solution, **103** was submitted to Suzuki–Miyaura coupling condition (Figure 15A, reaction parameters not fully disclosed). Alternatively, the 17-membered ring system in **105** could be formed by activation of **104** with pentafluorophenyl diphenylphosphinate (FDPP) in diluted DMF.

The synthesis of biphenomycin analogue **107** was reported by Carbonnelle et al. Instead of isolating the pinacolatoboron intermediate, the diiodinated ortho-substituted Compound **106** was converted to biaryl **107** in a tandem procedure with B_2pin_2 and $PdCl_2$(dppf) (Figure 15B) [161]. The efficiency of such a tandem reaction was, however, moderate and highly dependent on the concentration of the reaction, with significant formation of de-iodinated compound. The interest of the Zhu group in the use of Suzuki–Miyaura reactions for natural product synthesis was further illustrated by the synthesis of RP-66453 **108** [162], and the concise asymmetric total synthesis of biphenomycin B, wherein this cross-coupling served as the final step [107]. A different approach was reported by Waldmann et al. [163]. Here, the biaryl bridge in **108** was first prepared via intermolecular Suzuki–Miyaura of a pinacolatoboron- and iodo-precursor. The cyclization was then performed by lactamization, to again provide **108**. Owing to their widespread occurrence in nature [61] and the efficient access of biaryl-bridged peptides via Suzuki–Miyaura cross-couplings, multiple research teams focused their efforts on natural product synthesis of, among others, TMC-95A/B **109a,b** [164–168], arylomycins **110a,b** [169], complestatin [170,171] and Rubiyunnanin B [172], wherein the Suzuki–Miyaura reaction often serves as the disconnection site of the biaryl-bridge (Figure 15C).

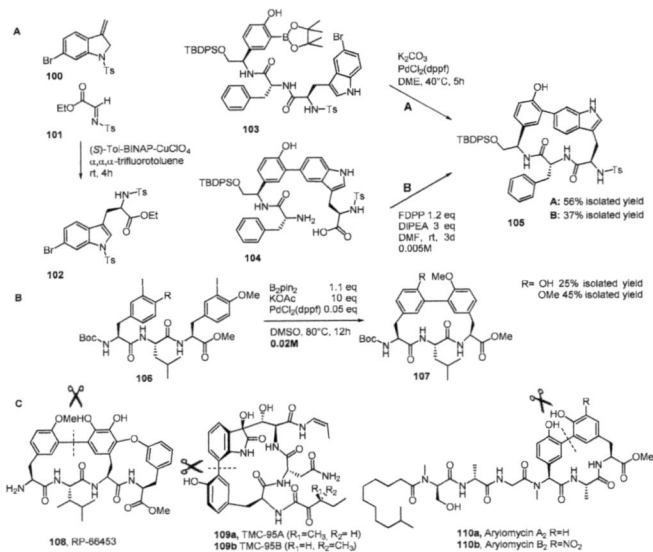

Figure 15. Cyclic peptide synthesis via Suzuki–Miyaura reactions. (**A**) Synthesis of a part of the ring system of complestatin (**B**), **105**. One-pot borylation - Suzuki–Miyaura on diiodinated substrate **106**. (**C**) Disconnection strategy for different natural products **108–110**, bearing a biaryl bridge.

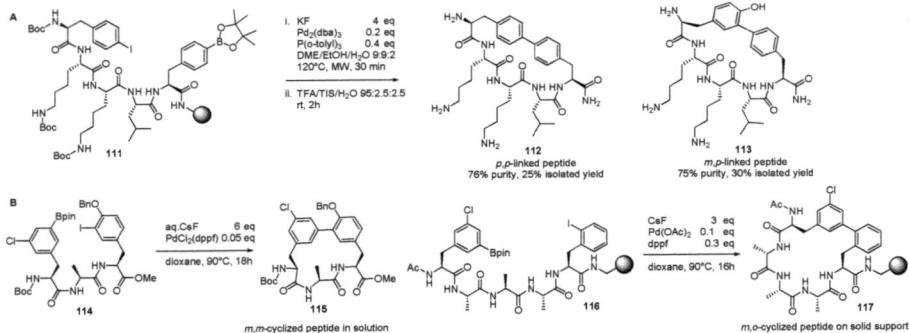

Figure 16. (**A**) Solid-phase synthesis of biaryl-linked peptide **112** with *p,p*-motif and **113** with *m,p*-motif. (**B**) Synthesis of *m,m*- and *m,o*-cyclic peptide **115** and **117** via respective solution-phase and solid-phase Suzuki–Miyaura.

Afonso et al. extended their work on cross-couplings of borylated peptides on solid support [158] through the cyclization of a linear peptide, for example, **111**, containing both the boronate and the halogenated aromatic amino acid [120]. Different ring sizes were obtained by varying both the length of the peptide sequence (three to eight residues) and the position of the halogen atom on the N-terminal amino acid (*p*-I–Phe, *m*-I–Tyr and 3-Br–His). The cyclization was carried out under microwave conditions and provided the expected biaryl cyclic peptides in high purity for both *p,p* and *m,p*-linked peptides as exemplified by, respectively, **112** and **113** in Figure 16A. An extension to their study was performed by alteration of the borylated residue [173]. The peptidyl resin containing protected 3-iodotyrosine was first borylated; and, after extension of the peptide chain, either protected 4-iodophenylalanine or 3-iodotyrosine was added to synthesize the *p,m*- and *m,m*-crosslinked biaryl peptides (not shown). Meyer et al. report the preparation of additional combinations of biaryl-bridged peptides [174]. Either a solid- or solution-phase approach allowed *m,m-*, *o,m-*, and *m,o-*bridged cyclic peptides, for example, **115** and **117**, through regiochemical variation of the aryl halide and arylboronate (Figure 16B). While solution-phase couplings were reported with $PdCl_2$(dppf) precatalyst, the solid-phase reactions were performed with $Pd(OAc)_2$ and dppf.

5. Protein Derivatization

Substrate complexity was pushed further and proteins bearing halogenated or borylated amino acid side chains were considered. Labeling of protein substrates through natural amino acids side chains is commonly carried out on cysteine, lysine or even aspartic acid side chains [31,175]. However, the abundance of nucleophilic groups in proteins hampers site-specific alteration of the substrates and emphasized the necessity for a chemoselective reaction toolbox. For this purpose, chemoselective bioorthogonal reactions have been explored and exploited for labeling substrates such as peptides and proteins, but these conversions have also been used for other purposes, namely to change the pharmacokinetic and pharmacodynamic properties of biologically relevant peptides and proteins [29–33]. Popular bioorthogonal reactions examples include the Staudinger ligation, click chemistry, cycloadditions, and transition-metal catalyzed C–C bond formation. Of interest to this work, the potential of palladium-mediated chemistry has even been extended to living cells [176–178]. However, in case of transformations on highly complex proteins, containing multiple complexation sites, superstoichiometrical quantities (often palladium >50 eq, boronic acid >500 eq) are required. Since the allowed palladium level thresholds in active pharmaceutical ingredients are very low and metal impurities can affect activity and toxicity studies of the target compounds, analysis and removal of palladium impurities is of utmost importance [179]. Due to the scale of all described reactions (R&D phase), recycling of the precious metal catalysts has not been studied in much detail when working on protein substrates. An exception to this can be found in the work of Dumas, which shows that inductively coupled plasma (ICP) with optical emission spectrometry (OES) can provide quantitative data on palladium content in protein samples after Suzuki–Miyaura reaction [34]. The addition of 3-mercaptopropionic acid has been shown to efficiently remove palladium from the protein surface [180]. Furthermore, recoverable palladium nanoparticles have been shown to possess excellent catalytic activity for the transformation of brominated or iodinated amino acids [145]. However, in view of future upscaling efforts, recyclability will have a major impact on newly developed methodologies [179,181]. Recent reviews on sustainable Suzuki–Miyaura reactions in aqueous media discuss the use of recoverable catalysts, yet these are limited to non-peptidic substrates [11,138,182]. In this respect, several recyclable catalysts [183–190], including silica- or polymer-supported catalysts and magnetically recoverable catalysts, are reported for biaryl formation and could potentially be implemented for transformations of peptide substrates.

Site-specific carbon–carbon bond formation on proteins was first realized by Kodama through Mizoroki–Heck and Sonogashira reactions on a His_6-fused Ras protein containing a 4-iodophenylalanine residue (iF32-Ras-His) [65,82]. The first example of a Suzuki–Miyaura reaction on protein substrates was reported by Brustad and coworkers via expansion of the genetic code, which allowed selective introduction of a 4-boronophenylalanine residue [83]. Subsequently, the Suzuki–Miyaura reaction induced the coupling of a fluorescent iodoaryl boron-dipyrromethene (bodipy-I) scaffold through use of Pd_2(dba)$_3$ complex in an aqueous buffered solution (20 mM EPPS) at 70 °C, though in a moderate conversion (ca. 30%) (Figure 17A). Both bodipy-I and Pd_2(dba)$_3$ were added in 20-fold excess with respect to the borylated protein. Chalker et al. previously showed that an aqueous catalyst solution of the sodium salt of 2-amino-4,6-dihydroxypyrimidine (ADHP, **80a**) and $Pd(OAc)_2$ catalyzed cross-couplings in excellent yields on C-terminally unprotected amino acids using only buffered aqueous conditions at 37 °C (vide supra, Section 2.4) [63]. Their elegant method was extended towards protein-based substrates by introduction of *p*-iodobenzyl cysteine (Pic) as replacement of a single cysteine residue in a subtilisin *Bacillus lentus* (SBL) mutant S156C

(SBL-156ArI, Figure 17B) [63]. To ensure fast couplings, an excess of the catalyst solution (50 eq) and boronic acid (500 eq) were employed. Further research by the Davis group also showed that when 4-iodophenylalanine is present in a His$_6$-tagged maltose binding protein (MBP), the ADHP-Pd(OAc)$_2$ combination was again perfectly suited for the envisaged derivatization [180]. Similarly, 50 eq of palladium catalyst and 680 eq of boronic acid were required for achieving complete conversion after 2 h at 37 °C. Reaction follow-up by mass spectrometry was however hampered by non-specific metal binding to the protein. Therefore, the addition of 3-mercaptopropionic acid scavenger was mandatory to remove palladium and re-emergence of proteinogenic ionic species during MS reaction analysis. Spicer et al. expanded the scope of the Suzuki–Miyaura reaction to living cell surface proteins [191]. Four mutant plasmids of the OmpC protein were co-transformed into *E. coli* cells to incorporate 4′-iodo-L-Phe, via amber stop codon suppression, in different loop regions of cell-surface channels. The Pd(OAc)$_2$.(ADHP)$_2$ catalyst system proved again appropriate for efficient labeling with fluorescent dyes on the outside of the cell-surface.

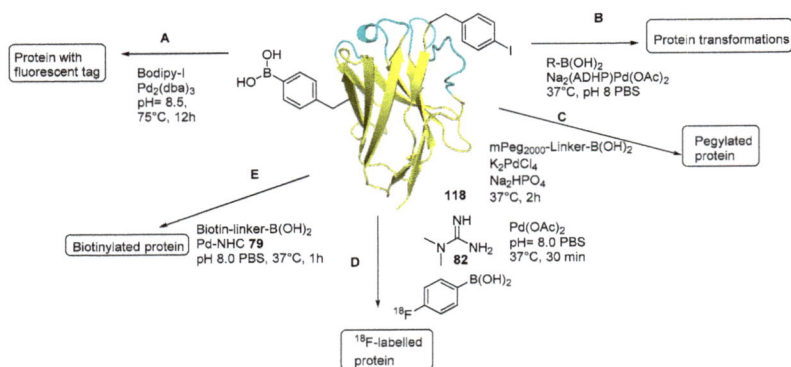

Figure 17. Schematic representation of different strategies for Suzuki–Miyaura reaction on protein substrates. Various proteins were utilized. Here, a common "dummy" substrate **118** is shown for illustrative purposes. (PBS = phosphate buffer).

Attachment of polyethylene glycol (PEG) chain to proteins (also denoted as the "PEGylation" process) in often used to improve stability and pharmacokinetic properties [192]. As a disadvantage, PEG-conjugates cope with a reduced biological activity which is related to the presence of multiple reactive sites at the level of amino acid side chain in the protein (usually lysine or cysteine residues). Therefore, the site-selective PEGylation via Suzuki–Miyaura reactions was envisaged as an alternative approach by Dumas (Figure 17C) [34]. A PEG unit (2 kDa or 20 kDa) was attached to two different iodinated proteins (Npβ-69pIPhe and SBL-156Pic). First, the derivatization of Boc-4-iodophenylalanine was unsuccessfully attempted with a monomethoxy PEG polymer phenylboronic acid (mPEG-PBA) reagent and ADHP as the ligand. Luckily, upon application of structurally related ligands such as 2-(N,N)-dimethyl-4,6-dihydroxypyrimidine (N,N-diMeADHP) and methylated guanidine derivatives, the derivatizations proceeded smoothly after 2 h at 37 °C. Intriguingly, reaction conversion also proceeded in >90% conversion when no ligand was added to reactions with K$_2$PdCl$_4$ as the palladium source. Converting these findings on the iodinated proteins showed that these "self-liganded" conditions resulted in high conversion for both Npβ-69pIPhe and SBL-156Pic with respectively 40 and 10 eq of Pd catalyst and 1000 eq of the mPEG$_{2k}$-PBA (Figure 17C). Application of the guanidine-based ligands was furthermore explored for the introduction of ^{18}F tags by Gao et al. [142]. With 1,1-dimethylguanidine **82** as a ligand, the first Pd-catalyzed introduction of ^{18}F, the most common radionuclide in PET imaging, was exemplified on SBL-156Pic (Figure 17D) with 250 eq of Pd-catalyst solution and 250 eq of labeled 4-fluorophenylboronic acid. Moreover, Ma and coworkers have shown that water-soluble NHCs, such as **79** (Figure 12), can be applied for protein derivatizations (Figure 17E) [135]. Pd-complexes with NHC-ligands bearing hydrophilic groups were first verified on *N*-Boc-4-iodophenylalanine under mild conditions (37 °C, 2–6 h). The feasibility of this methodology was transferred to a modified Bovine serum albumin (BSA) protein (not shown). In this study, the introduction of the halogen moiety was realized by derivatization of the lysine residues with (*p*-iodophenyl)methyl *p*-nitrophenyl carbonate. Incubation of the iodine-bearing protein *p*IPM-BSA with 100 eq of this Pd-NHC complex and 100 eq of biotin-B(OH)$_2$ yielded the desired derivatives (Figure 17E). Similarly, a lysozyme containing *p*-iodobenzoate (*p*IBZ-lyso) also underwent efficient cross-coupling with a PEG-bearing boronic acid (not shown) [135]. The utility of this catalytic system was even demonstrated to work on the cell-surface of live mammalian cells. Biotinylation of HeLa cells, pretreated with *N*-succinimidyl *p*-iodobenzoate, was achieved as detected by confocal microscopy. Interestingly, clearly, the high versatility of the Suzuki–Miyaura reaction is demonstrated by successful conversions on the abovementioned and highly challenging substrates.

6. The Suzuki–Miyaura Reaction as a Tool towards Improved Biological Activity

Most of the successful drugs can be subdivided in "small molecules" (MW < 500) and "biologicals" (MW > 5000). The latter class has emerged during the last decades due in part to an improved knowledge of recombinant protein expression and purification. These biomolecules often possess excellent biological activity and selectivity, but they are not amenable to oral bioavailability and come with a high production cost. Up until the past two decades, the large gap between these two classes had been less explored by the pharmaceutical industry. However, peptides of short and intermediate length (<50 AA) lie in the soft spot between small molecules and biologicals [45,193]. They still possess the high specificity for their biological target (which reduces the risk for off-target effects) and they can be made by synthetic organic chemistry. However, if we consider linear peptides bearing natural side chains, several inherent drawbacks arise. When comparing to typical "small molecules," disadvantages of these small peptides include poor metabolic stability, low (oral) bioavailability, and rapid clearance rates. In this regard, the Suzuki–Miyaura reaction provides a valuable bioorthogonal reaction to the medicinal chemist, as both the introduction of non-natural amino acids and a reduction of side chain flexibility can be realized via this reaction. Biaryl adducts have provided valuable SAR-information and proved to alter the selectivity, stability, and availability of the parent peptide. Selected examples to illustrate this are touched upon below.

The groups of Hagmann and Yang described the optimization of sulfonylated dipeptide inhibitors with subnanomolar inhibitory activity for integrin VLA-4 as illustrated by compound **119**. [194,195]. Highly potent analogues, for example, **120** and **121** (Figure 18A), were obtained when encompassing a biarylic C-terminal residue. Modification of the substituents on the distal ring of biphenylalanine altered the availability, half-life and clearance rate of the dipeptide ligands. Synthesis of these ligands was carried out by Suzuki–Miyaura reaction on the protected triflated tyrosine [194] or halogenated phenylalanine [195] with Pd(PPh$_3$)$_4$ in toluene and the corresponding boronic acids to yield **120** and **121**.

In analogy, dual $\alpha_4\beta_1/\alpha_4\beta_7$ integrin antagonists, for example, **122** and **124**, were designed by Sircar and Castanedo (Figure 18B) [196,197]. Sircar and coworkers prepared ligand **122** via Suzuki–Miyaura-catalyzed cross-coupling of triflated tyrosine or bromophenylalanine with Pd(PPh$_3$)$_4$ in DME [196]. Castanedo prepared the ligands via on-resin derivatization with PdCl$_2$(dppf) in NMP [197]. Exploration of the chemical space surrounding the distal ring of the biphenylalanine moiety indicated that substitution of the ortho-position had a beneficial effect on the inhibitory activity.

Short analogues of Glucagon-like peptide 1 (GLP-1) with potential antidiabetic activity were prepared by Mapelli (Figure 18C) [198]. The natural gastrointestinal peptide hormone exists predominantly as a 30 or 31 amino acid sequence with functional agonist activity at the GLP-1 G protein-coupled receptor. Endogenous GLP-1 is not suitable as a therapeutic agent due to its very short half-life time of ca. 2 min. A lead nonapeptide containing the critical residues for binding was extended with two biphenyl residues, to yield an 11-mer peptide **125**. The presence of two biphenylalanine moieties was found to be crucial for providing submicromolar agonist activity in the utilized cAMP accumulation assay. Introduction of small substituents on the distal rings of the biphenylalanines further improved the biological activity (EC_{50} decreased to 7 nM, not shown). Additionally, alterations to the initial 9-mer showed a significant improvement in activity (0.087 nM) and resulted in a nearly equipotent analogue **126**, as compared to the endogenous 30-mer, while prolonging the in vivo residence time to 20 h. Introduction of side chain-based macrocyclic constraints by lactam formation as illustrated in **127** or disulfide linkage in **128** produced agonists with stabilized secondary structural motifs (N-terminal β-turn and C-terminal helix) that differentially influenced affinity and agonist potency [199]. Both Mapelli and Hoang reported the synthesis of the unusual biaryl building blocks via attachment of 4-iodophenylalanine to the solid-support, followed by Suzuki–Miyaura reaction with the respective boronic acids and Pd(PPh$_3$)$_4$ or Pd(OAc)$_2$ and JohnPhos.

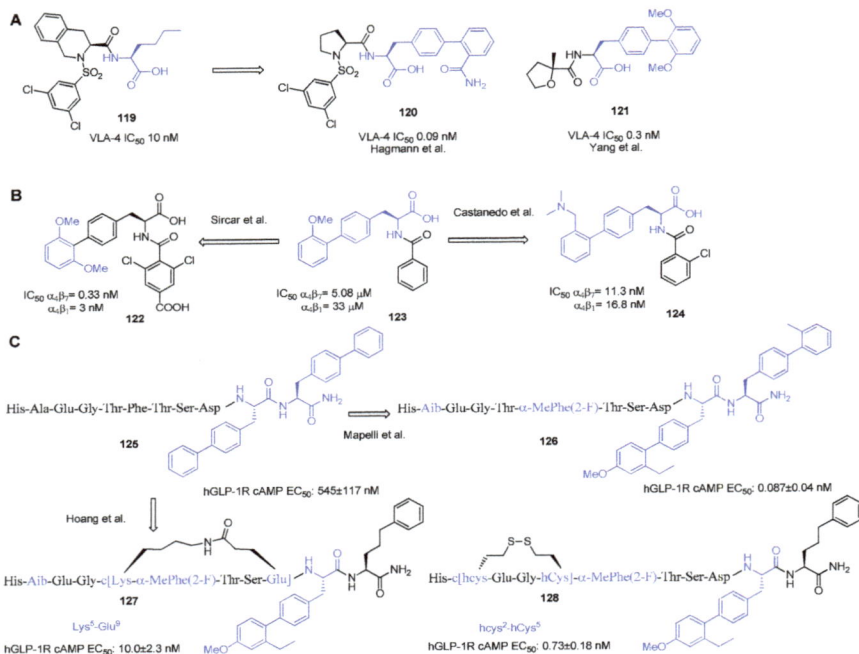

Figure 18. (**A**) Synthesis of dipeptide ligands **120** and **121** with inhibitory activity for integrin VLA-4. (**B**) Dual $\alpha_4\beta_1/\alpha_4\beta_7$ integrin antagonists **122** and **124**. (**C**) Short GLP-1 analogues **126–128** that are improved by amino acid substitution (**126**) and introduction of secondary motifs by backbone cyclization (**127** and **128**).

Cationic antimicrobial peptides (CAP) have been studied as a therapeutic alternative to the widespread bacterial resistance of common antibiotics. Typically, the sequences of these amphipathic peptides alternate between cationic residues and hydrophobic moieties. The positive influence of bulky hydrophobic groups has been described in these CAPs [200]. Evidently, biaryl containing peptides **129** and **130** were explored for their activity versus Gram-positive drug-resistance pathogenic bacteria (Figure 19A) [156,201]. In view of the development of novel antimicrobial peptides, Ng-Choi et al. incorporated 5-bromohistidine as a replacement of phenylalanine in undecapeptides at the 1- or 4-position (for example, **131**, Figure 19B) [202]. Derivatization of these building blocks was previously reported on solid-phase [159] and afforded a small library of biaryl peptides, which were tested for their biological activity in comparison with the lead peptide H–FKLFKKILKFL–NH$_2$. These arylhistidine-containing peptides displayed similar antibacterial and antifungal activity; however, a decreased hemolytic effect was observed. Hemolysis represents an important concern when developing CAPs.

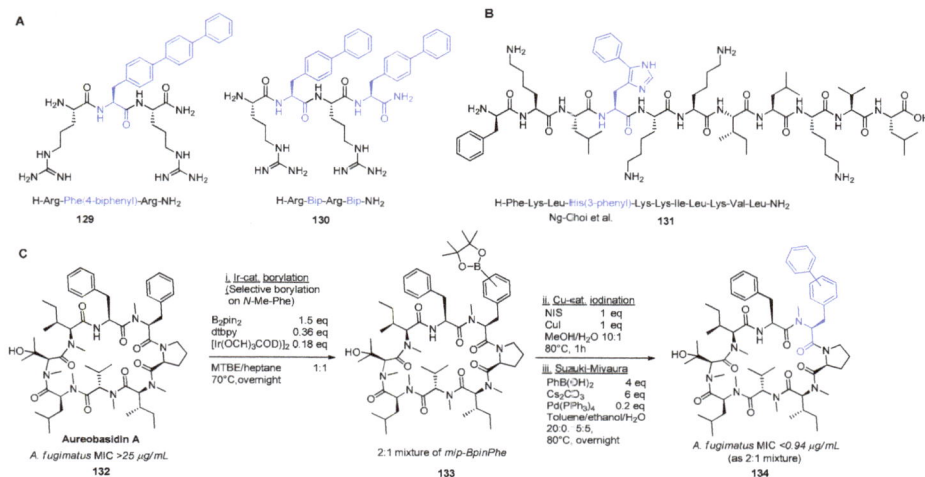

Figure 19. (**A**) Short cationic antimicrobial peptides **129** and **130** featuring, respectively, triaryl-and biarylphenylalanine (**B**) Improvement of the antimicrobial activity of undecapeptides by incorporation of 5-arylhistidine (**C**) Improvement of antifungal activity via borylation-Suzuki–Miyaura reaction on Aureobasidin A **134**.

The Suzuki–Miyaura reaction has also proven useful in the optimization of cyclic peptides. Wuts et al. explored analogues of Aurebasidin A (AbA) as antifungal agent to improve potency against a human pathogen (*Aspergillus fumigatus*) [203]. Selective borylation of the *N*-Me-Phe residue of **132** generated a 2:1 mixture of *m*- and *p*-substituted phenylalanine that was converted to iodophenylalanine and submitted to Suzuki–Miyaura catalysis (Figure 19C). Substrate **134** was found to be a potent ligand with an enhanced minimal inhibitory concentration (MIC) value, as compared to the native AbA.

7. Conclusions

In this review, we covered the initial developments of Suzuki–Miyaura cross-couplings on single amino acids in organic(/aqueous) media up to aqueous reaction conditions for the efficient derivatization of complex protein substrates. A plethora of unusual biaryl substituted amino acids was made accessible through this reaction, starting from either halogenated or borylated residues. Initial work was primarily focused on organic media with typical catalysts (Pd(PPh$_3$)$_4$, Pd(OAc)$_2$, Pd$_2$(dba)$_3$), to provide protected amino acids. Progress towards (pre)catalysts that are active, stable and more soluble in aqueous environments and subjected to mild conditions enabled the derivatization of more complex substrates. Especially, water-soluble pyrimidine ligands and hydrophilic analogues of classic ligands (PPh$_3$, NHCs, SPhos) have made transformations in aqueous media possible. This allowed the late-stage derivatization of peptides, both in solution- and solid-phase, for multiple applications, such as drug discovery and labeling efforts. Furthermore, the constrained biaryl-bridge motif was introduced in peptidic natural products and peptide substrates, and led to cyclic analogues. Transformations on protein substrates were realized as well, albeit superstoichiometrical quantities are required for derivatization of these highly complex substrates. Interestingly, recyclability of the palladium catalysts or recoverability of palladium is typically not discussed yet, due to the small scale on which the cross-couplings have been executed. Furthermore, catalyst development is still required to achieve higher substrate compatibility and a more broadly applicable catalyst system. Current limitations include the incompatibility of cross-couplings with alkylboronic acids and the limited use of chlorinated substrates. Nevertheless, it is clear that the Suzuki–Miyaura reaction is a powerful tool for the bioorthogonal transformation of amino acid substrates, peptides, and proteins. The incorporation of biaryl-bridged amino acids has been shown to significantly impact the biological activity of peptide sequences, and further development in the field of Suzuki–Miyaura catalysis will definitely have a positive impact on both synthetic methodology and pharmaceutical applications.

Acknowledgments: We gratefully acknowledge the IWT Flanders and Janssen Pharmaceutica for financial support of T.W. This work is supported by the Scientific Research Network (WOG) "Sustainable chemistry for the synthesis of fine chemicals" of the Research Foundation - Flanders (FWO).

Conflicts of Interest: The authors declare no conflict of interest.

References

1. Miyaura, N.; Suzuki, A. Palladium-Catalyzed Cross-Coupling Reactions of Organoboron Compounds. *Chem. Rev.* **2015**, *95*, 2457–2483. [CrossRef]
2. Suzuki, A. Cross-Coupling Reactions of Organoboranes: An Easy Way to Construct C-C Bonds (Nobel Lecture). *Angew. Chem. Int. Ed.* **2011**, *50*, 6722–6737. [CrossRef] [PubMed]
3. Genet, J.-P.; Savignac, M. Recent developments of palladium(0) catalyzed reactions in aqueous medium. *J. Organomet. Chem.* **1999**, *576*, 305–317. [CrossRef]
4. Lennox, J.J.; Lloyd-Jones, G.C. Preparation of Organotrifluoroborate Salts: Precipitation-Driven Equilibrium under Non-Etching Conditions. *Angew. Chem. Int. Ed.* **2012**, *51*, 9385–9388. [CrossRef] [PubMed]
5. Darses, S.; Genet, J.-P. Potassium Organotrifluoroborates: New Perspectives in Organic Synthesis. *Chem. Rev.* **2008**, *108*, 288–325. [CrossRef] [PubMed]
6. Knapp, D.M.; Gillis, E.P.; Burke, M.D. A General Solution for Unstable Boronic Acids: Slow-Release Cross-Coupling from Air-Stable MIDA Boronates. *J. Am. Chem. Soc.* **2009**, *131*, 6961–6963. [CrossRef] [PubMed]
7. Molander, G.A.; Ellis, N. Organotrifluoroborates: Protected Boronic Acids That Expand the Versatility of the Suzuki Coupling Reaction. *Acc. Chem. Res.* **2007**, *40*, 275–286. [CrossRef] [PubMed]
8. Lennox, J.J.; Lloyd-Jones, G.C. Selection of boron reagents for Suzuki–Miyaura coupling. *Chem. Soc. Rev.* **2014**, *43*, 412–443. [CrossRef] [PubMed]
9. Hall, D.G. Structure, properties, and preparation of boronic acid derivatives. overview of their reactions and applications. In *Boronic Acids: Preparation and Applications in Organic Synthesis and Medicine*; Hall, D.G., Ed.; Wiley-VCH Verlag GmbH & Co. KGaA: Weinheim, Germany, 2005; Volume 1, pp. 1–99.
10. Anderson, K.W.; Buchwald, S.L. General Catalysts for the Suzuki–Miyaura and Sonogashira Coupling Reactions of Aryl Chlorides and for the Coupling of Challenging Substrate Combinations in Water. *Angew. Chem. Int. Ed.* **2005**, *44*, 6173–6177. [CrossRef] [PubMed]
11. Polshettiwar, V.; Decottignies, A.; Len, C.; Fihri, A. Suzuki–Miyaura Cross-Coupling Reactions in Aqueous Media: Green and Sustainable Syntheses of Biaryls. *ChemSusChem* **2010**, *3*, 502–522. [CrossRef] [PubMed]
12. Shaugnessy, K.H. Cross-Coupling reactions in Aqueous Media. In *Cross-Coupling Reactions in Aqueous Media, in Palladium-Catalyzed Coupling Reactions: Practical Aspects and Future Developments*; Molnár, Á., Ed.; Wiley-VCH Verlag GmbH & Co. KGaA: Weinheim, Germany, 2013; Volume 1, pp. 235–286.
13. Rossi, R.; Bellina, F.; Lessi, M.; Manzini, C.; Marianetti, G.; Perego, L.A. Recent Applications of Phosphane-based Palladium Catalysts in Suzuki–Miyaura Reactions Involved in Total Syntheses of Natural Products. *Curr. Org. Chem.* **2015**, *19*, 1302–1409. [CrossRef]
14. Nicolaou, K.C.; Bulger, P.G.; Sarlah, D. Palladium-Catalyzed Cross-Coupling Reactions in Total Synthesis. *Angew. Chem. Int. Ed.* **2005**, *44*, 4442–4489. [CrossRef] [PubMed]
15. Ronson, T.O.; Taylor, R.J.K.; Fairlamb, I.J.S. Palladium-Catalysed macrocyclisations in the total synthesis of natural products. *Tetrahedron* **2015**, *71*, 989–1009. [CrossRef]
16. Heravi, M.M.; Hashemi, E. Recent applications of the Suzuki reaction in total synthesis. *Tetrahedron* **2012**, *68*, 9145–9478. [CrossRef]

17. Amann, N.; Wagenknecht, H.A. Preparation of pyrenyl-modified nucleosides via Suzuki–Miyaura cross-coupling reactions. *Synlett* **2002**, *5*, 687–691. [CrossRef]
18. Western, E.C.; Daft, J.R.; Johnson, E.M.; Gannett, P.M.; Shaughnessy, K.H. Efficient One-Step Suzuki Arylation of Unprotected Halonucleosides, Using Water-Soluble Palladium Catalysts. *J. Org. Chem.* **2003**, *68*, 6767–6774. [CrossRef] [PubMed]
19. Nencka, R.; Sinnaeve, D.; Karalic, I.; Martins, J.C.; Van Calenbergh, S. Synthesis of C-6-substituted uridine phosphonates through aerobic ligand-free Suzuki–Miyaura cross-coupling. *Org. Biomol. Chem.* **2010**, *8*, 5234–5246. [CrossRef] [PubMed]
20. Omumi, A.; Beach, D.G.; Baker, M.; Gabryelski, W.; Manderville, R.A. Postsynthetic Guanine Arylation of DNA by Suzuki–Miyaura Cross-Coupling. *J. Am. Chem. Soc.* **2011**, *133*, 42–50. [CrossRef] [PubMed]
21. Fresneau, N.; Hiebel, M.-A.; Agrofoglio, L.A.; Berteina-Raboin, S. Efficient Synthesis of Unprotected C-5-Aryl/Heteroaryl-2′-deoxyuridine via a Suzuki–Miyaura Reaction in Aqueous Media. *Molecules* **2012**, *17*, 14409–14417. [CrossRef] [PubMed]
22. De Ornellas, S.; Williams, T.J.; Baumann, C.G.; Fairlamb, I.J.S Catalytic C-H/C-X bond functionalisation of nucleosides, nucleotides, nucleic acids, amino acids, peptides and proteins. In *C–H and C–X Bond Functionalization*; Ribas, X., Ed.; Royal Society of Chemistry: Cambridge, UK, 2013; Volume 11, pp. 409–447.
23. Herve, G.; Sartori, G.; Enderlin, G.; Mackenzie, G.; Len, C. Palladium-catalyzed Suzuki reaction in aqueous solvents applied to unprotected nucleosides and nucleotides. *RSC Adv.* **2014**, *4*, 18558–18594. [CrossRef]
24. Carrow, B.P.; Hartwig, J.F. Distinguishing Between Pathways for Transmetalation in Suzuki–Miyaura Reactions. *J. Am. Chem. Soc.* **2011**, *133*, 2116–2119. [CrossRef] [PubMed]
25. Amatore, C.; Jutand, A.; Le Duc, G. Kinetic Data for the Transmetalation/Reductive Elimination in Palladium-Catalyzed Suzuki–Miyaura Reactions: Unexpected Triple Role of Hydroxide Ions Used as Base. *Chem. Eur. J.* **2011**, *17*, 2492–2503. [CrossRef] [PubMed]
26. Amatore, C.; Le Duc, G.; Jutand, A. Mechanism of Palladium-Catalyzed Suzuki–Miyaura Reactions: Multiple and Antagonistic Roles of Anionic "Bases" and Their Countercations. *Chem. Eur. J.* **2013**, *19*, 10082–10093. [CrossRef] [PubMed]
27. Lennox, A.J.J.; Lloyd-Jones, G.C. Organotrifluoroborate Hydrolysis: Boronic Acid Release Mechanism and an Acid-Base Paradox in Cross-Coupling. *J. Am. Chem. Soc.* **2012**, *134*, 7431–7441. [CrossRef] [PubMed]
28. Lennox, A.J.J.; Lloyd-Jones, G.C. Transmetalation in the Suzuki–Miyaura Coupling: The Fork in the Trail. *Angew. Chem. Int. Ed.* **2013**, *52*, 7362–7370. [CrossRef] [PubMed]
29. Lang, K.; Chin, J.W. Bioorthogonal Reactions for Labeling Proteins. *ACS Chem. Biol.* **2014**, *9*, 16–20. [CrossRef] [PubMed]
30. King, M.; Wagner, A. Developments in the Field of Bioorthogonal Bond Forming Reactions-Past and Present Trends. *Bioconj. Chem.* **2014**, *25*, 825–839. [CrossRef] [PubMed]
31. Spicer, C.D.; Davis, B.G. Selective chemical protein modification. *Nat. Commun.* **2014**, *5*, 4740–4753. [CrossRef] [PubMed]
32. Sletten, M.; Bertozzi, C.R. Bioorthogonal Chemistry: Fishing for Selectivity in a Sea of Functionality. *Angew. Chem. Int. Ed.* **2009**, *48*, 6974–6998. [CrossRef] [PubMed]
33. Chen, X.; Wu, Y.-W. Selective labelling of proteins. *Org. Biomol. Chem.* **2016**, *14*, 5417–5439. [CrossRef] [PubMed]
34. Dumas, A.; Spicer, C.D.; Gao, Z.; Takehana, T.; Lin, Y.A.; Yasukohchi, T.; Davis, B.G. Self-Liganded Suzuki–Miyaura Coupling for Site-Selective Protein PEGylation. *Angew. Chem. Int. Ed.* **2013**, *52*, 3916–3921. [CrossRef] [PubMed]
35. Keller, S.; Wage, T.; Hohaus, K.; Hölzer, E.E.; van Pée, K.-H. Purification and Partial Characterization of Tryptophan 7-Halogenase (PrnA) from *Pseudomonas fluorescens*. *Angew. Chem. Int. Ed.* **2000**, *39*, 2300–2302. [CrossRef]
36. Yeh, E.; Garneau, S.; Walsh, C.T. Robust in vitro activity of RebF and RebH, a two-component reductasehalogenase, generating 7-chlorotryptophan during rebeccamycin biosynthesis. *Proc. Natl. Acad. Sci. USA* **2005**, *102*, 3960–3965. [CrossRef] [PubMed]
37. Goss, R.J.M.; Newill, P.L.A. A Convenient Enzymatic Synthesis of L-halotryptophans. *Chem. Commun.* **2006**, *47*, 4924–4925. [CrossRef] [PubMed]
38. Smith, D.R.M.; Grüshow, S.; Goss, R.J.M. Scope and potential of halogenases in biosynthetic applications. *Curr. Opin. Chem. Biol.* **2013**, *17*, 276–283. [CrossRef] [PubMed]

39. Kaiser, P.F.; Churches, Q.I.; Hutton, C.A. Organoboron Reagents in the Preparation of Functionalized α-Amino Acids. *Aust. J. Chem.* **2007**, *60*, 799–810. [CrossRef]

40. Sivaev, I.B.; Bregadze, V.I. L-4-Boronophenylalanine (all around the one molecule). *ARKIVOC* **2008**, 47–61.

41. Moreira, I.S.; Fernandes, P.A.; Ramos, M.J. Hot spots—A review of the protein-protein interface determinant amino-acid residues. *Proteins: Struct. Funct. Genet.* **2007**, *68*, 803–812. [CrossRef] [PubMed]

42. Kotha, S.; Goyal, D.; Chavan, A.S. Diversity-Oriented Approaches to Unusual α-Amino Acids and Peptides: Step Economy, Atom Economy, Redox Economy, and Beyond. *J. Org. Chem.* **2014**, *78*, 12288–12313. [CrossRef] [PubMed]

43. Stevenazzi, A.; Marchini, M.; Sandrone, G.; Vergani, B.; Lattanzio, M. Amino acidic scaffolds bearing unnatural side chains: An old idea generates new and versatile tools for the life sciences. *Bioorg. Med. Chem. Lett.* **2014**, *24*, 5349–5356. [CrossRef] [PubMed]

44. Blaskovich, M.A.T. Unusual Amino Acids in Medicinal Chemistry. *J. Med. Chem.* **2016**, *59*, 10807–10836. [CrossRef] [PubMed]

45. Fosgerau, K.; Hoffmann, T. Peptide therapeutics: current status and future directions. *Drug Discov. Today* **2015**, *20*, 122–128. [CrossRef] [PubMed]

46. Adessi, C.; Soto, C. Converting a Peptide into a Drug: Strategies to Improve Stability and Bioavailability. *Curr. Med. Chem.* **2002**, *9*, 963–978. [CrossRef] [PubMed]

47. Van der Poorten, O.; Knuhtsen, A.; Pedersen, D.S.; Ballet, S.; Tourwé, D. Side Chain Cyclized Aromatic Amino Acids: Great Tools as Local Constraints in Peptide and Peptidomimetic Design. *J. Med. Chem.* **2016**, *59*, 10865–10890. [CrossRef] [PubMed]

48. Marsault, E.; Peterson, M.L. Macrocycles Are Great Cycles: Applications, Opportunities, and Challenges of Synthetic Macrocycles in Drug Discovery. *J. Med. Chem.* **2011**, *54*, 1961–2004. [CrossRef] [PubMed]

49. White, C.J.; Yudin, A.K. Contemporary strategies for peptide macrocyclization. *Nat. Chem.* **2011**, *3*, 509–524. [CrossRef] [PubMed]

50. Martí-Centelles, V.; Pandey, M.D.; Burguete, M.I.; Luis, S.V. Macrocyclization Reactions: The Importance of Conformational, Configurational, and Template-Induced Preorganization. *Chem. Rev.* **2015**, *115*, 8736–8834. [CrossRef] [PubMed]

51. Yudin, A.K. Macrocycles: Lessons from the distant past, recent developments, and future directions. *Chem. Sci.* **2015**, *6*, 30–49. [CrossRef]

52. Lau, Y.H.; de Andrade, P.; Wu, Y.; Spring, D.R. Peptide stapling techniques based on different macrocyclisation chemistries. *Chem. Soc. Rev.* **2015**, *44*, 91–102. [CrossRef] [PubMed]

53. Góngora-Benítez, M.; Tulla-Puche, J.; Albericio, F. Multifaceted Roles of Disulfide Bonds. Peptides as Therapeutics. *Chem. Rev.* **2014**, *114*, 901–926. [CrossRef] [PubMed]

54. Gleeson, E.C.; Roy Jackson, W.; Robinson, A.J. Ring-Closing metathesis in peptides. *Tetrahedron Lett.* **2016**, *57*, 4325–4333. [CrossRef]

55. Ahmad Fuaad, A.A.H.; Azmi, F.; Skwarczynski, M.; Toth, I. Peptide Conjugation via CuAAC 'Click' Chemistry. *Molecules* **2013**, *18*, 13148–13174. [CrossRef] [PubMed]

56. Dong, H.; Limberakis, C.; Liras, S.; Price, D.; James, K. Peptidic macrocyclization via palladium-catalyzed chemoselective indole C-2 arylation. *Chem. Commun.* **2012**, *48*, 11644–11646. [CrossRef] [PubMed]

57. Hopkins, B.A.; Smith, G.F.; Sciammetta, N. Synthesis of Cyclic Peptidomimetics via a Pd-Catalyzed Macroamination Reaction. *Org. Lett.* **2016**, *18*, 4072–4075. [CrossRef] [PubMed]

58. Makwana, K.M.; Mahalakshmi, R. Trp-Trp Cross-Linking: A Structure-Reactivity Relationship in the Formation and Design of Hyperstable Peptide β-Hairpin and α-Helix Scaffolds. *Org. Lett.* **2014**, *17*, 2498–2501. [CrossRef] [PubMed]

59. Mendive-Tapia, L.; Preciado, S.; Garciá, J.; Ramón, R.; Kielland, N.; Albericio, F.; Lavilla, R. New peptide architectures through C–H activation stapling between tryptophan-phenylalanine/tyrosine residues. *Nat. Comm.* **2015**, *6*, 7160. [CrossRef] [PubMed]

60. Lloyd-Williams, P.; Giralt, E. Atropisomerism, biphenyls and the Suzuki coupling: Peptide antibiotics. *Chem. Soc. Rev.* **2001**, *30*, 145–157. [CrossRef]

61. Feliu, L.; Planas, M. Cyclic Peptides Containing Biaryl and Biaryl Ether Linkages. *Int. J. Pept. Res. Ther.* **2005**, *11*, 53–97. [CrossRef]

62. Shaugnessy, K.H. Hydrophilic Ligands and Their Application in Aqueous-Phase Metal-Catalyzed Reactions. *Chem. Rev.* **2009**, *109*, 643–710. [CrossRef] [PubMed]

63. Chalker, J.M.; Wood, C.S.C.; Davis, B.G. A Convenient Catalyst for Aqueous and Protein Suzuki–Miyaura Cross-Coupling. *J. Am. Chem. Soc.* **2009**, *131*, 16346–16347. [CrossRef] [PubMed]

64. Deb Roy, A.; Goss, R.J.M.; Wagner, G.K.; Winn, M. Development of fluorescent aryltryptophans by Pd mediated cross-coupling of unprotected halotryptophans in water. *Chem. Commun.* **2008**, *39*, 4831–4833.

65. Kodama, K.; Fukuzawa, S.; Nakayama, H.; Sakamoto, K.; Kigawa, T.; Yabuki, T.; Matsuda, N.; Shirouzu, M.; Takio, K.; Yokoyama, S.; et al. Site-Specific Functionalization of Proteins by Organopalladium Reactions. *ChemBioChem* **2007**, *8*, 232–238. [CrossRef] [PubMed]

66. Jackson, R.F.W.; Wishart, N.; Wood, A.; James, K.; Wythes, M.J. Preparation of Enantiomerically Pure Protected 4-Oxo-α-amino Acids and 3-Aryl-α-amino Acids from Serine. *J. Org. Chem.* **1992**, *57*, 3397–3404. [CrossRef]

67. Jackson, R.F.W.; Moore, R.J.; Dexter, C.S. Concise Synthesis of Enantiomerically Pure Phenylalanine, Homophenylalanine, and Bishomophenylalanine Derivatives Using Organozinc Chemistry: NMR Studies of Amino Acid-Derived Organozinc Reagents. *J. Org. Chem.* **1998**, *63*, 7875–7884. [CrossRef]

68. Ross, A.J.; Lang, H.L.; Jackson, R.F.W. Much Improved Conditions for the Negishi Cross- Coupling of Iodoalanine Derived Zinc Reagents with Aryl Halides. *J. Org. Chem.* **2010**, *75*, 245–248. [CrossRef] [PubMed]

69. Ross, A.J.; Dreiocker, F.; Schäfer, M.; Oomens, J.; Meijer, A.J.H.M.; Pickup, B.T.; Jackson, R.F.W. Evidence for the Role of Tetramethylethylenediamine in Aqueous Negishi Cross-Coupling: Synthesis of Nonproteinogenic Phenylalanine Derivatives on Water. *J. Org. Chem.* **2011**, *76*, 1727–1734. [CrossRef] [PubMed]

70. Bittner, S.; Scherzer, R.; Harlev, E. The five bromotryptophans. *Amino Acids* **2007**, *33*, 19–42. [CrossRef] [PubMed]

71. Konda-Yamada, Y.; Okada, C.; Yoshida, K.; Umeda, Y.; Arima, S.; Sato, N.; Kai, T.; Takayanagi, H.; Harigaya, Y. Convenient synthesis of 7′- and 6′-bromo-D-tryptophan and their derivatives by enzymatic optical resolution using D-aminoacylase. *Tetrahedron* **2002**, *58*, 7851–7861. [CrossRef]

72. Kieffern, M.E.; Repka, L.M.; Reisman, S.E. Enantioselective Synthesis of Tryptophan Derivatives by a Tandem Friedel-Crafts Conjugate Addition/Asymmetric Protonation Reaction. *J. Am. Chem. Soc.* **2012**, *134*, 5131–5137. [CrossRef] [PubMed]

73. Blaser, G.; Sanderson, J.M.; Batsanov, A.S.; Howard, J.A.K. The facile synthesis of a series of tryptophan derivatives. *Tetrahedron Lett.* **2008**, *49*, 2795–2798. [CrossRef]

74. Winn, M.; Roy, A.D.; Grüschow, S.; Parameswaran, R.S.; Goss. R.J.M. A convenient one-step synthesis of L-aminotryptophans and improved synthesis of 5-fluorotryptophan. *Bioorg. Med. Chem. Lett.* **2008**, *18*, 4508–4510. [CrossRef] [PubMed]

75. Smith, D.R.M.; Willemse, T.; Gkotsi, D.S.; Schepens, W.; Maes, B.U.W.; Ballet, S.; Goss, R.J.M. The First One-Pot Synthesis of L-7-Iodotryptophan from 7-Iodoindole and Serine, and an Improved Synthesis of Other L-7-Halotryptophans. *Org. Lett.* **2014**, *16*, 2622–2625. [CrossRef] [PubMed]

76. Seibold, C.; Schnerr, H.; Rumpf, J.; Kunzendorf, A.; Hatscher, C.; Wage, T.; Ernyei, A.J.; Dong, C.; Naismith, J.H.; Van Pée, K.-H. A flavin-dependent tryptophan 6-halogenase and its use in modification of pyrrolnitrin biosynthesis. *Biocatal. Biotransformation* **2006**, *24*, 401–408. [CrossRef]

77. Zehner, S.; Kotzsch, A.; Bister, B.; Süssmuth, R.D.; Méndez, C.; Salas, J.A.; van Pée, K.-H. A Regioselective Tryptophan 5-Halogenase Is Involved in Pyrroindomycin Biosynthesis in *Streptomyces rugosporus* LL-42D005. *Chem. Biol.* **2005**, *12*, 445–452. [CrossRef] [PubMed]

78. Wang, L.; Brock, A.; Herberich, B.; Schultz, P.G. Expanding the Genetic Code of Escherichia coli. *Science* **2001**, *292*, 498–500. [CrossRef] [PubMed]

79. Xiao, H.; Schultz, P.W. At the Interface of Chemical and Biological Synthesis: An Expanded Genetic Code. *Cold Spring Harb. Perspect. Biol.* **2016**, *8*, 1–19. [CrossRef] [PubMed]

80. Liu, W.; Brock, A.; Chen, S.; Chen, S.; Schultz, P.G. Genetic incorporation of unnatural amino acids into proteins in mammalian cells. *Nat. Meth.* **2007**, *4*, 239–244. [CrossRef] [PubMed]

81. Liu, C.C.; Mack, A.V.; Brustad, E.M.; Mills, J.H.; Groff, D.; Smider, V.V.; Schultz, P.G. Evolution of Proteins with Genetically Encoded "Chemical Warheads". *J. Am. Chem. Soc.* **2009**, *131*, 9616–9617. [CrossRef] [PubMed]

82. Kodama, K.; Fukuzawa, S.; Nakayama, H.; Kigawa, T.; Sakamoto, K.; Yabuki, T.; Matsuda, N.; Shirouzu, M.; Takio, K.; Tachibana, K.; et al. Regioselective Carbon–Carbon Bond Formation in Proteins with Palladium Catalysis; New Protein Chemistry by Organometallic Chemistry. *ChemBioChem* **2006**, *7*, 134–139. [CrossRef] [PubMed]

83. Brustad, E.; Bushey, M.L.; Lee, J.W.; Groff, D.; Liu, W.; Schultz, P.G. A Genetically Encoded Boronate-Containing Amino Acid. *Angew. Chem. Int. Ed.* **2008**, *47*, 8220–8223. [CrossRef] [PubMed]

84. Casalnuovo, A.L.; Calabrese, J.C. Palladium-Catalyzed Alkylations in Aqueous Media. *J. Am. Chem. Soc.* **1999**, *112*, 4324–4330. [CrossRef]

85. Tilley, J.W.; Sarabu, R.; Wagner, R.; Mulkerins, K. Preparation of Carboalkoxyalkylphenylalanine Derivatives from Tyrosine. *J. Org. Chem.* **1990**, *55*, 906–910. [CrossRef]

86. Shieh, W.-C.; Carlson, J.A. A Simple Asymmetric Synthesis of 4-Arylphenylalanines via Palladium-Catalyzed Cross-Coupling Reaction of Arylboronic Acids with Tyrosine Triflate. *J. Org. Chem.* **1992**, *57*, 379–381. [CrossRef]

87. Burk, M.J.; Lee, J.R.; Martinez, J.P. A Versatile Tandem Catalysis Procedure for the Preparation of Novel Amino Acids and Peptides *J. Am. Chem. Soc.* **1994**, *116*, 10847–10848. [CrossRef]

88. Kotha, S.; Lahiri, K. A New Approach for Modification of Phenylalanine Peptides by Suzuki–Miyaura Coupling Reaction. *Bioorg. Med. Chem. Lett.* **2001**, *11*, 2887–2890. [CrossRef]

89. Wang, W.; Xiong, C.; Zhang, J.; Hruby, V.J. Practical, assymmetric synthesis of aromatic-substituted bulky and hydrophobic tryptophan and phenylalanine derivatives. *Tetrahedron* **2002**, *58*, 3101–3110. [CrossRef]

90. Espuña, G.; Arsequell, G.; Valencia, G.; Barluenga, J.; Alvarez-Gutiérrez, J.M.; Ballesteros, A.; González, J.M. Regioselective Postsynthetic Modification of Phenylalanine Side Chains of Peptides Leading to Uncommon *ortho*-Iodinated Analogues. *Angew. Chem. Int. Ed.* **2004**, *43*, 325–329. [CrossRef] [PubMed]

91. Limbach, M.; Löweneck, M.; Schreiber, J.V.; Frackenpohl, J.; Seebach, D. Synthesis of β^3-Homophenylalanine-Derived Amino Acids and Peptides by Suzuki Coupling in Solution and on Solid Support. *Helv. Chim. Acta* **2006**, *89*, 1427–1441. [CrossRef]

92. Knör, S.; Laufer, B.; Kessler, H. Efficient Enantioselective Synthesis of Condensed and Aromatic-Ring-Substituted Tyrosine Derivatives. *J. Org. Chem.* **2006**, *71*, 5625–5630. [CrossRef] [PubMed]

93. Cerezo, V.; Afonso, A.; Planas, M.; Feliu, L. Synthesis of 5-arylhistidines via a Suzuki–Miyaura cross-coupling. *Tetrahedron* **2007**, *63*, 10445–10453. [CrossRef]

94. Prieto, M.; Mayor, S.; Rodríguez, K.; Lloyd-Williams, P.; Giralt, E. Racemization in Suzuki Couplings: A Quantitative Study Using 4-Hydroxyphenylglycine and Tyrosine Derivatives as Probe Molecules. *J. Org. Chem.* **2007**, *72*, 1047–1050. [CrossRef] [PubMed]

95. Dumas, A.; Peramo, A.; Desmaële, D.; Couvreur, P. PLGA-PEG-supported Pd Nanoparticles as Efficient Catalysts for Suzuki–Miyaura Coupling Reactions in Water. *Chimia* **2016**, *70*, 252–257. [CrossRef] [PubMed]

96. Deb Roy, A.; Grüshow, S.; Cairns, N.; Goss, R.J.M. Gene Expression Enabling Synthetic Diversification of Natural Products: Chemogenetic Generation of Pacidamycin Analogs. *J. Am. Chem. Soc.* **2010**, *132*, 12243–12245. [PubMed]

97. Willemse, T.; Van Imp, K.; Vlijmen, H.W.T.; Schepens, W.; Goss, R.J.M.; Maes, B.U.W.; Ballet, S. Suzuki–Miyaura Diversification of Amino Acids and Dipeptides in Aqueous Media. *ChemCatChem* **2015**, *7*, 2055–2070. [CrossRef]

98. Frese, M.; Schnepel, C.; Minges, H.; Voß, H.; Feiner, R.; Sewald, N. Modular Combination of Enzymatic Halogenation of Tryptophan with Suzuki–Miyaura Cross-Coupling Reactions. *ChemCatChem* **2016**, *8*, 1799–1803. [CrossRef]

99. Maity, J.; Honcharenko, D.; Strömberg, R. Synthesis of fluorescent D-amino acids with 4-acetamidobiphenyl and 4-*N*,*N*-dimethylamino-1,8-naphthalimido containing side chains. *Tetrahedron Lett.* **2015**, *56*, 4780–4783. [CrossRef]

100. Qiao, J.X.; Fraunhoffer, K.J.; Hsiao, Y.; Li, Y.-X.; Wang, C.; Wang, T.C.; Poss, M.A. Synthesis of Fmoc-Protected Arylphenylalanines (Bip Derivatives) via Nonaqueous Suzuki–Miyaura Cross-Coupling Reactions. *J. Org. Chem.* **2016**, *81*, 9499–9506. [CrossRef] [PubMed]

101. Jana, R.; Pathak, T.P.; Sigman, M.S. Advances in Transition Metal (Pd,Ni,Fe)-Catalyzed Cross-Coupling Reactions Using Alkyl-organometallics as Reaction Partners. *Chem. Rev.* **2011**, *111*, 1417–1492. [CrossRef] [PubMed]

102. Yang, W.; Gao, X.; Wang, B. Boronic Acid Compounds as Potential Pharmaceutical Agents. *Med. Res. Rev.* **2003**, *23*, 346–368. [CrossRef] [PubMed]

103. Satoh, Y.; Gude, C.; Chan, K.; Firooznia, F. Synthesis of 4-Substitured Phenylalanine Derivatives by Cross-Coupling Reaction of *p*-Boronophenylalanine. *Tetrahedron. Lett.* **1997**, *38*, 7645–7648. [CrossRef]

104. Firooznia, F.; Gude, C.; Chan, K.; Marcopulos, N.; Satoh, Y. Enantioselective Synthesis of 4-Substituted Phenylalanines by Cross-Coupling Reactions. *Tetrahedron. Lett.* **1999**, *40*, 213–216. [CrossRef]

105. Iimura, S.; Wu, W. Palladium-catalyzed borylation of L-tyrosine triflate derivative with pinacolborane: practical route to 4-borono-L-phenylalanine (L-BPA) derivatives. *Tetrahedron Lett.* **2010**, *51*, 1353–1355. [CrossRef]

106. Jung, M.E.; Lazarova, T.I. New Efficient Method for the Total Synthesis of (S,S)-Isodityrosine from Natural Amino Acids. *J. Org. Chem.* **1999**, *64*, 2976–2977. [CrossRef] [PubMed]

107. Malan, C.; Morin, C. A Concise Preparation of 4-Borono-L-phenylalanine (L-BPA) from L-Phenylalanine. *J. Org. Chem.* **1998**, *63*, 8019–8020. [CrossRef]

108. Lépine, R.; Zhu, J. Microwave-Assisted Intramolecular Suzuki–Miyaura Reaction to Macrocycle, a Concise Asymmetric Total Synthesis of Biphenomycin B. *Org. Lett.* **2005**, *7*, 2981–2984. [CrossRef] [PubMed]

109. Wienhold, F.; Claes, D.; Graczyk, K.; Maison, W. Synthesis of Functionalized Benzoboroxoles for the Construction of Boronolectins. *Synthesis* **2011**, *24*, 4059–4067.

110. Kallepalli, V.A.; Shi, F.; Paul, S.; Onyeozili, E.N.; Maleczka, R.E.; Smith, M.R. Boc Groups as Protectors and Directors for Ir-Catalyzed C–H Borylation of Heterocycles. *J. Org. Chem.* **2009**, *74*, 9199–9201. [CrossRef] [PubMed]

111. Meyer, F.-M.; Liras, S.; Guzman-Perez, A.; Perrault, C.; Bian, J.; James, K. Functionalization of Aromatic Amino Acids via Direct C-H Activation: Generation of Versatile Building Blocks for Accessing Novel Peptide Space. *Org. Lett.* **2010**, *12*, 3870–3873. [CrossRef] [PubMed]

112. Audi, H.; Rémond, E.; Eymin, M.-J.; Tessier, A.; Malacea-Kabbara, R.; Jugé, S. Modular Hemisyntheses of Boronato- and Trifluoroborato-Substituted L-NHBoc Amino Acid and Peptide Derivatives. *Eur. J. Org. Chem.* **2013**, *2013*, 7960–7972. [CrossRef]

113. Bartolucci, S.; Bartoccini, F.; Righi, M.; Piersanti, G. Direct, Regioselective, and Chemoselective Preparation of Novel Boronated Tryptophans by Friedel-Crafts Alkylation. *Org. Lett.* **2011**, *14*, 600–603. [CrossRef] [PubMed]

114. Colgin, N.; Flinn, T.; Cobb, S.L. Synthesis and properties of MIDA boronate containing aromatic amino acids: New peptide building blocks. *Org. Biomol. Chem.* **2011**, *9*, 1864–1870. [CrossRef] [PubMed]

115. Gong, Y.; He, W. Direct Synthesis of Unprotected 4-Aryl Phenylalanines via the Suzuki Reaction under Microwave Irradiation. *Org. Lett.* **2002**, *4*, 3803–3805. [CrossRef] [PubMed]

116. Čapek, P.; Hocek, M. Efficient One-Step Synthesis of Optically Pure (Adenin-8-yl)phenylalanine Nucleosides. *Synlett* **2005**, *19*, 3005–3007.

117. Čapek, P.; Pohl, R.; Hocek, M. Cross-Coupling reactions of unprotected halopurine bases, nucleosides, nucleotides and nucleoside triphosphates with 4-boronophenylalanine in water. Synthesis of (purin-8-yl)- and (purin-6-yl)phenylalanines. *Org. Biomol. Chem.* **2006**, *4*, 2278–2284. [CrossRef] [PubMed]

118. Bartocinni, S.; Bartolucci, M.M.; Piersanti, G. A simple, modular synthesis of C4-substituted tryptophan derivatives. *Org. Biomol. Chem.* **2016**, *14*, 10095–10100. [CrossRef] [PubMed]

119. Loach, R.P.; Fenton, O.S.; Amaike, K.; Siegel, D.S.; Ozkal, E.; Movassaghi, M. C7-Derivatization of C3-Alkylindoles Including Tryptophans and Tryptamines. *J. Org. Chem.* **2014**, *79*, 11254–11263. [CrossRef] [PubMed]

120. Afonso, A.; Feliu, L.; Planas, M. Solid-Phase synthesis of biaryl cyclic peptides by borylation and microwave assisted intramolecular Suzuki–Miyaura reaction. *Tetrahedron* **2011**, *67*, 2238–2245. [CrossRef]

121. Campbell, A.D.; Raynham, T.M.; Taylor, R.J.K. The Synthesis of Novel Amino Acids v/a Hydroboration-Suzuki Cross Coupling. *Tetrahedron Lett.* **1999**, *40*, 5263–5266. [CrossRef]

122. Sabat, M.; Johnson, C.R. Synthesis of Unnatural Amino Acids via Suzuki Cross-Coupling of Enantiopure Vinyloxazolidine Derivatives. *Org. Lett.* **2000**, *2*, 1089–1092. [CrossRef] [PubMed]

123. Collier, P.N.; Campbell, A.D.; Patel, I.; Taylor, R.J.K. The direct synthesis of novel enantiomerically pure α-amino acids in protected form via Suzuki cross-coupling. *Tetrahedron Lett.* **2000**, *41*, 7115–7119. [CrossRef]

124. Rodríguez, A.; Miller, D.D.; Jackson, R.F.W. Combined application of organozinc chemistry and one-pot hydroboration-Suzuki coupling to the synthesis of amino acids. *Org. Biomol. Chem.* **2003**, *1*, 973–977. [CrossRef] [PubMed]

125. Reddy, V.J.; Chandra, J.S.; Reddy, M.V.R. Concise synthesis of ω-borono-α-amino acids. *Org. Biomol. Chem.* **2007**, *5*, 889–891. [CrossRef] [PubMed]

126. Harvey, J.E.; Kenworthy, M.N.; Taylor, R.J.K. Synthesis of non-proteinogenic phenylalanine analogues by Suzuki cross-coupling of a serine-derived alkyl boronic acid. *Tetrahedron Lett.* **2004**, *45*, 2467–2471. [CrossRef]

127. Zhang, Z.Q.; Yang, C.T.; Liang, L.J.; Xiao, B.; Lu, X.; Liu, J.-H.; Sun, Y.-Y.; Marder, T.B.; Fu, Y. Copper-Catalyzed/Promoted Cross-coupling of gem-Diborylalkanes with Nonactivated Primary Alkyl Halides: An Alternative Route to Alkylboronic Esters. *Org. Lett.* **2014**, *16*, 6342–6345. [CrossRef] [PubMed]

128. He, J.; Jiang, H.; Takise, R.; Zhu, R.-Y.; Chen, G.; Dai, H.-X.; Murali Dhar, T.G.; Shi, J.; Zhang, H.; Cheng, P.T.W.; et al. Ligand-Promoted Borylation of C(sp3)-H Bonds with Palladium(II) Catalysts. *Angew. Chem. Int. Ed.* **2016**, *55*, 785–789. [CrossRef] [PubMed]

129. Yoburn, J.C.; Van Vranken, D.L. Synthesis of Dityrosine Cross-Linked Peptide Dimers Using the Miyaura-Suzuki Reaction. *Org. Lett.* **2003**, *5*, 2817–2820. [CrossRef] [PubMed]

130. Hutton, C.A.; Skaff, O. A convenient preparation of dityrosine via Miyaura borylation-Suzuki coupling of iodotyrosine derivatives. *Tetrahedron Lett.* **2003**, *44*, 4895–4898. [CrossRef]

131. Kotha, S.; Shah, V.R.; Halder, S.; Vinodkumar, R.; Lahiri, K. Synthesis of bis-armed amino acid derivatives via the alkylation of ethyl isocyanoacetate and the Suzuki–Miyaura cross-coupling reaction. *Amino Acids* **2007**, *32*, 387–394. [CrossRef] [PubMed]

132. Prieto, M.; Mayor, S.; Lloyd-Williams, P.; Giralt, E. Use of the SPhos Ligand to Suppress Racemization in Arylpinacolboronate Ester Suzuki Couplings Involving r-Amino Acids. Synthesis of Biaryl Derivatives of 4-Hydroxyphenylglycine, Tyrosine, and Tryptophan. *J. Org. Chem.* **2009**, *74*, 9202–9205. [CrossRef] [PubMed]

133. Cho, J.H.; Prickett, C.D.; Shaughnessy, K.H. Efficient Sonogashira Coupling of Unprotected Halonucleosides in Aqueous Solvents Using Water-Soluble Palladium Catalysts. *Eur. J. Org. Chem.* **2010**, *2010*, 3678–3683. [CrossRef]

134. Roy, S.; Plenio, H. Sulfonated *N*-Heterocyclic Carbenes for Pd-Catalyzed Sonogashira and Suzuki–Miyaura Coupling in Aqueous Solvents. *Adv. Synth. Catal.* **2010**, *352*, 1014–1022. [CrossRef]

135. Ma, X.; Wang, H.; Chen, W. N-Heterocyclic Carbene-Stabilized Palladium Complexes as Organometallic Catalysts for Bioorthogonal Cross-Coupling Reactions. *J. Org. Chem.* **2014**, *79*, 8652–8658. [CrossRef] [PubMed]

136. Fleckenstein, C.; Roy, S.; Leuthäußer, S.; Plenio, H. Sulfonated N-heterocyclic carbenes for Suzuki coupling in water. *Chem. Commun.* **2007**, *27*, 2870–2872. [CrossRef] [PubMed]

137. Godoy, F.; Segarra, C.; Poyatos, M.; Peris, E. Palladium Catalysts with Sulfonate-Functionalized-NHC Ligands for Suzuki–Miyaura Cross-Coupling Reactions in Water. *Organometallics* **2011**, *30*, 684–688. [CrossRef]

138. Chatterjee, A.; Ward, T.R. Recent Advances in the Palladium Catalyzed Suzuki–Miyaura Cross-Coupling Reaction in Water. *Catal. Lett.* **2016**, *146*, 820–840. [CrossRef]

139. Lipshutz, B.H.; Ghorai, S. "Designer"-Surfactant-Enabled Cross-Couplings in Water at Room Temperature. *Aldrichim. Acta* **2012**, *45*, 3–16.

140. Li, J.-H.; Zhang, X.-D.; Xie, Y.-X. Efficient Pd(OAc)$_2$/pyrimidine catalytic system for Suzuki–Miyaura Cross-Coupling reaction. *Synlett* **2005**, *12*, 1897–1900. [CrossRef]

141. Li, S.; Lin, Y.; Cao, J.; Zhang, S. Guanidine/Pd(OAc)$_2$-Catalyzed Room Temperature Suzuki Cross-Coupling Reaction in Aqueous Media under Aerobic Conditions. *J. Org. Chem.* **2007**, *72*, 4067–4072. [CrossRef] [PubMed]

142. Gao, Z.; Gouverneur, V.; Davis, B.G. Enhanced Aqueous Suzuki–Miyaura Coupling Allows Site-Specific Polypeptide ^{18}F-Labeling. *J. Am. Chem. Soc.* **2013**, *135*, 13612–13615. [CrossRef] [PubMed]

143. Vilaró, M.; Arsequell, G.; Valencia, G.; Ballesteros, A.; Barluenga, J. Arylation of Phe and Tyr Side Chains of Unprotected Peptides by a Suzuki–Miyaura Reaction in Water. *Org. Lett.* **2008**, *10*, 3243–3245. [CrossRef] [PubMed]

144. Lin, Y.A.; Chalker, J.M.; Davis, B.G. Olefin Metathesis for Site-Selective Protein Modification. *ChemBioChem* **2009**, *10*, 959–969. [CrossRef] [PubMed]

145. Ai, H.-W.; Shen, W.; Brustad, E.; Schultz, P.G. Genetically Encoded Alkenes in Yeast. *Angew. Chem. Int. Ed.* **2010**, *49*, 935–937. [CrossRef] [PubMed]

146. Ojida, A.; Tsutsumi, H.; Kasagi, N.; Hamachi, I. Suzuki coupling for protein modification. *Tetrahedron Lett.* **2005**, *46*, 3301–3305. [CrossRef]

147. Lorsbach, B.A.; Kurth, M.J. Carbon–Carbon Bond Forming Solid-Phase Reactions. *Chem. Rev.* **1999**, *99*, 1549–1581. [CrossRef] [PubMed]

148. Colombel, V.; Presset, M.; Oehlrich, D.; Rombouts, F.; Molander, G.A. Synthesis and Reactivity of Solid-Supported Organotrifluoroborates in Suzuki Cross-Coupling. *Org. Lett.* **2012**, *14*, 1680–1683. [CrossRef] [PubMed]
149. Yokum, T.S.; Barany, G. Strategy in Solid-Phase Peptide Syntesis. In *Solid-Phase Synthesis: A Practical Guide*; Kates, S.A., Albericio, F., Eds.; Marcel Dekker, Inc.: New York, NY, USA, 2000; Volume 1, pp. 79–103.
150. Adams, J.H.; Cook, R.M.; Hudson, D.; Jammalamadaka, V.; Lyttle, M.H.; Songster, M.F. A Reinvestigation of the Preparation, Properties, and Applications of Aminomethyl and 4-Methylbenzhydrylamine Polystyrene Resins. *J. Org. Chem.* **1998**, *63*, 3706–3716. [CrossRef]
151. García-Martín, F.; Quintanar-Audelo, M.; García-Ramos, Y.; Cruz, L.J.; Gravel, C.; Furic, R.; Côté, S.; Tulla-Puche, J.; Albericio, F. ChemMatrix, a Poly(ethylene glycol)-Based Support for the Solid-Phase Synthesis of Complex Peptides. *J. Comb. Chem.* **2006**, *8*, 213–220. [CrossRef] [PubMed]
152. Colombo, A.; Fernàndez, J.-C.; de la Figuera, N.; Fernàndez-Forner, D.; Forns, P.; Albericio, F. Solid-Phase Preparation of a Library Based on a Phenylalanine Scaffold. *QSAR Comb. Sci.* **2005**, *24*, 913–922. [CrossRef]
153. Nielsen, T.E.; Le Quement, S.; Meldal, M. Solid-Phase synthesis of biarylalanines via Suzuki cross-coupling and intramolecular *N*-acyliminium Pictet-Spengler reactions. *Tetrahedron Lett.* **2005**, *46*, 7959–7962. [CrossRef]
154. Le Quement, S.; Nielsen, T.E.; Meldal, M. Solid-Phase Synthesis of Aryl-Substituted Thienoindolizines: Sequential Pictet-Spengler, Bromination and Suzuki Cross-Coupling Reactions of Thiophenes. *J. Comb. Chem.* **2008**, *10*, 447–455. [CrossRef] [PubMed]
155. Le Quement, S.; Ishoey, M.; Petersen, M.T.; Thastrup, J.; Hagel, G.; Nielsen, T.E. Solid-Phase Synthesis of Smac Peptidomimetics Incorporating Triazoloprolines and Biarylalanines. *ACS Comb. Sci.* **2011**, *13*, 667–675. [CrossRef] [PubMed]
156. Haug, D.E.; Stensen, W.; Svendsen, J.S. Application of the Suzuki–Miyaura cross-coupling to increase antimicrobial potency generates promising novel antibacterials. *Bioorg. Med. Chem. Lett.* **2007**, *17*, 2361–2364. [CrossRef] [PubMed]
157. Doan, N.-D.; Bourgault, S.; Létourneau, M.; Fournier, A. Effectiveness of the Suzuki–Miyaura Cross-Coupling Reaction for Solid-Phase Peptide Modification. *J. Comb. Chem.* **2008**, *10*, 44–51. [CrossRef] [PubMed]
158. Afonso, A.; Rosés, C.; Planas, M.; Feliu, L. Biaryl Peptides from 4-Iodophenylalanine by Solid-Phase Borylation and Suzuki–Miyaura Cross-Coupling. *Eur. J. Org. Chem.* **2010**, *2010*, 1461–1468. [CrossRef]
159. Cerezo, V.; Amblard, M.; Martinez, J.; Verdié, P.; Planas, M.; Feliu, L. Solid-phase synthesis of 5-arylhistidines via a microwave-assisted Suzuki–Miyaura cross-coupling. *Tetrahedron* **2008**, *64*, 10538–10545. [CrossRef]
160. Elder, A.M.; Rich, D.H. Two Syntheses of the 16- and 17-Membered DEF Ring Systems of Chloropeptin and Complestatin. *Org. Lett.* **1999**, *1*, 1443–1446. [CrossRef] [PubMed]
161. Carbonelle, A.-C.; Zhu, J. A Novel Synthesis of Biaryl-Containing Macrocycles by a Domino Miyaura Arylboronate Formation: Intramolecular Suzuki Reaction. *Org. Lett.* **2000**, *2*, 3477–3480. [CrossRef]
162. Boisnard, S.; Carbonelle, A.-C.; Zhu, J. Studies on the Total Synthesis of RP 66453: Synthesis of Fully Functionalized 15-Membered Biaryl-Containing Macrocycle. *Org. Lett.* **2001**, *3*, 2061–2064. [CrossRef] [PubMed]
163. Waldmann, H.; He, Y.-P.; Tan, H.; Arve, L.; Arndt, H.-D. Flexible total synthesis of biphenomycin B. *Chem. Commun.* **2008**, *43*, 5562–5564. [CrossRef] [PubMed]
164. Lin, S.; Danishefsky, S.J. The Total Synthesis of Proteasome Inhibitors TMC-95A and TMC-95B: Discovery of a New Method to Generate cis-Propenyl Amides. *Angew. Chem. Int. Ed.* **2002**, *41*, 512–515. [CrossRef]
165. Berthelot, A.; Piguel, S.; Le Dour, G.; Vidal, J. Synthesis of Macrocyclic Peptide Analogues of Proteasome Inhibitor TMC-95A. *J. Org. Chem.* **2003**, *68*, 9835–9838. [CrossRef] [PubMed]
166. Inoue, M.; Sakazaki, H.; Furuyama, H.; Hirama, M. Total Synthesis of TMC-95A. *Angew. Chem. Int. Ed.* **2003**, *42*, 2654–2657. [CrossRef] [PubMed]
167. Kaiser, M.; Siciliano, C.; Assfalg-Machleidt, I.; Groll, M.; Milbradt, A.; Moroder, L. Synthesis of a TMC-95A Ketomethylene Analogue by Cyclization via Intramolecular Suzuki Coupling. *Org. Lett.* **2003**, *5*, 3435–3437. [CrossRef] [PubMed]
168. Coste, A.; Bayle, A.; Marrot, J.; Evano, G. A Convergent Synthesis of the Fully Elaborated Macrocyclic Core of TMC-95A. *Org. Lett.* **2014**, *16*, 1306–1309. [CrossRef] [PubMed]
169. Dufour, J.; Neuville, L.; Zhu, J. Intramolecular Suzuki–Miyaura Reaction for the Total Synthesis of Signal Peptidase Inhibitors, Arylomycins A2 and B2. *Chem. Eur. J.* **2010**, *16*, 10523–10534. [CrossRef] [PubMed]

170. Jia, X.; Bois-Choussy, M.; Zhu, J. Synthesis of Diastereomers of Complestatin and Chloropeptin I: Substrate-Dependent Atropstereoselectivity of the Intramolecular Suzuki–Miyaura Reaction. *Angew. Chem. Int. Ed.* **2008**, *47*, 4167–4172. [CrossRef] [PubMed]

171. Wang, Z.; Bois-Choussy, M.; Jia, Y.; Zhu, J. Total Synthesis of Complestatin (Chloropeptin II). *Angew. Chem. Int. Ed.* **2010**, *49*, 2018–2022. [CrossRef] [PubMed]

172. Moschitto, M.J.; Lewis, C.A. Synthesis of the Rubiyunnanin B Core Aglycon. *Eur. J. Org. Chem.* **2016**, *2016*, 4773–4777. [CrossRef]

173. Afonso, A.; Cussó, O.; Feliu, L.; Planas, M. Solid-Phase Synthesis of Biaryl Cyclic Peptides Containing a 3-Aryltyrosine. *Eur. J. Org. Chem.* **2012**, *2012*, 6204–6211. [CrossRef]

174. Meyer, F.-M.; Collins, J.C.; Borin, B.; Bradow, J.; Liras, S.; Limberakis, C.; Mathiowetz, A.M.; Philippe, L.; Price, D.; Song, K.; et al. Biaryl-Bridged Macrocyclic Peptides: Conformational Constraint via Carbogenic Fusion of Natural Amino Acid Side Chains. *J. Org. Chem.* **2012**, *77*, 3099–3114. [CrossRef] [PubMed]

175. Antos, J.M.; Francis, M.B. Transition metal catalyzed methods for site-selective protein modification. *Curr. Opin. Chem. Biol.* **2006**, *10*, 253–262. [CrossRef] [PubMed]

176. Chankeshwara, S.V.; Indrigo, E.; Bradley, M. Palladium-mediated chemistry in living cells. *Curr. Opin. Chem. Biol.* **2014**, *21*, 128–135. [CrossRef] [PubMed]

177. Sasmal, P.K.; Streu, C.N.; Meggers, E. Metal Complex Catalysis in Living Biological Systems. *Chem. Commun.* **2013**, *49*, 1581–1587. [CrossRef] [PubMed]

178. Takaoka, Y.; Ojida, A.; Hamachi, I. Protein Organic Chemistry and Applications for Labeling and Engineering in Live-Cell Systems. *Angew. Chem. Int. Ed.* **2013**, *52*, 4088–4106. [CrossRef] [PubMed]

179. Garrett, C.E.; Prasad, K. The Art of Meeting Palladium Specifications in Active Pharmaceutical Ingredients Produced by Pd-Catalyzed Reactions. *Adv. Synth. Catal.* **2004**, *346*, 889–900. [CrossRef]

180. Spicer, C.D.; Davis, B.G. Palladium-Mediated site-selective Suzuki–Miyaura protein modification at genetically encoded aryl halides. *Chem. Commun.* **2011**, *47*, 1698–1700. [CrossRef] [PubMed]

181. Magano, J.; Dunetz, J.R. Large-Scale Applications of Transition Metal-Catalyzed Couplings for the Synthesis of Pharmaceuticals. *Chem. Rev.* **2011**, *111*, 2177–2250. [CrossRef] [PubMed]

182. Molnár, Á. Efficient, Selective, and Recyclable Palladium Catalysts in Carbon–Carbon Coupling Reactions. *Chem. Rev.* **2011**, *111*, 2251–2320. [CrossRef] [PubMed]

183. Lamblin, M.; Nasser-Hardy, L.; Hierso, J.-C.; Fouquet, E.; Felpin, F.-X. Recyclable Heterogeneous Palladium Catalysts in Pure Water: Sustainable Developments in Suzuki, Heck, Sonogashira and Tsuji-Trost Reactions. *Adv. Synth. Catal.* **2010**, *352*, 33–79. [CrossRef]

184. Zhang, Q.; Su, H.; Luo, J.; Wei, Y. Recyclable palladium(II) imino-pyridine complex immobilized on mesoporous silica as a highly active and recoverable catalyst for Suzuki–Miyaura coupling reactions in aqueous medium. *Tetrahedron* **2013**, *69*, 447–454. [CrossRef]

185. Shokouhimehr, M.; Lee, J.E.; Han, S.I.; Hyeon, T. Magnetically recyclable hollow nanocomposite catalysts for heterogeneous reduction of nitroarenes and Suzuki reactions. *Chem. Commun.* **2013**, *49*, 4779–4781. [CrossRef] [PubMed]

186. Shokouhimehr, M.; Kim, T.; Jun, S.W.; Shin, K.; Jang, Y.; Kim, B.H.; Kim, J.; Hyeon, T. Magnetically separable carbon nanocomposite catalysts for efficient nitroarene reduction and Suzuki reactions. *Appl. Catal. A* **2014**, *476*, 133–139. [CrossRef]

187. Choi, K.-H.; Shokouhimehr, M.; Sung, Y.-E. Heterogeneous Suzuki Cross-Coupling Reaction Catalyzed by Magnetically Recyclable Nanocatalyst. *Bull. Korean Chem. Soc.* **2013**, *34*, 1477–1480. [CrossRef]

188. Karimi, B.; Akhavan, P.F. Main-Chain NHC-palladium polymer as a recyclable self-supported catalyst in the Suzuki–Miyaura coupling of aryl chlorides in water. *Chem. Commun.* **2009**, *25*, 3750–3752. [CrossRef] [PubMed]

189. Lombardo, M.; Chiarucci, M.; Trombini, C. A recyclable triethylammonium ion-tagged diphenylphosphine palladium complex for the Suzuki–Miyaura reaction in ionic liquids. *Green Chem.* **2009**, *11*, 574–579. [CrossRef]

190. Edwards, G.A.; Trafford, M.A.; Hamilton, A.E.; Buxton, A.M.; Bardeaux, M.C.; Chalker, J.M. Melamine and Melamine-Formaldehyde Polymers as Ligands for Palladium and Application to Suzuki–Miyaura Cross-Coupling Reactions in Sustainable Solvents. *J. Org. Chem.* **2014**, *79*, 2094–2104. [CrossRef] [PubMed]

191. Spicer, C.D.; Triemer, T.; Davis, B.G. Palladium-Mediated Cell-Surface Labeling. *J. Am. Chem.Soc.* **2012**, *134*, 800–803. [CrossRef] [PubMed]

192. Alconcel, S.N.S.; Baas, A.S.; Maynard, H.D. FDA-Approved poly(ethylene glycol)–protein conjugate drugs. *Polym. Chem.* **2011**, *2*, 1442–1448. [CrossRef]

193. Craik, D.J.; Fairlie, D.P.; Liras, S.; Price, D. The future of peptide-based drugs. *Chem. Biol. Drug. Des.* **2013**, *81*, 136–147. [CrossRef] [PubMed]

194. Hagmann, W.K.; Durette, P.L.; Lanza, T.; Kevin, N.J.; de Laszlo, S.E.; Kopka, I.E.; Young, D.; Magriotis, P.A.; Li, B.; Lin, L.S.; et al. The Discovery of Sulfonylated Dipeptides as Potent VLA-4 Antagonists. *Bioorg. Med. Chem. Lett.* **2001**, *11*, 2709–2713. [CrossRef]

195. Yang, G.X.; Chang, L.L.; Truong, Q.; Doherty, G.A.; Magriotis, P.A.; de Laszlo, S.E.; Li, B.; MacCoss, M.; Kidambi, U.; Egger, L.A.; et al. *N*-Tetrahydrofuroyl-(L)-phenylalanine derivatives as potent VLA-4 antagonists. *Bioorg. Med. Chem. Lett.* **2002**, *12*, 1497–1500. [CrossRef]

196. Sircar, I.; Gudmundsson, K.S.; Martin, R.; Liang, J.; Nomura, S.; Jayakumar, H.; Teegarden, B.R.; Nowlin, D.M.; Cardarelli, P.M.; Mah, J.R.; et al. Synthesis and SAR of *N*-Benzcyl-L-Biphenylalanine Derivatives: Discovery of TR-14035, A Dual α4β1/α4β7 Integrin Antagonisty. *Bioorg. Med. Chem.* **2002**, *10*, 2051–2066. [CrossRef]

197. Castanedo, G.M.; Sailes, F.C.; Dubree, N.J.P.; Nicholas, J.B.; Caris, L.; Clark, K.; Keating, S.M.; Beresini, M.H.; Chiu, H.; Fong, S.; et al. Solid-Phase Synthesis of Dual α4β1/α4β7 Integrin Antagonists: Two Scaffolds with Overlapping Pharmacophores. *Bioorg. Med. Chem. Lett.* **2002**, *12*, 2913–2917. [CrossRef]

198. Mapelli, C.; Natarajan, S.I.; Meyer, J.-P.; Bastos, M.M.; Bernatowicz, M.S.; Lee, V.G.; Pluscec, J.; Riexinger, D.J.; Sieber-McMaster, E.S.; Constantine, K.L.; et al. Eleven Amino Acid Glucagon-like Peptide-1 Receptor Agonists with Antidiabetic Activity. *J. Med. Chem.* **2009**, *52*, 7788–7799. [CrossRef] [PubMed]

199. Hoang, H.N.; Song, K.; Hill, T.A.; Derksen, D.R.; Edmonds, D.J.; Kok, W.M.; Limberakis, C.; Liras, S.; Loria, P.M.; Mascitti, V.; et al. Short Hydrophobic Peptides with Cyclic Constraints Are Potent Glucagon-like Peptide-1 Receptor (GLP-1R) Agonists. *J. Med. Chem.* **2015**, *58*, 4080–4085. [CrossRef] [PubMed]

200. Haug, B.E.; Skar, M.L.; Svendsen, J.S. Bulky Aromatic Amino Acids Increase the Antibacterial Activity of 15-Residue Bovine Lactoferricin Derivatives. *J. Peptide Sci.* **2001**, *7*, 425–432. [CrossRef] [PubMed]

201. Lau, Q.Y.; Ng, F.M.; Cheong, J.W.D.; Yap, Y.Y.A.; Tan, Y.Y.F.; Jureen, R.; Hill, J.; Chia, C.S.B. Discovery of an ultra-short linear antibacterial tetrapeptide with anti-MRSA activity from a structure-activity relationship study. *Eur. J. Med. Chem.* **2015**, *105*, 138–144. [CrossRef] [PubMed]

202. Ng-Choi, I.; Soler, M.; Cerezo, V.; Badosa, E.; Montesinos, E.; Planas, M.; Feliu, L. Solid-Phase Synthesis of 5-Arylhistidine-Containing Peptides with Antimicrobial Activity Through a Microwave-Assisted Suzuki–Miyaura Cross-Coupling. *Eur. J. Org. Chem.* **2012**, *2012*, 4321–4332. [CrossRef]

203. Wuts, P.G.M.; Simons, L.J.; Metzger, B.P.; Sterling, R.C.; Slightom, J.L.; Elhammer, A.P. Generation of Broad-Spectrum Antifungal Drug Candidates from the Natural Product Compound Aureobasidin A. *ACS Med. Chem. Lett.* **2015**, *6*, 645–649. [CrossRef] [PubMed]

catalysts

MDPI

Review

The Use of Palladium on Magnetic Support as Catalyst for Suzuki–Miyaura Cross-Coupling Reactions

Magne O. Sydnes

Faculty of Science and Technology, University of Stavanger, NO-4036 Stavanger, Norway; magne.o.sydnes@uis.no; Tel.: +47-5183-1761

Academic Editor: Ioannis D. Kostas
Received: 6 December 2016; Accepted: 19 January 2017; Published: 23 January 2017

Abstract: The development of new solid supports for palladium has received a lot of interest lately. These catalysts have been tested in a range of cross-coupling reactions, such as Suzuki–Miyaura, Mizoroki-Heck, and Sonogashira cross-coupling reactions, with good outcomes. Attaching the catalyst to a solid support simplifies the operations required in order to isolate and recycle the catalyst after a reaction has completed. Palladium on solid supports made of magnetic materials is particularly interesting since such catalysts can be removed very simply by utilizing an external magnet, which withholds the catalyst in the reaction vessel. This review will showcase some of the latest magnetic solid supports for palladium and highlight these catalysts' performance in Suzuki–Miyaura cross-coupling reactions.

Keywords: palladium; magnetic support; Suzuki–Miyaura cross-coupling; catalyst recycling

1. Introduction

Since the discovery of the Suzuki–Miyaura cross-coupling reaction in 1979 (Scheme 1) [1–3], it has become one of the most utilized cross-coupling reactions both in industrial and academic research laboratories [3–7]. For the development of this reaction Suzuki was awarded the 2010 Nobel prize in chemistry, together with Negishi and Heck, the chemists behind the Negish coupling and Heck coupling, respectively [8–10]. Since the early reports by Suzuki and Miyaura on the cross-coupling reaction a tremendous development has taken place in terms of substrate flexibility and reaction conditions. A range of different catalysts and conditions are now available for the successful cross-coupling of halides, triflates, O-tosylates, and diazonium salts upon reaction with boronic acid, boronates, and trifluoroborates [3,7,11–15].

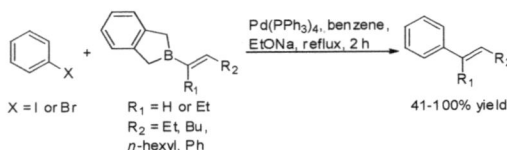

Scheme 1. Reaction conditions and substrates used in the first reported Suzuki–Miyaura cross-couplings [1].

Homogeneous palladium catalyst systems have, by far, been the most dominating catalysts utilized for this cross-coupling reaction [7]. In addition to the catalyst many reactions also require the addition of ligands in order to either promote a reaction that does not work or improve yields of sluggish reactions. However, lately there has been a growing interest for the use of heterogeneous catalysts for the Suzuki–Miyaura cross-coupling reaction, which can be seen by the increasing number of publications on the topic [16–18]. The growing interest is due to the many possibilities that heterogeneous catalysts open up for recovery and recycling of the metal catalyst.

Palladium on charcoal (Pd/C) was used for the first time as a catalyst for the Suzuki–Miyaura cross-coupling in 1994 [19]. The work by Marck et al. marked the beginning of the use of heterogeneous catalysts systems in this particular cross-coupling reaction. Since then it has been applied to an increasing number of reactions making Pd/C one of the most utilized catalyst supports for heterogeneous Suzuki–Miyaura cross-coupling reactions. Several review articles have discussed their use in cross-coupling reactions and Suzuki–Miyaura cross-couplings in particular [20–23]. Lately graphene has also been taken into use as a catalyst support for palladium [24], in addition to graphite oxide [25]. Although graphene, similarly to charcoal, is pure carbon, it adds new properties to the catalyst that are not seen with charcoal as the catalyst support.

Other types of carbon-containing materials are also frequently used as support for palladium nanoparticles (NPs). A good overview of naturally-occurring biopolymers used as catalyst supports was presented by Kumbhar and Salunkhe in 2015 [26]. For example, palladium on various types of tea leaves [27], cellulose [28,29], chitosan [30,31], pyrolysed whole plants [32], cotton [33], and filter paper [33] have been used with success in Suzuki–Miyaura cross-coupling reactions. Palladium on synthetic polymer films has also been utilized with very promising results both in view of the conversion, but also in terms of recyclability [34–36]. Chalcogenide nanoparticles are a support material that has been used as a catalyst support [37]. In addition, metal-organic frameworks (MOF) have been used in order to support palladium used for cross-coupling reactions. A recent review by Dhakshinamoorthy et al. describes the current status of this type of support materials for palladium [38]. The use of MOF as a catalyst support is also briefly discussed in a review describing recent advances and applications in boron chemistry [39].

Although many of the catalyst supports mentioned briefly above are suitable for recycling there are very few of them that can be easily recycled. One very good example of a catalyst that is easy to recycle is palladium on a polymer film support, the so called dip catalyst [34,35]. The polymer film with the catalyst can be easily added to the reaction mixture and when the reaction has reached completion the film can simply be removed from the reaction flask. A quick washing procedure then makes the catalyst ready for the next reaction.

The preparation of palladium on magnetic supports, predominately Fe_3O_4, with a range of different attachment strategies, has gained a lot of attention lately. The magnetic support makes it easy to isolate the catalyst by withholding it in the reaction flask by the use of an external magnet. The reaction mixture can then be removed and the catalyst can easily be washed preparing it for the next reaction. Naturally, catalysts that perform well over many runs with minimum loss of product yield are preferable. This review will discuss the developments made for the use of magnetic-supported catalysts for the Suzuki–Miyaura cross-coupling reaction since late 2013. For literature discussing the topic prior to 2013 the reader is referred to several reviews describing the topic [16,20,21].

2. Palladium on Magnetically Supports

Amongst the solid supports utilized for palladium, and other catalytic metals for that matter, magnetic particles have become quite popular. This is predominantly due to their easy removal from the reaction mixture after the reaction has reached completion [16]. Due to the magnetic beads, it is simple to apply an external magnetic field and decant the reaction mixture off (Figure 1). The catalyst can then be washed and used again in the next cross-coupling reaction. The ability for the various catalysts to tolerate many rounds of recycling then becomes an important parameter to consider.

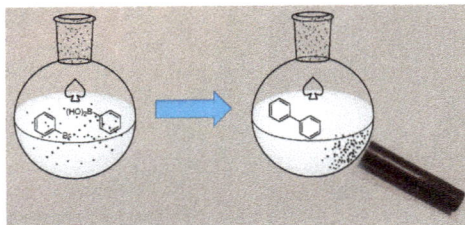

Figure 1. Application of an external magnetic field that collects the catalyst and withholds it in the reaction flask making it possible to decant the product mixture. The catalyst can then be washed and used in another cross-coupling reaction immediately afterwards.

2.1. The Magnetic Particle

A range of metals, Fe, Co, and Ni; alloys, FePt and CoPt; and ferrites, $CoFe_2O_4$, $MnFe_2O_4$, $CuFe_2O_4$, $CoFe_2O_4$, and $ZnFe_2O_4$, can be formed into magnetic nanoparticles (NPs) [40]. However, the majority of magnetic retractable NP catalysts in use are metal oxides, such as iron(II) oxide (FeO), magnetite (Fe_3O_4), and maghemite (γ-Fe_2O_3), due to their ease of formation. In the Suzuki–Miyaura cross-coupling the majority of catalyst systems on magnetic NPs are based on Fe_3O_4. Another very interesting feature with iron oxides (Fe_xO_y) is that they have been found to facilitate the reduction of oxidized palladium species onto the iron oxide in Suzuki–Miyaura cross-couplings and by such means capture leached palladium [41]. Palladium can be deposited directly on the iron oxide NP surface [41], or a thin coating can be applied often in the form of SiO_2 (Figure 2a), which functions as a handle for attaching organic ligands that hold palladium in place (Figure 2b). Furthermore, polystyrene-coated iron nanoparticles have also been utilized with good outcome [42].

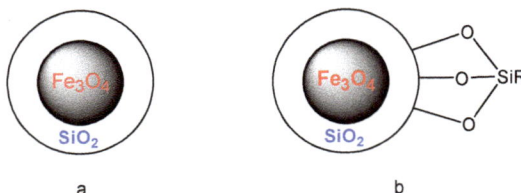

a   b

Figure 2. (a) Fe_3O_4 coated with SiO_2 is the predominant magnetic particle used in these catalyst systems; and (b) the finer details of how the ligand that holds palladium in place is attached to the magnetic particle.

2.2. Palladium Deposited Directly onto the Iron Oxide Nanoparticle

As just described in Section 2.1, one strategy for generating a magnetic retractable catalyst is to deposit the palladium directly onto the iron oxide nanoparticle. Several good examples of this strategy used for the Suzuki–Miyaura cross-coupling reaction have been reported over the last three years (Table 1).

In the work by Sun et al. Fe_3O_4@C-Pd@mSiO$_2$ was used in order to give the corresponding biaryls in good yield (Table 1, Entry 1) [43]. However, when 3-pyridyl boronic acid and 2-thiophenyl boronic acid was utilized the yields dropped to 19% and 38%, respectively. Park and co-workers obtained good, stable yields with Pd@Fe_3O_4@C as a catalyst (Table 1, Entry 2) [44]. The catalyst gave stable yields >99% (for PhX with PhB(OH)$_2$) over four rounds of recycling. Li et al. obtained predominantly higher yields than 90% for X = Br and I (Table 1, Entry 3) [45]. However, when X = Cl the yields dropped dramatically to around 50%. The catalyst, namely Fe_3O_4@C-Pd@mCeO$_2$, was recycled 10 times, which resulted in the yields dropping from 99% to 90% for the cross-coupling of 4-iodoanisole with phenylboronic acid.

Table 1. Use of palladium directly deposited onto iron oxide nanoparticles as a catalyst for Suzuki–Miyaura cross-coupling.

Entry	Catalyst	Conditions	R	ArB(OH)$_2$	X	Yield	Reference
1	Fe$_3$O$_4$@C-Pd@mSiO$_2$	ArB(OH)$_2$ (1.5 equiv.), ArX (1.0 equiv.), Cat. (1.5 mol% Pd), K$_2$CO$_3$, iPrOH, 70 °C, 6 h	H, 4-OMe, 4-NO$_2$, 4-COMe	Ph, 3-pyridyl, 2-thiophenyl, 1-naphthalene	I	73%–99% (for Ar = Ph) 19%–38% (for the heterocycles)	[43]
2	Pd@Fe$_3$O$_4$@C	ArB(OH)$_2$ (1.2 equiv.), ArX (1.0 equiv.), Cat. (1.0 mmol%), K$_2$CO$_3$, DMF/H$_2$O 4:1, 100 °C, 4 h	H, 4-F, 4-OMe, 2-Me, 4-COMe, 2-COMe	Ph (for all X), 4-Me-Ph, 4-CF$_3$-Ph	Cl, Br, OTf	>99% (for PhX with Ar = Ph) 45%–99%	[44]
3	Fe$_3$O$_4$@C-Pd@mCeO$_2$	ArB(OH)$_2$ (1.2 mmol), ArX (1.0 mmol), Cat. (Pd 3.05 wt%), K$_2$CO$_3$, EtOH/H$_2$O (1:1), 80 °C, 3 h	H, OMe, OEt, OCF$_3$, OH, NH$_2$, Me	Ph, 3-MePh, 4-MePh, 4-F, 4-OMe, 2-naphthyl	Cl, Br, I	68%–99% (X = Br and I) 50%–58% (X = Cl)	[45]
4	Fe$_3$O$_4$@Pd-OA [a]	ArB(OH)$_2$ (1.5 equiv.), ArX (1 equiv.), K$_3$PO$_4$, DMF, 115 °C, 5 h	4-OMe	Ph, 4-FPh	I	80%–94% [b]	[46]
5	Pd@ Fe$_3$O$_4$@ZnO	ArB(OH)$_2$ (1 equiv.), ArX (1 equiv.), Cat. (Pd 0.1 mol%), K$_2$CO$_3$, H$_2$O, 100 °C, 1–3 h (for X = Br, I), 7–14 h (for X = Cl)	4-Me, 4-COMe, 4-Cl, 4-F, 4-NO$_2$, 4-NH$_2$, 4-CN, 4-pyridin	Ph, 2-FPh, 3-FPh, 4-FPh, 4-EtPh	Cl, Br, I	70%–95% (X = Br and I) 25%–53% (X = Cl)	[47]
6	Fe$_3$O$_4$@SiO$_2$-Pd@mCeO$_2$	ArB(OH)$_2$ (1.2 equiv.), ArX (1 equiv.), Cat. (Pd 0.5 mmol%), K$_2$CO$_3$, EtOH/H$_2$O (9:1) (for Br, I), DMF/H$_2$O (9:1), 100 °C (for Cl), 3 h	H, 3-Me, 4-OMe, 4-OCF$_3$, 4-COCH$_3$	Ph, 3-MePh, 4-MePh, 4-OMePh, 4-FPh	Cl, Br, I	75%–99% (X = Br and I) 68%–88% (X = Cl)	[48]
7	Fe$_3$O$_4$@t-Pd@mCeO$_2$	ArB(OH)$_2$ (1.2 equiv.), ArX (1 equiv.), Cat. (Pd 0.5 mmols), K$_2$CO$_3$, EtOH/H$_2$O (9:1) (for Br, I), DMF/H$_2$O (9:1), 100 °C (for Cl), 3 h	H, 3-Me, 4-OMe, 4-OCF$_3$, 4-COCH$_3$	Ph, 3-MePh, 4-MePh, 4-OMePh, 4-FPh	Cl, Br, I	82%–99% (X = Br and I) 72%–92% (X = Cl)	[48]
8	Pd@Ni@CB CB = carbon black	ArB(OH)$_2$ (1.1 equiv.), ArX (1 equiv.), Cat. (Pd 0.1 mol%), K$_2$CO$_3$, EtOH/H$_2$O (1:1), 30 °C (for X = Br, I), 80 °C (for X = Cl), 45 min–1.5 h	H, 2-NO$_2$, 3-NO$_2$, 4-NO$_2$, 2-OMe, 3-OMe, 4-Me, 4-OH, 4-CN, 4-CHO	Ph	Cl, Br, I	79%–95% (X = Br and I) 5%–60% (X = Cl)	[49]

[a] OA = oleylamine; [b] Two examples were reported, Ar = Ph gave a yield of 80% and Ar = 4-FPh gave a yield of 94%.

The work by Strumia and co-workers predominantly focused on the development of the catalyst, however, the two examples of Suzuki–Miyaura cross-coupling reactions reported resulted in very good yields of the desired product (Table 1, Entry 4) [46]. Palladium on Fe_3O_4-ZnO nanoparticles were prepared by Hosseini-Sarvari et al. (Table 1, Entry 5) [47]. The catalyst system gave good yields for X = Br and I, but the yields with X = Cl was low (25%–53%) despite a prolonged reaction time. Ye and co-workers reported the use of two different magnetic palladium supports in their work (Table 1, Entries 6 and 7) [48]. $Fe_3O_4@h$-Pd@mCeO$_2$ was the preferable catalyst of choice, also giving good yields for the cross-coupling of aryl chlorides. Palladium on nickel and carbon black was used as a catalyst system by Wang and co-workers (Table 1, Entry 8) [49]. The catalyst gave good to excellent yields for X = Br and I, however, when the corresponding chloride was used, the yields dropped significantly. In their recycling experiments it was found that the product yield dropped by 18% over five runs.

Out of the catalysts presented in this section the work by Park and co-workers showed that the catalyst could be used four times, giving yields consistently at 99% [44], which is the most promising catalyst system in terms of recycling described in this section. However, in order to establish the full potential of the catalyst it should have been tested over several more cycles. There are several examples of catalysts that perform well over a few rounds of recycling before collapsing.

2.3. Iron Oxide with Ligands Holding Palladium

Another strategy, which is more commonly used, is to attach an organic substrate to the magnetic nanoparticle. A range of chelating agents have been attached to magnetic nanoparticles in order to chelate to palladium and, by such means, keep the catalyst attached to the magnetic particle. Several of these catalysts are depicted in Figure 3. All but one catalyst is based on Fe_3O_4 nanoparticles with γFe_2O_3 making up the last example. These catalysts have been utilized and put to the test in the Suzuki–Miyaura cross-coupling in addition to other cross-coupling reactions.

2.4. The Use of Palladium Chelated to Magnetic Nanoparticles

A general trend seen with all of the catalysts described in Figure 3, and further outlined in Table 2, is that the most reactive cross-coupling partners are the iodides, followed by bromides, with chlorides showing the lowest reactivity. The work by Zhang et al. utilized catalyst **1** for the formation of biaryls (Table 2, Entry 1) [50]. The catalyst gave constantly high yields for X = Br and I, however, when X = Cl the yields dropped dramatically. Somsook and co-workers utilized polyaniline attached to a magnetic nanoparticle in order to catalyze the Suzuki–Miyaura cross-coupling reaction between 1-bromo-4-methylbenzene and a range of aryl boronic acids (Table 2, Entry 2) [51]. The outcome of the reactions generally gave good yields with a few exceptions. Catalyst **2** showed high conversion and very good yields for aryl bromides, aryl iodides, and in the two examples that were performed with 1-chloro-4-nitrobenzene (Table 2, Entry 3) [52]. It would have been very interesting to see how this catalyst would perform in cross-coupling reactions between a broader range of aryl chlorides and more complex boronic acids. The catalyst also showed very good yields through 10 runs with only a 1% reduction in the isolated yield. Catalyst **3** gave good to excellent yields for aryl chlorides, bromides, and iodides (Table 2, Entry 4) [53]. The catalyst also maintained good catalytic activity over six runs both for cross-coupling of iodobenzene and 1-chloro-4-nitrobenzene with phenyl boronic acid.

Ma and co-workers prepared catalyst **4** and tested its performance in a range of Suzuki–Miyaura cross-coupling reactions (Table 2, Entry 5) [54]. The catalyst performed very well with aryl bromides and iodides in most cases, except for two examples, which contained a nitro group *ortho* to the halide. When aryl chlorides were tested under the same reaction conditions the yields dropped dramatically compared to the yields obtained with aryl bromides and iodides. In the work by Ghotbinejad et al. SPION-A-Pd(EDTA) (**5**) (EDTA = ethylenediaminetetraacetic acid) was used as catalyst under both conventional heating and ultrasound condition. The use of ultrasound reduced the reaction time from hours to minutes and generally also resulted in slightly better yields compared with the yields obtained when conventional heating was used (Table 2, Entry 6) [55].

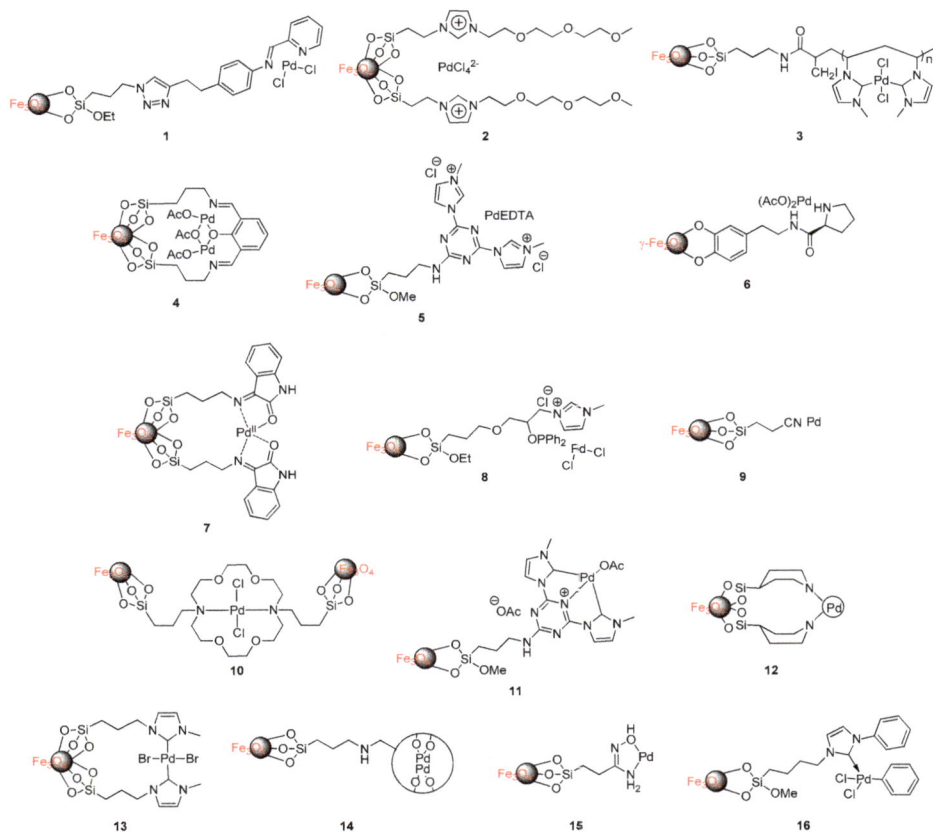

Figure 3. The structure of magnetically-supported palladium catalysts used in Suzuki–Miyaura cross-coupling reactions.

Catalyst **6** performed well in all the examples with aryl iodides take for the example where 1-iodo-4-nibrobenzene was cross-coupled with 2-thiophenyl boronic acid (Table 2, Entry 7) [56]. Out of the catalysts presented herein, catalyst **6** is the only one based on γ-Fe$_2$O$_3$ nanoparticles. Upon recycling of this catalyst the yields dropped from 99% to 90% over eight runs. Catalyst **7** prepared by Esmaeilpour and Javidi performed well for all three aryl halides with yields ranging from 70%–80% for the aryl chlorides tested (Table 2, Entry 8) [57]. The conditions reported by Biji and co-workers generally gave good yields for all three aryl halides when utilizing catalyst **8** (Table 2, Entry 9) [58]. Although the aryl chlorides only engaged in the cross-coupling reaction under quite forcing conditions (120 °C). Catalyst **9** worked excellently for aryl bromides and aryl iodides, however, the yields were disappointingly low in the examples with aryl chlorides (yields ranging from 20% to 25%) (Table 2, Entry 10) [59]. The catalyst prepared by Khakiani could be recycled seven times without significant loss in activity (the yield dropped by 1%–2%).

Table 2. Use of palladium catalyst chelated to magnetic nanoparticles as catalysts for Suzuki–Miyaura cross-coupling.

Entry	Catalyst #	Conditions	R	ArB(OH)$_2$	X	Yield	Reference
1	1	ArB(OH)$_2$ (1.2 equiv.), ArX (1.0 equiv.), Cat. (0.2 mol% Pd), K$_2$CO$_3$, EtOH/H$_2$O (1:1), 60 °C, 3–12 h	H, 4-Me, 4-COCH$_3$, 4-NO$_2$, 4-CF$_3$, 4-CN, 4-NH$_2$, 4-OH, 3-OMe, 3-CHO, 2-OMe	Ph, 4-MePh, 4-OMePh, 4-ClPh	Cl, Br, I	80%–99% (X = Br and I) 5%–12% (X = Cl)	[50]
2	Pd/Poly(*m*-ferrocenyl-aniline)	ArB(OH)$_2$ (1 equiv.), ArX (1 equiv.), Cat. (2.0 mol% Pd), KOH, toluene, reflux, 20 h	4-Me	Ph, 4-MePh, 3-MePh, 2-MePh, 4-CHOPh	Br	67%–98% 26% (for ArB(OH)$_2$ = 4-CHOPh)	[51]
3	2	ArB(OH)$_2$ (1.2 equiv.), ArX (1 equiv.), Cat. (0.025 mol% Pd), K$_2$CO$_3$, H$_2$O, 60–80 °C, 4–15 h (for Br, I), 24 h (for Cl) ^{a}	H, 4-NO$_2$, 4-OMe, 4-CHO, 4-OH, 4-NH$_2$, 3-OH, 2-CHO, 2-Me, 2-Et, 2,6-Me, 2-thiophenyl, 5-pyrimidinyl, 3-pyridyl, 2-pyridil	Ph, 4-MePh, 2-MePh, 3-pyridyl	Cl, Br, I (three examples with Cl)	72%–quant. (X = Br and I) 85%–96% (X = Cl)	[52]
4	3	ArB(OH)$_2$ (1.2 equiv.), ArX (1 equiv.), Cat. (0.0049 mol% Pd), K$_2$CO$_3$, NMP, 90 °C, 0.5–2.5 h (for X = Br or I), 2.5–9 h (for X = Cl)	H, 4-Me, 4-OMe, 4-CHO, 4-COCH$_3$, 4-CN, 4-NO$_2$, 4-Cl, 3-Me, 2-Me, 3-pyridyl, 5-pyrimidine, 3-thiophenyl	Ph,	Cl, Br, I	77%–95%	[53]
5	4	ArB(OH)$_2$ (1.5 equiv.), ArX (1 equiv.), Cat. (0.5 mol%), K$_2$CO$_3$, EtOH, 80 °C, 3–10 h	H, 4-Me, 2-Me, 4-OH, 2-NO$_2$	Ph, 4-ClPh (X = I)	Cl, Br, I	95%–99% (X = Br and I) 16%–64% (X = Cl)	[54]
6	5	ArB(OH)$_2$ (1.1 equiv.), ArX (1 equiv.), Cat. (0.003 mol% Pd), K$_2$CO$_3$, DMF/H$_2$O (1:2), 70 °C, 2–14 h Ultrasound applied power 160 W, 30 °C, 7–35 min	H, 4-Me, 4-OMe, 4-CHO, 4-Ac, 4-F	Ph, 4-OMePh	Br, I	82%–93% (70 °C, conventional heating) 87%–96% (ultrasound conditions)	[55]
7	6	ArB(OH)$_2$ (1.1 equiv.), ArX (1 equiv.), Cat. (0.1 mol%), Na$_2$CO$_3$, EtOH/H$_2$O (1:1). 80 °C, 30 min	H, 4-NO$_2$	4-COMe, 4-COOEt, 4-CHO, 4-ethenyl, 4-tolyl, 4-ethenyl, 2-furanyl, 2-thiophenyl	Cl, Br, I (only one example with X = Cl and Br)	60%–99% (for X = I)	[56]
8	7	ArB(OH)$_2$ (1.2 equiv.), ArX (1 equiv.), Cat. (0.3 mol%), K$_2$CO$_3$, NMP, 100 °C, 0.5–4.5 h (for Br, I), 4–12 h (for Cl)	H, 4-Me, 3-Me, 2-Me, 4-OMe, 4-NO$_2$, 2-Me-4-NO$_2$, 4-NH$_2$, 4-COMe, 4-CN, 1-naphthyl	Ph	Cl, Br, I	77%–96%	[57]
9	8	ArB(OH)$_2$ (1.5 equiv.), ArX (1 equiv.), Cat. (0.3 mol%), K$_2$CO$_3$, EtOH/H$_2$O (1:1), 30 °C (for X = I), 60 °C (for X = Br), 120 °C (for X = Cl), 1–24 h (for X = Br or I), 48 h (for X = Cl)	H, 4-OH, 4-OMe, 4-CHO, 4-NO$_2$, 4-CN, 4-COMe, 4-biphenyl, 1-naphthalen, 5-pyrimidine	Ph, 4-ClPh, 4-MePh, 3,5-diFPh, 2-FPh, 2-NO$_2$Ph, 1-naphthalene	Cl, Br, I	74%–98%	[58]
10	9	ArB(OH)$_2$ (1.1 equiv.), ArX (1 equiv.), Cat. (0.2 mol%), K$_2$CO$_3$, EtOH/H$_2$O, 20–25 °C, 0.2–6 h (for X = Br or I), 10–12 h (for X = Cl)	H, 4-Me, 4-COCH$_3$, 4-OMe, 4-Cl, 3-NO$_2$, 2-CHO, 1-naphthyl, 2-thienyl	Ph	Cl, Br, I	88%–98% (X = Br and I) 20%–25% (X = Cl)	[59]
11	Fe$_3$O$_4$@EDTA-PdCl$_2$	ArB(OH)$_2$ (1.1 equiv.), ArX (1 equiv.), K$_2$CO$_3$, TBAB^a, H$_2$O, 80 °C, 2–6 h	H, 3-Me, 3-CF$_3$, 1-naphthyl, 2-thienyl, (for X = Br) H, 4-Me, 4-OMe, 4-NO$_2$, 4-CHO (for X = I)	Ph	Br, I	76%–95%	[60]

Table 2. *Cont.*

$$R{-}\!\!\overset{\text{(HO)}_2\text{B}{-}}{\underset{X}{\bigcirc}} + \overset{\text{(HO)}_2\text{B}}{\bigcirc} \xrightarrow{\text{Conditions}} R{-}\overset{}{\bigcirc}{-}\bigcirc$$

Entry	Catalyst #	Conditions	R	ArB(OH)₂	X	Yield	Reference
12	**10**	ArB(OH)$_2$ (1.2 equiv.), ArX (1 equiv.), Cat. (0.5 mol%), Et$_3$N, TBAB ^{b}, DMF/H$_2$O (1:1), 75 °C, 1–6 h (X = Br and I), 24 h (X = Cl)	H, 4-Br, 2-thiophenyl, 1-naphthyl, 4-CN, 4-NO$_2$, 4-Me	Ph, 4-OMePh	Cl, Br, I	47%–94% (X = Br and I) 35%–37% (X = Cl)	[61]
13	Pd/Fe$_3$O$_4$/r-GO ^{c}	ArB(OH)$_2$ (1.5 equiv.), ArX (1.0 equiv.), Cat. (0.36 mol% Pd), K$_2$CO$_3$, H$_2$O, 80 °C, 15 min–2.5 h	H, 4-Me, 4-CN, 4-NO$_2$	Ph	Cl, Br, I	78%–99%	[62]
14	Fe$_3$O$_4$-NHC-Pd	ArB(OH)$_2$ (1.2 equiv.), ArX (1 equiv.), Cat. (0.02 mol%), K$_2$CO$_3$, EtOH/H$_2$O (3:1), 70 °C, 12 h	H, 4-Me, 4-OMe, 4-COMe, 4-CHO, 4-CO$_2$H, 4-CN, 4-NO$_2$	Ph, 4-MePh, naphthyl, 2,6-diMePh	Br	87%–99% 61% for 2,6-diMePh	[63]
15	**11**	ArB(OH)$_2$ (1.1 equiv.), ArX (1 equiv.), Cat. (0.002 mol%), K$_2$CO$_3$, DMF/H$_2$O (1:2), 70 °C, 4–14 h, microwave irradiation 200 W, 70 °C 2–7 min	H, 4-Me, 4-OMe, 4-CHO	Ph, 4-OMePh	Br, I	81%–91% (conventional heating) 85%–95% (microwave irradiation)	[64]
16	Fe$_3$O$_4$@PEG-iminophosphine-Pd	ArB(OH)$_2$ (1.2 equiv.), ArX (1 equiv.), Cat. (0.05 mol%), K$_2$CO$_3$, H$_2$O, 65 °C, 5–8 h (for Br, I) 24 h (for Cl)	4-Me, 3-Me, 2-Me, 4-OMe, 4-CH$_2$OH, 4-NO$_2$, 2-COMe, 4-F, 4-CN	Ph, 4-OMePh, 4-CF$_3$Ph	Cl (only one example), Br, I	89%–99% (X =Br and I) 88% (X =Cl, R = 4-NO$_2$, Ar = Ph)	[65]
17	**12**	ArB(OH)$_2$ (1.0 equiv.), ArX (1.0 equiv.), Cat. (0.37 mol%), Na$_2$CO$_3$, PEG, 100 °C, 20–150 min	H, 4-Me, 4-OMe, 4-CN, 4-CN, 4-Cl, 4-NO$_2$, 3-CHO, 3-CF$_3$,	Ph, 4-OMePh, 3,4-diFPh	Br, I	88%–98%	[66]
18	**13**	ArB(OH)$_2$ (1.2 equiv.), ArX (1 equiv.), Cat. (0.024 mol%), K$_2$CO$_3$, Tx ^{d} /THF (9:1), 65 °C, 30 h (for Br) 80 °C, 90 h (for Cl)	4-Me	Ph	Cl, Br	100% conv. ^{e}	[67]
19	**14**	ArB(OH)$_2$ (1.2 equiv.), ArX (1 equiv.), Cat. (0.03–0.30 mol%) ^{f} K$_2$CO$_3$, H$_2$O, reflux, 4–12 h (for X = Br or I), 24 h (for X = Cl)	H, 4-OMe, 4-CHO, 4-NO$_2$, 4-Cl, 2-Me, 5-pyrimidine,	Ph, 4-MePh, 4-OMePh, 3-NO$_2$Ph, 2,6-diMePh, 3,5-diFPh, 2-benzofuranyl, 2-naphtyl, 1-naphthyl	Cl, Br, I	75%–99% (X = Br and I) 36%–65% (X = Cl)	[68]
20	**15**	ArB(OH)$_2$ (1.2 equiv.), ArX (1 equiv.), Cat. (0.1 mol%), K$_2$CO$_3$, EtOH/H$_2$O (1:1), 25 °C, 1–5 h (for Br, I), 15–24 h (for Cl)	H, 4-Me, 4-OMe, 2-OMe, 4-F, 4-NO$_2$, 4-COMe	Ph	Cl, Br, I	85%–96% (X =Br and I) 60%–70% (X = Cl)	[69]
21	**16**	ArB(OH)$_2$ (1.1 equiv.), ArX (1 equiv.), Cat. (0.036 mol%), K$_2$CO$_3$, EtOH/H$_2$O (1:1), rt, 0.5–2 h	H,^{g} 4-OMe,^{g} 3-OMe, 2-OMe, 4-NO$_2$,^{g} 4-CHO 4-Cl, *m*-Cl, *o*-Cl (X = Br)	Ph	Cl, Br, I	75%–98% (X = Br and I) 65%–74% (X = Cl)	[70]
22	Pd/Fe$_3$O$_4$@PDA ^{h}	ArB(OH)$_2$ (1.1 equiv.), ArX (1 equiv.), Cat. (0.46 mol% for X = I and 0.27 for X = N$_2$$^+BF_4$$^-$), K$_2CO_3$, EtOH/H$_2$O (1:1) (for X = Br and I), 100 °C (for X = Br), 80 °C (for X = I), rt (for X = N$_2$$^+BF_4$$^-$), 10 h (for X = I) and 2–3.5 h (for X = N$_2$$^+BF_4$$^-$)	H (for X = I) 4-OMePh, 4-BrPh, 4-NO$_2$Ph (for X = N$_2$$^+BF_4$$^-$)		Br, I, N$_2$$^+BF_4$$^-$	77%–98%	[71]

^{a} Reaction conducted at 120 °C with the addition of 0.5 equiv. of TBAB and DMF/H$_2$O (1:1) as solvent; ^{b} TBAB = tetrabutylammonium bromide; ^{c} Pd/Fe$_3$O$_4$/reduced-graphene oxide nanohybrid; ^{d} Tx = aqueous solution 0.21 wt% of Triton TM X-405; ^{e} Conversion determined by GC; ^{f} Highest catalyst loading only used for X = Cl; ^{g} Reactions conducted with X = Cl; ^{h} PDA = polydopamine.

Heydari and co-workers utilized EDTA attached to the magnetic nanoparticle in order to coordinate to the palladium catalyst (Table 2, Entry 11) [60]. The catalyst gave good to excellent yields for the aryl bromides and iodides they tested, however, the yields dropped significantly when the catalyst was recycled (the yield dropped from 90% in run 1 to 80% in run 5). Catalyst **10** gave a mix of results, with yields ranging from poor to very good when tested by the Mobaraki group (Table 2, Entry 12) [61]. In particular, the aryl chlorides tested gave very poor yields. Palladium/Fe_3O_4/reduced-graphene oxide was used by Nasrabadi and co-workers with good outcome to catalyze the Suzuki–Miyaura cross-coupling between a range of aryl chlorides, bromides, and iodides (Table 2, Entry 13) [62]. The catalyst system also maintained its catalytic activity over eight runs with the yield only dropping by 2% from 97% in the first run to 95% in the eighth run.

Fe_3O_4-NHC-Pd was used by Wang et al. for a range of aryl bromides with a range of boronic acids (Table 2, Entry 14) [63]. The catalyst system was tested over 21 runs without loss of activity. The yields varied between 93% and 99% with 96% yield in the first run, and 97% yield in run number 21. Catalyst **11** gave very good yields for aryl bromides and iodides both under conventional heating and under microwave conditions (MW) (Table 2, Entry 15) [64]. The great advantage utilizing microwave conditions was that the reactions reached completion in a few minutes compared to hours under conventional conditions. In addition, the yields were slightly higher under MW conditions. The catalyst maintained most of its activity over seven rounds of recycling (yield dropped from 95% to 93%). Fe_3O_4@PEG-iminophosphine-Pd was used as catalyst by Lu and co-workers resulting in good to excellent yields of the desired biaryl system when utilizing all three aryl halides (Table 2, Entry 16) [65]. This catalyst system also showed good perseverance of activity over five runs with only a 2% drop in the isolated yield. Azadi and Ghorbani-Choghamarani developed catalyst **12**, which was put to the test in the Suzuki–Miyaura cross-coupling reaction with good to very good yields for both aryl bromides and aryl iodides (Table 2, Entry 17) [66].

Andrés and co-workers prepared catalyst **13** and tested it in two reactions, namely the cross-coupling of phenyl boronic acid with 1-chloro-4-methylbenzene and 1-bromo-4-methylbenzene (Table 2, Entry 18) [67]. In both cases they reported 100% conversion as judged by GC analysis. The β-cyclodextrine-based catalyst system performed well in Suzuki–Miyaura cross-couplings with aryl bromides and aryl iodides, however, the yields dropped dramatically for the corresponding aryl chlorides (Table 2, Entry 19) [68]. Catalyst **14** also showed a dramatic drop in activity over five runs with the isolated yield going down 16% over the five runs. Catalyst **15** performed very well when utilizing aryl bromides and aryl iodides, however, the yields dropped roughly 30% when the corresponding aryl chlorides were used (Table 2, Entry 20) [69]. A similar drop in yields was also experienced by Hajipour et al. when they tested catalyst **16** with aryl chlorides (Table 2, Entry 21) [70]. However, with the more active aryl halides, viz. bromide and iodide, the catalyst performed well.

Dubey and Kumar, in addition to aryl bromide and aryl iodide, also tested the corresponding diazonium tetrafluoroborate ($N_2^+BF_4^-$) as the coupling partner with boronic acid (Table 2, Entry 22) [71]. This is the only example where this coupling partner has been utilized in combination with a magnetically retrievable catalyst. The reaction time when diazonium tetrafluoroborate was utilized was significantly quicker than the reaction time obtained when Br and I were utilized. The reaction time was as short as 2–3.5 h at room temperature. The yields in all examples were generally good ranging from 77% to 98%.

As mentioned, the trend that is usually seen for aryl chlorides when homogenous catalysts are used is also present in the examples brought forward herein. As just mentioned, the reactivity is significantly lower resulting, in most cases, to much longer reaction times when aryl chlorides are used. The yields are, in most cases, also dramatically lower for aryl chlorides compared to the corresponding iodide or bromide counterpart, although some very nice exceptions can be found in the work of Andrés and co-workers (Table 2, Entry 18) [67].

2.5. Recycling

As seen in the examples discussed in the forgoing the majority of the catalyst systems give good yields for aryl bromides and aryl iodides, and if the reaction time is prolonged compared to the reaction time used for the aryl bromides and aryl iodides, a number of the aryl chlorides also result in good conversion. However, when it comes to the ability of the various catalysts to perform at the same high level of conversion over multiple cycles, the differences between them becomes obvious. While some catalysts manage to maintain a high level of conversion over numerous rounds of recycling, other catalysts drastically deteriorate after just a few rounds of recycling. In particular, two catalyst systems by Karimi et al. [52] and Lin and co-workers [63] are singled out as better than the rest with the catalysts used by the Lin group being outstanding with its 21 runs without the catalyst losing its activity.

2.6. Catalyst Leaching

The active form of the catalyst in heterogeneous catalysis is a topic that has been much debated over the years [72]. Is the reaction catalyzed by the solid catalyst or is it catalyzed by palladium leached out from the heterogeneous catalyst. Very few reports deal with this issue, however, the work by Azadi and Ghorbani-Choghamarani tested this in a very clever way [66]. By removing the catalyst with a magnet when the reaction had proceeded to 50% conversion, they could evaluate if the reaction proceeded after that point. Their data clearly showed that the reaction stopped after the catalyst was removed, determining that the active catalyst was the heterogeneous palladium catalyst.

3. Concluding Remarks

The results presented in this article clearly show that catalysts on magnetic particles are fully able to catalyze the Suzuki–Miyaura cross-coupling reaction in good yields over a reasonable reaction time. However, the substrates demonstrated thus far are relatively simple compared with some of the starting materials used in Suzuki–Miyaura cross-coupling reactions conducted under homogenous conditions. For example, the substrates during such cross-coupling reactions conducted during synthesis of complex natural products or pharmaceuticals are far more complex than the ones utilized during the examples shown herein for heterogeneous catalysis. The use of palladium on magnetic retractable catalyst supports would be of interest in these settings, in particular the latter, where there are strict regulations regarding leftover palladium in the product. In addition, this type of catalyst would become more generally attractive if they were commercially available.

Acknowledgments: The University of Stavanger (UiS), ToppForsk UiS, and the research program Bioactive is gratefully acknowledged for providing funding for my group.

Conflicts of Interest: The author declares no conflict of interest.

References

1. Miyaura, N.; Suzuki, A. Stereoselective synthesis of arylated (*E*)-alkenes by the reaction of alk-1-enylboranes with aryl halides in the presence of palladium catalyst. *J. Chem. Soc. Chem. Commun.* **1979**, *19*, 135–152. [CrossRef]
2. Miyaura, N.; Yamada, K.; Suzuki, A. A new stereospecific cross-coupling by the palladium-catalyzed reaction of 1-alkenylboranes with 1-alkenyl or 1-alkynyl halides. *Tetrahedron Lett.* **1979**, *20*, 3437–3440. [CrossRef]
3. Miyaura, N.; Suzuki, A. Palladium-catalyzed cross-coupling reactions of organoboron compounds. *Chem. Rev.* **1995**, *95*, 2457–2483. [CrossRef]
4. Miyaura, N. Cross-coupling reaction of organoboron compounds via base-assisted transmetalation to palladium(II) complexes. *J. Organomet. Chem.* **2002**, *653*, 54–57. [CrossRef]
5. Suzuki, A. Cross-coupling reactions via organoboranes. *J. Organomet. Chem.* **2002**, *653*, 83–90. [CrossRef]
6. Gildner, P.G.; Colacot, T.J. Reactions of the 21st Century: Two Decades of Innovative Catalyst Design for Palladium-Catalyzed Cross-Couplings. *Organometallics* **2015**, *34*, 5497–5508. [CrossRef]

7. Dobrounig, P.; Trobe, M.; Breinbauer, R. Sequential and iterative Pd-catalyzed cross-coupling reactions in organic synthesis. *Monatsh. Chem.* **2017**, *148*, 3–35. [CrossRef]

8. Suzuki, A. Cross-coupling reactions of organoboranes: An easy way to construct C–C Bonds (Nobel Lecture). *Angew. Chem. Int. Ed.* **2011**, *50*, 6723–6737. [CrossRef] [PubMed]

9. Negishi, E. Magical power of transition metals: Pas, present, and future (Nobel Lecture). *Angew. Chem. Int. Ed.* **2011**, *50*, 6738–6764. [CrossRef] [PubMed]

10. Seechurn, C.C.C.J.; Kitching, M.O.; Colacot, T.J.; Snieckus, V. Palladium-catalyzed cross-coupling: A historical contextual perspective to the 2010 Nobel Prize. *Angew. Chem. Int. Ed.* **2012**, *51*, 5062–5085. [CrossRef] [PubMed]

11. Makarasen, A.; Kuse, M.; Nishikawa, T.; Isobe, M. Substituent effect of imino-O-arenesulfonates, a coupling partner in Suzuki–Miyaura reaction for substitution of the pyrazine ring: Study for the synthesis of coelenterazine analogs. *Bull. Chem. Soc. Jpn.* **2009**, *82*, 870–878. [CrossRef]

12. Roglans, A.; Pla-Quintana, A.; Moreno-Mañas, M. Diazonium salts as substrates in palladium-catalyzed cross-coupling reactions. *Chem. Rev.* **2006**, *106*, 4622–4643. [CrossRef] [PubMed]

13. Taylor, J.G.; Moro, A.V.; Correia, C.R.D. Evolution and Synthetic applications of the Heck-Matsuda reaction: The return of arenediazonium salts to prominence. *Eur. J. Org. Chem.* **2011**, 1403–1428. [CrossRef]

14. Felpin, F.-X.; Nassar-Hardy, L.; Le Callonnec, F.; Fouquet, E. Recent advances in the Heck-Matsuda reaction in heterocyclic chemistry. *Tetrahedron* **2011**, *67*, 2815–2831. [CrossRef]

15. Mo, F.; Dong, G.; Zhang, Y.; Wang, J. Recent applications of arene diazonium salts in organic synthesis. *Org. Biomol. Chem.* **2013**, *11*, 1582–1593. [CrossRef] [PubMed]

16. Wang, D.; Astruc, D. Fast-growing field of magnetically recyclable nanocatalysts. *Chem. Rev.* **2014**, *114*, 6949–6985. [CrossRef] [PubMed]

17. Deraedt, C.; Astruc, D. "Homeopathic" palladium nanoparticle catalysis of cross carbon-carbon coupling reactions. *Acc. Chem. Res.* **2014**, *47*, 494–503. [CrossRef] [PubMed]

18. Paul, S.; Islam, M.M.; Islam, S.M. Suzuki–Miyaura reaction by heterogeneously supported Pd in water: Recent studies. *RSC Adv.* **2015**, *5*, 42193–42221. [CrossRef]

19. Marck, G.; Villiger, A.; Buchecker, R. Aryl couplings with heterogeneous palladium catalysts. *Tetrahedron Lett.* **1994**, *35*, 3277–3280. [CrossRef]

20. Sydnes, M.O. Recent Developments in the use of palladium on solid support in organic synthesis. *Curr. Org. Synth.* **2011**, *8*, 881–891. [CrossRef]

21. Sydnes, M.O. Use of nanoparticles as catalysts in organic synthesis—Cross-coupling reactions. *Curr. Org. Chem.* **2014**, *18*, 312–326. [CrossRef]

22. Zafar, M.N.; Mohsin, M.A.; Danish, M.; Nazar, M.F.; Murtaza, S. Palladium catalyzed Heck-Mizoroki and Suzuki–Miyaura coupling reactions (review). *Russ. J. Coord. Chem.* **2014**, *40*, 781–800. [CrossRef]

23. Felpin, F.-X.; Ayad, T.; Mitra, S. Pd/C: An old catalyst for new applications—Its use for the Suzuki–Miyaura reaction. *Eur. J. Chem.* **2006**, 2679–2690.

24. Fan, X.; Zhang, G.; Zhang, F. Multiple roles of graphene in heterogeneous catalysis. *Chem. Soc. Rev.* **2015**, *44*, 3023–3035. [CrossRef] [PubMed]

25. Santra, S.; Hota, P.K.; Bhattacharyya, R.; Bera, P.; Chosh, P.; Mandal, S.K. Palladium nanoparticles on graphite oxide: A recyclable catalyst for the synthesis of biaryl cores. *ACS Catal.* **2013**, *3*, 2776–2789. [CrossRef]

26. Kumbhar, A.; Salunkhe, R. Recent advances in biopolymer supported palladium in organic synthesis. *Curr. Org. Chem.* **2015**, *19*, 2075–2121. [CrossRef]

27. Lebaschi, S.; Hekmati, M.; Veisi, H. Green synthesis of palladium nanoparticles mediated by black tea leaves (*Camellia sinensis*) extract: Catalytic activity in the reduction of 4-nitrophenol and Suzuki–Miyaura coupling reaction under ligand-free conditions. *J. Colloid Interface Sci.* **2017**, *485*, 223–231. [CrossRef] [PubMed]

28. Xiao, J.; Lu, Z.; Li, Y. Carboxymethylcellulose-supported palladium nanoparticles generated in situ from palladium(II) carboxymethylcellulose: An efficient and reusable catalyst for Suzuki–Miyaura and Mizoroki-Heck reactions. *Ind. Eng. Chem. Res.* **2015**, *54*, 790–797. [CrossRef]

29. Chen, F.; Huang, M.; Li, Y. Synthesis of a novel cellulose microencapsulated palladium nanoparticle and its catalytic activities in Suzuki–Miyaura and Mizoroki-Heck reactions. *Ind. Eng. Chem. Res.* **2014**, *53*, 8339–8345. [CrossRef]

30. Naghipor, A.; Fakhri, A. Heterogeneous Fe$_3$O$_4$@chitosan-Schiff base Pd nanocatalyst: Fabrication, characterization and application as highly efficient and magnetically-recoverable catalyst for Suzuki–Miyaura and Heck-Mizoroki C–C coupling reactions. *Catal. Commun.* **2016**, *73*, 39–45. [CrossRef]

31. Jadhav, S.; Kumbhar, A.; Salunkhe, R. Palladium supported on silica-chitosan hybrid material (Pd-CS@SiO$_2$) for Suzuki–Miyaura and Mizoroki-Heck cross-coupling reactions. *Appl. Organometal. Chem.* **2015**, *29*, 339–345. [CrossRef]

32. Parker, H.L.; Rylott, E.L.; Hunt, A.J.; Dodson, J.R.; Taylor, A.F.; Bruce, N.C.; Clark, J.H. Supported palladium nanoparticles synthesized by living plants as a catalyst for Suzuki–Miyaura reactions. *PLoS ONE* **2014**, *9*, e87192. [CrossRef] [PubMed]

33. Nishikata, T.; Tsutsumi, H.; Gao, L.; Kojima, K.; Chikama, K.; Nagashima, H. Adhesive catalyst immobilization of palladium nanoparticles on cotton and filter Paper: Applications to reusable catalysts for sequential catalytic reactions. *Adv. Synth. Catal.* **2014**, *356*, 951–960. [CrossRef]

34. Hariprasad, E.; Radhakrishnan, T.P. Palladium nanoparticle-embedded polymer thin film "dip catalyst" for Suzuki–Miyaura reaction. *ACS Catal.* **2012**, *2*, 1179–1186. [CrossRef]

35. Rao, V.K.; Radhakrishnan, T.P. Hollow bimetallic nanoparticles generated in situ inside a polymer thin film: fabrication and catalytic application of silver-palladium-poly(vinyl alcohol). *J. Mater. Chem. A* **2013**, *1*, 13612–13618. [CrossRef]

36. Oliveira, D.G.M.; Alvarenga, G.; Scheeren, C.W.; Rosa, G.R. Densenvovimento de reator tipo «dip catalyst» para filmes poliméricos contendo nanoparticulas de metais de transição. *Quim. Nova* **2014**, *37*, 1401–1403.

37. Kumar, A.; Rao, G.K.; Kumar, S.; Singh, A.K. Formation and role of palladium chalcogenide and other species in Suzuki–Miyaura and Heck C–C coupling reactions catalyzed with palladium(II) complexes of organochalcogen ligands: realities and speculations. *Organometallics* **2014**, *33*, 2921–2943. [CrossRef]

38. Dhakshinamoorthy, A.; Asiri, A.M.; Garcia, H. Metal-organic frameworks catalyzed C–C and C-heteroatom coupling reactions. *Chem. Soc. Rev.* **2015**, *44*, 1922–1947. [CrossRef] [PubMed]

39. Zhu, Y.; Hosmane, N.S. Nanocatalysis: Recent advances and applications in boron chemistry. *Coord. Chem. Rev.* **2015**, *293–294*, 357–367. [CrossRef]

40. Wang, D.; Deraedt, C.; Ruiz, J.; Astruc, D. Magnetic and dendritic catalysts. *Acc. Chem. Res.* **2015**, *48*, 1871–1880. [CrossRef] [PubMed]

41. Yao, Y.; Patzig, C.; Hu, Y.; Scott, R.W.J. In Situ X-ray absorption spectroscopic study of Fe@FexOy/Pd and Fe@FexOy/Cu nanoparticle catalysts prepared by galvanic exchange reactions. *J. Phys. Chem. C* **2015**, *119*, 21209–21218. [CrossRef]

42. Wittmann, S.; Majoral, J.-P.; Grass, R.N.; Stark, W.J.; Reiser, O. Carbon coated magnetic nanoparticles as supports in microwave-assisted palladium catalyzed Suzuki-Miyarura coupling. *Green Process. Synth.* **2012**, *1*, 275–279.

43. Sun, Z.; Yang, J.; Wang, J.; Li, W.; Kaliaguine, S.; Hou, X.; Deng, Y.; Zhao, D. A versatile designed synthesis of magnetically separable nano-catalysts with well-defined core-shell nanostructures. *J. Mater. Chem. A* **2014**, *2*, 6071–6074. [CrossRef]

44. Woo, H.; Lee, K.; Park, J.C.; Park, K.H. Facile synthesis of Pd/Fe$_3$O$_4$/charcoal bifunctional catalyst with high metal loading for high product yields in Suzuki–Miyaura coupling reactions. *New J. Chem.* **2014**, *38*, 5626–5632. [CrossRef]

45. Li, Y.; Zhang, Z.; Shen, J.; Ye, M. Hierarchical nanospheres based on Pd nanoparticles dispersed on carbon coated magnetic cores with a mesoporours ceria shell: A highly intergrated multifunctional catalyst. *Dalton Trans.* **2015**, *44*, 16592–16601. [CrossRef] [PubMed]

46. Cappelletti, A.L.; Uberman, P.M.; Martín, S.E.; Sateta, M.E.; Troiani, H.E.; Sánchez, R.D.; Carbonio, R.E.; Strumia, M.C. Synthesis, characterization, and nanocatalysis application of core-shell superparamagnetic nanoparticles of Fe$_3$O$_4$@Pd. *Aust. J. Chem.* **2015**, *68*, 1492–1501. [CrossRef]

47. Hosseini-Sarvari, M.; Khanivar, A.; Moeini, F. Palladium immobilized on Fe$_3$O$_4$/ZnO nanoparticles: A novel magnetically recyclable catalyst for Suzuki–Miyaura and Heck reactions under ligand-free conditions. *J. Iran. Chem. Soc.* **2016**, *13*, 45–53. [CrossRef]

48. Li, Y.; Zhang, Z.; Fan, T.; Li, X.; Dong, P.; Baines, R.; Shen, J.; Ye, M. Magnetic core-shell to yolk-shell structures in palladium-catalyzed Suzuki–Miyaura reactions: Heterogeneous versus homogeneous nature. *ChemPlusChem* **2016**, *81*, 564–573. [CrossRef]

49. Xia, J.; Fu, Y.; He, G.; Sun, X.; Wang, X. Core-shell-like Ni-Pd nanoparticles supported on carbon black as a magnetically separable catalyst for green Suzuki–Miyaura coupling reactions. *Appl. Catal. B* **2017**, *200*, 39–46. [CrossRef]

50. Zhang, Q.; Su, H.; Luo, J.; Wei, Y. "Click" magnetic nanoparticle-supported palladium catalyst: A phosphine-free, highly efficient and magnetically recoverable catalyst for Suzuki–Miyaura coupling reactions. *Catal. Sci. Technol.* **2013**, *3*, 235–243. [CrossRef]

51. Chaicharoenwimolkul, L.; Chairam, S.; Namkajorn, M.; Khamthip, A.; Kamonsatikul, C.; Tewasekson, U.; Jindabot, S.; Pon-On, W.; Somsook, E. Effect of ferrocene substituents and ferricinium additive on the properties of polyaniline derivatives and catalytic activities of palladium-doped poly(m-ferrocenylaniline)-catalyzed Suzuki–Miyaura cross-coupling reactions. *J. Appl. Polym. Sci.* **2013**, *130*, 1489–1497. [CrossRef]

52. Karimi, B.; Mansouri, F.; Vali, H. A highly water-dispersible/magnetically separable palladium catalyst based on a Fe_3O_4@SiO_2 anchored TEG-imidazolium ionic liquid for the Suzuki–Miyaura coupling reaction in water. *Green Chem.* **2014**, *16*, 2587–2596. [CrossRef]

53. Esmaeilpour, M.; Javidi, J.; Dodeji, F.N.; Hassannezhad, H. Fe_3O_4@SiO_2-polymer-imid-Pd magnetic porous nanosphere as magnetically separable catalyst for Mizoroki-Heck and Suzuki–Miyaura coupling ractions. *J. Iran. Chem. Soc.* **2014**, *11*, 1703–1715. [CrossRef]

54. Le, X.; Dong, Z.; Jin, Z.; Wang, Q.; Ma, J. Suzuki–Miyaura cross-coupling reactions catalyzed by efficient and recyclable Fe_3O_4@SiO_2@SiO_2-Pd(II) catalyst. *Catal. Commun.* **2014**, *53*, 47–52. [CrossRef]

55. Ghotbinejad, M.; Khosropour, A.R.; Mohammadpoor-Baltork, I.; Moghadam, M.; Tangestaninejad, S.; Mirkhani, V. Ultrasound-assisted C–C coupling reactions catalyzed by unique SPION-A-Pd(EDTA) as a robust nanocatalyst. *RSC Adv.* **2014**, *4*, 8590–8596. [CrossRef]

56. Nehlig, E.; Waggeh, B.; Millot, N.; Lalatonne, Y.; Motte, L.; Guénin, E. Immobilized Pd on magnetic nanoparticles bearing proline as a highly efficient and retrievable Suzuki–Miyaura catalyst in aqueous media. *Dalton Trans.* **2015**, *44*, 501–505. [CrossRef] [PubMed]

57. Esmaeilpour, M.; Javidi, J. Magnetically-recoverable Schiff Base Complex of Pd(II) immobilized on Fe_3O_4@SiO_2 nanoparticles: An efficient catalyst for Mizoroki-Heck and Suzuki–Miyaura coupling reactions. *J. Chin. Chem. Soc.* **2015**, *62*, 614–626. [CrossRef]

58. Gholinejad, M.; Razeghi, M.; Ghaderi, A.; Biji, P. Palladium supported on phosphinite functionalized Fe_3O_4 nanoparticles as a new magnetically separable catalyst for Suzuki–Miyaura coupling reactions in aqueous media. *Catal. Sci. Technol.* **2016**, *6*, 3117–3127. [CrossRef]

59. Khakiani, B.A.; Pourshamsian, K.; Veisi, H. A highly stable and efficient magnetically recoverable and reusable Pd nanocatalyst in aqueous media heterogeneously catalyzed Suzuki C–C cross-coupling reactions. *Appl. Organometal. Chem.* **2015**, *29*, 259–265. [CrossRef]

60. Azizi, K.; Ghonchepour, E.; Karimi, M.; Heydari, A. Encapsulation of Pd(II) into superparamagnetic nanoparticles grafted with EDTA and their catalytic activity towards reduction of nitroarenes and Suzuki–Miyaura coupling. *Appl. Organometal. Chem.* **2015**, *29*, 187–194. [CrossRef]

61. Mowassagh, B.; Takallou, A.; Mobaraki, A. Magnetic nanoparticle-supported Pd(II)-cryptand 22 complex: An efficient and reusable heterogeneous precatalyst in the Suzuki–Miyaura coupling and the formation of aryl-sulfur bonds. *J. Mol. Catal. A Chem.* **2015**, *401*, 55–65. [CrossRef]

62. Hoseini, S.J.; Heidari, V.; Nasrabadi, H. Magnetic Pd/Fe_3O_4/reduced-graphene oxide nanohybrid as an efficient and recoverable catalyst for Suzuki–Miyaura coupling reaction in water. *J. Mol. Catal. A Chem.* **2015**, *396*, 90–95. [CrossRef]

63. Wang, Z.; Yu, Y.; Zhang, Y.X.; Li, Z.; Qian, H.; Lin, Z.Y. A magnetically separable palladium catalyst containing a bulky N-heterocyclic carbine ligand for the Suzuki–Miyaura reaction. *Green Chem.* **2015**, *17*, 413–420. [CrossRef]

64. Ghotbinejad, M.; Khosropour, A.R.; Mohammadpoor-Baltork, I.; Moghadam, M.; Tangestaninejad, S.; Mirkhani, V. SPIONs-bis(NHC)-palladium(II): A novel, powerful and efficient catalyst for Mizoroki-Heck and Suzuki–Miyaura C–C coupling reactions. *J. Mol. Catal. A Chem.* **2014**, *385*, 78–84. [CrossRef]

65. Liu, X.; Zhao, X.; Lu, M. A highly water-dispersible and magnetically separable palladium catalyst based on functionalized poly(ethylene glycol)-supported iminophosphine for Suzuki–Miyaura coupling in water. *Appl. Organometal. Chem.* **2015**, *29*, 419–424. [CrossRef]

66. Azadi, G.; Ghorbani-Choghamarani, A. Immobilized palladium on modified magnetic nanoparticles and study of its catalytic activity in Suzuki–Miyaura C–C coupling reaction. *Appl. Organometal. Chem.* **2016**, *30*, 360–366. [CrossRef]

67. Martinez-Olid, F.; Andrés, R.; de Jesus, E.; Flores, J.C.; Gómez-Sal, P.; Heuze, K.; Vellutini, L. Magnetically recoverable catalysts base don mono- or bis-(NHC) complexes of palladium for the Suzuki–Miyaura reaction in aqueous media: Two NHC-Pd linkages are better than one. *Dalton Trans.* **2016**, *45*, 11633–11638. [CrossRef] [PubMed]

68. Salemi, H.; Kaboudin, B.; Kazemi, F.; Yokomatsu, T. Highly water-dispersible magnetite nanoparticle supported-palladium-β-cyclodextrin as an efficient catalyst for Suzuki–Miyaura and Sonogashira coupliong reactions. *RSC Adv.* **2016**, *6*, 52656–52664. [CrossRef]

69. Ghorbani-Vaghei, R.; Hemmati, S.; Hekmati, M. Pd immobilized on modified magnetic Fe_3O_4 nanoparticles: Magnetically recoverable and reusable Pd nanocatalyst for Suzuki–Miyaura coupling reactions and Ullmann-type N-arylation of indoles. *J. Chem. Sci.* **2016**, *128*, 1157–1162. [CrossRef]

70. Hajipour, A.R.; Tadayoni, N.S.; Khorsandi, Z. Magnetic iron oxide nanoparticles-N-heterocyclic carbine-palladium(II): A new, efficient and robust recyclable catalyst for Mizoroki-Heck and Suzuki–Miyaura coupling reactions. *Appl. Organometal. Chem.* **2016**, *30*, 590–595. [CrossRef]

71. Dubey, A.V.; Kumar, A.V. A biomimetic magnetically recoverable palladium nanocatalyst for the Suzuki cross-coupling reaction. *RSC Adv.* **2016**, *6*, 46864–46870. [CrossRef]

72. Phan, N.T.S.; Van Der Sluys, M.; Jones, C.W. On the nature of the active species in palladium catalyzed Mizoroki-Heck and Suzuki–Miyaura couplings—Homogeneous or heterogeneous catalysis, a critical review. *Adv. Synth. Catal.* **2006**, *348*, 609–679. [CrossRef]

catalysts

MDPI

Communication

Intramolecular Transfer of Pd Catalyst on Carbon–Carbon Triple Bond and Nitrogen–Nitrogen Double Bond in Suzuki–Miyaura Coupling Reaction

Takeru Kamigawara, Hajime Sugita, Koichiro Mikami, Yoshihiro Ohta and Tsutomu Yokozawa *

Department of Materials and Life Chemistry, Kanagawa University, 3-27-1 Rokkakubashi, Kanagawa-ku, Yokohama 221-8686, Japan; katze102@icloud.com (T.K.); h.sugita1992@gmail.com (H.S.); koichiro.mikami@sagami.or.jp (K.M.); y-ohta0112@kanagawa-u.ac.jp (Y.O.)
* Correspondence: yokozt01@kanagawa-u.ac.jp; Tel.: +81-45-481-5661

Academic Editor: Ioannis D. Kostas
Received: 24 March 2017; Accepted: 20 June 2017; Published: 23 June 2017

Abstract: Intramolecular transfer of t-Bu$_3$P-ligated Pd catalyst on a carbon–carbon triple bond (C≡C) and nitrogen–nitrogen double bond (N=N) was investigated and compared with the case of a carbon–carbon double bond (C=C), which is resistant to intramolecular transfer of the Pd catalyst. Suzuki–Miyaura coupling reaction of equimolar 4,4′-dibromotolan (**1a**) or 4,4′-dibromoazobenzene (**1b**) with 3-isobutoxyphenylboronic acid (**2**) was carried out in the presence of t-Bu$_3$P-ligated Pd precatalyst **3** and KOH/18-crown-6 as a base at room temperature. In both cases, the diphenyl-substituted product was selectively obtained, indicating that the Pd catalyst walked from one benzene ring to the other through the C≡C or N=N bond after the first substitution with **2**. Taking advantage of this finding, we conducted unstoichiometric Suzuki–Miyaura polycondensation of 1.3 equiv. of **1** and 1.0 equiv. of phenylenediboronic acid (ester) **6** in the presence of **3** and CsF/18-crown-6 as a base, obtaining high-molecular-weight conjugated polymer with a boronic acid (ester) moiety at both ends, contrary to the Flory principle.

Keywords: palladium catalyst; Suzuki coupling; catalyst transfer; conjugated polymer; unstoichometric polycondensation

---

1. Introduction

Suzuki–Miyaura coupling reaction is a powerful protocol for the synthesis of polyarylenes containing π-conjugated polymers [1]. We have found that t-Bu$_3$PPd(Ar)Br [2,3] initiates chain-growth Suzuki–Miyaura coupling polymerization of haloarylboronic acid (ester) as an AB type monomer to afford well-defined polyfluorene [4], poly(p-phenylene) [5], and poly(hexylthiophene) [6], and other researchers have obtained well-defined poly(phenanthrene) [7] and poly(fluorene-*alt*-benzothiadiazole) [8]. The chain-growth polymerization progresses via intramolecular transfer of the Pd catalyst to the terminal C–X (X = halogen) bond after reductive elimination of polymer–Pd–ArX. Therefore, these types of chain-growth polymerizations, including Kumada–Tamao and other coupling polymerizations, are known as catalyst-transfer condensation polymerizations (CTCPs) [9,10].

When this t-Bu$_3$PPd(0) catalyst, which has a propensity for intramolecular catalyst transfer on a π-electron face, was used for Suzuki–Miyaura coupling polymerization of dibromoarene and arenyldiboronic acid ester (AA + BB polycondensation), high-molecular-weight π-conjugated polymer with a boronate moiety at both ends was obtained, even though excess dibromoarene was used [11]. This unstochiometric polycondensation behavior is accounted for by successive substitution of the bromides in dibromoarene with arenyldiboronic acid ester or oligomers having boronate moieties at both ends through intramolecular transfer of the Pd catalyst on the π face of dibromoarene. However,

Suzuki–Miyaura coupling reaction of 4,4'-dibromostilbene with phenylboronic acid in the presence of *t*-Bu₃PPd(0) catalyst did not selectively afford diphenyl-substituted stilbene, implying that the Pd catalyst did not walk from one benzene ring to the other through the carbon–carbon double bond (C=C) in the stilbene after the first substitution of 4,4'-dibromostilbene with phenylboronic acid [12]. We found that this failure of intramolecular catalyst transfer is due to bimolecular intermolecular transfer of *t*-Bu₃PPd(0) on the C=C of dibromostilbene to the C=C of another dibromostilbene. However, this intermolecular transfer of the catalyst could be suppressed by introduction of alkoxy groups at the ortho positions of the C=C.

We were next interested in whether or not the Pd catalyst undergoes intramolecular catalyst transfer on other multiple bonds. In the present study, we investigated catalyst transfer on a carbon–carbon triple bond (C≡C) and nitrogen–nitrogen double bond (N=N) by means of Suzuki–Miyaura coupling reaction of 4,4'-dibromotolan (**1a**) or 4,4'-dibromoazobenzene (**1b**) with phenylboronic acid **2** in the presence of *t*-Bu₃PPd(0) precatalyst **3** [13] (Scheme 1). If the catalyst undergoes intramolecular transfer on the multiple bond X in **1** and then inserts itself into the C–Br bond after the first substitution, the main product would be disubstituted **5**. On the other hand, if the catalyst diffuses into the reaction mixture after the first substitution, the main product would be monosubstituted **4**. Bielawski and coworkers have demonstrated intramolecular catalyst transfer of *t*-Bu₃PPd(0) catalyst on C≡C in Stille CTCP of 4-(tributylstannylethynyl)-2,5-dialkoxyiodobenzene as an AB monomer [14]. However, this catalyst transfer on C≡C might be dependent on the steric effect of the tributylstannyl and/or alkoxy groups, as in the case of catalyst transfer on C=C in *o*-alkoxy-substituted stilbene. In the present work, we found that *t*-Bu₃PPd(0) catalyst undergoes intramolecular catalyst transfer on C≡C and N=N even if the benzene rings adjacent to these multiple bonds are unsubstituted, in contrast to the case of catalyst transfer on C=C. To demonstrate effective catalyst transfer on C≡C and N=N, we further conducted unstoichiometric Suzuki–Miyaura polycondensation of excess **1** and 1.0 equiv. of phenylenediboronic acid (ester) in the presence of **3**, obtaining π-conjugated polymer with a boronate moiety at both ends.

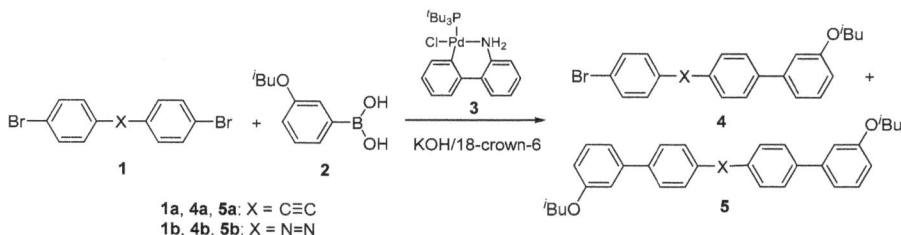

Scheme 1. Suzuki–Miyaura coupling reaction of **1** with **2** in the presence of **3** to examine intramolecular catalyst transfer on the multiple bond X in **1**.

2. Results and Discussion

2.1. Suzuki–Miyaura Coupling Reaction of **1** with **2**

Intramolecular catalyst transfer on C≡C was first investigated by Suzuki–Miyaura coupling reaction of equimolar **1a** with **2** in the presence of 5 mol % of *t*-Bu₃PPd(0) precatalyst **3** and KOH/18-crown-6 (Table 1). In order to determine the product ratio of monosubstituted **4a** to disubstituted **5a** by means of gas chromatography (GC), authentic **4a** and **5a** were prepared by the reaction of equimolar **1a** and **2** in the presence of (Ph₃P)₄Pd, which has no propensity for intramolecular catalyst transfer on a π face, and by the reaction of **1a** and 2 equiv. of **2** in the presence of **3**, respectively.

Table 1. Suzuki–Miyaura coupling reaction of 1 with 2 in the presence of 3 [1].

Entry	1	Additive	Conv. of 1 (%) [2]	Yield of 4 and 5 (%) [2]	4:5 [2]
1	1a	-	74	70	6:94
2	1a	styrene	80	77	74:26
3	1a	stilbene	74	66	23:77
4	1a	phenylacetylene	97	56	35:65
5	1a	tolan	69	66	8:92
6	1a	1-pheynyl-1-propyne	67	62	10:90
7 [3]	1b	-	61 [4]	52 [4]	11:89 [4]

[1] Reaction of 1 and 2 was carried out in the presence of 5.0 mol % of 3, KOH (4.5 equiv.)/18-crown-6 (8.0 equiv.), and 4 equiv. of additive in THF ($[1]_0 = 1.60 \times 10^{-2}$ M) and water (THF/water = 26.7/1, *v/v*) at rt for 2 h; [2] Determined by GC; [3] Reaction was carried out for 24 h; [4] Determined from the ^{1}H NMR spectra.

In the reaction of **1a** with **2**, disubstituted **5a** was obtained with high selectivity (Entry 1). Furthermore, the product ratio of **5a** was consistently more than 90% from the beginning of the reaction (Figure S1). This is important, because if **5a** was selectively formed due to higher reactivity of **4a** than **1a**—i.e., not via the catalyst-transfer mechanism—then **4a** should be accumulated at the beginning of the reaction, in which the concentration of **1a** is much higher than that of **4a** [15]. Therefore, these results indicate that intramolecular transfer of *t*-Bu$_3$PPd(0) catalyst on C≡C took place, in contrast to the case of the reaction of 4,4'-dibromostilbene with **2** in the presence of *t*-Bu$_3$PPd(0) catalyst, which afforded mainly monosubstituted stilbene [12]. We next examined the effects of additives to see whether intramolecular catalyst transfer on C≡C is disturbed by additives containing C=C or C≡C. If the affinity of an additive for *t*-Bu$_3$PPd(0) is stronger than that of C≡C, the additive would trap the Pd catalyst during intramolecular catalyst transfer on **1a**, resulting in a decrease of the ratio of disubstituted **5a**. In the case of styrene as an additive, the ratio of **5a** was drastically decreased to 26% (Entry 2), whereas stilbene resulted in decrease to 77% (Entry 3). It turned out that terminal C=C disturbed intramolecular catalyst transfer on C≡C more strongly than did internal C=C. As for additives containing C≡C, phenylacetylene also induced a decrease of the ratio of **5a** (entry 4), whereas tolan and 1-phenyl-1-propyne did not affect the intramolecular transfer of the catalyst, and **5a** was selectively formed (Entries 5 and 6). Consequently, compounds bearing internal C≡C, including substrate **1a**, turned out not to disturb intramolecular catalyst transfer on C≡C.

We next investigated intramolecular catalyst transfer on N=N. Suzuki–Miyaura coupling reaction of 4,4'-dibromoazobenzene (**1b**) with **2** was similarly carried out. Products **4b** and **5b** were isolated by preparative HPLC, and we confirmed that aromatic proton signals of **1b**, **4b**, and **5b** appeared separately in ^{1}H NMR spectra (Figure S2). The same reaction was conducted again in the presence of 1,4-bis(hexyloxy)benzene as an internal standard, and conversion of **1b** and the product ratio of **4b** to **5b** were determined by ^{1}H NMR spectroscopy. The ratio of **4b** to **5b** was 11:89, indicating that intramolecular catalyst transfer also took place on N=N (Entry 7), although the ratio of **5b** was slightly decreased as compared to the case of reaction of **1a** with **2** (Entry 1).

The observation that *t*-Bu$_3$PPd(0) catalyst underwent intramolecular catalyst transfer on C≡C and N=N, in contrast to the case of C=C, might be accounted for by lower stability of the η^2 coordination complexes of these multiple bonds to *t*-Bu$_3$PPd, compared with the η^2 complex of C=C. As the η^2 complex becomes less stable, the chance of bimolecular ligand exchange of η^2 complex between multiple bonds would be decreased. The observation that intramolecular catalyst transfer on C≡C is disturbed by additives containing C=C implies that the η^2 complex of C=C is more stable than that of C≡C. In addition, density functional theory (DFT) calculations were performed to estimate the stability of the η^2 coordination complexes of these multiple bonds with *t*-Bu$_3$PPd. The values of stabilization energy of complexation of stilbene, tolan, and azobenzene with *t*-Bu$_3$PPd were −27.9, −25.6, and −20.8 kcal/mol, respectively (Figures S3–S5). These results support our interpretation, although calculations for model PdPH$_3$ systems with ethylene and acetylene indicated that the Pd–C bonds are stronger in the acetylene complex than in the ethylene complex [16,17].

2.2. Unstoichiometric Suzuki–Miyaura Polycondensation

2.2.1. Polycondensation of Dibromotolan and Phenylenediboronic Acid (Ester)

Since we found that t-Bu$_3$PPd(0) catalyst undergoes intramolecular catalyst transfer on C≡C, we conducted unstoichiometric Suzuki–Miyaura polycondensation of 1.3 equiv. of **1a** and 1.0 equiv. of phenylenediboronic acid **6a** in the presence of Pd catalyst **3** and CsF/18-crown-6 as a base at room temperature to afford polymer with M_n of 5500 Da and M_w/M_n of 3.95 (Scheme 2). The matrix-assisted laser desorption ionization time-of-flight (MALDI-TOF) mass spectrum of the polymer showed one series of peaks due to polymer with a boronic acid moiety at both ends (Figure 1). These results indicated that the polymerization involved intramolecular catalyst transfer on C≡C-containing monomer **1a**.

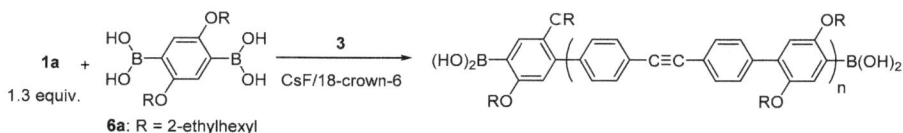

Scheme 2. Unstoichiometric Suzuki–Miyaura polycondensation of **1a** with **6a** in the presence of **3**.

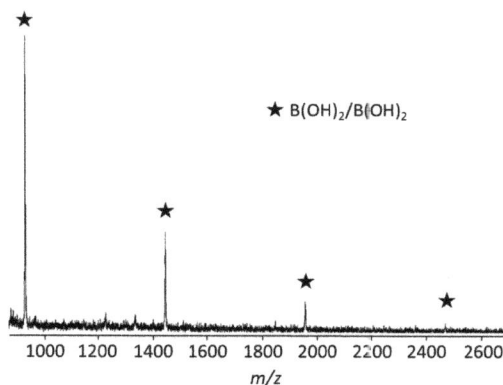

Figure 1. MALDI-TOF MS spectrum of polymer obtained by the polymerization of 1.3 equiv. of **1a** and 1.0 equiv. of **6a** in the presence of 2.0 mol % of **3**, CsF (5 equiv.), and 18-crown-6 (8 equiv.) in THF ($[\mathbf{6a}]$ = 8.33 × 10^{-3} M) and water (THF/water = 30/1, *v/v*) at rt for 15 h, followed by quenching with 1 M hydrochloric acid.

The polymerization of alkoxy-substituted dibromotolan **1c** and alkoxy-substituted phenylene diboronic acid ester **6b** was also conducted under similar conditions (Scheme 3). The molecular weight increased with reaction time and reached M_n = 10,500 Da (M_w/M_n = 2.77) (Figure 2a), even though 1.3 equiv. of **1c** was used. The MALDI-TOF mass spectrum of polymer obtained at 24 h similarly showed major peaks due to polymer with a boronic acid ester moiety at both ends (BPin/BPin). Minor peaks in the lower-molecular-weight region are assignable to BPin/H and BPin/B(OH)$_2$, which would be formed by hydrolysis of polymer–Pd–Br and the pinacol boronate moiety, respectively, by hydrochloric acid used to quench the polymerization. Formation of higher-molecular-weight polymer, in contrast to the polymerization of unsubstituted tolan **1a** with **6a**, is presumably accounted for by a higher propensity for intramolecular catalyst transfer on **1c** than **1a**, as in the case of Suzuki–Miyaura coupling reaction of dibromostilbene with phenylboronic acid: alkoxy groups at the ortho position of C=C of stilbene promoted intramolecular catalyst transfer, as mentioned in the Introduction.

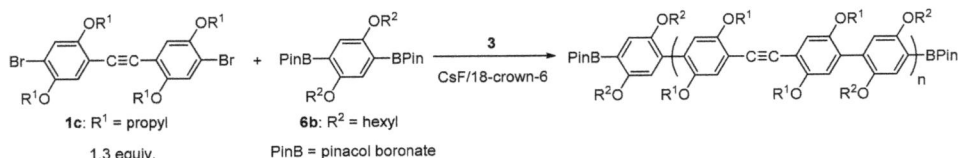

Scheme 3. Unstoichiometric Suzuki–Miyaura polycondensation of **1c** with **6b** in the presence of **3**.

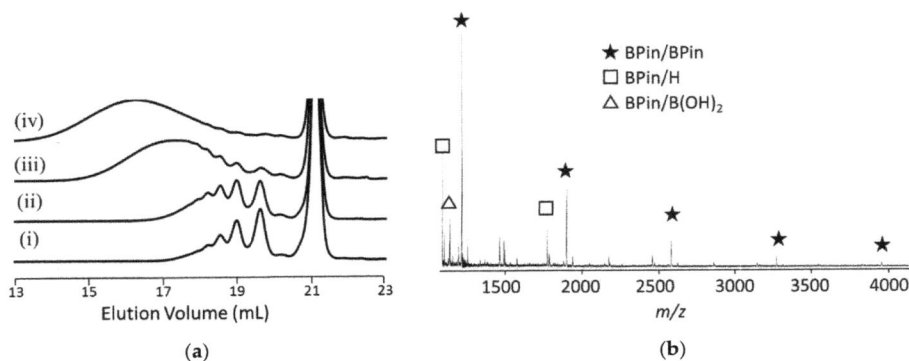

(a) (b)

Figure 2. Unstoichiometric Suzuki–Miyaura polycondensation of 1.3 equiv. of **1c** with 1.0 equiv. of **6b** in the presence of 2.0 mol % of **3**, CsF (5.0 equiv.), and 18-crown-6 (8.0 equiv.) in THF ([6b] = 8.33 × 10^{-3} M) and water (THF/water = 30/1, *v/v*) at rt: (a) GPC profiles of polymer obtained at (i) 1 h (M_n = 2470 Da, M_w/M_n = 1.29), (ii) 5 h (M_n = 3000 Da, M_w/M_n = 1.39), (iii) 24 h (M_n = 6640 Da, M_w/M_n = 2.03), (iv) 48 h (M_n = 10,500 Da, M_w/M_n = 2.77; (b) MALDI-TOF mass spectra of polymer obtained at 24 h (M_n = 6640 Da, M_w/M_n = 2.03).

2.2.2. Polycondensation of Dibromoazobenzene and Phenylenediboronic Acid Ester

We applied intramolecular catalyst transfer on N=N, as mentioned in Section 2.1, for unstoichiometric polycondensation of 1.3 equiv. of dibromoazobenzene **1b** and 1.0 equiv. of phenylenediboronic acid ester **6b** in the presence of Pd catalyst **3** (Scheme 4). The polymerization proceeded homogeneously, but polymer was precipitated at 42 h. Accordingly, the THF-soluble part of the polymer obtained at 42 h was analyzed. The GPC elution curve of the obtained polymer showed M_n = 5500 Da and M_w/M_n = 1.44, and the polymer ends were confirmed to be BPin/BPin by MALDI-TOF mass spectrometry (Figure 3). Therefore, it turned out that polycondensation involving intramolecular catalyst transfer on **1b** proceeded, although high-molecular-weight polymer was not obtained due to the low solubility of the polymer in THF.

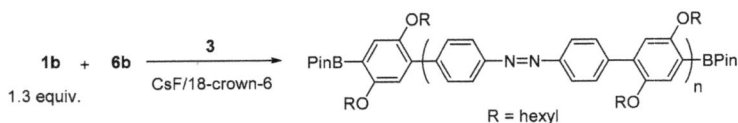

Scheme 4. Unstoichiometric Suzuki–Miyaura polycondensation of **1b** with **6b** in the presence of **3**.

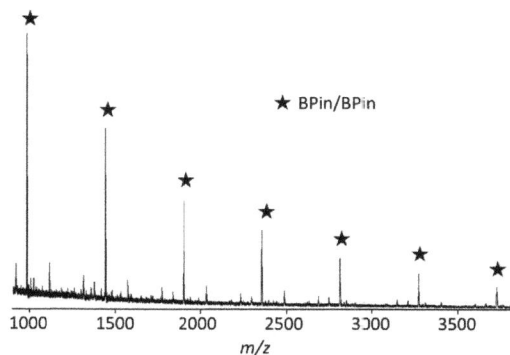

Figure 3. MALDI-TOF MS spectrum of polymer obtained by the polymerization of 1.3 equiv. of **1b** and 1.0 equiv. of **6b** in the presence of 2.0 mol % of **3**, CsF (5 equiv.), and 18-crown-6 (8 equiv.) in THF ([**6b**] = 8.33×10^{-3} M) and water (THF/water = 30/1, v/v) at rt for 42 h, followed by quenching with 1 M hydrochloric acid.

In order to obtain more soluble polymer, polymerization of **1b** with **6a**, bearing a 2-ethylhexyloxy group, was carried out. The polymerization proceeded homogeneously, and the molecular weight was increased to 8300. However, the molecular weight distribution was exceptionally broad (M_w/M_n = 12.4); the reason for this is not clear at present. Probably due to this broad molecular weight distribution, the MALDI-TOF mass spectrum did not show clear peaks in the region of molecular weight higher than 2000 Da.

3. Materials and Methods

3.1. Materials

All starting materials were purchased from commercial suppliers (TCI, Aldrich, Wako, and Kanto, Tokyo, Japan) and used without further purification. Dry tetrahydrofuran (THF, stabilizer-free, Kanto, Tokyo, Japan) and distilled water (Wako, Tokyo, Japan) were used as received. Dibromoazobenzene **1b** [18] and phenylenediboronic acid **6a** [11] were prepared according to the literature.

3.2. Measurements

^{1}H and ^{13}C NMR spectra were obtained on JEOL ECA-500 (JEOL, Tokyo, Japan and ECA-600 spectrometers (JEOL, Tokyo, Japan). The internal standard for ^{1}H NMR spectra in CDCl$_3$ was tetramethylsilane (0.00 ppm) and the internal standard for ^{13}C NMR spectra in CDCl$_3$ was the midpoint of CDCl$_3$ (77.0 ppm). IR spectra were recorded on a JASCO FT/IR-410 (JASCO, Tokyo, Japan). All melting points were measured with a Yanagimoto hot stage melting point apparatus (Yanaco, Tokyo, Japan) without correction. GC was performed on a Shimadzu GC-14B gas chromatograph (Shimazu, Kyoto, Japan) equipped with a Shimadzu fused silica capillary column CBP1-W12–100 (12 m length, 0.53 mm i.d.) and a flame ionization detector (FID). Isolation of **4b** and **5b** was carried out on a Japan Analytical Industry LC908-C60 recycling preparative HPLC (eluent, CHCl$_3$) (JAI, Tokyo, Japan) with two JAIGEL columns (1H-40 and 2H-40). Column chromatography was performed on silica gel (Kieselgel 60, 230–400 mesh, Merck, Darmstadt, Germany) with a specified solvent. The M_n and M_w/M_n values of polymer were measured on a Tosoh HLC-8020 gel permeation chromatography (GPC) unit (eluent, THF; calibration, polystyrene standards) (Tosoh, Yamaguchi, Japan) with two TSK-gel columns (2 × Multipore H$_{XL}$-M). MALDI-TOF mass spectra were recorded on a Shimadzu/Kratos AXIMA-CFR plus (Shimadzu, Kyoto, Japan) in the reflectron ion mode (laser λ = 337 nm). DCTB (trans-2-[3-(4-*tert*-butylphenyl)-2-methyl-2-propenylidene]malononitrile) was used as the matrix for the MALDI-TOF mass measurements.

3.3. Synthesis of Monosubstituted Tolan **4a**

Transfer of reagents and withdrawal of small aliquots of the reaction mixture for analysis were carried out via a syringe from a three-way stopcock under a stream of nitrogen. A two-necked round-bottomed flask was equipped with a three-way stopcock and a dimroth condenser. Dibromotolan **1a** (134.9 mg, 0.40 mmol), phenylboronic acid **2** (81.4 mg, 0.42 mmol), (PPh$_3$)$_4$Pd (23.3 mg, 0.020 mmol), K$_3$PO$_4$ (370.9 mg, 1.75 mmol), and 18-crown-6 (846 mg, 3.26 mmol) were placed in the flask, and the flask was flushed with argon. Dry THF (10.0 mL) and distilled water (0.60 mL) were added to the flask via a syringe, and the reaction mixture was refluxed for 3 h. The reaction was quenched with 1 M hydrochloric acid, and the mixture was extracted with CHCl$_3$. The organic layer was washed with water and dried over anhydrous MgSO$_4$. The solvent was removed under reduced pressure, and the residue was purified by means of column chromatography (SiO$_2$, hexane) to afford **4a** as a pale yellow viscous liquid (111.3 mg, 69%); mp 96–99 °C.

^{1}H NMR (500 MHz, CDCl$_3$) δ 7.58 (s, 4H), 7.50 (d, J = 8.6 Hz, 2H), 7.40 (d, J = 8.6 Hz, 2H), 7.35 (t, J = 8.0 Hz, 1H), 7.18 (d, J = 6.9 Hz, 1H), 7.13 (s, 1H), 6.90 (dd, J = 5.7 Hz and 2.3 Hz, 1H), 3.78 (d, J = 6.9 Hz, 2H), 2.14-2.10 (m, 1H), 1.05 (d, J = 6.3 Hz, 6H); ^{13}C NMR (126 MHz, CDCl$_3$) δ 159.7, 141.8, 140.9, 132.0, 129.8, 127.3, 122.3, 119.3, 113.4, 90.0, 75.6, 30.9, 28.3, 19.3; IR (KBr) 3460, 2908, 1606, 1471, 1350, 1206, 1105, 965, 776, 690, 513 cm^{-1}.

3.4. Synthesis of Disubstituted Tolan **5a**

Transfer of reagents and withdrawal of small aliquots of the reaction mixture for analysis were carried out via a syringe from a three-way stopcock under a stream of nitrogen. Dibromotolan **1a** (134.3 mg, 0.40 mmol), **2** (157.7 mg, 0.81 mmol), **3** (10.2 mg, 0.020 mmol), KOH (105.4 mg, 1.88 mmol), and 18-crown-6 (861.0 mg, 3.26 mmol) were placed in a flask equipped with a three-way stop cock, and the atmosphere in the flask was replaced with argon. Dry THF (10.0 mL) and distilled water (0.60 mL) were added to the flask via a syringe, and the mixture was degassed under reduced pressure and filled with argon. The reaction mixture was stirred at room temperature for 2 h. The reaction was quenched with hydrochloric acid, and the mixture was extracted with CHCl$_3$. The organic layer was washed with water and dried over anhydrous MgSO$_4$. The solvent was removed under reduced pressure, and the residue was purified by means of column chromatography (SiO$_2$, hexane) to afford **5a** as a dark yellow viscous solid (184.1 mg, 96%); mp 130–134 °C.

^{1}H NMR (500 MHz, CDCl$_3$) δ 7.60 (s, 8H), 7.35 (t, J = 7.7 Hz, 2H), 7.18 (d, J = 7.7 Hz, 2H), 7.15 (s, 2H), 6.90 (dd, J = 6.0 and 2.0 Hz, 2H), 3.79 (d, J = 6.6 Hz, 4H), 2.13-2.10 (m, 2H), 1.04 (d, J = 6.9 Hz, 12H); ^{13}C NMR (126 MHz, CDCl$_3$) δ 167.7, 152.2, 147.2, 144.7, 130.8, 124.5, 115.5, 114.2, 83.9, 69.8, 29.7, 24.8; IR (KBr) 3461, 2907, 1606, 1522, 1295, 1206, 1050, 1013, 873, 830, 776, 690, 538 cm^{-1}.

3.5. Synthesis of Mono- and Disubstituted Azobenzene **4b** and **5b**

Transfer of reagents and withdrawal of small aliquots of the reaction mixture for analysis were carried out via a syringe from a three-way stopcock under a stream of nitrogen. Dibromoazobenzene **1b** (50.9 mg, 0.15 mmol), **2** (30.6 mg, 0.16 mmol), **3** (4.2 mg, 0.016 mmol), KOH (37.8 mg, 0.68 mmol), and 18-crown-6 (318.4 mg, 1.2 mmol) were placed in a flask equipped with a three-way stop cock, and the atmosphere in the flask was replaced with argon. Dry THF (6.0 mL) and distilled water (0.36 mL) were added to the flask via a syringe, and the mixture was degassed under reduced pressure and filled with argon. The reaction mixture was stirred at room temperature for 24 h. The reaction was quenched with hydrochloric acid, and the mixture was extracted with CHCl$_3$. The organic layer was washed with water and dried over anhydrous MgSO$_4$. The solvent was removed under reduced pressure, and the residue was purified by means of HPLC to afford **4b** as a pale yellow viscous solid (10.4 mg, 17%) and **5b** as a pale yellow viscous solid (22.3 mg, 31%).

4b: mp 89–95 °C; ^{1}H NMR (500 MHz, CDCl$_3$) δ 7.98 (d, *J* = 8.6 Hz, 2H), 7.82 (d, *J* = 8.9 Hz, 2H), 7.74 (d, *J* = 8.6 Hz, 2H), 7.66 (d, *J* = 8.6 Hz, 2H), 7.37 (t, *J* = 8.0 Hz, 1H), 7.24 (d, *J* = 7.7 Hz, 1H), 7.20 (s, 1H), 6.94 (dd, *J* = 7.7 and 1.7 Hz, 1H), 3.80 (d, *J* = 6.6 Hz, 1H), 2.16–2.10 (m, 1H), 1.06 (d, *J* = 6.6 Hz, 6H), ^{13}C NMR (126 MHz, CDCl$_3$) δ 159.7, 151.9, 143.6, 141.6, 129.8, 127.8, 123.3, 119.4, 113.8, 113.6, 74.5, 28.3, 19.3, IR (KBr) 3460, 2911, 1599, 1469, 1397, 1217, 1108, 965, 840, 610, 552 cm^{-1}.

5b: mp 195–198 °C; ^{1}H NMR (500 MHz, CDCl$_3$) δ 8.00 (d, *J* = 8.6 Hz, 4H), 7.75 (d, *J* = 8.3 Hz, 4H), 7.37 (t, *J* = 8.0 Hz, 2H), 7.24 (d, *J* = 8.3 Hz, 2H), 7.20 (s, 2H), 6.93 (dd, *J* = 6.3 and 2.0 Hz, 2H), 3.81 (d, *J* = 6.3 Hz, 4H), 2.17–2.11 (m, 2H), 1.06 (d, *J* = 6.6 Hz, 12H), ^{13}C NMR (126 MHz, CDCl$_3$) δ 159.7, 151.6, 132.3, 129.9, 128.3, 127.8, 124.3, 119.4, 113.9, 113.6, 77.3, 28.3, 19.3, IR (KBr) 3448, 2924, 2362, 1719, 1571, 1473, 1281, 1065, 1006, 835, 657, 577 cm^{-1}.

3.6. Suzuki–Miyaura Coupling Reaction of **1a** with **2**

Transfer of reagents and withdrawal of small aliquots of the reaction mixture for analysis were carried out via a syringe from a three-way stopcock under a stream of nitrogen. Dibromotolan **1a** (0.100 mmol), **2** (0.105 mmol), **3** (0.005 mmol), KOH (0.5 mmol), 18-crown-6 (0.80 mmol), and 1,4-bis(hexyloxy)benzene (0.040 mmol) as an internal standard substance were placed in the flask, and the atmosphere in the flask was replaced with argon. Dry THF (4.0 mL), distilled water (0.40 mL), and additive (none or 0.100 mmol) were added to the flask via a syringe, and the mixture was degassed under reduced pressure and filled with argon. The reaction mixture was stirred at room temperature for 2 h, and the reaction was quenched with 1 M hydrochloric acid. The mixture was extracted with CHCl$_3$, and the organic layer was subjected to GC analysis for estimation of conversion of **1a** and the product ratio of **4a** to **5a**.

3.7. Suzuki–Miyaura Coupling Reaction of **1b** with **2**

Transfer of reagents and withdrawal of small aliquots of the reaction mixture for analysis were carried out via a syringe from a three-way stopcock under a stream of nitrogen. Dibromoazobenzene **1b** (50.9 mg, 0.15 mmol), **2** (30.6 mg, 0.16 mmol), **3** (4.2 mg, 0.008 mmol), KOH (37.8 mg, 0.68 mmol), 18-crown-6 (318.4 mg, 1.2 mmol), and 1,4-bis(hexyloxy)benzene (27.2 mg, 0.016 mmol) as an internal standard substance were placed in the flask, and the atmosphere in the flask was replaced with argon. Dry THF (6.0 mL) and distilled water (0.36 mL) were added to the flask via a syringe, and the mixture was degassed under reduced pressure and filled with argon. The reaction mixture was stirred at room temperature for 24 h, and the reaction was quenched with 1 M hydrochloric acid. The mixture was extracted with CHCl$_3$ and dried over MgSO$_4$. The solvent was removed under reduced pressure. The whole mixture was dissolved in CDCl$_3$ and then subjected to ^{1}H NMR analysis to determine conversion of **1b** and the product ratio of **4b** to **5b**.

3.8. Polycondensation of **1c** and **6b**

Transfer of reagents and withdrawal of small aliquots of the reaction mixture for analysis was carried out via a syringe from a three-way stopcock under a stream of nitrogen. Dibromo monomer **1c** (36.89 mg, 0.065 mmol), phenylenediboronic acid pinacol ester **6b** (26.91 mg, 0.051 mmol), CsF (34.1 mg, 0.22 mmol), 18-crown-6 (119.1 mg, 0.45 mmol), and **3** (1.4 mg, 0.027 mmol) were placed in the flask, and the atmosphere in the flask was replaced with argon. Dry THF (6.0 mL) and distilled water (0.2 mL) were added to the flask via a syringe, and the mixture was degassed under reduced pressure and filled with argon. The reaction mixture was stirred at room temperature for two days, and the reaction was quenched with 1 M hydrochloric acid. The mixture was extracted with CHCl$_3$, and the organic layer was washed with water, then dried over anhydrous MgSO$_4$. Concentration under reduced pressure gave residue, which was purified by preparative HPLC to give polymer (32.7 mg, 94%).

4. Conclusions

We have demonstrated that *t*-Bu$_3$PPd(0) catalyst undergoes intramolecular transfer on C≡C and N=N, in contrast to the case of C=C, by means of Suzuki–Miyaura coupling reaction of equimolar dibromotolan **1a** or dibromoazobenzene **1b** with phenylboronic acid **2** in the presence of *t*-Bu$_3$P-ligated Pd precatalyst **3** and by unstoichiometric polycondensation of **1** and phenyleneboronic acid (ester) **6** with **3**. In the former reaction, successive disubstitution of **1** with **2** selectively proceeded, and the latter polycondensation yielded high-molecular-weight conjugated polymer with a boronic acid (ester) moiety at both ends, even if excess dibromo monomer **1** was used. We believe these findings provide useful information about the requirements for Pd catalyst transfer on π-conjugated faces, as well as affording entry into a new class of π-conjugated polymer architectures via functionalization of chain-end boronate moieties.

Supplementary Materials: The following details are available online at www.mdpi.com/2073-4344/7/7/195/s1, Synthesis of mono- and disubtituted tolan **4a** and **5a**, Synthesis of mono- and disubstituted azobenzene **4b** and **5b**, The product ratio of **5a** to **4a** and **5a** as a function of conversion of **1a**, Estimation of product ratio of **4b** to **5b** by ^{1}H NMR spectra, and DFT calculations for complexation of diphenylacetylene, stilbene, and azobenene with PdP(*t*-Bu)$_3$.

Acknowledgments: This study was supported by a Grant in Aid (No. 15H03819) for Scientific Research from the Japan Society for the Promotion of Science (JSPS) and by the MEXT-Supported Program for the Strategic Research Foundation at Private Universities (No. S1311032), 2013-2018.

Author Contributions: H.S. and T.Y. conceived and designed the experiments; T.K. performed the experiments; K.M. and T.Y. analyzed the data; Y.O. contributed reagents/materials/analysis tools; T.Y. wrote the paper.

Conflicts of Interest: The authors declare no conflict of interest.

References

1. Sakamoto, J.; Rehahn, M.; Wegner, G.; Schlüter, A.D. Suzuki polycondensation: Polyarylenes à la carte. *Macromol. Rapid Commun.* **2009**, *30*, 653–687. [CrossRef] [PubMed]
2. Stambuli, J.P.; Bühl, M.; Hartwig, J.F. Synthesis, characterization, and reactivity of monomeric, arylpalladium halide complexes with a hindered phosphine as the only dative ligand. *J. Am. Chem. Soc.* **2002**, *124*, 9346–9347. [CrossRef] [PubMed]
3. Stambuli, J.P.; Incarvito, C.D.; Buhl, M.; Hartwig, J.F. Synthesis, structure, theoretical studies, and ligand exchange reactions of monomeric, T-shaped arylpalladium(II) halide complexes with an additional, weak agostic interaction. *J. Am. Chem. Soc.* **2004**, *126*, 1184–1194. [CrossRef] [PubMed]
4. Yokoyama, A.; Suzuki, H.; Kubota, Y.; Ohuchi, K.; Higashimura, H.; Yokozawa, T. Chain-growth polymerization for the synthesis of polyfluorene via Suzuki–Miyaura coupling reaction from an externally added initiator unit. *J. Am. Chem. Soc.* **2007**, *129*, 7236–7237. [CrossRef] [PubMed]
5. Yokozawa, T.; Kohno, H.; Ohta, Y.; Yokoyama, A. Catalyst-transfer Suzuki–Miyaura coupling polymerization for precision synthesis of poly(*p*-phenylene). *Macromolecules* **2010**, *43*, 7095–7100. [CrossRef]
6. Yokozawa, T.; Suzuki, R.; Nojima, M.; Ohta, Y.; Yokoyama, A. Precision synthesis of poly(3-hexylthiophene) from catalyst-transfer Suzuki–Miyaura coupling polymerization. *Macromol. Rapid Commun.* **2011**, *32*, 801–806. [CrossRef] [PubMed]
7. Verswyvel, M.; Hoebers, C.; De Winter, J.; Gerbaux, P.; Koeckelberghs, G. Study of the controlled chain-growth polymerization of poly(3,6-phenanthrene). *J. Polym. Sci. Part A: Polym. Chem.* **2013**, *51*, 5067–5074. [CrossRef]
8. Elmalem, E.; Kiriy, A.; Huck, W.T.S. Chain-growth Suzuki polymerization of n-type fluorene copolymers. *Macromolecules* **2011**, *44*, 9057–9061. [CrossRef]
9. Yokozawa, T.; Ohta, Y. Transformation of step-growth polymerization into living chain-growth polymerization. *Chem. Rev.* **2016**, *116*, 1950–1968. [CrossRef] [PubMed]
10. Yokozawa, T.; Ohta, Y. Chapter 1 controlled synthesis of conjugated polymers in catalyst-transfer condensation polymerization: Monomers and catalysts. In *Semiconducting Polymers: Controlled Synthesis and Microstructure*; The Royal Society of Chemistry: Cambridge, UK, 2017; pp. 1–37.

11. Nojima, M.; Kosaka, K.; Kato, M.; Ohta, Y.; Yokozawa, T. Alternating intramolecular and intermolecular catalyst-transfer Suzuki–Miyaura condensation polymerization: Synthesis of boronate-terminated π-conjugated polymers using excess dibromo monomers. *Macromol. Rapid Commun.* **2016**, *37*, 79–85. [CrossRef] [PubMed]

12. Nojima, M.; Ohta, Y.; Yokozawa, T. Structural requirements for palladium catalyst transfer on a carbon–carbon double bond. *J. Am. Chem. Soc.* **2015**, *137*, 5682–5685. [CrossRef] [PubMed]

13. Zhang, H.-H.; Xing, C.-H.; Tsemo, G.B.; Hu, Q.-S. *t*-Bu₃P-coordinated 2-phenylaniline-based palladacycle complex as a precatalyst for the Suzuki cross-coupling polymerization of aryl dibromides with aryldiboronic acids. *ACS Macro Lett.* **2013**, *2*, 10–13. [CrossRef] [PubMed]

14. Kang, S.; Ono, R.J.; Bielawski, C.W. Controlled catalyst transfer polycondensation and surface-initiated polymerization of *p*-phenyleneethynylene-based monomer. *J. Am. Chem. Soc.* **2013**, *135*, 4984–4987. [CrossRef] [PubMed]

15. Bryan, Z.J.; Hall, A.O.; Zhao, C.T.; Chen, J.; McNeil, A.J. Limitations of using small molecules to identify catalyst-transfer polycondensation reactions. *ACS Macro Lett.* **2016**, *5*, 69–72. [CrossRef]

16. Zenkina, O.V.; Gidron, O.; Shimon, L.J.W.; Iron, M.A.; van der Boom, M.E. Mechanistic aspects of aryl–halide oxidative addition, coordination chemistry, and ring-walking by palladium. *Chem. Eur. J.* **2015**, *21*, 16113–16125. [CrossRef] [PubMed]

17. Zenkina, O.; Altman, M.; Leitus, G.; Shimon, L.J.W.; Cohen, R.; van der Boom, M.E. From azobenzene coordination to aryl-halide bond activation by platinum. *Organometallics* **2007**, *26*, 4528–4534. [CrossRef]

18. Ma, H.; Li, W.; Wang, J.; Xiao, G.; Gong, Y.; Qi, C.; Feng, Y.; Li, X.; Bao, Z.; Cao, W.; et al. Organocatalytic oxidative dehydrogenation of aromatic amines for the preparation of azobenzenes under mild conditions. *Tetrahedron* **2012**, *68*, 8358–8366. [CrossRef]

catalysts

MDPI

Communication

Aziridine- and Azetidine-Pd Catalytic Combinations. Synthesis and Evaluation of the Ligand Ring Size Impact on Suzuki-Miyaura Reaction Issues

Hamza Boufroura, Benjamin Large, Talia Bsaibess, Serge Perato, Vincent Terrasson, Anne Gaucher and Damien Prim *

Institut Lavoisier de Versailles, UMR CNRS 8180, Université de Versailles Saint-Quentin-en-Yvelines, 45 Avenue des Etats-Unis, 78035 Versailles CEDEX, France; hamza.boufroura@uvsq.fr (H.B.); benjamin.large@uvsq.fr (B.L.); talia.bsaibess@uvsq.fr (T.B.); serge.perato@u-psud.fr (S.P.); vincent.terrasson@hotmail.fr (V.T.); anne.gaucher@uvsq.fr (A.G.)
* Correspondence: damien.prim@uvsq.fr; Tel.: +33-013-925-4455

Academic Editor: Ioannis D. Kostas
Received: 29 November 2016; Accepted: 10 January 2017; Published: 13 January 2017

Abstract: The synthesis of new vicinal diamines based on aziridine and azetidine cores as well as the comparison of their catalytic activities as ligand in the Suzuki-Miyaura coupling reaction are described in this communication. The synthesis of three- and four-membered ring heterocycles substituted by a methylamine pendant arm is detailed from the parent nitrile derivatives. Complexation to palladium under various conditions has been examined affording vicinal diamines or amine-imidate complexes. The efficiency of four new catalytic systems is compared in the preparation of variously substituted biaryls. Aziridine- and azetidine-based catalytic systems allowed Suzuki-Miyaura reactions from aryl halides including chlorides with catalytic loadings until 0.001% at temperatures ranging from 100 °C to r.t. The evolution of the Pd-metallacycle ring strain moving from azetidine to aziridine in combination with a methylamine or an imidate pendant arm impacted the Suzuki-Miyaura reaction issue.

Keywords: aziridine; azetidine; vicinal diamine; imidate; ligand; Suzuki-Miyaura coupling

1. Introduction

The Suzuki-Miyaura reaction is "an easy way for C–C bonding" [1]. This statement is especially worthwhile in modern organic synthesis. Indeed, the advent of the Suzuki-Miyaura reaction represents a critical step in this field. This reaction has become for decades a major methodological tool available to chemists to build molecular architectures especially based on a biaryl scaffold. The biaryl motif can be found in numerous natural products, agrochemicals, drugs, polymers, ligands and thus triggered the attention of the scientific community for a wide range of applications [2–9]. If the formation of a biphenyl motif belongs nowadays to a textbook knowledge, the past decade has witnessed spectacular innovations and improvements such as ligandless transformations [10], supported or nanostructured catalytic systems [11–13], performing additives [14,15], neat and/or aqueous conditions [2,13,15], MW activations [10] for example.

Although the presence of a ligand is not strictly mandatory, increase of selectivity, the use of poorly reactive chlorides, room temperature conditions remains challenging and usually requires a ligand [11,13,15,16]. Additionally, preformed and geometrically constrained catalytic systems appears as crucial features beneficial to a catalytic activity enhancement [8,17]. If phosphorus-based catalytic systems dominated this area during the last decades, nitrogen-containing ligands revealed recently useful and appear as a pertinent alternative. In this context, ligands incorporating pyridyl-based scaffolds as well as combinations of pyridines linked to flexible alkylamine arms or rigid cycloalkylamines such as piperidines have been successfully developed in recent years [17–23]. In this communication, we describe new ligands based on strained aza-heterocycles such as aziridines and azetidines. We focused on two strategies using these small aza-heterocycles. The first one involves a combination of these small aza-heterocycles and a flexible methylamine side chain while the second requires a more rigid pendant arm based on an imidate moiety (Figure 1).

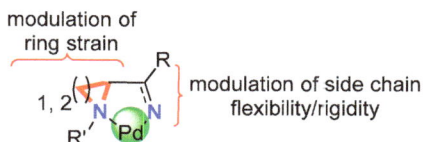

Figure 1. Aziridine- and azetidine-based catalytic systems. Modulation of ring and side chain strain.

We expect the modulation of the ring and side chain strains to impact the catalytic properties of these new ligands families. The abilities of the latter to promote couplings especially involving aromatic chlorides will be examined. In addition, room temperature conditions will further be prioritized in the context of the synthesis of variously substituted biphenyls.

2. Results and Discussion

2.1. Synthesis of Catalytic System

2.1.1. Synthesis of Ligand Precursors

Cyanoazetidine **1** and cyanoaziridine **2** have been easily obtained in two and three steps respectively starting from commercially available reactants as reported earlier [24,25]. The further step was to prepare vicinal diamines **3** and **4**. Both diamines were obtained using a two-step sequence, first an addition of phenylmagnesium bromide followed by a reduction with sodium borohydride. According to recent literature [26–28] reduction afforded a 9/1 ratio of diastereomers. In the absence of X-ray data, a relative stereochemistry can only be hypothesized. In good agreement with literature data [26–28], the assumed *anti* selectivity, might results from a hydride attack on the less strained Re face of the imine moiety leading as depicted in Scheme 1. The proposed chelate model could result either from an intramolecular hydrogen bond between the tertiary amine and imino moieties or from the formation of a five-membered Mg-chelate after addition of the Grignard reagent. The probable major anti diastereomer was in each case isolated after silica gel chromatography purification in 56% and 69% yields respectively (Scheme 1).

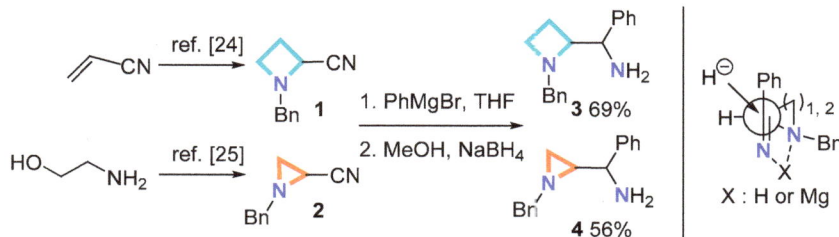

Scheme 1. Synthesis of aziridine and azetidine ligand precursors.

2.1.2. Synthesis of Pd Complexes

The next step was the preparation of Pd complexes starting from both nitrile and diamine precursors.

- Synthesis of vicinaldiamine-based Pd complexes

For this purpose, vicinal diamines **3** and **4** were reacted with Na$_2$PdCl$_4$ in freshly distilled MeOH for 20 h at room temperature. Expected complexes **A** and **B** were easily obtained after filtration and successive washings with cold MeOH, ether and pentane in 94% and 83% yields respectively (Scheme 2).

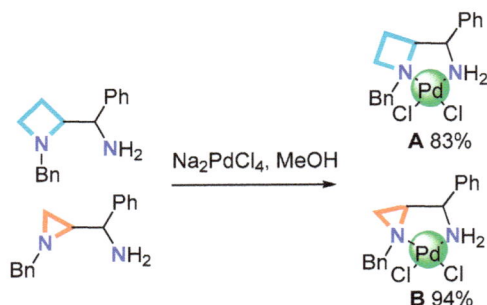

Scheme 2. Preparation of diamine-Pd complexes **A** and **B**.

- Synthesis of heterocycle/amidate-based Pd complexes

Amidate-Pd complexes were prepared under similar conditions and isolated by simple filtration and washings as shown in Scheme 3. Dual complexation of the nitrile group and the heterocyclic nitrogen atom to palladium in MeOH allowed the activation of the nitrile group and the formation of the imidate fragment through nucleophilic addition of one equivalent of methanol moiety. Bidendate complexes **C** and **D** display characteristic chemical shifts in ^{1}H and ^{13}C NMR corresponding to the presence of an imidate fragment as shown in Scheme 3. Methoxy groups appear at 3.87 and 3.65 ppm in ^{1}H NMR and the sp^2 carbon appear at 175.3 and 177.3 ppm in ^{13}C NMR. Both novel complexes were obtained in 70% and 63% yields respectively (Scheme 3).

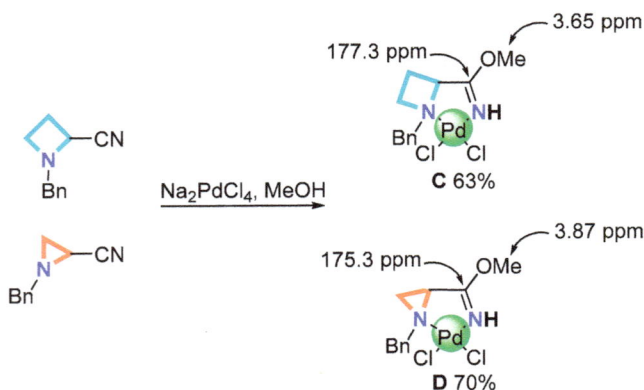

Scheme 3. Preparation of amine/imidate-Pd complexes **C** and **D** and characteristic chemical shifts in ^{1}H and ^{13}C NMR spectra.

Within these series, complexes appeared more stable than the corresponding ligands. Especially the aziridine **3** displays averaged stability, we assume to arise from ring opening without careful storage under inert conditions. Complexes display high stability for months. No modification of aspect, ^{1}H NMR spectra or catalytic activities was noticed.

2.2. Evaluation of Catalytic Properties

Catalytic properties of complexes **A**, **B**, **C** and **D** were next evaluated. The coupling between 4-bromonitrobenzene and 4-tolylboronic acid was chosen as the model coupling reaction (Scheme 4). Gratifyingly, running the reaction at room temperature for 10 h, using 1% of catalytic loading and base/solvent combination (Cs$_2$CO$_3$ and DMF/H$_2$O) as already reported for benzhydrylamines [20–24], allowed us to identify 4-nitro-4'-methyl biphenyl **5** in the crude material ^{1}H NMR spectra. Various conversions were observed ranging from 10% to 50% evidencing that our catalytic systems were able to trigger Suzuki-Miyaura coupling at room temperature.

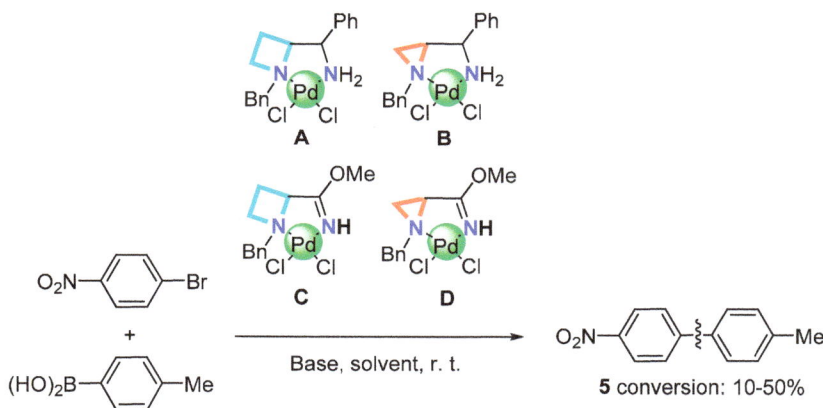

Scheme 4. Catalysts **A**–**D** applied to a model Suzuki reaction.

- Optimisation of catalytic conditions

The next step was to adjust the base/solvent combination. Among several base (Na$_2$CO$_3$, NaHCO$_3$, K$_3$PO$_4$, CsF) and solvent (THF, dioxane, toluene/ethanol/H$_2$O) combination tests, Cs$_2$CO$_3$ as the base and DMF/H$_2$O (9/1) as the solvent, gave the best conversions.

All catalytic systems were further evaluated and compared each other as shown in Figure 2. In the context of the synthesis of biphenyl derivative **5**, iodo-, bromo- and chloro-precursors were independently reacted with 4-tolylboronic acid using **A**, **B**, **C** and **D** catalysts, Cs$_2$CO$_3$ as the base and DMF/H$_2$O (9/1) as the solvent at various catalytic loadings. Results obtained with 1% and 0.1% of catalyst loading at room temperature are compared in Figure 2a,b respectively and results obtained with 0.01% and 0.001% of catalyst loading at 100 °C are compared in Figure 2c,d respectively.

Figure 2a shows obtained results when using 1% of catalyst loading for 16 h at room temperature. Iodides readily react affording the expected biphenyl motif in 93% to quantitative Yields whatever the catalyst **A**, **B**, **C** or **D**. The use of bromides and chlorides allowed us to clearly differentiate the efficiency of both catalyst families. Indeed, diamine-based catalysts **A** and **B** and amine/imidate catalysts **C** and **D** behaved differently. Best results were obtained using the more strained catalysts **C** and **D**. If transformation of arylchloride using **D** as the catalyst led to 41% in 16 h at room temperature, an extended reaction time to 48 h at room temperature allowed us to reach 94% yield.

A similar trend was observed with 0.1% catalytic loading for 24 h at room temperature (Figure 2b). Amine/imidate complexes **C** and **D** revealed superior to diamine complexes **A** and **B**. Aryliodides were transformed up to quantitative yields at room temperature. An expected decrease of catalytic activity is observed when moving to bromides and chloride. At this loading, complex **D** that displays an enhanced ring strain and a more rigid side chain led to best results. Again, an extended reaction time to 48 h improved the yield of biphenyl 5% to 58% at room temperature.

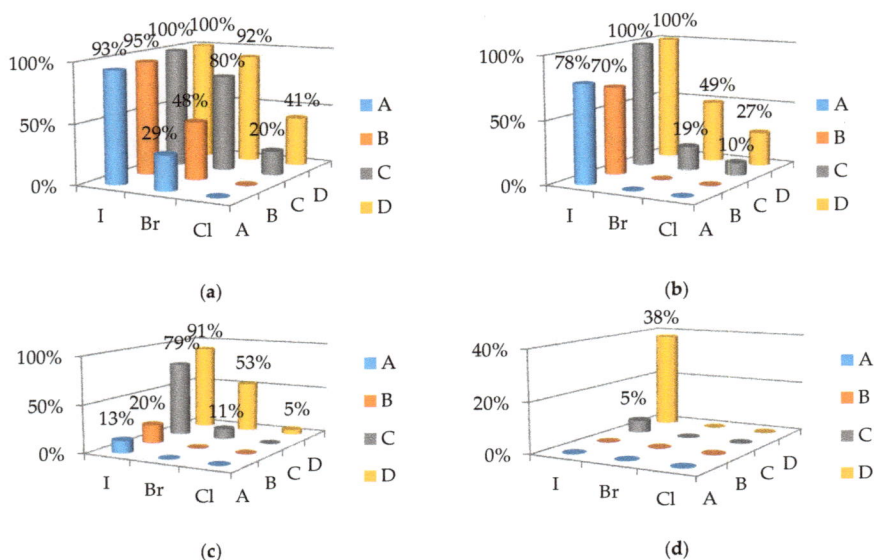

Figure 2. Comparison of catalytic system: (**a**) Catalytic loading 1%, 16h, r.t.; (**b**) Catalytic loading 0.1%, 24 h, r.t.; (**c**) Catalytic loading 0.01%, 24 h, 100 °C; (**d**) Catalytic loading 0.001%, 24 h, 100 °C.

At lower catalytic loadings such as 0.01% and 0.001% (Figure 2c,d) reactions were run at 100 °C. Again both complex families were compared and the amine/imidate **C** and **D** were found superior to the diamine complexes **A** and **B**. Again combination of ring and side chain enhanced strain in catalyst **D** revealed beneficial to catalytic activity. Indeed, the use of **D** at 0.01%, allowed reaching 91% yield for aryliodides and 53% for bromides. In contrast, chlorides exhibit only poor conversion. At 0.001% at 100 °C only aryl iodides react affording the expected biphenyl motif in 38% yield. In contrast, using bromides and chlorides analogues showed no reaction.

In fact, complex **D** was found superior to all others at room temperature until 0.1% loading and at 100 °C at lower loadings. Better results obtained using complex **D**, may result from the joint presence of an enhanced ring strain and a more rigid side chain which both likely play a crucial role in the transmetalation step of the catalytic cycle and the final product releasing step.

- Application to various substrates

Complex **D** was next evaluated in several Suzuki-Miyaura reactions. tab:catalysts-07-00027-t001 gathers results obtained using various combinations of electronic effects for both aryl halides and boronic acids. As shown, chlorides bearing electron withdrawing groups are transformed in fair to high yields at room temperature using catalytic loadings ranging from 1% to 0.1% (entries 1–6, 9, 12–13). Reactions using boronic acids substituted by electron withdrawing groups such as NO_2-phenyl, CH_3CO-phenyl or sterically hindered substituents required the use of bromides in order to obtain the desired biphenyl with fair to excellent yields (entries 10–11, 14–16).

Phenyl halides substituted by electron donating groups such as MeO-phenyl and HO-phenyl (entries 17–21) also required the use of bromides at catalytic loadings from 0.1% to 1% and/or increase of temperature from room temperature to 100 °C respectively. Finally, 2,2′-biphenyl (entries 22–24) could be obtained in high yields (89%–90%) at 100 °C using either chlorides or bromides at 0.1% and 0.01% catalytic loading.

Table 1. Synthesis of biaryls using catalyst **D**, Cs_2CO_3 as the base and DMF/H_2O (9/1) as the solvent.

Entry	Halide Fragment	Boronic Acid Fragment	Halide	Cat. Load.	Conditions	Yield (%)
1	NC—⬡—	—⬡(OMe)(OMe)	Cl	0.1%	r.t., 24 h	28
2			Cl	1%	r.t., 18 h	88
3	NC—⬡—	—⬡(Me)	Cl	0.1%	r.t., 24 h	21
4			Cl	1%	r.t., 24 h	69 [29]
5	O₂N—⬡—	—⬡(OMe)(OMe)	Cl	0.1%	r.t., 24 h	21
6			Cl	1%	r.t., 24 h	60
7			Br	0.01%	100 °C, 24 h	89
8			Br	0.1%	100 °C, 6 h	93 [a] [30]
9	OHC—⬡—⬡(NO₂)		Cl	1%	r.t., 24 h	30
10			Br	0.1%	r.t., 24 h	43
11			Br	1%	r.t., 24 h	61
12	OHC—⬡—⬡(Me)		Cl	0.1%	r.t., 24 h	16
13			Cl	1%	r.t., 24 h	45
14			Br	0.1%	r.t., 24 h	58
15	⬡—⬡(C=O)		Br	0.1%	r.t., 24 h	52
16			Br	1%	r.t., 24 h	81 [31]
17	MeO—⬡—⬡		Cl	1%	100°C, 18 h	32
18			Br	1%	r.t., 24 h	39 [32]
19	MeO—⬡—⬡(OHC)		Cl	1%	100 °C, 18 h	25
20			Br	1%	100 °C, 18 h	57 [33]
21	HO—⬡—⬡		Br	0.1%	r.t., 18 h	45
22	HO—⬡—⬡		Cl	1%	r.t., 24 h	30
23			Cl	0.1%	100 °C, 6 h	89
24			Br	0.01%	100 °C, 24 h	90

[a] A similar yield of 87% has been obtained using an analogue of catalyst **C** under similar reaction conditions [27].

3. Materials and Methods

3.1. Materials

Unless otherwise noted, all starting materials were obtained from commercial suppliers and used without purification. Petroleum ether was distilled under Argon. NMR spectra were recorded on a 300 MHz and 200 MHz Brucker spectrometers (Bruker BioSpin GmbH, Rheinstetten, Germany). Chemical shifts were reported in ppm relative to the residual solvent peak (7.27 ppm for $CHCl_3$) for ^1H spectra and (77.00 ppm for $CDCl_3$) for ^{13}C spectra. High Resolution Mass spectroscopy data were recorded on an Autospec Ultima (Waters/Micromass) device (Waters, Gyancourt, France,) with a resolution of 5000 RP at 5%. Thin-layer chromatography (TLC) was carried out on aluminium sheets precoated with silica gel 60 F254. Column chromatography separations were performed using silica gel (0.040–0.060 mm). (*N*-benzyl)-2-cyanoazetidine **1** and (*N*-benzyl)-2-cyanoaziridine **2** have been prepared according to references [24,25].

3.2. Methods

3.2.1. General Procedure for Addition/Reduction Sequence

The phenylmagnesium chlorde (2 mmol) was added to a solution of 2-cyanoderivative **1** or **2** (1 mmol) in dry THF (10 mL) at 0 °C under argon. After stirring for 20 min, MeOH (10 mL) and $NaBH_4$ (1.2 mmol) were successively added. After a further 1 h, the reaction was quenched with saturated aqueous NH_4Cl solution (5 mL), and extracted with EtOAc (3 × 10 mL). The combined organic extracts were washed with brine, dried with magnesium sulfate and concentrated under reduced pressure. Amines **3** and **4** were purified by silica gel column chromatography using cyclohexane/Et_2O 1:1 as the eluant.

3.2.2. General Complexation Procedure

To a stirred solution of ligand 1–4 (0.25 mmol) in 5 mL of freshly distilled MeOH was added Na_2PdCl_4 (74 mg, 0.25 mmol). The mixture was stirred at room temperature for 1 to 16 h and filtered over a celite pad. The filtrate was removed by evaporation under vacuum. The residue was then purified over silica gel pad eluting first with cyclohexane/EtOAc 7:3 to remove traces of free ligand, then with EtOAc for ligands **3** and **4** and with AcOEt/MeOH 95:5 for ligands **1** and **2**.

3.2.3. General Suzuki Coupling Procedure

To a stirred solution of aromatic halide (0.5 mmol), boronic acid (0.6 mmol) and Cs_2CO_3 (407 mg, 1.25 mmol) in 1 mL of DMF/H_2O (95:5) was added the palladium complex as a solid or in solution in DMF/H_2O (95:5). The mixture was stirred at room temperature or 100 °C (refer to tab:catalysts-07-00027-t001). 10 mL of EtOAc and 10 mL of water were then added and the aqueous phase was extracted with EtOAc (3 × 5 mL). The combined organic layers were dried ($MgSO_4$), filtered and concentrated under vacuum, the crude product was purified by flash chromatography on silica gel to give the biaryl product.

4. Conclusions

Two new families of ligands based on an aziridine and azetidine core have been developed. Starting from the parent aziridine- and azetidine-nitriles, vicinal diamines and amine/imidate palladium complexes were obtained respectively using a nucleophilic addition/reduction-complexation sequence or a direct complexation in methanol. Evaluation and comparison of the catalytic activities of four complexes are described. The amine/imidate family **C** and **D** was found superior to the diamine analogues family **A** and **B**. Within the amine/imidate family best results were achieved using complex **D** that combines both enhanced ring strain and side chain rigidity. The aziridine-imidate complex **D** proved to be efficient for the synthesis of various substituted biphenyls. Catalyst **D** allowed iodides to react at 100 °C and catalytic loadings as low as 0.001%. In addition, bromides are able to be used as partners in couplings at 100 °C and loadings of 0.01%. Finally, **D** catalyzes the reaction of chlorides at room temperature using catalytic loadings ranging from 1% to 0.1%.

Supplementary Materials: The following are available online at www.mdpi.com/2073-4344/7/1/27/s1, experimental procedures for the preparation of precursors, catalytic systems and Suzuki couplings as well as analytical data for new compounds.

Acknowledgments: University of Versailles St Quentin, MENRT-France, ANR (ANR-11-BS07-030-01), IDEX (ANR-10-IDEX-0003-02) and LabEx CHARMMMAT (ANR-11-LABEX-0039) are gratefully acknowledged for financial supports and grants (HB, BL, SP, VT).

Author Contributions: H.B., B.L., T.B., S.P. and V.T. conceived and performed the experiments; A.G. and D.P. analyzed the data and wrote the paper.

Conflicts of Interest: The authors declare no conflict of interest.

References

1. Suzuki, A. Cross-coupling reactions of organoboranes: An easy way to construct C–C bonds (Nobel Lecture). *Angew. Chem. Int. Ed.* **2011**, *50*, 6722–6737. [CrossRef] [PubMed]

2. Chatterjee, A.; Ward, T.R. Recent advances in the palladium catalyzed Suzuki-Miyaura cross-coupling reaction in water. *Catal. Lett.* **2016**. [CrossRef]

3. Zhang, D.; Wang, Q. Palladium catalyzed asymmetric Suzuki-Miyaura coupling reactions to axially chiral birayl compounds: Chiral ligands and recent advances. *Coord. Chem. Rev.* **2015**, *286*, 1–16. [CrossRef]

4. Maluenda, I.; Navarro, O. Recent developments in the Suzuki-Miyaura reaction: 2010–2014. *Molécules* **2015**, *20*, 7528–7557. [CrossRef] [PubMed]

5. Han, F.-Y. Transition-metal-catalyzed Suzuki-Miyaura cross-coupling reactions: A remarkable advance from palladium to nickel catalysts. *Chem. Soc. Rev.* **2013**, *42*, 5270–5298. [CrossRef] [PubMed]

6. Kumar, A.; Kumar Rao, G.; Kumar, S.; Singh, A.K. Organosulphur and related ligands in Suzuki-Miyaura C–C coupling. *Dalton Trans.* **2013**, *42*, 5200–5223. [CrossRef] [PubMed]

7. Rossi, R.; Bellina, F.; Lessi, M. Selective palladium-catalyzed Suzuki–Miyaura reactions of polyhalogenated heteroarenes. *Adv. Synth. Catal.* **2012**, *354*, 1181–1255. [CrossRef]

8. Li, H.; Seechurn, C.C.C.J.; Colacot, T.J. Development of preformed Pd catalysts for cross-coupling reactions, beyond the 2010 Nobel prize. *ACS Catal.* **2012**, *2*, 1147–1164. [CrossRef]

9. Liu, S.-Y.; Li, H.-Y.; Shi, M.-M.; Jiang, H.; Hu, X.-L.; Li, W.-Q.; Fu, L.; Chen, H.-Z. Pd/C as a clean and effective heterogeneous catalyst for C–C couplings toward highly pure semiconducting polymers. *Macromolecules* **2012**, *45*, 9004–9009. [CrossRef]

10. Arvela, R.K.; Leadbeater, N.E.; Sangi, M.S.; Williams, V.A.; Granados, P.; Singer, R.D. A reassessment of the transition-metal free Suzuki-type coupling methodology. *J. Org. Chem.* **2005**, *70*, 161–168. [CrossRef] [PubMed]

11. Jawale, D.V.; Gravel, E.; Boudet, C.; Shah, N.; Geertsen, V.; Li, H.; Namboothiri, I.N.N.; Doris, E. Room temperature Suzuki coupling of aryl iodides, bromides and chlorides using a heterogeneous carbon nanotube-palladium nanohybrid catalyst. *Catal. Sci. Technol.* **2015**, *5*, 2388–2392. [CrossRef]

12. Hong, M.C.; Choi, M.C.; Chang, Y.W.; Lee, Y.; Kim, J.; Rhee, H. Palladium nanoparticles on thermoresponsive hydrogels and their application as recyclable Suzuki-Miyaura coupling reaction catalysts in water. *Adv. Synth. Catal.* **2012**, *354*, 1257–1263. [CrossRef]

13. Maegawa, T.; Kitamura, Y.; Sako, S.; Udzu, T.; Sakurai, A.; Tanaka, A.; Kobayashi, Y.; Endo, K.; Bora, U.; Kurita, T.; et al. Heterogeneous Pd/C-catalyzed ligand free, room-temperature Suzuki-Miyaura coupling reactions in aqueous media. *Chem Eur. J.* **2007**, *13*, 5937–5943. [CrossRef] [PubMed]

14. Liu, D.-X.; Gong, W.-J.; Li, H.-X.; Gao, J.; Li, F.-L.; Lang, J.-P. Palladium(II)-catalyzed Suzuki-Miyaura reactions of arylboronic acid with aryl halide in water in the presence of 4-(benzylthio)-*N*,*N*,*N*-trimethylbenzenammonium chloride. *Tetrahedron* **2014**, *70*, 3385–3389. [CrossRef]

15. Lipshutz, B.H.; Petersen, T.B.; Abela, A.R. Room-temperature Suzuki-Miyaura couplings in water facilitated by non-ionic amphiphiles. *Org. Lett.* **2008**, *10*, 1333–1336. [CrossRef] [PubMed]

16. Tang, Y.; Zeng, Y.; Hu, Q.; Huang, F.; Jin, L.; Mo, W.; Sun, N.; Hu, B.; Shen, Z.; Hu, X.; et al. Efficient catalyst for both Suzuki and Heck cross-coupling reactions: Synthesis and catalytic behaviour of geometry-constrained iminopyridylpalladium chlorides. *Adv. Synth. Catal.* **2016**, *358*, 2642–2651. [CrossRef]

17. Yang, J.; Liu, S.; Zheng, J.-F.; Zhou, J. Room-temperature Suzuki-Miyaura coupling of heteroaryl chlorides and tosylates. *Eur. J. Org. Chem.* **2012**, 6248–6259. [CrossRef]

18. Najera, C.; Gil-Molto, J.; Karlström, S. Suzuki-Miyaura and related cross-coupling in aqueous solvents catalyzed by di(2-pyridyl)methylamine-palladium dichloride complexes. *Adv. Synth. Catal.* **2004**, *346*, 1798–1811. [CrossRef]

19. Puget, B.; Roblin, J.-P.; Prim, D.; Troin, Y. New 2-(2-pyridyl)piperidines: Synthesis, complexation of palladium and catalytic activity in Suzuki reaction. *Tetrahedron Lett.* **2008**, *49*, 1706–1709. [CrossRef]

20. Terrasson, V.; Prim, D.; Marrot, J. *N*-Heterocyclic benzhydrylamines as New *N*,*N*-Bidentate ligands in palladium complexes: Synthesis, characterization and catalytic activity. *Eur. J. Inorg. Chem.* **2008**, 2739–2745.

21. Gunawan, M.-A.; Qiao, C.; Abrunhosa-Thomas, I.; Puget, B.; Roblin, J.-P.; Prim, D.; Troin, Y. Simple pyridylmethylamines: Efficient and robust *N*,*N*-ligands for Suzuki-Miyaura coupling reactions. *Tetrahedron Lett.* **2010**, *51*, 5392–5394. [CrossRef]

22. Grach, G.; Pieters, G.; Dinut, A.; Terrasson, V.; Medimagh, R.; Bridoux, A.; Razafimahaleo, V.; Gaucher, A.; Marque, S.; Marrot, J.; et al. N-Heterocyclic pyridylmethylamines: Synthesis, complexation, molecular structure, and application to asymmetric Suzuki-Miyaura and oxidative coupling reactions. *Organometallics* **2011**, *30*, 4074–4086. [CrossRef]

23. Requet, A.; Yalgin, H.; Prim, D. Convenient and rapid strategies towards 6-(hetero)arylpyridylmethylamines. *Tetrahedron Lett.* **2015**, *56*, 1378–1382. [CrossRef]

24. Couty, T.; David, O.; Larmanjat, B.; Marrot, J. Strained azetidinium ylides: New reagents for cyclopropanation. *J. Org. Chem.* **2007**, *72*, 1058–1061. [CrossRef] [PubMed]

25. Ayi, A.I.; Guedj, R. Reaction of hydrogen fluoride in pyridine solution with cis-cyano-2-and cis-amido-2-aziridines. Preparation of β-fluoro-α-amino acids and esters by means of acidic hydrolysis and alcoholysis of β-fluoro-α-amino nitriles and/or β-fluoro-α-amino acid amides. *J. Chem. Soc. Perkin Trans.* **1983**, 2045–2051. [CrossRef]

26. Keller, L.; Vargas-Sanchez, M.; Prim, D.; Couty, F.; Evano, G.; Marrot, J. Azetidines as ligands in the palladium (II) complexes series. *J. Organomet. Chem.* **2005**, *690*, 2306–2311. [CrossRef]

27. Pieters, G.; Puget, B.; Terrasson, V.; Roblin, J.-P.; Gaucher, A.; Marque, S.; Prim, D.; Troin, Y. On the robustness of methylamines-Pd catalytic systemsin the Suzuki reaction: Compromise examples between synthesis and catalysis. *Rev. Chim. (Bucarest)* **2010**, *61*, 825–827.

28. Besev, M.; Engman, L. Diastereocontrol by a hydroxyl auxiliary in the synthesis of pyrrolidines via radical cyclization. *Org. Lett.* **2002**, *4*, 3023–3025. [CrossRef] [PubMed]

29. Kataoka, N.; Shelby, Q.; Stambuli, J.P.; Hartwig, J.F. Air stable, sterically hindered ferrocenyl dialkylphosphines for palladium-catalyzed $C-C$, $C-N$, and $C-O$ bond-forming cross-couplings. *J. Org. Chem.* **2002**, *67*, 5553–5566. [CrossRef] [PubMed]

30. Yanagisawa, S.; Sudo, T.; Noyori, R.; Itami, K. Direct $C-H$ Arylation of (Hetero)arenes with Aryl Iodides via Rhodium Catalysis. *J. Am. Chem. Soc.* **2006**, *128*, 11748–11749. [CrossRef] [PubMed]

31. Tao, B.; Boykin, D.W. Simple amine/Pd(OAc)$_2$-catalyzed Suzuki coupling reactions of aryl bromides under mild aerobic conditions. *J. Org. Chem.* **2004**, *69*, 4330–4335. [CrossRef] [PubMed]

32. Mino, T.; Shirae, Y.; Saito, T.; Sakamoto, M.; Fujita, T. Palladium-catalyzed Sonogashira and Hiyama reactions using phosphine-free hydrazone ligands. *J. Org. Chem.* **2006**, *71*, 9499–9502. [CrossRef] [PubMed]

33. Mendes Da Silva, J.F.; Perez, A.F.Y.; Pinto de Almeida, N. An efficient and new protocol for phosphine-free Suzuki coupling reaction using palladium-encapsulated and air-stable MIDA boronates in an aqueous medium. *RSC Adv.* **2014**, *4*, 28148–28155. [CrossRef]

catalysts

MDPI

Article

Suzuki-Miyaura C-C Coupling Reactions Catalyzed by Supported Pd Nanoparticles for the Preparation of Fluorinated Biphenyl Derivatives

Roghayeh Sadeghi Erami [1,2], Diana Díaz-García [1], Sanjiv Prashar [1],
Antonio Rodríguez-Diéguez [3], Mariano Fajardo [1], Mehdi Amirnasr [2] and
Santiago Gómez-Ruiz [1,*]

[1]   Departamento de Biología y Geología, Física y Química Inorgánica, ESCET, Universidad Rey Juan Carlos,
       Calle Tulipán s/n, E-28933 Móstoles (Madrid), Spain; romochem@gmail.com (R.S.E.);
       dianadiazgarcia2@gmail.com (D.D.-G.); sanjiv.prashar@urjc.es (S.P.); mariano.fajardo@urjc.es (M.F.)
[2]   Department of Chemistry, Isfahan University of Technology, Isfahan 84156-83111, Iran; amirnasr@cc.iut.ac.ir
[3]   Departamento de Química Inorgánica, Universidad de Granada, 18071 Granada, Spain; antonio5@ugr.es
*   Correspondence: santiago.gomez@urjc.es; Tel.: +34-914-888-507

Academic Editor: Ioannis D. Kostas
Received: 20 December 2016; Accepted: 24 February 2017; Published: 28 February 2017

Abstract: Heterogeneous recyclable catalysts in Suzuki-Miyaura C-C coupling reactions are of great interest in green chemistry as reusable alternatives to homogeneous Pd complexes. Considering the interesting properties of fluorinated compounds for the pharmaceutical industry, as precursors of novel materials, and also as components of liquid crystalline media, this present study describes the preparation of different fluorinated biphenyl derivatives by Suzuki-Miyaura coupling reactions catalyzed by a heterogeneous system (G-COOH-Pd-10) based on Pd nanoparticles supported onto COOH-modified graphene. The catalytic activity of the hybrid material G-COOH-Pd-10 has been tested in Suzuki-Miyaura C–C coupling reactions observing excellent versatility and good conversion rates in the reactions of phenylboronic acid, 4-vinylphenylboronic acid, 4-carboxyphenylboronic acid, and 4-fluorophenylboronic acid with 1-bromo-4-fluorobenzene. In addition, the influence of the arylbromide has been studied by carrying out reactions of 4-fluorophenylboronic acid with 1-bromo-2-fluorobenzene, 1-bromo-3-fluorobenzene, 1-bromo-4-fluorobenzene, 2-bromo-5-fluorotoluene, and 2-bromo-4-fluorotoluene. Finally, catalyst recyclability tests show a good degree of reusability of the system based on G-COOH-Pd-10 as the decrease in catalytic activity after five consecutive catalytic cycles in the reaction of 1-bromo-4-fluorobenzene with 4-florophenylboronic acid at 48 hours of reaction is lower than 8% while in the case of reactions at three hours the recyclability of the systems is much lower.

Keywords: Pd nanoparticles; C-C coupling; fluorinated compounds; graphene; supported catalysts; Suzuki-Miyaura reactions

1. Introduction

Palladium-catalyzed reactions have been one of the main methods for C-C cross-coupling processes [1]. In particular, Suzuki-Miyaura are one of the most widely used reactions for the preparation of biphenyl derivatives [2]. These reactions have been principally carried out using homogeneous catalysts based on simple or sophisticated Pd complexes [2–5]. However, the current needs of industry and the search for greener alternatives to these catalyst are pushing the development of new heterogeneous and recyclable systems [6]. These heterogeneous C-C coupling catalysts are based on either supported palladium complexes [7,8] or supported palladium nanoparticles [9].

Most of the studied palladium-based heterogeneous catalytic systems have shown lower efficiencies and catalytic activities than homogeneous counterparts [9]. Nevertheless, current research advances of the scientific community have led to the development of highly active, reusable, and robust heterogeneous systems. The majority of these systems are based on palladium nanoparticles (PdNPs) which take advantage of their interesting properties, such as high surface area and high catalytic activity [9].

In addition, supporting PdNPs on different nanostructured materials enhances the recyclability properties and facilitates the separation of the products which are usually dissolved in the reaction mixture [10]. Thus, the ongoing research in this topic is very intensive because there is still much work that needs to be done to improve the catalytic performance of the supported systems. Therefore, many groups are working on supporting Pd nanoparticles onto mesoporous silica [11,12], alumina [13,14], graphene [15,16], modified graphene [17,18], graphene oxide [19], graphite oxide [20], reduced graphene oxide [21], or other carbon-based materials [22], for example, for the development of novel catalytic systems. However, with supported catalysts considerable work still needs to be carried out in order to increase the versatility of the reagents and products of the C-C coupling reactions.

In this context, our group has decided to study the preparation of different fluorinated biphenyl derivatives by Suzuki-Miyaura coupling reactions. In general, fluorinated compounds, although generally viewed as mostly inert because of their lack of chemical reactivity [23], may have biological activity which could be of interest in different therapies. For example simple and accepted compounds, such as Prozac[TM], Redux[TM], or 5-fluorouracyl, are fluorinated compounds with anti-depressant, anti-obesity, and anticancer properties, respectively. Furthermore, there is a long list of fluorine-containing drugs that have been introduced to the market during last two decades [24]. The incorporation of fluorine in drugs normally improves their metabolic stability and impedes the oxidative attack of cytochrome P450 enzymes, thus improving their activity in vivo [25]. Pharmaceutical use is not the unique application of F-containing organic derivatives, for example, fluorination compounds are used to improve material properties opening new fields of research [26]. In particular, fluorinated biaryl derivatives are highly suitable as components of liquid crystalline media [27] and, in the form of ethers, have also recently been considered as pro-drug scaffolds employing the chemical-microbial approach [28].

We have only found in the literature a few examples reporting the preparation of fluorinated biaryl derivatives via C-C coupling catalytic reactions. Almost all of these reports described homogeneous systems based on Pd-complexes as catalysts [29–35], while only one study was carried out using Pd nanoparticles [36]. Therefore, we report here the synthesis and characterization of palladium nanoparticles supported onto commercial graphene modified with COOH groups and the study of the application of this composite material in Suzuki-Miyaura C-C coupling heterogeneous catalytic reactions in the formation of fluorinated biphenyls. We have studied different parameters for this reaction, including the recyclability of the catalytic systems. Furthermore, we have developed a new quantification method, as an alternative to gas chromatography (GC) or high-performance liquid chromatography (HPLC) for the fluorinated products of the catalytic reactions based on a simple ^{19}F-NMR study using an internal standard method.

2. Results and Discussion

2.1. Synthesis and Characterization of the Supported PdNPs

Supported palladium nanoparticles were prepared by the reaction of commercial COOH-modified graphene (G-COOH) with different amounts of [PdCl$_2$(cod)] in toluene for 48 hours. The supported PdNPs were presumably formed via the reduction of [PdCl$_2$(cod)] as was previously reported by our group for silica- and titania-based materials [12,14,37,38]. This synthetic method requires the reduction of the organometallic palladium complex which is achieved by toluene and the carbon atoms at the surface of the graphene which presumably act as the reducing agents. In addition, an agglomeration of the graphene layers occurs giving rise to the formation of a hybrid material that consists mainly of a graphite support in a mixture of phases and impregnated palladium nanoparticles.

This reaction was repeated using different amounts of the organometallic Pd precursor [PdCl$_2$(cod)] to study the Pd loading on the materials which was determined by X-ray fluorescence (XRF) analysis. Thus, a theoretical amount of 5 wt %, 10 wt %, and 15 wt %. Pd was used for the reactions to give the materials G-COOH-Pd-5, G-COOH-Pd-10, and G-COOH-Pd-15, respectively. After analysis of the materials by XRF, the incorporation of palladium to the material was 3.06 wt %, 7.93 wt %, and 11.20 wt % Pd. for G-COOH-Pd-5, G-COOH-Fd-10, and G-COOH-Pd-15, respectively. Therefore, the higher Pd efficacy was achieved for G-COOH-Pd-10 (79.3%) while in the case of G-COOH-Pd-5 and G-COOH-Pd-15 this value was 61.2% and 74.6%, respectively (Table 1 and Table S1 of Supplementary Material). The differences in the incorporaticn rate are not very high. Other materials have shown that the loading capacity follows a logarithmic tendency as the material saturates and limits the reduction of the Pd(II) complex to palladium nanoparticles [12]. This may be the reason for the lower Pd incorporation rate found for G-COOH-Pd-15. In general, these materials showed higher loading capacities than other silica-based materials reported previously by our group [12], similar Pd incorporation to that found for other C-based materials [15–20] and slightly lower than in the case of alumina submicronic particles [14].

Table 1. Theoretical and experimental Pd (wt %) quantity and Pd incorporation rate (%) in materials G-COOH-Pd-5, G-COOH-Pd-10, and G-COOH-Pd-15.

Material	Theoretical Pd Quantity (wt %)	Experimental Pd Quantity (wt %)	Incorporation Rate (%)
G-COOH-Pd-5	5	3.06	61.2
G-COOH-Pd-10	10	7.93	79.3
G-COOH-Pd-15	15	11.20	74.6

Bearing in mind that the most effective incorporation of Pd in the material was achieved for G-COOH-Pd-10, this catalytic system was selected for the catalytic studies and for further characterization. G-COOH-Pd-10 was then studied by transmission electronic microscopy (TEM). The TEM image of G-COOH-Pd-10 (Figure 1a) shows that this material contains Pd nanoparticles that can be easily observed as black dots impregnated on the external surface of the carbon-based support materials. The obtained PdNPs have a poorly defined shape with a size of 14.6 ± 1.4 nm. Pd nanoparticle size distribution (Figure 1b) has been calculated by using the software ImageJ (1.51j, Wayne Rasband, National Institutes of Health, MD, USA, 2017) [39] and a subsequent Gaussian fit using Origin (OriginLab 8.0, Northampton, MA, USA, 2009). In addition, the TEM image (Figure 1) shows that the larger palladium particles are formed by clusters of small Pd nanoparticles (for additional images see Figures S1–S3 of the Supplementary Material).

In addition, the material was characterized by Fourier-transformed infrared spectroscopy (FT-IR), N$_2$ adsorption-desorption isotherms (BET), and X-ray diffraction (XRD). The FT-IR spectrum shows the expected signals for the O-H vibration, stretching band of the C=O and vibration of the C-O bond of the COOH groups at ca. 2900, 1600–1750, and 1100 cm^{-1}, respectively (Figure S4 of the Supplementary Material). The characterization by N$_2$ adsorption-desorption isotherms (BET method), showed a specific surface area of material G-COOH-Pd-10 of 4.1 m^2/g and an irregular pore size distribution of the material. The measurements showed type III isotherms (Figure 2) according to the IUPAC classification [40] which is indicative of non-porous materials with low affinity adsorbent-adsorbate. The measured surface areas by BET studies is much lower than the theoretical surface area limit of graphene which is estimated to be ca. 2600 m^2/g. Therefore, material G-COOH-Pd-10 in dry state is probably affected by the stacking of the graphene sheets, causing a decrease in its surface area. This also indicates that the nitrogen gas of the BET analysis does not easily penetrate the graphene layers of the material giving type III isotherms of low surface area. The partial stacking of layers in the dry state, while measuring the textural properties of the materials giving low surface area, has been previously observed in similar systems [19].

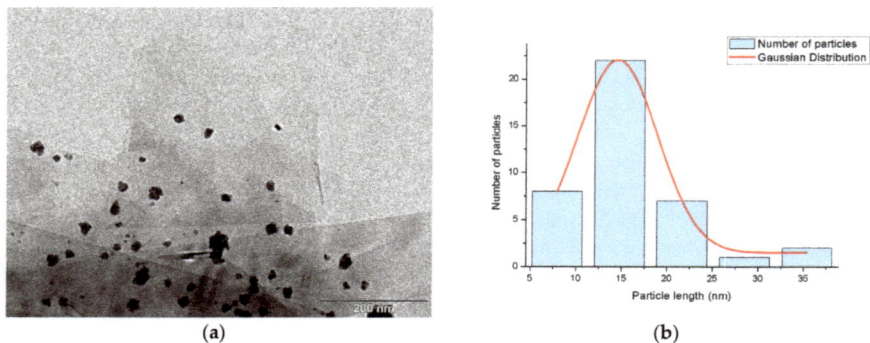

(a)   (b)

Figure 1. (a) Transmission electronic microscopy (TEM) image of the material G-COOH-Pd which consists of Pd nanoparticles supported on a graphite support in a mixture of phases; and (b) Pd-particle size Gaussian distribution.

In spite of the partial stacking of the layers detected in the BET analysis, an interesting difference in the adsorptive parameters of the material after Pd-functionalization was observed, namely, a decrease in the surface area, a slight increase in the Barnett, Joyner and Halenda (BJH) adsorption or desorption cumulative volume of pores and a slight decrease in the BJH adsorption or desorption cumulative surface area of pores (Table S2 and Figure S5 of the Supplementary Material). These changes indicate the impregnation of the Pd nanoparticles on the external surface which decreases the surface area and slightly increases the estimated pore volume. This reveals that the Pd nanoparticles perturb the pure stacking of the graphene layers because the impregnation of the particles in the material increases the distance between layers due to the intercalation of the metal nanoparticles.

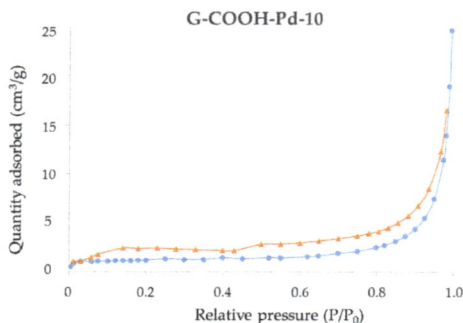

Figure 2. N_2 adsorption-desorption isotherm of material G-COOH-Pd-10.

Finally, the material G-COOH-Pd-10 was characterized by powder XRD to confirm the presence of Pd nanoparticles. The XRD pattern (Figure S6 of the Supplementary Material) shows the peaks corresponding to a mixture of carbon-based materials with graphite as a broad peak at a 2θ of ca. $26°$ (indicating the partial agglomeration of graphene layers to graphite) and the peaks assigned to the Pd nanoparticles at $39°$, $43°$, $67°$, $78°$, and $84°$ corresponding to the Miller planes (111), (200), (220), (311), and (222), respectively. This confirms, therefore, the presence of Pd nanoparticles impregnated onto the carbon-based material, as was previously observed in TEM images.

2.2. Catalytic Study

2.2.1. Determination of the Optimal Conditions and Influence of Different Boronic Acids

The heterogeneous catalyst G-COOH-Pd-10 was tested in four coupling reactions of 1-bromo-4-fluorobenzene as aryl bromide with boronic acids with different substituents, namely, phenylboronic acid, 4-vinylphenylboronic acid, 4-carboxyphenylboronic acid and 4-fluorophenylboronic acid and (Scheme 1, reactions a–d, respectively).

The reaction conditions were determined previously by our group using analogous supported catalysts based on palladium nanoparticles and silica or alumina [12,14]. Thus, all of the catalytic tests of this study were carried out using a DMF/H₂O (95:5) mixture as solvent, K_2CO_3 as the base, and two different temperatures (70 °C and 110 °C). The reactions were carried out at different time intervals of 3, 8, 24, and 48 h, in order to determine the kinetic parameters of each reaction.

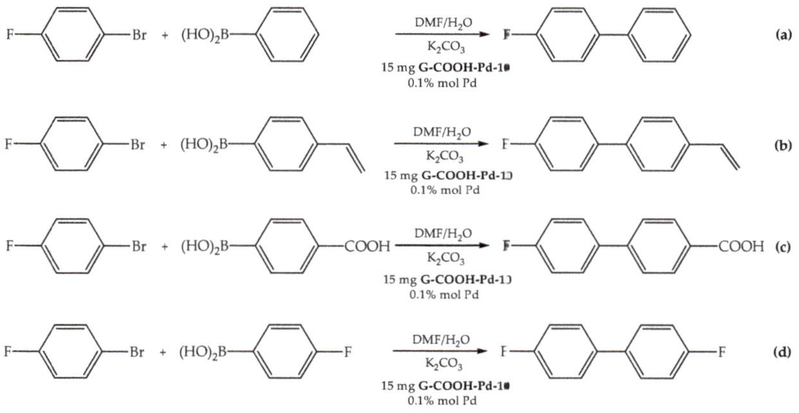

Scheme 1. Reaction of 1-bromo-4-fluorobenzene with (**a**) phenylboronic acid; (**b**) 4-vinylphenylboronic acid; (**c**) 4-carboxyphenylboronic acid; and (**d**) 4-fluorophenylboronic acid catalyzed by G-COOH-Pd-10.

The results obtained in the C-C cross-coupling reactions of Scheme 1 at different time intervals are given in Table 2 and presented in Figure 3. In general, the increase of the temperature from 70 to 110 °C results in higher conversion percentages as expected for this kind of reaction.

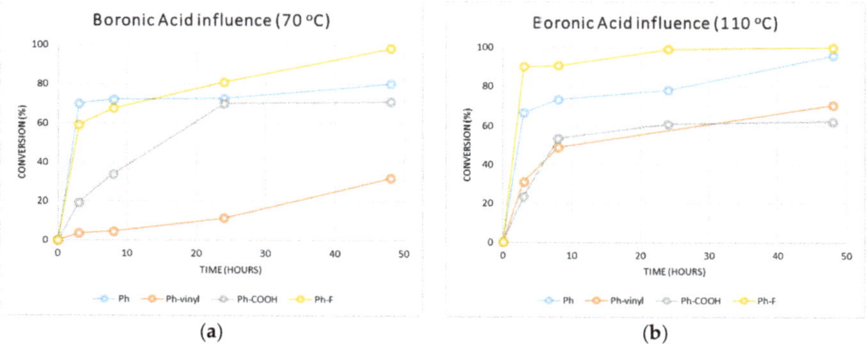

Figure 3. Conversion vs. time in the reaction of different boronic acids and 1-bromo-4-fluorobenzene catalyzed by G-COOH-Pd-10: (**a**) at 70 °C; and (**b**) at 110 °C.

However, these results are in contrast with the previous study of Pd-supported nanoparticles using mesoporous silica-based materials such as MSU-2 or SBA-15 [12] in which a decrease in the catalytic activity was observed when increasing the temperature, due to a higher mobility of the nanoparticles which increase the aggregation of the catalytic centers, thus, decreasing the catalytic activity.

Table 2. Bromide conversions in C-C coupling reactions using 1-bromo-4-flurobenzene and different boronic acids catalyzed by G-COOH-Pd-10.

Time (h)	T (°C)	Bromide Conversion (%)	TON [a]	TOF [b] (h^{-1})	Reaction
3		70	157	52	
8	70	72	161	20	
24		73	162	7	
48		80	179	4	F—⟨⟩—Br + (HO)₂B—⟨⟩
3		67	149	50	
8	110	73	164	21	
24		78	175	7	
48		96	215	5	
3		3	8	3	
8	70	5	10	1	
24		11	26	1	
48		32	71	2	F—⟨⟩—Br + (HO)₂B—⟨⟩
3		31	69	23	
8	110	49	109	14	
24		1			
48		71	158	3	
3		19	43	15	
8	70	34	76	10	
24		70	157	7	
48		71	159	3	F—⟨⟩—Br + (HO)₂B—⟨⟩—COOH
3		24	53	18	
8	110	54	120	15	
24		61	136	6	
48		62	139	3	
3		59	132	44	
8	70	67	151	19	
24		81	181	8	
48		98	219	5	F—⟨⟩—Br + (HO)₂B—⟨⟩—F
3		90	201	67	
8	110	91	202	25	
24		99	221	9	
48		100	223	5	

[a] Turnover number. [b] Turnover frequency. [1] Not analyzed.

In addition, one can clearly observe that the most effective boronic acid in terms of conversion of 1-bromo-4-fluorobenzene at both studied temperatures is 4-fluorophenylboronic acid, which leads to almost complete halide conversion. The second most active boronic acid at both temperatures is phenylboronic acid which, at 3 h and 8 h reaction time at 70 °C, is even slightly more active than 4-fluorophenylboronic acid. Both reactions reach almost the maximum level of halide conversion after just 3–8 h of reaction and the conversion does not increase significantly from 8 to 48 h. Thus, the highest TOF value (67.1 h^{-1}) found in these studies was observed for the reaction using 4-fluorophenylboronic acid at 110 °C after 3 h of reaction. In the case of phenylboronic acid, the TOF values at 70 or 110 °C were of ca. 50 h^{-1}. In all cases, these TOF values were much higher than in the case of similar systems based on alumina and Pd nanoparticles [14] or mesoporous silica and palladium nanoparticles [12]. In addition, they are in the same range, if not somewhat higher, than those described for homogeneous systems based on palladium complexes used for the preparation of fluorinated biaryls [29–35].

The reactions using the other two boronic acids, 4-carboxyphenylboronic acid, and 4-vinylphenylboronic acid, showed less conversion of the halide. It appears that the reaction with 4-vinylphenylboronic acid is more temperature-sensitive than in the case of 4-carboxyphenylboronic acid as the conversions for 4-vinylphenylboronic acid at 110 °C are much higher than at 70 °C. However, in the case of 4-carboxyphenylboronic acid this increase is not as high. The reaction using 4-carboxyphenylboronic acid seems to be more effective than that of 4-vinylphenylboronic acid but not comparable to phenylboronic acid or 4-fluorophenylboronic acids. The differences in the activity are presumably due to the difference in the electronic properties of the substituents. Thus, –F and –COOH have an electron-withdrawing inductive effect –*I*, while vinyl group has an electron-releasing effect +*I*. Therefore, as many mechanistic studies have proven that the only reactions involving boronic acids occur at a significant rate, is between the neutral boronic acid and oxo-palladium species [41]. It seems that the activation of the boronic group is faster in the case of F- and COOH- substituted phenylboronic acids, and this results in a higher activity.

2.2.2. Influence of the Fluorinated Aryl Bromide

In view of the interesting catalytic properties when using 4-fluorophenylboronic acid (which was superior to that of phenylboronic acid and 4-carboxyphenylboronic acid), this reagent was selected and used with other fluorinated bromoaryls, namely, 1-bromo-2-fluorobenzene, 1-bromo-3-fluorobenzene, 2-bromo-5-fluorotoluene, and 2-bromo-4-fluorotoluene (Scheme 2a–d, respectively), for the formation of different difluorinated biphenyls.

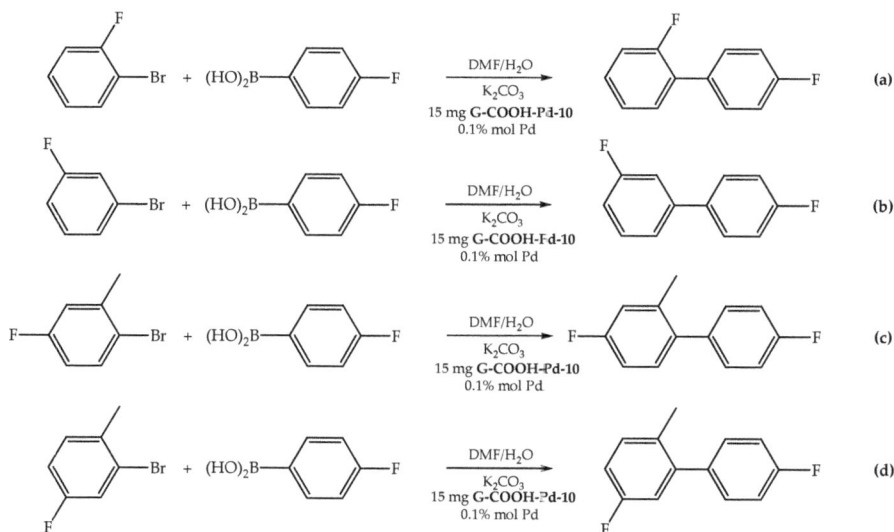

Scheme 2. Reaction of 4-fluorophenylboronic acid with (**a**) 1-bromo-2-fluorobenzene; (**b**) 1-bromo-3-fluorobenzene; (**c**) 2-bromo-5-fluorotoluene catalyzed; and (**d**) 2-bromo-4-fluorotoluene by G-COOH-Pd-10.

The results obtained in the C-C cross-coupling reactions of Scheme 2 at different time intervals are given in Table 3 and represented in Figure 4. As occurred in the previous study, concerning the influence of boronic acids, the increase of the temperature from 70 °C to 110 °C again results in higher conversion percentages of the bromides. In addition, the results show that, except for 1-bromo-4-fluorotoluene and 2-bromo-4-fluorotoluene at 110 °C, the reactions achieve almost the maximum conversion between 3 and 8 h. In the case of 1-bromo-2-fluorobenzene, 1-bromo-3-fluorobenzene, and 1-bromo-4-fluorobenzene at both 70 °C or 110 °C and 3 h of reaction, the obtained TOF values were between ca. 44 and 67 h^{-1}. In contrast, in the case of 2-bromo-5-fluorotoluene and 2-bromo-4-fluorotoluene at 70 °C or 110 °C and 3 h of reaction, the TOF values were lower, indicating an inferior activity when using this substituted bromide.

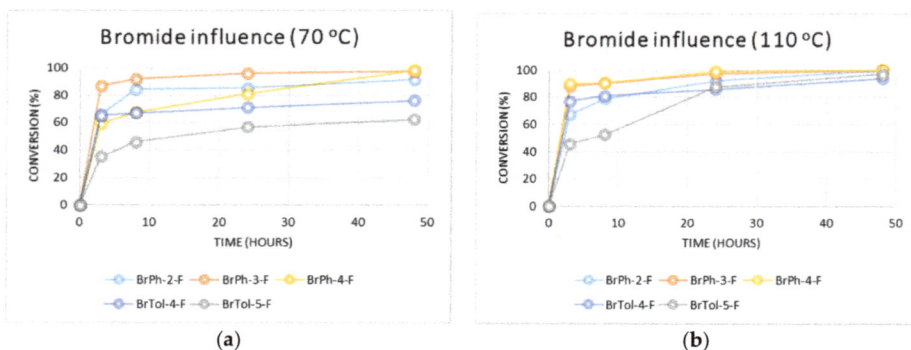

Figure 4. Conversion vs. time in the reaction of different fluorinated arylbromides and 4-fluorophenylboronic acid catalyzed by G-COOH-Pd-10: (**a**) at 70 °C; and (**b**) at 110 °C.

It seems clear that the position of the fluorine substituent at the phenyl ring does not have a remarkable influence in the catalytic activity. Thus, the steric effect does not seem to be determinant for their reactivity. However, in the case of 1-bromo-4-fluorotoluene and 2-bromo-4-fluorotoluene, the incorporation of the methyl group in *ortho*-position to the bromine atom, results in a decrease of the catalytic activity due to both the steric hindrance and the electronic +*I* effect of the methyl group in the bromide, which decrease the catalytic activity.

Finally, we have carried out the reaction between 4-fluorobenzeneboronic acid and 1-chloro-4-fluorobenzene observing halide conversion after 48 h of ca. 12%. However, from the reaction mixture we were unable to isolate the coupling product observing various unidentified F-containing compounds. This indicates the limited applicability in coupling reactions of these systems when starting from arylchlorides.

2.2.3. Recyclability Tests

It is well known that one of the most important advantages of heterogeneous catalytic systems is the possibility of recovery and recyclability. Thus, a series of catalytic tests were carried out to determine the degree of loss of activity of the synthesized catalyst G-COOH-Pd-10 after several consecutive catalytic cycles. The studied recyclability tests were performed using similar experimental conditions, but tested in up to five consecutive catalytic cycles. The selected reagents for this study of recyclability were 1-bromo-4-fluorobenzene and 4-fluorophenylboronic acid (highest TOF values). After each catalytic cycle, the catalyst was centrifuged and washed with water and diethylether, dried under vacuum, and then used in the subsequent catalytic test.

Table 3. Halide conversions in C-C coupling reactions using 4-fluorophenylboronic acid and different fluorinated arylbromides catalyzed by G-COOH-Pd-10.

Time (h)	T (°C)	Bromide Conversion (%)	TON	TOF (h⁻¹)	Reaction
3	70	66	147	49	
8		84	189	24	
24		85	191	8	
48		91	204	4	
3	110	67	151	50	
8		79	177	22	
24		92	207	9	
48		100	224	5	
3	70	87	195	65	
8		92	206	26	
24		96	215	9	
48		97	218	5	
3	110	88	197	66	
8		90	202	25	
24		98	218	9	
48		100	224	5	
3	70	35	78	26	
8		46	102	13	
24		57	127	5	
48		62	140	3	
3	110	46	102	34	
8		53	117	15	
24		88	197	8	
48		97	217	5	
3	70	65	149	50	
8		67	151	19	
24		71	159	6	
48		76	169	3	
3	110	77	172	57	
8		81	182	23	
24		86	194	8	
48		94	210	4	

For the studied reaction, a very low loss of activity (less than 10%) of the catalyst G-COOH-Pd-10 was observed after the five catalytic cycles (Figure 5). It is important to note that these systems based on palladium nanoparticles may lead to agglomeration of the nanoparticles after the first or subsequent cycles because of the mobility of the palladium nanoparticles at high temperatures. This effect has previously been observed in similar hybrid systems based on mesoporous silica [12] and graphene oxide [16], but not significantly in alumina-based materials [14], or graphene oxide when using microwaves [16]. In the case of G-COOH-Pd-10 the very small loss of activity suggests that there is no formation of big clusters of Pd nanoparticles as this would lead to deactivation of the catalyst.

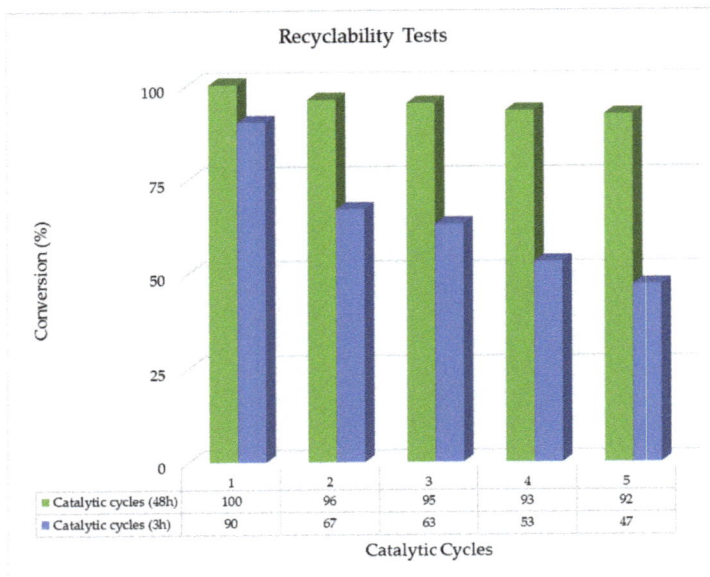

Figure 5. Recyclability tests of the consecutive reactions of 1-bromo-4-fluorobenzene and 4-fluorophenylboronic acid at 110 °C for 48 h (green) or 3 h (blue) catalyzed by G-COOH-Pd-10.

We have carried out additional recyclability tests using a reaction time of 3 h. We have observed a progressive loss of activity in the consecutive tests at 3 h from 90% to 47% after the fifth cycle (Figure 5). This did not happen at 48 h probably because the reaction time is higher and, therefore, compensates for the loss of activity.

Thus, in order to determine the reason of the loss of activity, we measured the Pd leaching in the solution by XRF although we were unable to determine Pd in concentrated solutions. However, when we carried out a TEM measurement of the material after the fifth catalytic cycle (Figure 6), we observed the formation of clusters of palladium nanoparticles which agglomerate and, therefore, cause a decrease in the catalytic activity.

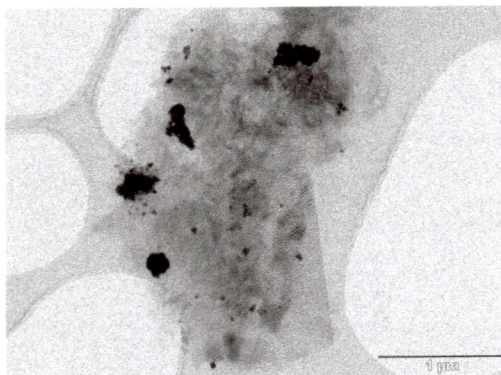

Figure 6. Transmission electronic microscopy (TEM) image of G-COOH-Pd-10 after five consecutive catalytic cycles.

3. Materials and Methods

3.1. General Conditions

All manipulations were performed under dry nitrogen gas using standard Schlenk techniques and a dry box. Solvents were distilled from the appropriate drying agents and degassed before use. Graphene modified with COOH groups UGRAY^{TM}-COOH (Graphene-carboxyl, G-COOH) was purchased from United Nanotech (Karnataka, India) and was used as purchased, after a simple dehydration process (see Section 3.3.2.). Water (resistance 18.2 MΩ·cm) used in the study was obtained from a Millipore Milli-Q-System (Billerica, MA, USA).

3.2. General Remarks on the Characterization of the Materials

X-ray diffraction (XRD) patterns of the hybrid materials were obtained on a Philips Diffractometer model PW3040/00 X'Pert MPD/MRD at 45 KV and 40 mA, using a wavelength Cu Kα (λ = 1.5418 Å). Pd wt % determination by X-ray fluorescence was carried out with a X-ray fluorescence spectrophotometer Philips MagiX with an X-ray source of 1 kW and a Rh anode using a helium atmosphere. The quantification method is capable of analyzing from 0.0001% to 100% Pd. N$_2$ gas adsorption-desorption isotherms were performed using a Micromeritics ASAP 2020 analyzer (Micromeritics, Norcross, GA., USA). Conventional transmission electron microscopy (TEM) was carried out on a TECNAI 20 Philips unit (Philips, Eindhoven, The Netherlands), operating at 200 kV.

3.3. Preparation of the Hybrid Materials G-COOH-Pd-X

3.3.1. Preparation of the Palladium Precursor

PdCl$_2$ (1.53 g, 8.6 mmol) was dissolved in 6 mL of concentrated HCl. The cooled solution was diluted with 150 mL of absolute ethanol and passed through a filter paper; the residue and filter paper were then washed with 2 × 10 mL of ethanol. Afterwards, 1,5-cyclooctadiene (2.5 mL, 20.4 mmol) was added to the resulting solution under stirring. The color of the solution turned from brown to orange and the solid product precipitated immediately. The reaction was stirred for an additional 20 min and then filtered and the yellow-orange solid washed with diethylether (3 × 10 mL). The final product was dried under vacuum overnight giving 2.32 g (16.4 mmol) of [PdCl$_2$(cod)] (yield: 97%).

3.3.2. Dehydration of G-COOH

In order to reduce the quantity of physisorbed solvents or water on the external surface area of G-COOH, activation by dehydration of the corresponding material was carried out at 150 °C under vacuum.

3.3.3. Pd-Loading Study

Functionalization of G-COOH has been studied using different quantities of Pd precursor [PdCl$_2$(cod)] and one gram of G-COOH. Table S1 of the supplementary material shows the quantity of palladium complex and G-COOH employed in each reaction.

The general procedure for the preparation of the Pd-supported nanoparticles was carried out using a similar method to that published by our group for titanium oxide-based materials [12,14,37,38]: In summary, the corresponding amount of G-COOH (1.0 g) and [PdCl$_2$(cod)] (141.2 mg, for a theoretical Pd loading of 5 wt %) were added to a Schlenk tube and dried under vacuum for 1 h at room temperature. Subsequently, 50 mL of toluene (THF) was added under an inert atmosphere. The reaction mixture was then heated to 110 °C and stirred for 48 h. The resulting material was isolated by filteration and washed with toluene, water and diethylether (2 × 50 mL each). The material was dried under vacuum for 12 h to remove all trace of solvents.

3.4. Catalytic Study

3.4.1. Determination of the Optimal Conditions and Influence of the Boronic Acid

The study was focused on the reaction of 1-bromo-4-fluorobenzene with different boronic acids, two different temperatures were tested (70 and 110 °C) and the reactions were monitored after 3, 8, 24, and 48 h reaction time. The different boronic acids used in the reaction were phenylboronic acid, 4-vinylphenylboronic acid, 4-carboxyphenylboronic acid and 4-fluorophenylboronic acid (Scheme 1a–d, respectively). The reactions were performed under identical conditions in order to facilitate a subsequent analysis of the results. In all the reactions, the limiting reagent was 1-bromo-4-fluorobenzene, the molar ratio between the halide and the boronic acid was 1:1.2, the molar ratio between the halide and the base (K_2CO_3) was 1:2 and the amount of catalyst was in all cases 15 mg. All the reactions were carried out using degassed solvents and under a nitrogen atmosphere to achieve higher final conversions [42].

Reaction of 1-bromo-4-fluorobenzene with phenylboronic acid derivatives: a stock solution of 1-bromo-4-fluorobenzene (262 mg, 1.5 mmol) in a mixture of degassed solvents ($DMF:H_2O$ 95:5, 15 mL) was carried out under nitrogen in a Schlenk tube. Subsequently, Schlenk flasks were filled with either phenylboronic acid (36.6 mg, 0.300 mmol), 4-vinylphenylboronic acid (44.4 mg, 0.300 mmol), 4-carboxyphenylboronic acid (49.8 mg, 0.300 mmol), or 4-fluorophenylboronic acid (42.0 mg, 0.300 mmol), K_2CO_3 (69.1 mg, 0.5 mmol), and Pd catalyst (15 mg G-COOH-Pd-10, 1.18 mg of Pd, $1.11 \cdot 10^{-3}$ mmol Pd, 0.44% molar ratio bromide:Pd, 0.1% mol Pd in the reaction). Three vacuum/N_2 cycles (10 min/1 min) were carried out to remove oxygen from the reaction atmosphere and adsorbed water from the solids. Afterwards, 2.5 mL (0.25 mmol) of the stock solution of 1-bromo-4-fluorobenzene were transferred under N_2 to the Schlenk flasks containing the solid mixtures. The suspension was then heated to the corresponding temperature (70 °C or 110 °C) using a condenser and stirred for 3, 8, 24, or 48 h. After this time, the solution was cooled to room temperature and filtered over a nylon filter (0.4 μm).

3.4.2. Quantification of the Conversion of 1-Bromo-4-Fluorobenzene

For the quantification of the conversion of 1-bromo-4-fluorobenzene, 0.3 mL of each resulting solution of the reactions was mixed together with 0.3 mL of a standard 2.5 wt % solution of 4-fluorobenzophenone in deuterated acetone. An additional NMR tube (blank) was prepared by mixing 0.3 mL of the original solution of 1-bromo-4-fluorobenzene (262 mg, 1.5 mmol) in a mixture of degassed solvents ($DMF:H_2O$ 95:5, 15 mL) with 0.3 mL of a standard 1.0% v/v solution of 4-fluorobenzophenone in deuterated acetone. The quantification was carried out by comparison of the ratio of the integral of the signal of the fluorine atom in 1-bromo-4-fluorobenzene (δ −111.7 ppm) with that of the standard (4-fluorobenzophenone, δ −102.9 ppm) before and after the reaction. The chosen internal standard was 4-fluorobenzophenone (final concentration of 0.75% v/v), an inert compound with respect to reagents and products and that has a very different chemical shift to that of the other compounds. The solution samples were always in the range of the concentration of 1-bromo-4-fluorobenzene of the calibration curve. The calibration curve was prepared using a concentration of 1-bromo-4-fluorobenzene in the range 0.005–0.1 M. Using this method, the molar quantification of the conversion of 1-bromo-4-fluorobenzene was carried out. For an example of the ^{19}F NMR spectra of the reaction between 1-bromo-4-fluorobenzene and phenylboronic acid see Figures S7–S12 of the Supplementary Material.

3.4.3. Isolation of the Coupling Products

For the isolation of the product, the solvent of the filtrated solution after 48 h of reaction was eliminated under vacuum and the solid or oily residue was dissolved in $CDCl_3$ and analyzed by ^1H, and ^{19}F NMR spectroscopy. For further details of all the synthesized fluorinated biaryl derivatives see section Spectroscopic Data (^1H and ^{19}F NMR) of the Supplementary Material.

3.4.4. Reactions of Different Fluorinated Aryl Bromides with 4-Fluorophenylboronic Acid

In order to determine the influence of different fluorinated aryl bromides in the catalytic C-C coupling reaction, additional reactions of 1-bromo-3-fluorobenzene, 1-bromo-2-fluorobenzene, 2-bromo-5-fluorotoluene, or 2-bromo-4-fluorotoluene with 4-fluorophenylboronic acid were carried out (Scheme 2a–d), following a similar procedure to that described in the Section 3.4.2 using temperatures of 70 °C and 110 °C and reaction times of 3, 8, 24, or 48 h. In this case, the quantification was carried out by comparison of the ratio of the integral of the signal of the fluorine atom in the ^{19}F NMR spectrum of 1-bromo-3-fluorobenzene (δ −106.9 ppm), 1-bromo-2-fluorobenzene (δ −104.4 ppm), 2-bromo-5-fluorotoluene (δ −112.0 ppm), or 2-bromo-4-fluorotoluene (δ −112.0 ppm) with that of the standard (4-fluorobenzophenone, δ 102.9 ppm) before and after the reaction. For an example of the ^{19}F NMR spectra of the reaction between 4-fluorophenylboronic acid and 1-bromo-2-fluorobenzene see Figure S13 of the Supplementary Material.

3.4.5. Studies of the Catalyst Recyclability

Additional catalytic tests were carried out to determine the loss of activity of the catalyst after several catalytic cycles. Recyclability tests were carried out using similar experimental procedures but tested in up to five consecutive catalytic cycles. The reactions were performed on a larger scale, but under the same experimental conditions. The tests were conducted using a higher starting amount of catalyst (75 mg) in order to be able to carry out up to five catalytic cycles. After each cycle, the catalyst was centrifuged and washed with water (2 × 100 mL) and diethylether (2 × 30 mL).

4. Conclusions

In this work we have synthesized a hybrid heterogeneous catalyst based on Pd-supported nanoparticles onto COOH-modified graphite support in a mixture of phases (G-COOH-Pd-10). This material has shown interesting catalytic behavior in the Suzuki-Miyaura coupling reaction of fluorinated aryls with TOF values of up to 67 h^{-1} which are higher than those found for reported homogeneous palladium complexes in similar reactions for the formation of fluorinated biaryls. In addition, the recyclability of the system G-COOH-Pd-10 at short 3 h reaction time shows a progressive loss of activity in the catalytic tests from 90% to 47% after the fifth cycle. However, this material showed a good degree of recyclability with a low deactivation of less than 8% after up to five catalytic cycles after 48 hours, probably due to the longer reaction time compensating for the loss of activity.

Supplementary Materials: The following are available online at www.mdpi.com/2073-4344/7/3/76/s1. Table S1: Experimental quantities of reagents for the Pd loading study; Table S2: Adsorptive parameters of the materials G-COOH and G-COOH-Pd-10; Figure S1: TEM image of G-COOH showing the single layer of graphene; Figure S2: TEM image of a cluster of agglomerated Pd nanoparticles; Figure S3: TEM image showing the impregnation of a cluster of Pd nanoparticles at the edge of the graphene layer; Figure S4: FT-IR spectrum of G-COOH-Pd-10; Figure S5: Nitrogen adsorption desorption isotherm of G-COOH; Figure S6: XRD of the material G-COOH-Pd-10; Figure S7: Comparison of the ^{19}F NMR spectra of the reaction between 1-bromo-4-fluorobenzene and phenylboronic acid catalyzed by G-COOH-Pd-10 in the presence of a constant quantity of standard (4-fluorobenzophenone) at different reaction time periods; Figure S8: ^{19}F NMR spectrum of the starting solution of 1-bromo-4-fluorobenzene in the presence of a constant quantity of standard (4-fluorobenzophenone) (0 hours); Figures S9–S12: ^{19}F NMR spectrum of the reaction between 1-bromo-4-fluorobenzene and phenylboronic acid catalyzed by G-COOH-Pd-10 after 3, 8, 24, and 48 hours of reaction in the presence of a constant quantity of standard (4-fluorobenzophenone), Figure S13: Comparison of the ^{19}F NMR spectra of the reaction between 1-bromo-2-fluorobenzene and 4-fluorophenylboronic acid catalyzed by G-COOH-Pd-10 in the presence of a constant quantity of standard (4-fluorobenzophenone) at different reaction time periods. In addition, the spectroscopic data (^{1}H and ^{19}F NMR) of all the fluorinated biaryl derivatives are included.

Acknowledgments: We gratefully acknowledge financial support from the Ministerio de Economía y Competitividad, Spain (Grant no. CTQ2015-66164-R). We would also like to thank Universidad Rey Juan Carlos and Banco de Santander for supporting our Research Group of Excellence QUINANOAP. We would also like to thank Isfahan University of Technology for the partial financial support of the research stay of R.S.E. Finally, we thank Sandra Carralero and Carmen Forcé for their valuable advice with NMR experiments.

Author Contributions: S.G-R. conceived and designed the experiments; R.S.E. and D.D.-G. performed the experiments; S.G-R., R.S.E., D.D.-G., S.P., M.A., A.R.-D., and M.F. analyzed the data and contributed with different analysis tools; finally, S.G-R., R.S.E., D.D.-G., and S.P. wrote the paper.

Conflicts of Interest: The authors declare no conflict of interest. The funding sponsors had no role in the design of the study; in the collection, analyses, or interpretation of data; in the writing of the manuscript, and in the decision to publish the results.

References

1. Seechurn, C.C.C.J.; Kitching, M.O.; Colacot, T.J.; Snieckus, V. Palladium-Catalyzed Cross-Coupling: A Historical Contextual Perspective to the 2010 Nobel Prize. *Angew. Chem. Int. Ed.* **2012**, *51*, 5062–5085. [CrossRef] [PubMed]
2. Maluenda, I.; Navarro, H. Recent Developments in the Suzuki-Miyaura Reaction: 2010–2014. *Molecules* **2015**, *20*, 7528–7557. [CrossRef] [PubMed]
3. Miyaura, N.; Suzuki, A. Palladium-Catalyzed Cross-Coupling Reactions of Organoboron Compounds. *Chem. Rev.* **1995**, *95*, 2457–2483. [CrossRef]
4. Nicolaou, K.C.; Bulger, P.G.; Sarlah, D. Palladium-Catalyzed Cross-Coupling Reaction in Total Synthesis. *Angew. Chem. Int. Ed.* **2005**, *44*, 4442–4489. [CrossRef] [PubMed]
5. Kotha, S.; Lahiri, K.; Kashinath, D. Recent Applications of the Suzuki-Miyaura Cross-Coupling Reaction in Organic Synthesis. *Tetrahedron* **2002**, *58*, 9633–9695. [CrossRef]
6. Schlögl, R. Heterogeneous Catalysis. *Angew. Chem. Int. Ed.* **2015**, *54*, 3465–3520. [CrossRef] [PubMed]
7. Polshettiwar, V.; Len, C.; Fihri, A. Silica-supported palladium: Sustainable catalysts for cross-coupling reactions. *Coord. Chem. Rev.* **2009**, *253*, 2599–2626. [CrossRef]
8. Kann, N. Recent Applications of Polymer Supported Organometallic Catalysts in Organic Synthesis. *Molecules* **2010**, *15*, 6306–6331. [CrossRef] [PubMed]
9. Bej, A.; Ghosh, K.; Sarkar, A.; Knight, D.W. Palladium nanoparticles in the catalysis of coupling reactions. *RSC Adv.* **2016**, *6*, 11446–11453. [CrossRef]
10. Pérez-Lorenzo, M. Palladium Nanoparticles as Efficient Catalysts for Suzuki Cross-Coupling Reactions. *J. Phys. Chem. Lett.* **2012**, *3*, 167–174. [CrossRef]
11. Parlett, C.M. A.; Bruce, D.W.; Hondow, N.S.; Newton, M.A.; Lee, A.F.; Wilson, K. Mesoporous Silicas as Versatile Supports to Tune the Palladium-Catalyzed Selective Aerobic Oxidation of Allylic Alcohols. *ChemCatChem* **2013**, *5*, 939–950. [CrossRef]
12. Balbín, A.; Gaballo, F.; Ceballos-Torres, J.; Prashar, S.; Fajardo, M.; Kaluderovic, G.N.; Gómez-Ruiz, S. Dual application of Pd nanoparticles supported on mesoporous silica SBA-15 and MSU-2: Supported catalysts for C–C coupling reactions and cytotoxic agents against human cancer cell lines. *RSC Adv.* **2014**, *4*, 54775–54787. [CrossRef]
13. Kumar, A.P.; Kumar, B.P.; Kumar, A.B.V.K.; Huy, B.T.; Lee, Y.-I. Preparation of palladium nanoparticles on alumina surface by chemical co-precipitation method and catalytic applications. *Appl. Surface Sci.* **2013**, *265*, 500–509. [CrossRef]
14. Hossain, A.M.S.; Balbín, A.; Erami, R.S.; Prashar, S.; Fajardo, M.; Gómez-Ruiz, S. Synthesis and study of the catalytic applications in C–C coupling reactions of hybrid nanosystems based on alumina and palladium nanoparticles. *Inorg. Chim. Acta* **2017**, *455*, 645–652. [CrossRef]
15. Siamaki, A.R.; Khder, A.E.R.S.; Abdelsayed, V.; El-Shall, M.S.; Gupton, B.F. Microwave-assisted synthesis of palladium nanoparticles supported on graphene: A highly active and recyclable catalyst for carbon–carbon cross-coupling reactions. *J. Catal.* **2011**, *279*, 1–11. [CrossRef]
16. Gómez-Martínez, M.; Buxaderas, E.; Pastor, I.M.; Alonso, D.A. Palladium nanoparticles supported on graphene and reduced graphene oxide as efficient recyclable catalyst for the Suzuki–Miyaura reaction of potassium aryltrifluoroborates. *J. Mol. Catal. A: Chem.* **2015**, *404–405*, 1–7. [CrossRef]
17. Joshi, H.; Sharma, K.N.; Sharma, A.K.; Singh, A.K. Palladium–phosphorus/sulfur nanoparticles (NPs) decorated on graphene oxide: Synthesis using the same precursor for NPs and catalytic applications in Suzuki–Miyaura coupling. *Nanoscale* **2014**, *6*, 4588–4597. [CrossRef] [PubMed]
18. Movahed, S.K.; Esmatpoursalmani, R.; Bazgir, A. N-Heterocyclic carbene palladium complex supported on ionic liquid-modified graphene oxide as an efficient and recyclable catalyst for Suzuki reaction. *RSC Adv.* **2014**, *4*, 14586–14591. [CrossRef]

19. Yamamoto, S.-I.; Kinoshita, H.; Hashimoto, H.; Nishina, Y. Facile preparation of Pd nanoparticles supported on single-layer graphene oxide and application for the Suzuki–Miyaura cross-coupling reaction. *Nanoscale* **2014**, *6*, 6501–6505. [CrossRef] [PubMed]

20. Rumi, L.; Scheuermann, G.M.; Mülhaupt, R.; Bannwarth, W. Palladium Nanoparticles on Graphite Oxide as Catalyst for Suzuki-Miyaura, Mizoroki-Heck, and Sonogashira Reactions. *Helv. Chim. Acta* **2011**, *94*, 966–976. [CrossRef]

21. Lin, J.; Mei, T.; Lv, M.; Zhang, C.; Zhao, Z.; Wang, X. Size-controlled PdO/graphene oxides and their reduction products with high catalytic activity. *RSC Adv.* **2014**, *4*, 29563–29570. [CrossRef]

22. Hattori, T.; Tsubone, A.; Sawama, Y.; Monguchi, Y.; Sajiki, H. Palladium on Carbon-Catalyzed Suzuki-Miyaura Coupling Reaction Using an Efficient and Continuous Flow System. *Catalysts* **2015**, *5*, 18–25. [CrossRef]

23. Groult, H.; Leroux, F.; Tressaud, A. *Modern synthesis Processes and Reactivity of Fluorinated Compounds*; Elsevier: London, UK, 2017.

24. Wang, J.; Sánchez-Roselló, M.; Aceña, J.L.; del Pozo, C.; Sorochinsky, A.E.; Fustero, S.; Soloshonok, V.A.; Liu, H. Fluorine in Pharmaceutical Industry: Fluorine-Containing Drugs Introduced to the Market in the Last Decade (2001–2011). *Chem. Rev.* **2014**, *114*, 2432–2506. [CrossRef] [PubMed]

25. Murphy, C.D. Drug metabolism in microorganisms. *Biotechnol. Lett.* **2015**, *37*, 19–28. [CrossRef] [PubMed]

26. Berger, R.; Resnati, G.; Metrangolo, P.; Weberd, E.; Hulliger, J. Organic fluorine compounds: A great opportunity for enhanced materials properties. *Chem. Soc. Rev* **2011**, *40*, 3496–3508. [CrossRef] [PubMed]

27. Coates, D.; Greenfield, S.; Smith, G.; Chambers, M.K.; Kurmeier, H.-A.; Dorsch, D. Fluorinated biphenyl derivatives. Merck, Darmstadt, Germany. WO1990015115, 1990.

28. Hampton, A.S.; Mikulski, L.; Palmer-Brown, W.; Murphy, C D.; Sandford, G. Evaluation of fluorinated biphenyl ether pro-drug scaffolds employing the chemical-microbial approach. *Bioorg. Med. Chem. Lett.* **2016**, *26*, 2255–2258. [CrossRef] [PubMed]

29. Fang, X.; Huang, Y.; Chen, X.; Lin, X.; Bai, Z.; Huang, K.-W.; Yuan, Y.; Weng, Z. Preparation of fluorinated biaryls through direct palladium-catalyzed coupling of polyfluoroarenes with aryltrifluoroborates. *J. Fluorine Chem.* **2013**, *151*, 50–57. [CrossRef]

30. Wei, Y.; Su, W. Pd(OAc)$_2$-Catalyzed Oxidative C-H/C-H Cross-Coupling of Electron-Deficient Polyfluoroarenes with Simple Arenes. *J. Am. Chem. Soc.* **2010**, *132*, 16377–16379. [CrossRef] [PubMed]

31. Li, H.; Liu, J.; Sun, C.-L.; Li, B.-J.; Shi, Z.-J. Palladium-Catalyzed Cross-Coupling of Polyfluoroarenes with Simple Arenes. *Org. Lett.* **2011**, *13*, 276–279. [CrossRef] [PubMed]

32. Liu, N.; Liu, C.; Jin, Z. Green synthesis of fluorinated biaryl derivatives via thermoregulated ligand/palladium-catalyzed Suzuki reaction. *J. Organomet. Chem.* **2011**, *696*, 2641–2647. [CrossRef]

33. Liu, Y.; Wang, J. Synthesis of 4-Substituted Styrene Compounds via Palladium-Catalyzed Suzuki Coupling Reaction Using Free Phosphine Ligand in Air. *Synth. Commun* **2010**, *40*, 196–205. [CrossRef]

34. Wang, Z.-Y.; Ma, Q.-M.; Lia, R.-H.; Shao, L.-X. Palladium-catalyzed Suzuki–Miyaura coupling of aryl sulfamates with arylboronic acids. *Org. Biomol. Chem.* **2013**, *11*, 7899–7906. [CrossRef] [PubMed]

35. Wang, Z.-Y.; Chen, G.-Q.; Shao, L.-X. N-Heterocyclic Carbene—Palladium(II)—1-Methylimidazole Complex-Catalyzed Suzuki—Miyaura Coupling of Aryl Sulfonates with Arylboronic Acids. *J. Org. Chem.* **2012**, *77*, 6608–6614. [CrossRef] [PubMed]

36. Kurscheid, B.; Belkoura, L.; Hoge, B. Air-Stable and Catalytically Active Phosphinous Acid Transition-Metal Complexes. *Organometallics* **2012**, *31*, 1329–1334. [CrossRef]

37. Lázaro-Navas, S.; Prashar, S.; Fajardo, M.; Gómez-Ruiz, S. Visible light-driven photocatalytic degradation of the organic pollutant methylene blue with hybrid palladium–fluorine-doped titanium oxide nanoparticles. *J. Nanopart. Res.* **2015**, *17*, 94. [CrossRef]

38. Rico-Oller, B.; Boudjemaa, A.; Bahruji, H.; Kebir, M.; Prashar, S.; Bachari, K.; Fajardo, M.; Gómez-Ruiz, S. Photodegradation of organic pollutants in water and green hydrogen production via methanol photoreforming of doped titanium oxide nanoparticles. *Sci. Total Environ.* **2016**, *563–564*, 921–932. [CrossRef] [PubMed]

39. Schneider, C.A.; Rasband, W.S.; Eliceiri, K.W. NIH Image to ImageJ: 25 years of image analysis. *Nat. Methods* **2012**, *9*, 671–675. [CrossRef] [PubMed]

40. Sing, K.S.W.; Everett, D.H.; Haul, R.A.W.; Moscou, L.; Pierotti, R.A.; Rouquerol, J.; Siemieniewska, T. Reporting physisorption data for gas/solid systems with special reference to the determination of surface area and porosity (Recommendations 1984). *Pure Appl. Chem.* **1985**, *57*, 603–619. [CrossRef]
41. Lennox, A.J. J.; Lloyd-Jones, G.C. Selection of boron reagents for Suzuki–Miyaura coupling. *Chem. Soc. Rev.* **2014**, *43*, 412–443. [CrossRef] [PubMed]
42. Trilla, M.; Borja, G.; Pleixats, R.; Man, M.W.C.; Bied, C.; Moreau, J.J.E. Recoverable Palladium Catalysts for Suzuki–Miyaura Cross-Coupling Reactions Based on Organic-Inorganic Hybrid Silica Materials Containing Imidazolium and Dihydroimidazolium Salts. *Adv. Synth. Catal.* **2008**, *350*, 2566–2574. [CrossRef]

catalysts

MDPI

Article

Graphene Oxide-Supported Oxime Palladacycles as Efficient Catalysts for the Suzuki–Miyaura Cross-Coupling Reaction of Aryl Bromides at Room Temperature under Aqueous Conditions

Melania Gómez-Martínez, Alejandro Baeza * and Diego A. Alonso *

Departamento de Química Orgánica and Instituto de Síntesis Orgánica (ISO), Facultad de Ciencias, Universidad de Alicante. Apdo. 99, E-03080 Alicante, Spain; melania.gomez@ua.es
* Correspondence: alex.baeza@ua.es (A.B.); diego.alonso@ua.es (D.A.A.);
    Tel.: +34-965-902-888 (A.B.); +34-965-909-841 (D.A.A.)

Academic Editor: Ioannis D. Kostas
Received: 25 January 2017; Accepted: 16 March 2017; Published: 22 March 2017

Abstract: Palladacycles are highly efficient precatalysts in cross-coupling reactions whose immobilization on carbonaceous materials has been hardly studied. Herein, we report a detailed study on the synthesis and characterization of new oxime palladacycle-graphene oxide non-covalent materials along with their catalytic activity in the Suzuki–Miyaura reaction. Catalyst **1-GO**, which has been fully characterized by ICP, XPS, TGA, and UV-Vis analyses has been demonstrated to be an efficient catalyst for the Suzuki–Miyaura coupling between aryl bromides and arylboronic acids using very low catalyst loadings (0.002 mol % of Pd) at room temperature under aqueous conditions.

Keywords: graphene oxide; Suzuki–Miyaura coupling; palladacycle; aryl bromide; palladium; aqueous conditions

---

1. Introduction

Palladium-catalyzed cross-coupling reactions occupy a predominant place in the arsenal of synthetic chemists [1–9]. Carbometallated Pd(II) compounds, especially the highly active palladacycles [10–13], have emerged as very promising catalysts for C–C bond forming reactions. These complexes usually involve very low catalyst loadings, minimizing the cost-effective impact of the expensive palladium. However, the product contamination problems associated with this toxic metal have converted the immobilization of this type of well-defined complex into an attractive strategy to develop sustainable catalytic processes. Until now, different palladacycles have been immobilized on inorganic and organic supports through covalent or non-covalent interactions [11,12]. Among them, silica-based supports [14–24], organic polymers [19,25–37], monolithic supports [38], magnetic nanoparticles [39,40], macrocyclic molecules [41], and Montmorillonite [42,43], are the studied systems. Although these catalysts can be generally reused, a progressive deactivation caused by complex degradation is usually observed, with the reaction products generally being vulnerable to palladium contamination [44]. Also, contribution to the catalysis of leached species, from the solid into the solution, is usually observed.

During the last years, carbon-derived materials have been designed as interesting supports for palladium nanoparticles due to their chemical stability as well as large specific surface area. This type of catalyst has been mainly used in the Suzuki–Miyaura cross-coupling reaction [45–62]. However, to the best of our knowledge, only very recently has some attention been paid to the immobilization of palladacycles to graphene materials, with two studies having been reported so far using covalent immobilization strategies [63,64].

Very recently, we have shown that palladium nanoparticles (Pd NPs), supported on graphene and reduced graphene oxide, efficiently catalyze the Suzuki–Miyaura coupling between aryl bromides and potassium aryltrifluoroborates under aqueous and low loading conditions (0.1 mol % Pd) employing conventional or microwave heating [65]. Based on the experience and knowledge of our research group with oxime palladacycles as highly efficient precatalysts in cross-coupling reactions [66–68], herein we report a detailed study on the synthesis and characterization of new oxime palladacycle-graphene oxide non-covalent materials along with their catalytic activity in the Suzuki–Miyaura reaction.

2. Results and Discussion

We initially chose GO as support for oxime palladacycles **1–3** (Figure 1) since this carbon material shows interesting hydrophilicity and stability properties in liquid phase reactions due to the abundant oxygen-containing functional groups on its surface. A non-covalent immobilization strategy, through hydrogen-bond and π-stacking interactions between the pre-formed palladacycles and the functionalized carbon support was selected to prepare the new catalytic materials. This approach should integrally preserve the palladacycle structure, avoiding an uncontrolled generation of other Pd(II) or Pd(0) species, as demonstrated by other authors when using a post-palladation strategy [17]. Furthermore, we expect the carbonaceous support to act as a reservoir and stabilizing medium for the in situ generated Pd NPs, which could extend the catalyst lifetime.

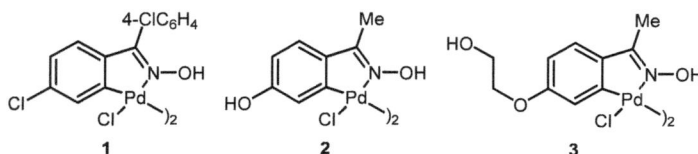

Figure 1. Oxime palladacycles employed in this study.

Preliminary studies on the immobilization of the palladacycles over **GO** were carried out with complex **1**. The preparation procedures tested to obtain the **1-GO** catalyst are indicated in Scheme 1. Basically, they consisted of the synthesis of the corresponding oxime palladacycle **1** from 4,4′-dichlorobenzophenone [69] and a subsequent immobilization of this complex over GO [70] in THF or water as solvents (Scheme 1). As depicted, sonication of the palladacycle in the presence of **GO** for 1 h (methods B and C) significantly improved the content of the immobilized complex on the carbon surface as determined by Inductively Coupled Plasma Optical Emission Spectrometry ICP-OES, being the highest (2 wt % Pd) when using THF as solvent (method B). Finally, following B conditions but starting from 4-hydroxyacetophenone and 1-[4-(2-hydroxyethoxy)phenyl]ethan-1-one, the supported catalysts **2-GO** (2.72 wt % Pd by ICP-OES) and **3-GO** (2.98 wt % Pd by ICP-OES) were also prepared (Scheme 1).

X-ray photoelectron spectroscopy (XPS) was used to analyze the elemental composition of the surface of the different catalysts. Interestingly, XPS analysis of the palladacycles **1–3** and all the supported catalyst batches showed all the binding energies associated with the oxime palladacycle structure (see SI). This point confirmed the stability of the cyclopalladated complexes during the immobilization process to **GO**. In fact, in the case of the preparation of **1-GO**, no structural changes were observed by ^{1}H-NMR analysis in the eluted palladacycle after the immobilization process (see SI). Generally, none of the prepared catalysts contained palladium oxides or palladium(0) species on their surface according to the XPS analyses. Indeed, in isolated cases we could identify small amounts of the precursor Li_2PdCl_4 by XPS analysis, easily removed by simple washings with water. On the other hand, in the case of using water as solvent (method C) for the preparation of **1-GO**, we also observed small amounts (~8%) of a non-identified oxidized palladium specie (336.0 and 341.1 eV by XPS, see SI).

Catalysts **2017**, 7, 94

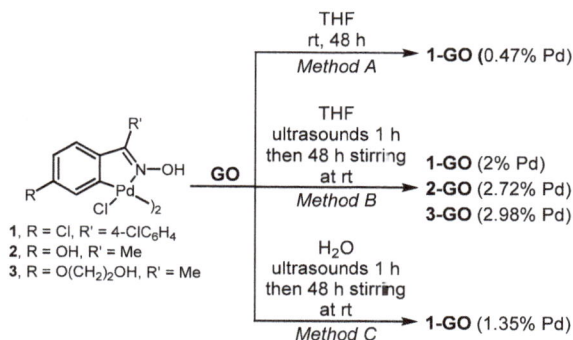

Scheme 1. Synthesis of supported oxime palladacycles.

Thus, performing the palladation reaction using THF as solvent before immobilization onto GO led to the integral cyclopalladated species being supported. In fact, Pd(0) has been previously detected when anchoring oxime palladacycles to different supports such as, 3-hydroxypropyltriethoxysilyl-functionalized MCM-41 [17] and 3-(aminomethyl)pyridine-functionalized graphene oxide [63].

Figures 2 and 3 show the XPS spectra of **1** and **1-GO** where the spectroscopic signature of palladium, carbon, oxygen, and nitrogen atoms has been determined [71–73] Regarding palladium (Figure 3a), the Pd3d XPS spectrum for **1** and **1-GO** showed the corresponding binding energies at 337.70 (3d5/2) and 343.02 (3d3/2) eV and 337.62 (3d5/2) and 342.83 (3d3/2) eV, respectively. These energies were assigned to the Pd(II) from the oxime palladacycle. On the other hand, O1s XPS analysis for catalyst **1** showed three different components related to the C=N–OH bond at 531.76 eV, the O–H bond at 533.5 eV and the C–O bond at 532.20 eV (Figure 3b). The higher content of the O–H and C–O groups in **1-GO** (532.20 and 533.32 eV, respectively) can be associated with the corresponding groups on the graphene oxide surface. The carbonaceous material was also responsible for the higher content of the C=C/C–H aromatic ring binding energies (284.60 and 286.71 eV) in the C1s XPS spectrum of **1-GO** when compared with **1** (Figure 3c). The N1s XPS spectra of **1** and **1-GO** showed the N=C bonding energies at 400.36 and 399.93 eV, respectively. Furthermore, the peaks corresponding to the N–Pd moiety appeared at 402.86 eV for 1 and 401.93 eV for **1-GO** (Figure 3d). Finally, the Cl2p XPS spectra of **1** was analyzed showing the corresponding binding energies at 198.6 eV (2p), 199.9 (2p3/2) and 200.3 eV (2p1/2). These energies were assigned to the Cl–Pd bonding energy from the oxime palladacycle **1**. The band at 200.1 8 eV corresponded to the 2p3/2 aromatic–Cl bonding energy (Figure 3e) [63].

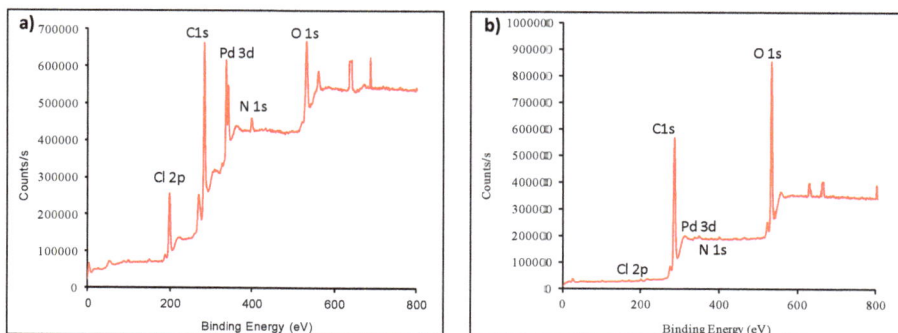

Figure 2. X-ray photoelectron spectroscopy (XPS) full spectra of: (a) **1**; (b) **1-GO**.

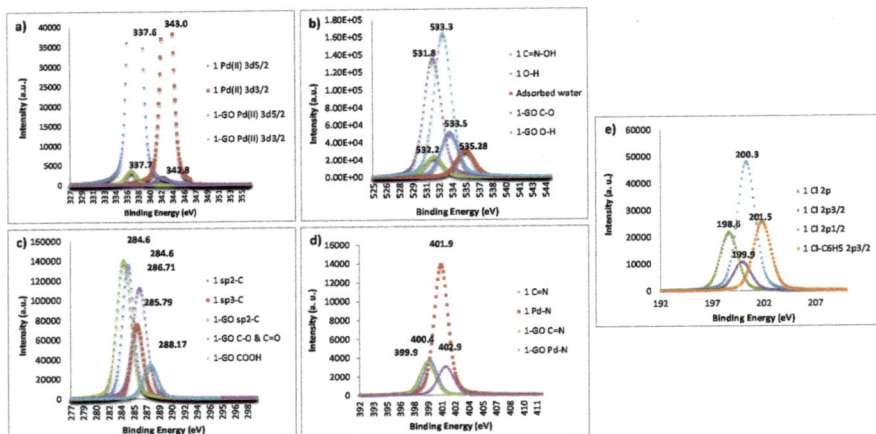

Figure 3. XPS spectra of **1** and **1-GO**: (**a**) Pd3d; (**b**) O1s; (**c**) C1s; (**d**) N1s; (**e**) Cl2p, and deconvoluted peaks (%) for Pd3d, C1s, O1s, N1s and Cl2p core levels.

Thermogravimetric analysis (TGA) of **1** (line blue), GO (black), and **1-GO** (line red) was next performed (Figure 4). As shown, palladacycle **1** presents a 39% mass loss below 300 °C which corresponds with the degradation of the 4,4′-dichlorobenzophenone oxime. On the other hand, regarding the support, GO exhibited around a 13% mass loss below 100 °C and almost a 27% loss at 210 °C resulting from the removal of the labile oxygen-containing functional groups and H_2O. Finally, **1-GO** exhibited a 28% mass loss at 200 °C due to the loss of the carbonaceous oxygenated species from GO and the loss of the organic ligand of the palladacycle. A final and significant common drop in mass was observed for the three materials around at 925 °C connected with the pyrolysis of the carbon skeletons (Figure 4). According to the TGA, the Pd content for catalyst **1-GO** (measured over different material batches) was between 1.35% to 2% which slightly differs from the content obtained from the more accurate ICP-OES technique (1.10% Pd).

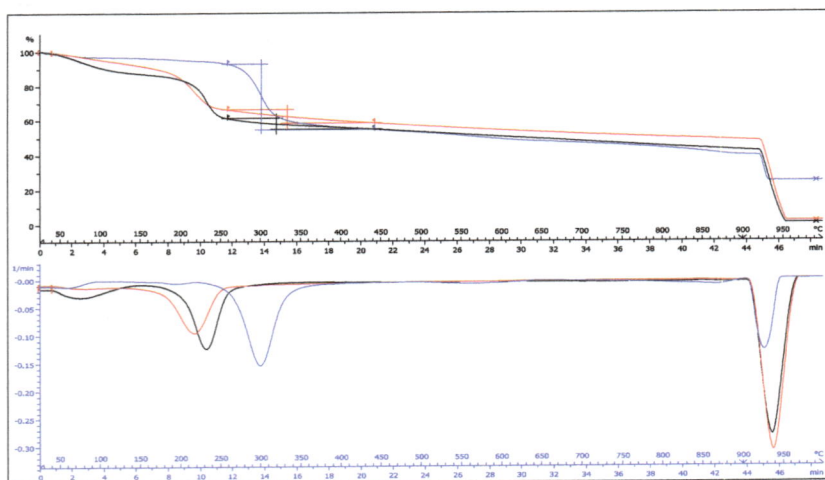

Figure 4. Thermogravimetric analysis (TGA) of **1** (line blue), GO (black), and **1-GO** (line red).

Catalyst **1-GO** was also characterized by solid UV-vis spectroscopy. As depicted in Figure 5, 4,4′-dichlorobenzophenone oxime (EtOH solution) showed the characteristic absorption C band at 265 nm. This absorption band was slightly blue shifted to 261 nm for palladacycle **1**, which also exhibited the specific metal-to-ligand charge transfer band at about 357 nm. As shown in the UV-vis absorption spectra, graphene oxide exhibited two absorption peaks, a maximum at 230 nm corresponding to π/π^* transitions of aromatic C–C bonds, and a shoulder at 293 nm attributed to n/π^* transitions of C=O bonds [74]. Finally, in the UV-vis spectrum of **1-GO**, the absorption C band from the palladacycle at 262 nm could be detected (Figure 5).

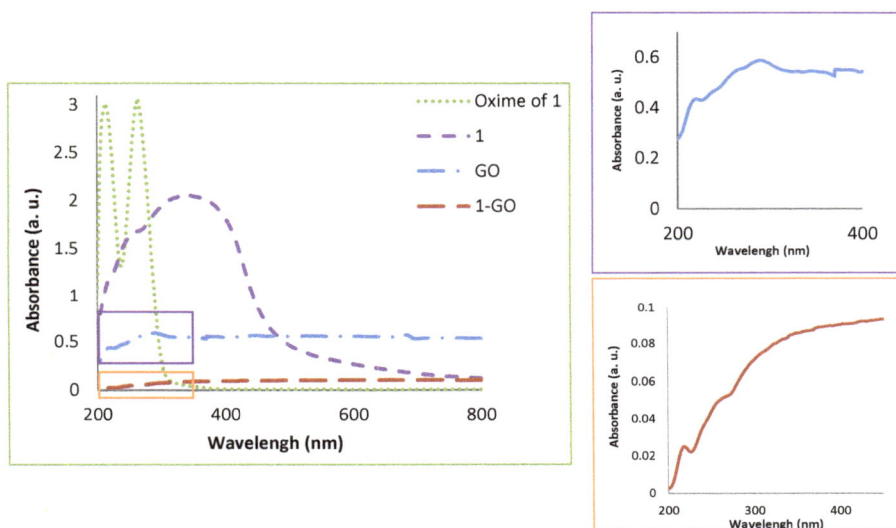

Figure 5. UV-Vis spectra of: 4,4′-dichlorobenzophenone oxime (EtOH solution), palladacycle **1** (solid state), GO (solid state), and **1-GO** (solid state).

Palladacycles **1** and **2** have been demonstrated by our group to be very active precatalysts in the Suzuki–Miyaura reaction under organic [75] and aqueous conditions [76]. On the other hand, the amphiphilic character of GO (hydrophilic edges and hydrophobic basal plane) [77] should be beneficial in aqueous processes, converting this material in a potential phase-transfer catalyst. Thus, the synthesized materials **1-GO**, **2-GO**, and **3-GO** were tested as precatalysts in the aqueous Suzuki–Miyaura cross-coupling between 4-bromoanisole and phenylboronic acid at room temperature under low loading conditions (0.02 mol % Pd) (Table 1). Initially, we demonstrated that GO (2 wt %) was not active in the cross-coupling process (Table 1, entry 1). As depicted in entries 2–4, the non-supported palladacycles **1–3** afforded 4-methoxybiphenyl (**4**) with high yields after 20 h (83, 80, and 99%, respectively). Similar results were obtained for the supported catalysts **1-GO** (85%) and **3-GO** (82%) in the cross-coupling reaction, while catalyst **2-GO** afforded **4** in a lower 69% conversion (Table 1, entries 5–7). In order to compare catalytic activities under the studied reaction conditions, Herrmann's catalyst (**5**) (Herrmann's catalyst was purchased from Aldrich (Madrid, Spain)) was also supported in GO following immobilization method B (Scheme 1) to afford catalyst **5-GO** with a 3.22% of Pd content according to the ICP-OES analysis (see SI for the synthesis and characterization data). As depicted in entries 8 and 9, both **5** and **5-GO** afforded 4-methoxybiphenyl with good conversion, showing similar catalytic activities than the rest of the tested palladacyclic catalysts.

Table 1. Suzuki–Miyaura reaction. Catalyst study.

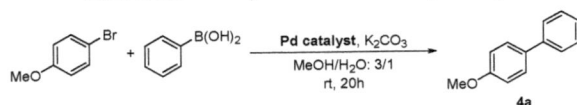

Entry	Pd Catalyst	Mol % Pd	4a (Conv., %) [1]
1	GO	-	0
2	1	0.02	83
3	2	0.02	80
4	3	0.02	99
5	1-GO	0.02	85
6	2-GO	0.02	69
7	3-GO	0.02	86
8	5	0.02	80
9	5-GO	0.02	81
10	PdNPs-G	0.02	30
11	PdNPs-rGO	0.02	40
12	1-GO	0.002	99 (73)
13	2-GO	0.002	45
14	3-GO	0.002	33
15	5-GO	0.002	15

[1] Reaction conversion towards **4** determined by GC analysis. Isolated yield in brackets.

The activity of immobilized palladium nanoparticles on graphene nanoplatelets (**PdNPs-G**) and reduced graphene oxide (**PdNPs-rGO**) as a catalyst was also evaluated in the model Suzuki–Miyaura coupling. As previously described, these supported catalysts, obtained from NanoInnova Technologies S.L., have shown good activity in the Suzuki cross-coupling of potassium aryltrifluoroborates with aryl halides [65]. However, under the studied conditions, these materials afforded 4-methoxybiphenyl (**4**) in a 30 and 40% conversion, respectively (Table 1, entries 10 and 11).

In view of the obtained results, it became evident that a reduction in the catalyst loading was mandatory to determine the most active catalyst in the model Suzuki cross-coupling. Thus, supported catalysts **1-GO**, **2-GO**, **3-GO**, and **5-GO** were tested reducing the catalyst loading down to 0.002 mol % of Pd in order to test the limits of the new materials. As depicted in Table 1 (entries 12–15), only oxime-palladacycle-derived catalyst **1-GO** was active enough to give 4-methoxybiphenyl in a good 73% isolated yield at room temperature.

Next, we performed a substrate scope using catalyst **1-GO** under the optimized reaction conditions (Table 2). Initially, we checked the reactivity of different boron-derived nucleophiles with 4-bromoanisole (Table 2, entries 1–4), none of them being as active as the phenylboronic acid. Then, we confirmed the activity of the catalytic system using deactivated aryl bromides as electrophiles, since not only 4-bromoanisole but also 4-bromophenol afforded, after reaction with phenylboronic acid, the corresponding [1,1'-biphenyl]-4-ol in a 61% yield (Table 2, entry 5). Both the activated 4-bromoacetophenone and the neutral bromobenzene reacted with phenylboronic acid and 4-tolylboronic acid to afford biphenyls **4c** and **4d** in 66 and 94% yields, respectively. On the other hand, as depicted in entries 8 and 9, **1-GO** seemed to be very sensitive towards the steric hindrance of the reactants since a 42% yield was obtained for the reaction between 2-bromotoluene and phenylboronic acid (entry 9), while no reaction was observed when 4-bromoanisole reacted with (2,6-dimethylphenyl)boronic acid (Table 2, entry 9). Regarding the electronic nature of the nucleophile, the electron-rich 4-tolylboronic acid afforded **4g** in a 68% yield after reaction with 4-bromoanisole (Table 2, entry 10). On the contrary, the electron-poor 4-(trifluoromethyl)phenylboronic acid showed, as expected, a very low reactivity with the same deactivated electrophile giving biphenyl **4h** in a 12% yield (Table 2, entry 11). Finally, we could also perform a Suzuki alkenylation of 4-bromoanisole with styrylboronic acid, which afforded compound **4i** in a 35% yield (Table 2, entry 12).

Table 2. Suzuki coupling catalyzed by **1-GO** [1].

Entry	R	R′	BX	Product	Yield (%) [2]	TON
1	4-OMe	H	B(OH)$_2$		73	36,500
2	4-OMe	H	BF$_3$K		33	16,500
3	4-OMe	H	B(OCMe$_2$)$_2$	**4a**	27	13,500
4	4-OMe	H	B(OCOCH$_2$)$_2$NMe		5	-
5	4-OH	H	B(OH)$_2$	**4b**	61	30,500
6	4-Ac	H	B(OH)$_2$	**4c**	66	33,000
7	H	4-Me	B(OH)$_2$	**4d**	94	47,000
8	2-Me	H	B(OH)$_2$	**4e**	42	21,000
9	4-OMe	2,6-(Me)$_2$	B(OH)$_2$	**4f**	5	-
10	4-OMe	4-Me	B(OH)$_2$	**4g**	68	34,000

Table 2. *Cont.*

Entry	R	R'	BX	Product	Yield (%) [2]	TON
11	4-OMe	4-CF$_3$	B(OH)$_2$	**4h**	12	6000
12	4-OMe	-[3]	B(OH)$_2$	**4i**	35	17,500

[1] Reaction conditions: ArBr (1.25 mmol, 1 eq.), ArB(OH)$_2$ (1.56 mmol, 1.25 eq.), K$_2$CO$_3$ (2.5 mmol, 2 eq.), **1-GO** (2×10^{-3} mol % Pd); [2] Isolated yields after flash chromatography; [3] Styrylboronic acid was used as nucleophile.

With respect to the active species involved in the Suzuki reaction employing **1-GO** as a pre-catalyst, it is well-known that oxime palladacycles act as stable precursors to highly active and truly Pd(0) catalyst, usually generated after transmetallation by the activated boron nucleophile followed by a reductive elimination reaction of the aryl-ligated palladacycle [78]. The presence of water in the reaction medium, as demonstrated by Blackmond [79], would also play a key role in accelerating the initial formation of the monomeric active catalytic species. As commented above, palladium nanoparticles on graphene nanoplatelets (**PdNPs-G**, average nanoparticle size: 4.50 nm) and reduced graphene oxide (**PdNPs-rGO**, average nanoparticle size: 10.9 nm) have shown lower activity in the model Suzuki reaction under the optimized reaction conditions, probably due to the agglomeration of the nanoparticles under the reaction conditions. The slow release of the active Pd species from the oxime palladacycle in **1-GO**, would account for the better activity of this material in the Suzuki reaction.

A simple recovery and reuse of transition-metal immobilized catalytic systems is highly desirable from both economic and environmental points of view. Therefore, a study of **1-GO** recyclability was carried out on the model reaction, i.e., the coupling between 4-bromoanisole and phenylboronic acid using 0.1 mol % Pd as catalyst loading. After each cycle, the catalyst was easily separated from the reaction mixture by washing the crude reaction with a EtOAc/H$_2$O: 1/1 solvent mixture and subsequent centrifugation (see SI, for details). As depicted in Scheme 2, catalyst **1-GO** showed good catalytic activity for the two first reactions runs, decreasing the conversion of the reaction from the third cycle.

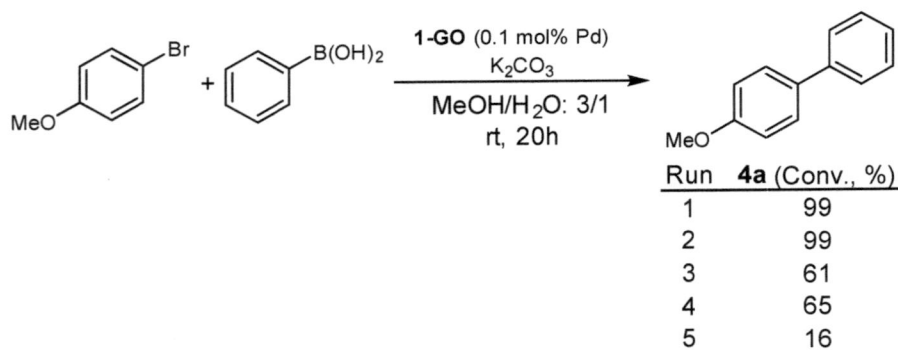

Run	**4a** (Conv., %)
1	99
2	99
3	61
4	65
5	16

Scheme 2. 1-GO recyclability and yields in the Suzuki–Miyaura coupling in different reaction cycles.

Figure 6 shows the reasons why **1-GO** undergoes a loss of catalytic activity along the successive reaction cycles. The Pd3d X-ray photoelectron spectroscopy (XPS) spectrum (Figure 6a) indicated the presence of two different Pd(II) entities. The component at the higher binding energy was assigned to oxime palladacycle **1**, while the component at the lower binding energy [336.2 (Pd3d5/2) and 341.8 (Pd3d3/2) eV] was assigned to PdO due to the surface oxidation of the Pd NPs generated from the palladacycle during the catalytic cycles [80]. C1s XPS analysis also showed a marked decrease of the oxygen content in **1-GO** after the fifth run: Fresh **1-GO** (O/C: 0.714); **1-GO** after the fifth cycle (O/C: 0.371). The low catalytic activity of palladium oxide, especially being agglomerated as demonstrated by the TEM analysis (Figure 6b), would account for the observed decrease of the catalytic activity of the **1-GO** system. Also, inductively coupled plasma mass spectrometry (ICP-OES) analysis of the washings after the fifth run showed 67 ppb of Pd leaching.

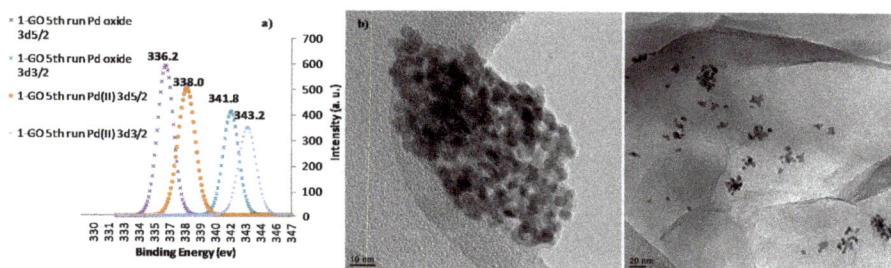

Figure 6. Analysis of **1-GO** after five reaction cycles: (**a**) XPS Pd spectrum; (**b**) TEM images.

Oxime palladacycles are considered reservoirs of highly active palladium nanoparticles [66–68]. Usually, PdNps supported on graphene materials are prepared by reduction of a Pd salt or Pd catalyst in the presence of the corresponding carbonaceous support [45–62]. Then, we next turned our attention to the reduction of **1** with NaBH$_4$ and the catalytic activity of the supported PdNPs was thus obtained. When unsupported palladacycle **1** was reduced with NaBH$_4$ in the presence of **GO** (Scheme 3, Eq. a) [81], a carbonaceous material [**Pd(0)-rGO**] with an 8.40% Pd content (by ICP-OES analysis) was obtained. XPS analysis of this material showed a 75.8% content of Pd(0), which unfortunately suffered agglomeration during the reduction process as demonstrated by TEM analysis (Scheme 3 Eq. a). On the other hand, the reduction of **1-GO** with NaBH$_4$ (Scheme 3, Eq. b) afforded a material [**Pd(II)-rGO**] which, according to ICP-OES analysis, contained a 5.97% Pd. XPS analysis of [**Pd(II)-rGO**] demonstrated the stability of palladacycle under these reduction conditions since 76.8% of **1** was not reduced. Both materials suffered a strong decrease of oxygen content as a consequence of the reduction of **GO** (O/C: 0.655) to **rGO** (O/C: 0.237–0.369).

We also studied the activity of **Pd(0)-rGO** and **Pd(II)-rGO** in the model Suzuki reaction between 4-bromoanisole and phenylboronic acid under the optimized reaction conditions (0.002 mol % Pd) (Scheme 4). As expected, **Pd(0)-rGO** was catalytically inactive in the process. On the other hand, **Pd(II)-rGO** afforded 4a in a 40% yield (Scheme 4). This result ratified the important amphiphilic effect of **GO** on the Suzuki coupling carried out under aqueous conditions. Palladacycle 1 was also immobilized over rGO following Method B (see SI), but the material thus obtained (1-rGO) only contained a 0.035% of Pd, according to the ICP-OES analysis.

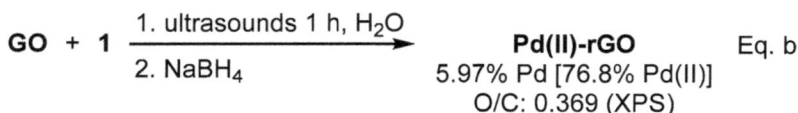

Scheme 3. Palladacycle reduction studies in the presence of **GO**.

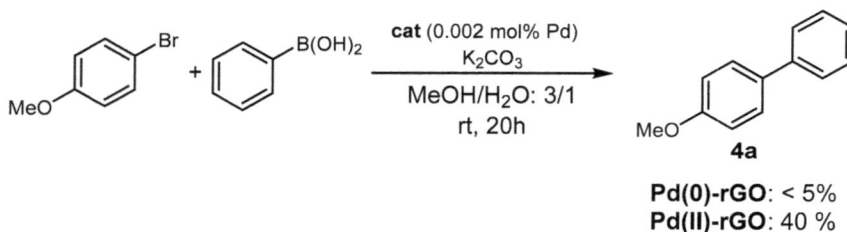

Pd(0)-rGO: < 5%
Pd(II)-rGO: 40 %

Scheme 4. Suzuki coupling using **rGO** as a catalyst support.

3. Materials and Methods

3.1. General

Unless otherwise noted, all commercial reagents and solvents were used without further purification. ^{1}H-NMR (300 MHz) and ^{13}C-NMR (75 MHz) spectra were obtained on a Bruker AC-300, using CDCl$_3$ as solvent and TMS (0.003%) as reference, unless otherwise stated. Low-resolution mass spectra (MS) were recorded in the electron impact mode (EI, 70 eV, He as carrier phase) using an Agilent 5973 Network Mass Selective Detector spectrometer, being the samples introduced through a GC chromatograph Agilent 6890N equipped with a HP-5MS column [(5%-phenyl)–methylpolysiloxane; length 30 m; ID 0.25 mm; film 0.25 mm]. Analytical TLC was performed on Merck aluminum sheets with silica gel 60 F254. Silica gel 60 (0.04–0.06 mm) was employed for flash chromatography. The conversion of the reactions was determined by GC analysis on an Agilent 6890 N Network GC system. Centrifugations were carried out in a Hettich centrifuge (Universal 320, 6000 rpm, 15 min). ICP-MS analyses were performed on an Agilent 7700x equipped with HMI (high matrix introduction) and He mode ORS as standard. Elemental analyses were determined with a CHNS elemental micro analyzer with Micro detection system TruSpec LECO. X-ray Powder Diffraction (XRD) was performed in a Bruker D8-Advance with mirror Goebel (non-planar samples) with high temperature Chamber (up to 900 °C), with a generator of x-ray KRISTALLOFLEX K 760-80F (power: 3000 W, voltage: 20–60 KV and current: 5–80 mA) with a tube of RX with copper anode. X-ray photoelectron spectroscopy (XPS) was performed in a VG-Microtech Mutilab 3000 equipment equipped with a hemispherical electron analyzer with 9 channeltrons (with energy of passage of (2–200 eV) and an X-ray radiation source with Mg and Al anodes. Transmission electron microscopy (TEM) was performed in JEOL Model JEM-2010. This microscope features an OXFORD X-ray detector model INCA Energy TEM 100 for microanalysis (EDS). The image acquisition camera is of the brand GATAN model ORIUS SC600. It is mounted on the axis with the microscope at the bottom and is integrated into the GATAN Digital Micrograph 1.80.70 image acquisition and processing program for GMS 1.8.0. The supported catalyst were sonicated in a ultrasons P-Selecta (360 W). Solid state UV-Vis spectroscopy was performed in a JASCO V-670 dual-beam UV-Vis/NIR spectrometer covering the wavelength range from 190 to 2700 nm. The equipment has a single monochromator with double netting, one for the UV-Vis region (1200 grooves/mm) and one for the NIR region (300 grooves/mm). The detectors are a photomultiplier tube for the UV-Vis region and a PbS detector for the NIR region. The switching of both the detectors and the networks is automatically effected at a wavelength set by the user between 750 and 900 nm. The sources used are a deuterium lamp (190 to 350 nm) and a halogen lamp (330 to 2700 nm). UV-Vis spectroscopy analysis of oxime precursor of catalyst **1** was performed in ethanol solution (1.04 mg of oxime in 20 mL of ethanol aproxymately) in a SHIMAZU UV-1603 spectrophotometer covering the wavelength range from 190 to 2700 nm.

3.2. Synthesis of Oxime Palladacycle 1

A MeOH (0.8 mL) solution of the oxime derived from 4,4'-dichlorobenzophenone (1 eq), sodium acetate (61.70 mg, 0.752 mmol, 1 eq) and Li_2PdCl_4 (1.5 mL of a 0.5 M solution in MeOH, 1eq) was stirred at room temperature under argon atmosphere for 4 days. Then, water was added to precipitate the corresponding palladium complex and subsequently, the palladacycle was filtered and washed with water and hexane. Finally, the catalyst was dried under reduced pressure overnight giving oxime palladacycle **1** as a yellow solid (47% yield).

3.3. Synthesis of 1-GO

A 10 mL glass vessel was charged with **GO** (200 mg) and **1** (45.89 mg). Then, anhydrous THF (4 mL) was added and the reaction mixture was sonicated for 1 h. Afterwards, the reaction was stirred at room temperature for 48 h. After this time, the catalyst was submitted to four washing (THF, 10 mL)-centrifugation (6000 rpm, 20 min) cycles, with the solvent being eliminated after each cycle with a syringe equipped with a 4 mm/0.2 μm PTFE syringe filter. The residual solvent was completely removed under reduced pressure affording **1-GO** (2% Pd by ICP-OES).

3.4. Typical Procedure for the Suzuki–Miyaura Reaction

A 10 mL glass vessel was charged with **1-GO** (0.002 mol % Pd, 0.198 mg), 4-bromoanisole (156.5 μL, 1.25 mmol, 1 eq), phenylboronic acid (190.20 mg, 1.56 mmol, 1.25 eq), K_2CO_3 (345.50 mg, 2.5 mmol, 2 eq) and $MeOH/H_2O$: 3/1 (3 mL). The vessel was sealed with a pressure cap and the mixture was stirred at room temperature for 20 h. Then, H_2O (5 mL) and EtOAc (5 mL) were added and the liquid mixture was filtered with cotton and extracted with EtOAc (3 × 10 mL). The organic layers were dried over $MgSO_4$ and concentrated under reduced pressure. The crude residue was purified by flash chromatography (silica gel, Hexane/EtOAc: 95/5) to obtain 0.167 g of pure **4a** (73% yield).

3.5. Typical Procedure for the Recovery of the Catalyst in the Suzuki–Miyaura Reaction

Once the reaction was finished, the mixture was suspended and stirred for 15 min in a 10 mL mixture of $EtOAc/H_2O$: 1/1. Then, this mixture was centrifuged (6000 rpm, 20 min) and the solvent was eliminated using a syringe equipped with 4 mm/0.2 μm PTFE syringe filter. The washing/centrifugation sequence was repeated four additional times until no product was detected in the liquid phase by thin layer chromatography. The residual solvent was completely removed under reduced pressure affording the supported palladacycle catalyst which was directly used in the same tube with fresh reagents for the next run. This procedure was repeated for every cycle and the conversion of the reaction was determined by GC using decane as internal standard.

4. Conclusions

We have synthesized and characterized new non-covalent supported oxime palladacycles on graphene oxide as efficient catalysts for the Suzuki–Miyaura coupling between aryl bromides and arylboronic acids. Catalyst **1-GO** resulted very active for this process under aqueous conditions at room temperature under very low catalyst loadings (0.002 mol % Pd). Catalyst **1-GO** can be recovered and reused with loss of catalytic activity after the second cycle due to the oxidation and metal agglomeration processes.

Supplementary Materials: The following are available online at www.mdpi.com/2073-4344/7/3/94/s1.

Acknowledgments: Financial support from the University of Alicante (UAUSTI16-03, UAUSTI16-10, VIGROB-173), the Spanish Ministerio de Economía, Industria y Competitividad (CTQ2015-66624-P) is acknowledged.

Author Contributions: Melania Gómez-Martínez performed the synthetic works. Alejandro Baeza and Diego A. Alonso designed the experiments of the project and supervised the whole studies reported in the manuscript. Melania Gómez-Martínez, Alejandro Baeza and Diego A. Alonso wrote the manuscript.

Conflicts of Interest: The authors declare no conflict of interest.

References

1. Diederich, F.; Stang, P.J. *Metal-Catalyzed Cross-Coupling Reactions*; Wiley-VCH: Weinheim, Germany, 1998.
2. Beller, M.; Bolm, C. *Transition Metals for Organic Synthesis. Building Blocks and Fine Chemicals*; Wiley-VCH: Weinheim, Germany, 1998.
3. Negishi, E. *Handbook of Organopalladium Chemistry for Organic Synthesis*; Wiley-Interscience: New York, NY, USA, 2002.
4. Miyaura, N. *Cross-Coupling Reactions. A Practical Guide*; Springer: Berlin, Germany, 2002.
5. Tsuji, J. *Palladium Reagents and Catalysts. Innovations in Organic Synthesis*; Wiley: Chichester, UK, 2004.
6. Beller, M.; Bolm, C. *Transition Metals for Organic Synthesis. Building Blocks and Fine Chemicals*, 2nd ed.; Wiley-VCH: Weinheim, Germany, 2004.
7. Diederich, F.; Stang, P.J. *Metal-Catalyzed Cross-Coupling Reactions*; Wiley-VCH: Weinheim, Germany, 2004.
8. Farina, V.; Miyaura, N.; Buchwald, S.L. Special Issue on Cross-Coupling Chemistry. *Adv. Synth. Catal.* **2004**, *346*, 1505–1879.
9. Negishi, E. Transition Metal-Catalyzed Organometallic Reactions that Have Revolutionized Organic Synthesis. *Bull. Chem. Soc. Jpn.* **2007**, *80*, 233–257. [CrossRef]
10. Dupont, J.; Pfeffer, M. *Palladacycles: Synthesis, Characterization and Applications*; Wiley-VCH: Weinheim, Germany, 2008.
11. Ratti, R. Palladacycles—Versatile Catalysts for Carbon-Carbon Coupling Reactions. *Can. Chem. Trans.* **2014**, *2*, 467–488.
12. Das, P.; Linert, W. Schiff base-derived homogeneous and heterogeneous palladium catalysts for the Suzuki–Miyaura reaction. *Coord. Chem. Rev.* **2016**, *311*, 1–23. [CrossRef]
13. Nájera, C. Oxime-Derived Palladacycles: Applications in Catalysis. *ChemCatChem* **2016**, *8*, 1865–1881. [CrossRef]
14. Bedford, R.B.; Cazin, C.S.J.; Hursthouse, M.B.; Light, M.E.; Pike, K.J.; Wimperis, S. Silica-supported imine palladacycles—Recyclable catalysts for the Suzuki reaction? *J. Organomet. Chem.* **2001**, *633*, 173–181. [CrossRef]
15. Baleizão, C.; Corma, A.; García, H.; Leyva, A. An oxime-carbapalladacycle complex covalently anchored to silica as an active and reusable heterogeneous catalyst for Suzuki cross-coupling in water. *Chem. Commun.* **2003**, 606–607. [CrossRef]
16. Yu, K.; Sommer, W.; Weck, M.; Jones, C.W. Silica and polymer-tethered Pd–SCS-pincer complexes: Evidence for precatalyst decomposition to form soluble catalytic species in Mizoroki–Heck chemistry. *J. Catal.* **2004**, *226*, 101–110. [CrossRef]
17. Venkatesan, C.; Singh, A.P. Synthesis and characterization of carbometallated palladacycles over 3-hydroxypropyltriethoxysilyl-functionalized MCM-41. *J. Catal.* **2004**, *227*, 148–163. [CrossRef]

18. Baleizão, C.; Corma, A.; García, H.; Leyva, A. Oxime Carbapalladacycle Covalently Anchored to High Surface Area Inorganic Supports or Polymers as Heterogeneous Green Catalysts for the Suzuki Reaction in Water. *J. Org. Chem.* **2004**, *69*, 439–446. [CrossRef] [PubMed]
19. Corma, A.; Das, D.; García, H.; Leyva, A. A periodic mesoporous organosilica containing a carbapalladacycle complex as heterogeneous catalyst for Suzuki cross-coupling. *J. Catal.* **2005**, *229*, 322–331. [CrossRef]
20. Mohamed, M.F.; Neverov, A.A.; Brown, R.S. An Immobilized Ortho-Palladated Dimethylbenzylamine Complex as an Efficient Catalyst for the Methanolysis of Phosphorothionate Pesticides. *Inorg. Chem.* **2009**, *48*, 1183–1191. [CrossRef] [PubMed]
21. Li, G.; Yang, H.; Lia, W.; Zhang, G. Rationally designed palladium complexes on a bulky N-heterocyclic carbene-functionalized organosilica: An efficient solid catalyst for the Suzuki–Miyaura coupling of challenging aryl chlorides. *Green Chem.* **2011**, *13*, 2939–2947. [CrossRef]
22. Farsadpour, S.; Ghoochany, L.T.; Shylesh, S.; Dörr, G.; Seifert, A.; Ernst, S.; Thiel, W.R. A Covalently Supported Pyrimidinylphosphane Palladacycle as a Heterogenized Catalyst for the Suzuki-Miyaura Cross Coupling. *ChemCatChem* **2012**, *4*, 401–407. [CrossRef]
23. Lu, F.-H.; Yue, P.; Wang, X.-R.; Lu, Z.-L. Synthesis and immobilization of oxime-derived palladacycles as effective and reusable catalysts for the degradation of phosphorothionate pesticides. *Inorg. Chem. Commun.* **2013**, *34*, 19–22. [CrossRef]
24. Liu, H.; Li, T.; Xue, X.; Xua, W.; Wu, Y. The mechanism of a self-assembled Pd(ferrocenylimine)-Si compound-catalysed Suzuki coupling reaction. *Catal. Sci. Technol.* **2016**, *6*, 1667–1676. [CrossRef]
25. Lin, C.-A.; Luo, F.-T. Polystyrene-supported recyclable palladacycle catalyst for Heck, Suzuki and Sonogashira reactions. *Tetrahedron Lett.* **2003**, *44*, 7565–7568. [CrossRef]
26. Bedford, R.B.; Coles, S.J.; Hursthouse, M.B.; Scordia, V.J.M. Polystyrene-supported dicyclohexylphenylphosphine adducts of amine- and phosphite-based palladacycles in the Suzuki coupling of aryl chlorides. *Dalton Trans.* **2005**, 991–995. [CrossRef] [PubMed]
27. Luo, F.-T.; Xue, C.; Ko, S.-L.; Shao, Y.-D.; Wu, C.-J.; Kuo, Y.-M. Preparation of polystyrene-supported soluble palladacycle catalyst for Heck and Suzuki reactions. *Tetrahedron* **2005**, *61*, 6040–6045. [CrossRef]
28. Corma, A.; Garcia, H.; Leyva, A. Comparison between polyethylenglycol and imidazolium ionic liquids as solvents for developing a homogeneous and reusable palladium catalytic system for the Suzuki and Sonogashira coupling. *Tetrahedron* **2005**, *61*, 9848–9854. [CrossRef]
29. Corma, A.; Garcia, H.; Leyva, A. Polyethyleneglycol as scaffold and solvent for reusable C–C coupling homogeneous Pd catalysts. *J. Catal.* **2006**, *240*, 87–99. [CrossRef]
30. Solodenko, W.; Mennecke, K.; Vogt, C.; Gruhl, S.; Kirschning, A. Polyvinylpyridine, a Versatile Solid Phase for Coordinative Immobilisation of Palladium Precatalysts—Applications in Suzuki-Miyaura Reactions. *Synthesis* **2006**, 1873–1881.
31. Hershberger, J.C.; Zhang, L.; Lu, G.; Malinakova, H.C. Polymer-Supported Palladacycles: Efficient Reagents for Synthesis of Benzopyrans with Palladium Recovery. Relationship among Resin Loading, Pd:P Ratio, and Reactivity of Immobilized Palladacycles. *J. Org. Chem.* **2006**, *71*, 231–235. [CrossRef] [PubMed]
32. Mennecke, K.; Solodenko, W.; Kirschning, A. Carbon-Carbon Cross-Coupling Reactions under Continuous Flow Conditions Using Poly(vinylpyridine) Doped with Palladium. *Synthesis* **2008**, 1589–1599. [CrossRef]
33. Alacid, E.; Najera, C. Kaiser oxime resin-derived palladacycle: A recoverable polymeric precatalyst in Suzuki–Miyaura reactions in aqueous media. *J. Organomet. Chem.* **2009**, *694*, 1658–1665. [CrossRef]
34. Islam, M.; Mondal, P.; Roy, A.S.; Tuhina, K. Catalytic oxidation of organic substrates using a reusable polystyrene-anchored orthometallated palladium (II) complex. *J. Appl. Polym. Sci.* **2010**, *118*, 52–62. [CrossRef]
35. Yang, Y.-C.; Toy, P.H. Self-Supported Ligands as a Platform for Catalysis: Use of a Polymeric Oxime in a Recyclable Palladacycle Precatalyst for Suzuki–Miyaura Reactions. *Synlett* **2014**, *25*, 1319–1324.
36. Cho, H.-J.; Jung, S.; Kong, S.; Park, S.-J.; Lee, S.-M.; Lee, Y.-S. Polymer-Supported Electron-Rich Oxime Palladacycle as an Efficient Heterogeneous Catalyst for the Suzuki Coupling Reaction. *Adv. Synth. Catal.* **2014**, *356*, 1056–1064. [CrossRef]
37. Sudheendran, M.; Eitel, S.H.; Naumann, S.; Buchmeiser, M.R.; Peters, R. Heterogenization of ferrocene palladacycle catalysts on ROMP-derived monolithic supports and application to a Michael addition. *New J. Chem.* **2014**, *38*, 5597–5607. [CrossRef]

38. Karami, K.; Najvani, S.D.; Naeini, N.H.; Hervés, P. Palladium particles from oxime-derived palladacycle supported on Fe$_3$O$_4$/oleic acid as a catalyst for the copper-free Sonogashira cross-coupling reaction. *Chin. J. Catal.* **2015**, *36*, 1047–1053. [CrossRef]

39. Gholinejad, M.; Razeghi, M.; Najera, C. Magnetic nanoparticles supported oxime palladacycle as a highly efficient and separable catalyst for room temperature Suzuki-Miyaura coupling reaction in aqueous media. *RSC Adv.* **2015**, *5*, 49568–49576. [CrossRef]

40. Karami, K.; Haghighat Naeini, N. Palladium nanoparticles supported on cucurbit[6]uril: An efficient heterogeneous catalyst for the Suzuki reaction under mild conditions. *Appl. Organomet. Chem.* **2015**, *29*, 33–39. [CrossRef]

41. Singh, V.; Ratti, R.; Kaur, S. Synthesis and characterization of recyclable and recoverable MMT-clay exchanged ammonium tagged carbapalladacycle catalyst for Mizoroki–Heck and Sonogashira reactions in ionic liquid media. *J. Mol. Catal. A Chem.* **2011**, *334*, 13–19. [CrossRef]

42. Karami, K.; Hashemi, S.; Dinari, M. Investigation of catalytic properties of two new orthopalladated complexes supported on montmorillonite: Synthesis, characterization and application in aerobic oxidation of alcohols. *Appl. Organomet. Chem.* **2017**, in press. [CrossRef]

43. Hübner, S.; de Vries, J.G.; Farina, V. Why Does Industry Not Use Immobilized Transition Metal Complexes as Catalysts? *Adv. Synth. Catal.* **2016**, *358*, 3–25. [CrossRef]

44. Scheuermann, G.M.; Rumi, L.; Steurer, P.; Bannwarth, W.; Mülhaupt, R. Palladium Nanoparticles on Graphite Oxide and Its Functionalized Graphene Derivatives as Highly Active Catalysts for the Suzuki-Miyaura Coupling Reaction. *J. Am. Chem. Soc.* **2009**, *131*, 8262–8270. [CrossRef] [PubMed]

45. Siamaki, A.R.; Khder, A.E. R.S.; Abdelsayed, V.; El-Shall, M.S.; Gupton, B.F. Microwave-assisted synthesis of palladium nanoparticles supported on graphene: A highly active and recyclable catalyst for carbon–carbon cross-coupling reactions. *J. Catal.* **2011**, *279*, 1–11. [CrossRef]

46. Rumi, L.; Scheuermann, G.M.; Mülhaupt, R.; Bannwarth, W. Palladium Nanoparticles on Graphite Oxide as Catalyst for Suzuki-Miyaura, Mizoroki-Heck and Sonogashira Reactions. *Helv. Chim. Acta* **2011**, *94*, 966–976. [CrossRef]

47. Xiang, G.; He, J.; Li, T.; Zhuang, J.; Wang, X. Rapid preparation of noble metal nanocrystalsvia facile coreduction with graphene oxide and their enhanced catalytic properties. *Nanoscale* **2011**, *3*, 3737–3742. [CrossRef] [PubMed]

48. Machado, B.F.; Serp, P. Graphene-based materials for catalysis. *Catal. Sci. Technol.* **2012**, *2*, 54–75. [CrossRef]

49. Moussa, S.; Siamaki, A.R.; Gupton, B.F.; El-Shall, M.S. Pd-Partially Reduced Graphene Oxide Catalysts (Pd/PRGO): Laser Synthesis of Pd Nanoparticles Supported on PRGO Nanosheets for Carbon-Carbon Cross Coupling Reactions. *ACS Catal.* **2012**, *2*, 145–154. [CrossRef]

50. Ioni, Y.V.; Lyubimov, S.E.; Korlyukov, A.A.; Antipin, M.Y.; Davankov, V.A.; Gubin, S.P. Activity of palladium nanoparticles on graphene oxide in the Suzuki-Miyaura reaction. *Russ. Chem. Bull., Int. Ed.* **2012**, *61*, 1825–1827. [CrossRef]

51. Hu, J.; Wang, Y.; Han, M.; Zhou, Y.; Jiang, X.; Sun, P. A facile preparation of palladium nanoparticles supported on magnetite/s-graphene and their catalytic application in Suzuki-Miyaura reaction. *Catal. Sci. Technol.* **2012**, *2*, 2332–2340. [CrossRef]

52. Qu, K.; Wu, L.; Ren, J.; Qu, X. Natural DNA-Modified Graphene/Pd Nanoparticles as Highly Active Catalyst for Formic Acid Electro-Oxidation and for the Suzuki Reaction. *ACS Appl. Mater. Inter.* **2012**, *4*, 5001–5009. [CrossRef] [PubMed]

53. Shang, N.; Feng, C.; Zhang, H.; Gao, S.; Tang, R.; Wang, C.; Wang, Z. Suzuki-Miyaura reaction catalyzed by graphene oxide supported palladium nanoparticles. *Catal. Commun.* **2013**, *40*, 111–115. [CrossRef]

54. Shang, N.; Gao, S.; Feng, C.; Zhang, H.; Wang, C.; Wang, Z. Graphene oxide supported N-heterocyclic carbene-palladium as a novel catalyst for the Suzuki-Miyaura reaction. *RSC Adv.* **2013**, *3*, 21863–21868. [CrossRef]

55. Hoseini, S.J.; Dehghani, M.; Nasrabadi, H. Thin film formation of Pd/reduced-graphene oxide and Pd nanoparticles at oil-water interface, suitable as effective catalyst for Suzuki-Miyaura reaction in water. *Catal. Sci. Technol.* **2014**, *4*, 1078–1083. [CrossRef]

56. Yamamoto, S.-I.; Kinoshita, H.; Hashimoto, H.; Nishina, Y. Facile preparation of Pd nanoparticles supported on single-layer graphene oxide and application for the Suzuki-Miyaura cross-coupling reaction. *Nanoscale* **2014**, *6*, 6501–6505. [CrossRef] [PubMed]

57. Joshi, H.; Sharma, K.N.; Sharma, A.K.; Singh, A.K. Palladium-phosphorus/sulfur nanoparticles (NPs) decorated on graphene oxide: Synthesis using the same precursor for NPs and catalytic applications in Suzuki-Miyaura coupling. *Nanoscale* **2014**, *6*, 4588–4597. [CrossRef] [PubMed]

58. Bai, C.; Zhao, Q.; Li, Y.; Zhang, G.; Zhang, F.; Fan, X. Palladium Complex Immobilized on Graphene Oxide as an Efficient and Recyclable Catalyst for Suzuki Coupling Reaction. *Catal. Lett.* **2014**, *144*, 1617–1623. [CrossRef]

59. Park, J.H.; Raza, F.; Jeon, S.-J.; Kim, H.-I.; Kang, T.W.; Yim, S.; Kim, J.-H. Recyclable N-heterocyclic carbene/palladium catalyst on graphene oxide for the aqueous-phase Suzuki reaction. *Tetrahedron Lett.* **2014**, *55*, 3426–3430. [CrossRef]

60. Movahed, S.K.; Esmatpoursalmani, R.; Bazgir, A. N-Heterocyclic carbene palladium complex supported on ionic liquid-modified graphene oxide as an efficient and recyclable catalyst for Suzuki reaction. *RSC Adv.* **2014**, *4*, 14586–14591. [CrossRef]

61. Lin, J.; Mei, T.; Lv, M.; Zhang, C.; Zhao, Z.; Wang, X. Size-controlled PdO/graphene oxides and their reduction products with high catalytic activity. *RSC Adv.* **2014**, *4*, 29563–29570. [CrossRef]

62. Pérez-Mayoral, E.; Calvino-Casilda, V.; Soriano, E. Metal-supported carbon-based materials: Opportunities and challenges in the synthesis of valuable products. *Catal. Sci. Technol.* **2016**, *6*, 1265. [CrossRef]

63. Fath, R.H.; Hoseini, S.J. Covalently cyclopalladium(II) complex/reduced-graphene oxide as the effective catalyst for the Suzuki-Miyaura reaction at room temperature. *J. Organomet. Chem.* **2017**, *828*, 16–23. [CrossRef]

64. Xue, Z.; Huang, P.; Li, T.; Qin, P.; Xiao, D.; Liu, M.; Chen, P.; Wu, Y. A novel "tunnel-like" cyclopalladated arylimine catalyst immobilized on graphene oxide nano-sheet. *Nanoscale* **2017**, *9*, 781–791. [CrossRef] [PubMed]

65. Gómez-Martínez, M.; Buxaderas, E.; Pastor, I.M.; Alonso, D.A. Palladium nanoparticles supported on graphene and reduced graphene oxide as efficient recyclable catalyst for the Suzuki-Miyaura reaction of potassium aryltrifluoroborates. *J. Mol. Catal. A: Chem.* **2015**, *404*, 1–7.

66. Alonso, D.A.; Botella, L.; Nájera, C.; Pacheco, M.C. Synthetic Applications of Oxime-Derived Palladacycles as Versatile Catalysts in Cross-Coupling Reactions. *Synthesis* **2004**, 1713–1718.

67. Alacid, E.; Alonso, D.A.; Botella, L.; Nájera, C.; Pacheco, M.C. Oxime palladacycles revisited: Stone-stable complexes nonetheless very active catalysts. *Chem. Rec.* **2006**, *6*, 117–132. [CrossRef] [PubMed]

68. Alonso, D.A.; Nájera, C. Oxime-derived palladacycles as source of palladium nanoparticles. *Chem. Soc. Rev.* **2010**, *39*, 2891–2902. [CrossRef] [PubMed]

69. Alonso, D.A.; Nájera, C.; Pacheco, M.C. Oxime Palladacycles: Stable and Efficient Catalysts for Carbon—Carbon Coupling Reactions. *Org. Lett.* **2000**, *2*, 1823–1826. [CrossRef] [PubMed]

70. GO Was Supplied by Nanoinnova Technologies S.L. Available online: http://www.nanoinnova.com/Product/Details/24 (accessed on 21 March 2017).

71. Singh, G.; Botcha, V.C.; Sutar, D.S.; Talwar, S.S.; Srinivasa, R.S.; Major, S.S. Graphite mediated reduction of graphene oxide monolayer sheets. *Carbon* **2015**, *95*, 843–851. [CrossRef]

72. Yu, B.; Wang, X.; Qian, X.; Xing, W.; Yang, H.; Ma, L.; Lin, Y.; Jiang, S.; Song, L.; Hu, Y.; et al. Functionalized graphene oxide/phosphoramide oligomer hybrids flame retardant prepared via in situ polymerization for improving the fire safety of polypropylene. *RSC Adv.* **2014**, *4*, 31782–31794. [CrossRef]

73. Lin, Y.; Pan, X.; Qi, W.; Zhang, B.; Su, D.S.J. Nitrogen-doped onion-like carbon: A novel and efficient metal-free catalyst for epoxidation reaction. *Mater. Chem. A* **2014**, *2*, 12475–12483. [CrossRef]

74. Paredes, J.I.; Villar-Rodil, S.; Martínez-Alonso, A.; Tascón, J.M.D. Au/graphene hydrogel: Synthesis, characterization and its use for catalytic reduction of 4-nitrophenol. *Langmuir* **2008**, *24*, 10560–10564. [CrossRef] [PubMed]

75. Alonso, D.A.; Nájera, C.; Pacheco, M.C. Highly Active Oxime-Derived Palladacycle Complexes for Suzuki—Miyaura and Ullmann-Type Coupling Reactions. *J. Org. Chem.* **2002**, *67*, 5588–5594. [CrossRef] [PubMed]

76. Botella, L.; Nájera, C. A Convenient Oxime-Carbapalladacycle-Catalyzed Suzuki Cross-Coupling of Aryl Chlorides in Water. *Angew. Chem. Int. Ed.* **2002**, *41*, 179–181. [CrossRef]

77. Kim, J.; Cote, L.J.; Kim, F.; Yuan, W.; Shull, K.R.; Huang, J. Graphene Oxide Sheets at Interfaces. *J. Am. Chem. Soc.* **2010**, *132*, 8180–8186. [CrossRef] [PubMed]

78. Bedford, R.B.; Cazin, C.S.J.; Coles, S.J.; Gelbrich, T.; Horton, P.N.; Hursthouse, M.B.; Light, M.E. High-Activity Catalysts for Suzuki Coupling and Amination Reactions with Deactivated Aryl Chloride Substrates: Importance of the Palladium Source. *Organometallics* **2003**, *22*, 987–999. [CrossRef]

79. Rosner, T.; Le Bars, J.; Pfaltz, A.; Blackmond, D.G. Kinetic Studies of Heck Coupling Reactions Using Palladacycle Catalysts: Experimental and Kinetic Modeling of the Role of Dimer Species. *J. Am. Chem. Soc.* **2001**, *123*, 1848–1855. [CrossRef] [PubMed]

80. Wang, L.J.; Zhang, J.; Zhao, X.; Xu, L.L.; Lyu, Z.Y.; Laia, M.; Chen, W. Palladium nanoparticle functionalized graphene nanosheets for Li–O_2 batteries: Enhanced performance by tailoring the morphology of the discharge product. *RSC Adv.* **2015**, *5*, 73451–73456. [CrossRef]

81. Chua, C.K.; Pumera, M. Chemical reduction of graphene oxide: A synthetic chemistry viewpoint. *Chem. Soc. Rev.* **2014**, *43*, 291–312. [CrossRef] [PubMed]

catalysts

Article

Plant Extract Mediated Eco-Friendly Synthesis of Pd@Graphene Nanocatalyst: An Efficient and Reusable Catalyst for the Suzuki-Miyaura Coupling

Mujeeb Khan [1], Mufsir Kuniyil [1], Mohammed Rafi Shaik [1], Merajuddin Khan [1,*],
Syed Farooq Adil [1], Abdulrahman Al-Warthan [1], Hamad Z. Alkhathlan [1], Wolfgang Tremel [2],
Muhammad Nawaz Tahir [2] and Mohammed Rafiq H. Siddiqui [1,*]

[1] Department of Chemistry, College of Science, King Saud University, P.O. 2455, Riyadh 11451, Saudi Arabia;
kmujeeb@ksu.edu.sa (Muj.K.); mufsir@gmail.com (Muf.K.); rafiskm@gmail.com (M.R.S.);
sfadil@ksu.edu.sa (S.F.A.); awarthan@ksu.edu.sa (A.A.-W.); khathlan@ksu.edu.sa (H.Z.A.)
[2] Institute of Inorganic and Analytical Chemistry, Johannes Gutenberg-University of Mainz, Duesbergweg,
55128 Mainz, Germany; tremel@uni-mainz.de (W.T.); tahir@uni-mainz.de (M.N.T.)
* Correspondence: mkhan3@ksu.edu.sa (Me.K.); rafiqs@ksu.edu.sa (M.R.H.S.);
Tel.: +966-11-467-5910 (Me.K.); +966-1-467-6082 (M.R.H.S.)

Academic Editor: Ioannis D. Kostas
Received: 27 November 2016; Accepted: 5 January 2017; Published: 9 January 2017

Abstract: Suzuki-Miyaura coupling reaction catalyzed by the palladium (Pd)-based nanomaterials is one of the most versatile methods for the preparation of biaryls. However, use of organic solvents as reaction medium causes a big threat to environment due to the generation of toxic byproducts as waste during the work up of these reactions. Therefore, the use of water as reaction media has attracted tremendous attention due to its environmental, economic, and safety benefits. In this study, we report on the synthesis of green Pd@graphene nanocatalyst based on an in situ functionalization approach which exhibited excellent catalytic activity towards the Suzuki–Miyaura cross-coupling reactions of phenyl halides with phenyl boronic acids under facile conditions in water. The green and environmentally friendly synthesis of Pd@graphene nanocatalyst (PG-HRG-Pd) is carried out by simultaneous reduction of graphene oxide (GRO) and $PdCl_2$ using *Pulicaria glutinosa* extract (PGE) as reducing and stabilizing agent. The phytomolecules present in the plant extract (PE) not only facilitated the reduction of $PdCl_2$, but also helped to stabilize the surface of PG-HRG-Pd nanocatalyst, which significantly enhanced the dispersibility of nanocatalyst in water. The identification of PG-HRG-Pd was established by various spectroscopic and microscopic techniques, including, high-resolution transmission electron microscopy (HRTEM), X-ray diffraction (XRD), ultraviolet–visible spectroscopy (UV-Vis), Fourier transform infrared spectroscopy (FT-IR), and Raman spectroscopy. The as-prepared PG-HRG-Pd nanocatalyst demonstrated excellent catalytic activity towards the Suzuki-Miyaura cross coupling reactions under aqueous, ligand free, and aerobic conditions. Apart from this the reusability of the catalyst was also evaluated and the catalyst yielded excellent results upon reuse for several times with marginal loss of its catalytic performance. Therefore, the method developed for the green synthesis of PG-HRG-Pd nanocatalyst and the eco-friendly protocol used for the Suzuki coupling offers a mild and effective substitute to the existing protocols and may significantly contribute to the endeavors of green chemistry.

Keywords: green synthesis; plant extract; palladium; graphene and Suzuki-Miyaura coupling

1. Introduction

Catalysis plays a very important role in many aspects of life, ranging from academic research to the chemical industry [1]. In the absence of catalysts, manufacturing of several essential commodities would not be possible at affordable prices—such as medicines, polymers, paints, lubricants, and various types of fine chemicals [2,3]. So far, the field of catalysis has made remarkable advances in the development of several catalytic processes for various important organic transformations including C–H bond activations, chemoselective oxidation and reductions, asymmetric hydrogenations, oxidative aminations, etc. [4,5]. Apart from these catalytic transformations, the cross-coupling reactions for the creation of C–C bonds, such as Heck coupling, Stille, Sonogashira, Kumada, and Suzuki-Miyaura coupling reactions have gained immense interest in several fields [6]. These reactions have been extensively applied for the vast industrial organic transformations, including for the production of agrochemicals, pharmaceuticals, and various other fine chemicals [7].

Among various coupling reactions, the Suzuki–Miyaura coupling has become the most powerful tool for the formation of C–C bonds, which is a transition-metal-catalyzed cross-coupling between an organoboron compound and an organic (pseudo) halide [8]. The Suzuki–Miyaura couplings have gained immense popularity due to their mild reaction conditions, wide range of functional group tolerance, high stability, and easy availability of organoboron reagents [9]. Although, several transitional-metal-based catalysts have been investigated for the Suzuki–Miyaura cross-coupling but palladium-based heterogeneous catalysts have so far gained great success, due to their robust ligand design [10–12]. Particularly, due to the rapid advancement of nanotechnology, Pd nanoparticle (NP)-based heterogeneous nanocatalysts have emerged as excellent substitutes to other organometallic-based conventional materials, which are found to be robust and possessing high surface area and that were used as catalyst support for Suzuki–Miyaura cross-coupling [13,14].

The nano-sized Pd-based catalysts enhances the surface area many folds which results in increasing the contact between substrate and catalyst, and their insolubility in reaction solvents facilitates effortless separation [15]. In order to further enhance the efficiency, cost effectiveness of Pd-based nanocatalysts, different types of support materials are required which facilitate stabilization and homogeneous dispersion of nanocatalyst [16]. So far, silica-based support materials, due to their excellent stability and porosity, have been commonly applied for the Suzuki–Miyaura cross-couplings reactions [17]. Apart from this, many carbon-based support materials have also been applied, as their high specific surface area, superior electronic conductivity and excellent stability significantly improve the activity and stability of Pd nanocatalysts [18]. Recently, graphene, as a new promising candidate among various carbon materials, has attracted tremendous attention of scientists and technologists due to its remarkable properties [19]. Owing to its 2D planar structure, excellent conductivity and large theoretical specific surface area of 2630 $m^2 \cdot g^{-1}$, graphene has garnered an incredible reputation as a support material in catalysis [20].

The synthesis of composite, combining the intrinsic properties of both Pd and graphene in the form of nanocomposite, may collectively contribute to the further enhancement of the catalytic properties of hybrid nanocatalysts [21]. The dispersion of Pd NPs on the surface of graphene is usually carried out either via post immobilization (ex situ hybridization), which involves the mixing of separate solutions of graphene and presynthesized Pd NPs or by the in-situ crystallization of NPs [22,23]. The in-situ binding is carried out via simultaneous reduction of Pd salts and graphene oxide (GRO) with several reduction techniques including chemical, thermal or electrochemical reduction [24–26]. So far, majority of Pd-graphene (Pd@graphene nanocomposite) is prepared by in-situ chemical reduction of Pd salts and GRO, which usually involves potentially hazardous and toxic reagents, starting materials, solvents, and stabilizers [27,28]. Recently, atomic layer deposition technique was used for the preparation of Pd/graphene nanocomposites [29,30], which give enhanced control over particle size and dispersion, however these techniques cannot be carried out easily in all labs and will be more difficult to scale up.

Due to growing environmental concerns, the involvement of the concepts of the green chemistry, which are designed to eliminate or reduce harmful chemicals or chemical processes that are a threat to the environment have become imminent in the field of nanocatalysis [31,32]. Therefore, applying green methods for the synthesis of catalysts and using sustainable catalytic processes may tremendously enhance the energy efficiency, environmental friendliness, and several economic benefits for the large scale industrial processes [33]. Recently, the trends of applying green reductants—such as microorganisms, plant extracts (PEs), and amino acids—in the field of nanotechnology have garnered significant popularity [34]. Particularly, the acceptance of PEs as suitable alternatives to the chemical reductants has been steadily growing in recent years, due to their easy availability and cost effectiveness [35,36]. To date, PEs have been used for the preparation of different metallic NPs and also for the reduction of GRO, however, they are rarely applied for the in-situ preparation of graphene-metallic NP-based nanocomposites [37–39]. For instance, in our previous study, Pd@graphene nanocomposites were prepared in a facile method by the simultaneous reduction of GRO and PdCl$_2$ using *Salvadora persica* L. (miswak) root extract (RE) as bioreductant [38]. The as-prepared nanocomposites demonstrated excellent catalytic activities towards the selective oxidation of alcohols.

Herein, we demonstrate the synthesis of Pd@graphene nanocomposites using *Pulicaria glutinosa* extract (PGE) as reducing agent. The reducing ability of PGE that possesses rich phenolic contents has been already tested for the efficient synthesis of various metallic NPs [40]. Apart from this, the PGE has also been applied for the effective reduction of GRO and stabilization of HRG [41]. In view of the excellent reducing properties of the PGE for nanoparticles as well as GRO, we contemplated the study of one-step preparation of nanocomposite involving metallic NPs and GRO via simultaneous reduction. In continuation of our previous study, the PGE is now being used as a green reductant for the preparation of catalytically active Pd@graphene nanocatalyst (Scheme 1). The PGE not only acted as a reducing agent but also functioned as an in-situ functionalizing ligand to facilitate the dispersion of Pd NPs onto surface of HRG nanosheets. During this study, the catalytic activity of the Pd@graphene nanocomposites towards Suzuki-Miyura coupling reactions were tested and compared with the activity of previously prepared Pd NPs with the same PGE. Although, in both cases comparable catalytic activity is obtained, however, in the latter scenario a higher amount of Pd nanocatalyst was required to obtained similar catalytic activity. Furthermore, detailed kinetic study of catalytic coupling reactions and reusability of the catalyst was also performed. Pd@graphene nanocomposites and the products of the catalytic reactions were characterized using various microscopic and analytical techniques including gas chromatography (GC), XRD, FT-IR, UV-Vis, and HRTEM.

Scheme 1. Schematic illustration of the green synthesis of Pd@graphene nanocatalyst (PG-HRG-Pd) using an aqueous extract of the *P. glutinosa* and their catalytic activity for the Suzuki Miyaura coupling.

2. Results and Discussion

The preparation of Pd@graphene nanocatalyst is performed in a single step via concurrent reduction of GRO and PdCl$_2$ under mild conditions using *P. glutinosa* extract (PGE) as both reducing and stabilizing agent. To begin with, PGE was added in the required amount to the aqueous mixture of GRO and PdCl$_2$. The resulting mixture was stirred under reflux conditions until a clear color change from dark brown to black was observed, which indicated the formation of Pd@graphene nanocatalyst (PG-HRG-Pd). Furthermore, to study the effect of the amount of Pd NPs on the catalytic activities of PG-HRG-Pd nanocatalyst, two different samples including PG-HRG-Pd-1 and PG-HRG-Pd-2 were prepared by taking 50 wt % and 75 wt % of PdCl$_2$ with that of GRO, respectively, however, to monitor the influence of precursors on the density and distribution of Pd NPs on graphene support, the amount of PGE was not changed during this process.

Most of the oxygen-containing functional groups are removed during the reduction of GRO which affects the dispersing properties of HRG severely and makes it difficult to use HRG for further processing. To counter this, various additional stabilizing agents and surfactants are usually added for the dispersion of HRG, which facilitates better dispersibility of the resulting mixture. However, the PGE which is used here as a reducing agent is already known to possess various phytomolecules such as terpenoids, flavonoids, etc., which are rich in oxygen containing functional groups. These functional groups not only facilitated the simultaneous reduction of GRO and PdCl$_2$ but also helped to stabilize PG-HRG-Pd nanocatalyst by sticking on the surface of the resulting material. The oxygen containing functional groups of various phytomolecules of residual PGE, which remained attached on the surface of the HRG nanosheets, extend into the solution to provide electrostatic repulsion that stabilizes the suspension. Among these functional groups, several carboxyl and hydroxyl groups are also present, which may act as the active sites for the adsorption of Pd ions and also facilitate nucleation and growth of Pd NPs.

2.1. TEM and EDX Analysis

The presence of residual phytomolecules on the surface of PG-HRG-Pd not only enhances the aqueous dispersibility of the resulting composite, but also facilitates the homogeneous distribution of Pd NPs on the surface of HRG nanosheets. Since the nucleation of Pd starts directly on to the bound phytomolecules, it results in highly dense coating of Pd NPs on the surface of HRG sheets as shown in transmission electron micrographs (Figure 1). Detailed size distribution analysis of Pd NPs using 50 wt % ratio of Pd precursor is difficult to comment on as the particles are mostly anisotropic. The most excited morphology among metal nanoparticles is two-dimensional (2D) nano-plates (triangle or hexagonal shape). As shown in HRTEM images, the lower concentration of Pd precursor also results mostly in triangle shape Pd NPs. Here, the average edge length of Pd triangle is ~15–18 nm. The *fcc* structure of the Pd phase is confirmed by measuring the lattice *d*-spacing on top of one of the triangle shape HRTEM image which is 2.4 Å corresponding to (111) plane. However, upon using a higher concentration of Pd precursor, the morphology became isotropic with particle sizes of ~7–8 nm but it further improved the density and distribution of Pd nanoparticles onto graphene. It has provided an efficient support for the homogenous distribution of Pd NPs. In the former case where less concentration of Pd precursor is used, it could be the result of fast depletion of precursor to give rise to Ostwald ripening, resulting in bigger and anisotropic particles. This is also evidenced by very small particles in the background of bigger particles attached to support material (HRG). Whereas, in the latter case where more precursor salt is used, there is enough precursor to grow the nuclei beyond a critical radius. However, it is worth mentioning that these nanoparticles are polycrystalline with domains exposed on the surface having d-spacing (2.23 Å) corresponding to the (111) plane and also d-spacing 1.89 Å related to (200) of cubic Pd phase. The composition of the product is confirmed by energy-dispersive X-ray spectroscopy (EDX) analysis (Figure 2). Apart from the signal of Pd, the oxygen and carbon can also be seen in the EDX spectrum of PG-HRG-Pd-1, which can be attributed to the oxygen-containing functional groups of the residual phytomolecules.

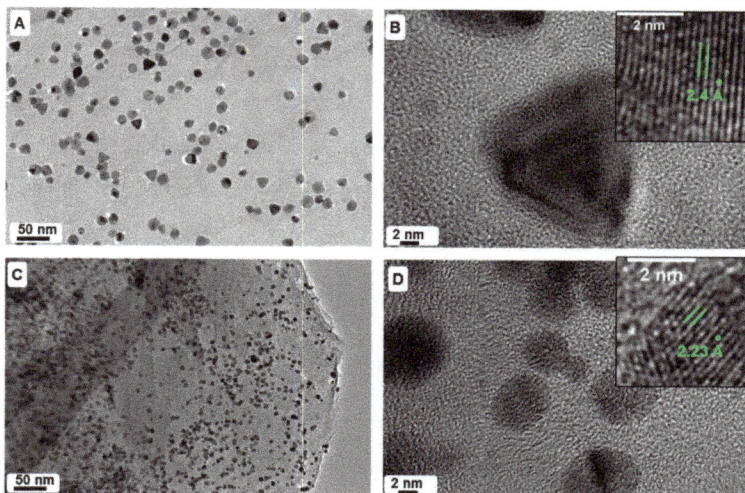

Figure 1. Transmission electron microscope (TEM) and high-resolution (HRTEM) images of the Pd@graphene nanocatalyst (PG-HRG-Pd-1) (**A**) overview PG-HRG-Pd (50 wt % ratio of Pd precursor); (**B**) HRTEM image confirming the triangular shape of Pd nanoparticles with distance between two Pd lattice fringes is 2.4 Å (inset) corresponding to (111) crystal planes of face-centered cubic (*fcc*) Pd crystals; (**C**) overview PG-HRG-Pd (75 wt % ratio of Pd precursor) showing the enhanced density of Pd nanoparticles and (**D**) high-resolution (HRTEM) image of the corresponding sample confirming the spherical morphology and also *fcc* structure of as synthesized Pd NPs (inset).

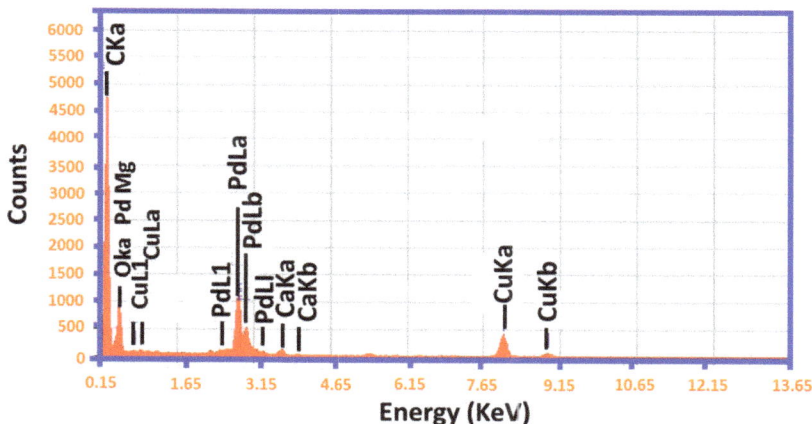

Figure 2. Energy dispersive X-ray spectrum (EDX) of as-synthesized Pd@graphene nanocatalyst (PG-HRG-Pd-1) confirming the composition of product.

2.2. Raman Spectroscopy

Raman spectroscopy was applied to monitor the reduction of GRO. Raman spectra of GRO and PG-HRG-Pd-1 nanocatalyst are displayed in Figure 3. Pristine graphene consists of two different characteristic signals, with the G and D bands situated at 1575 cm^{-1} and 1350 cm^{-1}, respectively. However, after oxidation, the sp^2 character in graphene is destroyed and various defects are formed in GRO, due to which the characteristic bands of graphene in this case are shifted to 1602 and 1340 cm^{-1}, respectively. Notably, these bands are relocated closer to their ideal positions after the reduction with PGE in the Raman spectrum of PG-HRG-Pd-1, i.e., the G band is shifted from 1602 to 1592 cm^{-1}, whereas the D band is relocated from 1340 to 1336 cm^{-1}. The emergence of some visible changes in the Raman spectra of GRO after reduction with PGE clearly reflects the formation of PG-HRG-Pd-1.

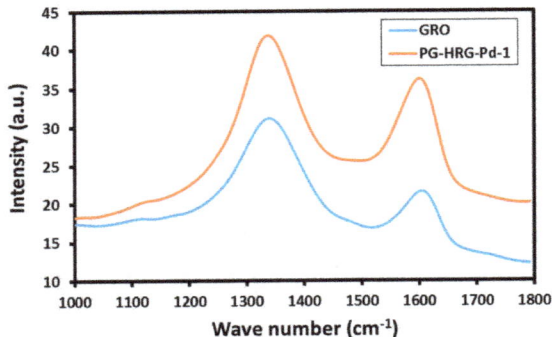

Figure 3. Raman spectra of graphene oxide (GRO) and plant extract-mediated Pd@graphene nanocatalyst (PG-HRG-Pd-1).

2.3. XRD Analysis

XRD analysis was used to confirm the phase and crystallinity of PG-HRG-Pd-1. For this purpose, XRD diffractogram of pristine graphite, GRO, and PG-HRG-Pd-1 were measured as shown in Figure 4. Typically, graphite consists of an intense reflection at 26.4° (002), which is moved to the lower brag angle in case of GRO at 10.9° corresponding to (001) plane with d-spacing of 0.81 nm, due to the addition of various oxygen containing functional groups between the graphite layers during the oxidation process [42]. Notably, in the XRD pattern after reduction, i.e., of PG-HRG-Pd-1, a broad reflection appeared at ~22.4° (002), whereas the reflection corresponding to (001) plane of GRO at 10.9° disappeared. This clearly points towards the destruction of the regular layered structure of GRO and the formation of a few layer stacked HRG sheets in PG-HRG-Pd-1 after the reduction with PGE. Besides, the XRD spectrum of PG-HRG-Pd-1 also contains several reflections at 40.02° (111), 46.49° (200), 68.05° (220), 81.74° (311), and 86.24° (222), which can be attributed to the face centered cubic (*fcc*) structure of Pd NPs on the surface of HRG (JCPDS: 87-0641, space group: Fm3m (225)).

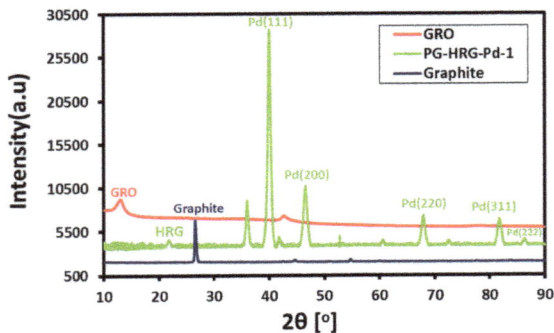

Figure 4. XRD (X-ray diffraction) diffractograms of graphite, graphene oxide (GRO), and plant extract-mediated Pd@graphene nanocatalyst (PG-HRG-Pd-1).

2.4. UV-Vis Spectral Analysis

UV spectroscopy is a very useful tool for investigating the formation of nanocatalysts and also to confirm the presence of plant extract as stabilizing agent on the surface of PG-HRG-Pd nanocatalyst. The UV spectra of PdCl$_2$, pure PGE, GRO, and PG-HRG-Pd nanocatalyst are displayed in Figure 5. Typically, GRO exhibits two absorption bands at ~230 and ~301 nm, while PdCl$_2$ possesses a distinguished absorption band at ~420 nm. The absence of these peaks in the UV spectrum of PG-HRG-Pd evidently confirms the concurrent reduction of both GRO and PdCl$_2$ and also confirms the formation of PG-HRG-Pd nanocatalyst. Notably, the UV absorption spectrum of PG-HRG-Pd exhibits two weak absorption bands at ~280 and ~325 nm which closely resemble the absorption peaks of PGE. This clearly suggests the presence of residual phytomolecules of PGE on the surface of PG-HRG-Pd nanocatalyst as stabilizing agent. This is further confirmed by FT-IR analysis by measuring the FT-IR spectra of pure PGE, GRO, and PG-HRG-Pd nanocatalyst as shown in Figure 5. Typically, the FT-IR signals belonging to the GRO either disappear or their intensities are significantly reduced after the reduction process, which in turn confirms the formation of HRG. For instance, in the FT-IR spectrum of PG-HRG-Pd the peaks at ~1630 cm^{-1} (for C=C stretching), ~1740 cm^{-1} (for C=O stretching) corresponding to the GRO disappear, whereas the intensities of some of the distinguished peaks are relatively decreased, such as a broad band at around 3440 cm^{-1} for hydroxyl groups, which points towards the reduction of GRO to HRG in PG-HRG-Pd. Upon further analysis, it was revealed that the IR spectrum of PG-HRG-Pd also bear close resemblance to the FT-IR spectrum of pure PGE. The FT-IR spectrum of the PGE exhibits signals at 3746 and 3410 cm^{-1} belonging to the OH group and a C−H peak at 2943 cm^{-1}. Apart from this, the peaks around 1753, 1622, and 1407 cm^{-1} are the typical aromatic peaks for the C−H, C−C, and C−O stretching, respectively. Whereas, the peaks corresponding to the C−O stretching of carboxylic acids, alcohols, ether, and ester groups appeared at 1264 and 1077 cm^{-1}. Expectedly, most of these peaks of PGE are also found in the FT-IR spectrum of PG-HRG-Pd, either on the same position or with slight shifts. This clearly suggests that PGE not only acted as a reducing agent but also functioned as a stabilizing agent by attaching to the surface of HRG sheets.

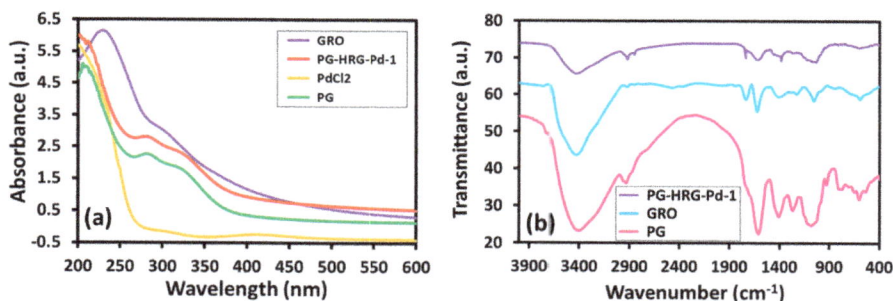

Figure 5. (a) UV-Vis absorption spectra of graphene oxide (GRO), Pd@graphene nanocatalyst (PG-HRG-Pd-1), palladium chloride (PdCl$_2$), and PGE; (b) FT-IR spectra of graphene oxide (GRO), Pd@graphene nanocatalyst (PG-HRG-Pd-1) and the PGE.

2.5. Suzuki Reaction Catalyzed by Pd@graphene Nanocatalyst

Suzuki coupling is one of the most useful methods for the synthesis of biaryls and alkene derivatives [43]. These reactions are usually carried out by using various Pd-based catalysts in different organic solvents [44]. Typically, such types of carbon–carbon couplings, including Suzuki reactions, are performed in a mixture of an organic solvent and an aqueous inorganic base, under inert conditions [45,46]. However, the concerns of the environmental impact of various chemical products and the chemical processes by which they are produced have been mounting rapidly in recent years due to global warming. Since, majority of the chemical wastes from a reaction mixture corresponds to the solvents. Therefore, the use of water as a reaction medium, which is considered a benign solvent due to its natural abundance and physiological compatibility, may potentially contribute towards the serious efforts of reducing the environmental impacts of the organic reactions [47]. In this study, the catalytic activity of the as-synthesized PG-HRG-Pd-1 nanocatalyst was evaluated in the Suzuki Miyaura coupling of bromobenzene with phenylboronic acid in water. For this purpose, the Pd@graphene nanocomposite (PG-HRG-Pd) obtained by the simultaneous reduction of both $PdCl_2$ and GRO using PGE as reducing agent was used as a catalyst. The as-prepared nanocatalyst effectively catalyzed the reaction of various aryl halides, including, chloro, bromo, iodobenzene, chloro-benzophenone, and bromo-acetophenone with benzeneboronic acid in water containing sodium lauryl sulfate and K_3PO_4 under aerobic conditions to obtain biaryls (cf. Figure 6). The variation in the substrate structure was carried out in order to understand the catalytic performance of the nanocatalyst in the presence of different ring substituents.

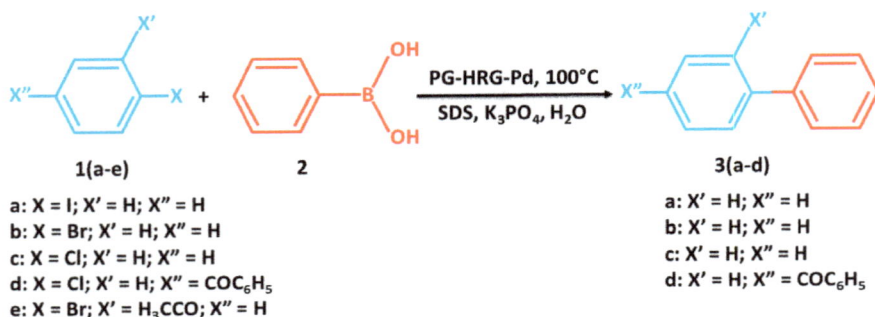

1(a-e)

a: X = I; X' = H; X" = H
b: X = Br; X' = H; X" = H
c: X = Cl; X' = H; X" = H
d: X = Cl; X' = H; X" = COC_6H_5
e: X = Br; X' = H_3CCO; X" = H

3(a-d)

a: X' = H; X" = H
b: X' = H; X" = H
c: X' = H; X" = H
d: X' = H; X" = COC_6H_5

Figure 6. Schematic representation of the Suzuki reaction of iodobenzene, bromobenzene, and chlorobenzene, with phenylboronic acid under aqueous conditions.

The comparison between the catalytic performance of the two nanocatalysts, i.e., PG-HRG-Pd-1 and PG-HRG-Pd-2, has been found to be negligible whereas the catalytic performance varied extensively in the presence of different substituents. The results reveal that iodo substituted aryl halide yields the coupled product at the fastest reaction rate within 30 min for both the catalysts, however for the bromo- and chloro-substituted substrates yield ~90% and ~54% products. The 4-Chlorobenzophenone yielded least conversion product of 38%, while the 2'-Bromoacetophenone did not yielded any coupling product. This could be due to the increasing anionic nature among halogens down the group, which induces the difference in the rate of the reaction, while in the case of 2'-Bromoacetophenone, the steric hindrance from the acetate group could be responsible for not yielding any coupling product. The kinetics of the reaction were carried out using the gas chromatography (GC), which was studied by collecting the reaction mixture at equal intervals of time and quenched immediately. The graphical representation of the data is presented in the Figure 7.

The reusability of the nanocatalyst was also studied in order to find out if there is depreciation in the catalytic performance with consecutive reuse. When the catalyst was reused under similar reaction conditions, it was observed that there was a slight decrease in the percentage product formation upon consecutive reuse, which is very much unlike the results obtained when the same studies were carried out using Pd NPs prepared without graphene support [48]. Comparisons of the results obtained from reusability studies of the catalysts Pd NPs and PG-HRG-Pd-1, revealed that the depreciation of catalytic performance found by employing the PG-HRG-Pd-1, is much lower than Pd NPs wherein the catalyst loses about catalytic performance yielding an ~50% conversion product after three consecutive re-use, while PG-HRG-Pd-1 yields an ~77% conversion product after five consecutive re-uses. The graphical representation of the results obtained after several reuses of catalyst is given in Figure 8. A leaching study was carried out in order to ascertain the leaching of Pd NPs during the catalytic process; the product was subjected to ICP-MS studies. It was found that there is 0% percent Pd NPs in the reaction mixture. This indicates that the Pd NPs formed are well supported on the HRG, hence there is no leaching of Pd NPs, which in turn results in the excellent reusability of the prepared catalyst.

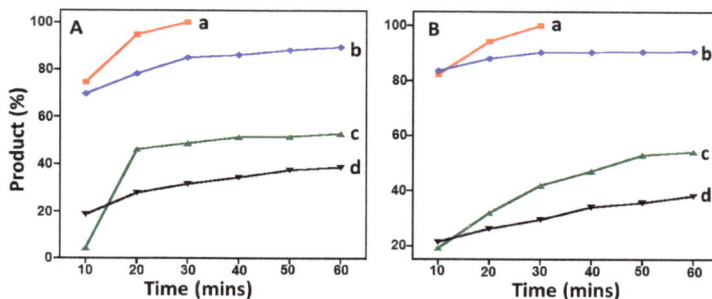

Figure 7. Time dependent conversion efficiency of the Suzuki reaction employing (**A**) PG-HRG-Pd-1 and (**B**) PG-HRG-Pd-2 for various substrates (a) iodobenzene, (b) bromobenzene, (c) chlorobenzene, and (d) chloro benzophenone with phenylboronic acid under aqueous and aerobic conditions determined by GC analysis.

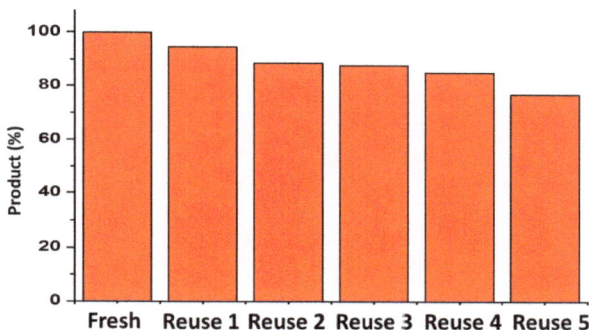

Figure 8. Graphical representation of conversion product obtained by the reuse of catalyst for the catalytic conversion of iodobenzene.

The results obtained in the present study were compared with our previously reported study [48], where in the catalyst was prepared employing the same green reducing agent without the presence of highly reduced graphene. It was observed that the prepared nanocatalyst containing graphene uses 2.5 mol % Pd NPs which is half the amount used for the previously reported, i.e., 5 mol %. However, to better understand the results obtained from both the studies, the turn over number (TON) values for both the catalysts are calculated. Upon comparison, it was observed that the TON value obtained for Pd NPs was found to be 20, while it was found to be ~74 and ~52 for the catalysts PG-HRG-Pd-1 and PG-HRG-Pd-2, respectively. A graphical representation of the comparative TON values is given in Figure 9.

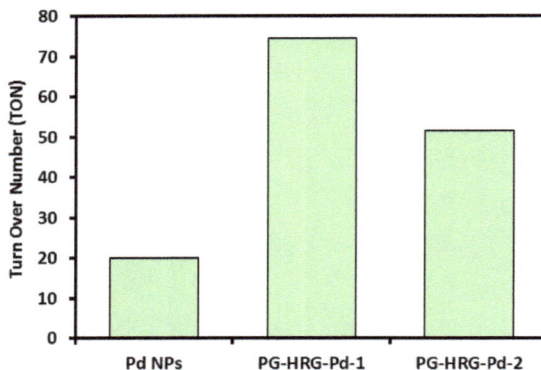

Figure 9. Graphical illustration of comparative turn over number (TON) values.

3. Materials and Methods

3.1. Materials

Natural graphite powder (99.999%, 200 mesh) was purchased from Alfa Aesar (Tewksbury, MA, USA); Palladium (II) Chloride (PdCl$_2$ 99.99%), NaBH$_4$ (96%), concentrated sulfuric acid (H$_2$SO$_4$ 98%), potassium permanganate (KMnO$_4$ 99%), sodium nitrate (NaNO$_3$, 99%), hydrogen peroxide (H$_2$O$_2$, 30 wt %), bromobenzene (99.5%), sodium dodecyl sulfate (98%), phenyl boronic acid (95%), tripotassium phosphate (98%), and all organic solvents were obtained from Aldrich Chemicals (St. Louis, MI, USA) and were used directly without further purification. The details about the procedure of the collection of *P. glutinosa* and the preparation of extract are given in our previous study [40].

3.2. Methods

3.2.1. Preparation of Pd@graphene Nanocomposite (PG-HRG-Pd-1)

Graphite oxide (GO) used for the preparation of PG-HRG-Pd-1 was synthesized according to our previously reported method [41]. Initially, as-prepared graphite oxide or GO (200 mg) was dispersed in 40 mL of distilled water (DW) water and sonicated for 30 min to obtain graphene oxide (GRO) sheets. The resulting suspension was taken in a round bottom flask, to which 100 mg (0.563 mmol) of PdCl$_2$ was added. The flask was mounted with a cooling condenser, which was heated to 100 °C. Subsequently, 10 mL of an aqueous solution of the PGE (0.1 g·mL^{-1}) was added to the suspension, which was then allowed to stir for 24 h at 98 °C. Afterwards, the resulting black powder of (PG-HRG-Pd-1) was collected by filtration, and further washed with DI water several times to remove excess PGE residue and redistributed into water for sonication. This suspension was centrifuged at 4000 rpm for another 30 min. The final product was collected by vacuum filtration and dried in vacuum.

3.2.2. Suzuki Reaction Catalyzed by Pd@graphene Nanocatalyst

In a typical experiment, a mixture of sodium dodecyl sulfate (144 mg, 0.5 mmol), tripotassium phosphate (K$_3$PO$_4$, 399 mg), phenylboronic acid (146 mg, 1.2 mmol) and deionized water (20 mL) was taken in a 100 mL round bottom flask. Halobenzene (1.0 mmol) was added to this mixture under stirring, followed by the as-prepared PG-HRG-Pd-1 nanocatalyst (5 mol %, 5.32 mg). The mixture was stirred at 100 °C in an oil bath for 5 min and then extracted with ethyl acetate (3 × 20 mL). The combined organic extract was dried over anhydrous sodium sulfate (Na$_2$SO$_4$), and the resulting mixture was analyzed by gas chromatography (GC). In order to identify the product obtained from the catalytic reaction, the as-obtained mixture was crystallized from ethanol. The coupling product obtained from using bromobenzene as starting material was isolated using column chromatography which was found to be a white powder, identified as biphenyl using ^{1}H and ^{13}C solution nuclear magnetic resonance spectroscopy (NMR) and mass spectroscopy. ^{1}H-NMR δ 8.25 (d, J = 8.3 Hz, 4H, C–CH, next to ipso), 7.25–7.26 (m, 6H, remaining protons of phenyl ring); ^{13}C-NMR δ 141.3 (2C, C–C, ipso), 128.8 (4C, CH–CH), 127.3 (2C, CH–CH, edge carbons), 127.2 (4C, C–CH, next to ipso); electron impact-mass spectrometry(EIMS) m/z 154 (M$^+$). The progress of reaction employing other substrates was monitored using gas chromatography (GC), however the coupling product was not isolated. NMR spectra's obtained are provided as Supplementary information as Figures S1 and S2.

3.3. Characterization of Catalysts

3.3.1. Transmission Electron Microscopy (TEM)

Transmission electron microscopy (TEM) and high-resolution transmission electron microscopy (HRTEM) images were obtained using a JEOL JEM 1101 (USA) instrument. The samples for the TEM measurements were prepared by suspending the composites in ethanol and were drop-cast onto a carbon-coated 200-mesh copper grid and subsequently dried at room temperature.

3.3.2. X-ray Powder Diffraction (XRD)

X-ray powder diffraction (XRD) measurements were performed on an Altima IV (Rigaku, Tokyo, Japan) instrument, which is equipped with a Cu Ka radiation source.

3.3.3. UV-Vis Spectroscopy

UV-Vis measurements were conducted on a Perkin Elmer lambda 35 (USA) UV-vis spectrophotometer (Perkin Elmer, Waltham, MA, USA). The analysis was performed in quartz cuvettes using DI water as a reference solvent. Stock solutions of PG-HRG-Pd-1 and GRO for the UV measurements were prepared by dispersing 5 mg of sample in 10 mL of DI water and sonicating for 30 min. The UV samples of GRO and PG-HRG-Pd-1 were prepared by diluting 1 mL of stock solution with 9 mL of water.

3.3.4. Fourier Transform Infrared Spectrometer (FT-IR)

Fourier transform infrared spectrometer (FT-IR) spectra were measured on a Perkin-Elmer 1000 (Perkin Elmer, Waltham, MA, USA) Fourier transform infrared spectrometer.

4. Conclusions

In summary, we have developed a green and facile method for the preparation of Pd@graphene nanocomposites and investigated their catalytic application towards the Suzuki-Miyura coupling reactions. This graphene-based novel catalyst was prepared in a single step by the spontaneous reduction of both GRO and PdCl$_2$ using *P. glutinosa* extract as reducing agent. The resultant nanocatalyst showed homogeneous distribution of Pd NPs on the surface of HRG with excellent dispersion properties, due to the presence of residual phytomolecules as stabilizing ligands. The enhanced dispersibility has enabled PG-HRG-Pd nanocatalysts to be utilized as effective catalysts for various Suzuki-Miyura coupling reactions in aqueous solution. These features, together with the ease and greenness of the synthetic process may promote the suitability of the method for large-scale production of efficient catalysts for various important organic transformations including the Suzuki-Miyura couplings. HRG support plays an excellent role due to which there is no leaching of the catalyst during the reaction, which in turn results in excellent reusability of the catalyst. Comparative studies of the TON values revealed that the prepared catalyst PG-HRG-Pd-1 is the better-performing catalyst when compared to Pd NPs prepared employing the same procedure.

Supplementary Materials: The following are available online at www.mdpi.com/2073-4344/7/1/20/s1, Figure S1: ^{1}H-NMR spectra of the Suzuki–Miyaura Coupling product. Figure S2: ^{13}C-NMR spectra of the Suzuki–Miyaura Coupling product.

Acknowledgments: This project was supported by King Saud University, Deanship of Scientific Research, College of Science, Research Centre.

Author Contributions: Muj.K., S.F.A., and Me.K. designed the project. Muj.K., S.F.A., Me.K., and M.R.S. helped to draft the manuscript. Me.K. and H.Z.A. carried out the preparation of plant extract and characterization of plant extract material. Muf.K. and M.R.S. carried out the experimental part and some part of characterization. A.A.-W., W.T., M.N.T., and M.R.H.S provided scientific guidance for successful completion of the project and also helped to draft the manuscript. All authors read and approved the final manuscript.

Conflicts of Interest: The authors declare that they have no conflict of interests.

References

1. Trost, B.M. Atom economy—A challenge for organic synthesis: Homogeneous catalysis leads the way. *Angew. Chem. Int. Ed. Engl.* **1995**, *34*, 259–281. [CrossRef]
2. Czaja, A.U.; Trukhan, N.; Müller, U. Industrial applications of metal–organic frameworks. *Chem. Soc. Rev.* **2009**, *38*, 1284–1293. [CrossRef] [PubMed]
3. Alemán, J.; Cabrera, S. Applications of asymmetric organocatalysis in medicinal chemistry. *Chem. Soc. Rev.* **2013**, *42*, 774–793. [CrossRef] [PubMed]
4. Chng, L.L.; Erathodiyil, N.; Ying, J.Y. Nanostructured catalysts for organic transformations. *Acc. Chem. Res.* **2013**, *46*, 1825–1837. [CrossRef] [PubMed]
5. Maki-Arvela, P.I.; Simakova, I.L.; Salmi, T.; Murzin, D.Y. Production of lactic acid/lactates from biomass and their catalytic transformations to commodities. *Chem. Rev.* **2013**, *114*, 1909–1971. [CrossRef] [PubMed]
6. Chu, C.K.; Liang, Y.; Fu, G.C. Silicon–carbon bond formation via nickel-catalyzed cross-coupling of silicon nucleophiles with unactivated secondary and tertiary alkyl electrophiles. *J. Am. Chem. Soc.* **2016**, *138*, 6404–6407. [CrossRef] [PubMed]
7. Ruiz-Castillo, P.; Buchwald, S.L. Applications of palladium-catalyzed C–N cross-coupling reactions. *Chem. Rev.* **2016**, *116*, 12564–12649. [CrossRef] [PubMed]
8. Han, F.-S. Transition-metal-catalyzed Suzuki-Miyaura Cross-Coupling reactions: A remarkable advance from palladium to nickel catalysts. *Chem. Soc. Rev.* **2013**, *42*, 5270–5298. [CrossRef] [PubMed]
9. Miyaura, N.; Suzuki, A. Palladium-catalyzed cross-coupling reactions of organoboron compounds. *Chem. Rev.* **1995**, *95*, 2457–2483. [CrossRef]
10. Deraedt, C.; Astruc, D. "Homeopathic" palladium nanoparticle catalysis of cross carbon–carbon coupling reactions. *Acc. Chem. Res.* **2013**, *47*, 494–503. [CrossRef] [PubMed]
11. Hattori, T.; Tsubone, A.; Sawama, Y.; Monguchi, Y.; Sajiki, H. Palladium on carbon-catalyzed Suzuki-Miyaura coupling reaction using an efficient and continuous flow system. *Catalysts* **2015**, *5*, 18–25. [CrossRef]
12. Gniewek, A.; Ziółkowski, J.J.; Trzeciak, A.M.; Zawadzki, M.; Grabowska, H.; Wrzyszcz, J. Palladium nanoparticles supported on alumina-based oxides as heterogeneous catalysts of the Suzuki-Miyaura reaction. *J. Catal.* **2008**, *254*, 121–130. [CrossRef]
13. Kalidindi, S.B.; Jagirdar, B.R. Nanocatalysis and prospects of green chemistry. *ChemSusChem* **2012**, *5*, 65–75. [CrossRef] [PubMed]
14. Manabe, K. Palladium catalysts for cross-coupling reaction. *Catalysts* **2015**, *5*, 38–39. [CrossRef]
15. Wang, J.; Gu, H. Novel metal nanomaterials and their catalytic applications. *Molecules* **2015**, *20*, 17070–17092. [CrossRef] [PubMed]
16. Ohtaka, A.; Okagaki, T.; Hamasaka, G.; Uozumi, Y.; Shinagawa, T.; Shimomura, O.; Nomura, R. Application of "boomerang" linear polystyrene-stabilized Pd nanoparticles to a series of C–C coupling reactions in water. *Catalysts* **2015**, *5*, 106–118. [CrossRef]
17. Gautam, P.; Dhiman, M.; Polshettiwar, V.; Bhanage, B.M. KCC-1 supported palladium nanoparticles as an efficient and sustainable nanocatalyst for carbonylative Suzuki-Miyaura Cross-Coupling. *Green Chem.* **2016**, *18*, 5890–5899. [CrossRef]
18. Adib, M.; Karimi-Nami, R.; Veisi, H. Palladium NPs supported on novel imino-pyridine-functionalized MWCNTs: Efficient and highly reusable catalysts for the Suzuki-Miyaura and Sonogashira coupling reactions. *New J. Chem.* **2016**, *40*, 4945–4951. [CrossRef]

19. Raccichini, R.; Varzi, A.; Passerini, S.; Scrosati, B. The role of graphene for electrochemical energy storage. *Nat. Mater.* **2015**, *14*, 271–279. [CrossRef] [PubMed]

20. Fan, X.; Zhang, G.; Zhang, F. Multiple roles of graphene in heterogeneous catalysis. *Chem. Soc. Rev.* **2015**, *44*, 3023–3035. [CrossRef] [PubMed]

21. Khan, M.; Tahir, M.N.; Adil, S.F.; Khan, H.U.; Siddiqui, M.R.H.; Al-warthan, A.A.; Tremel, W. Graphene based metal and metal oxide nanocomposites: Synthesis, properties and their applications. *J. Mater. Chem. A* **2015**, *3*, 18753–18808. [CrossRef]

22. Metin, Ö.; Kayhan, E.; Özkar, S.; Schneider, J.J. Palladium nanoparticles supported on chemically derived graphene: An efficient and reusable catalyst for the dehydrogenation of ammonia borane. *Int. J. Hydrogen Energy* **2012**, *37*, 8161–8169. [CrossRef]

23. Xu, W.; Wang, X.; Zhou, Q.; Meng, B.; Zhao, J.; Qiu, J.; Gogotsi, Y. Low-temperature plasma-assisted preparation of graphene supported palladium nanoparticles with high hydrodesulfurization activity. *J. Mater. Chem.* **2012**, *22*, 14363–14368. [CrossRef]

24. Zhang, Y.; Shu, H.; Chang, G.; Ji, K.; Oyama, M.; Liu, X.; He, Y. Facile synthesis of palladium–graphene nanocomposites and their catalysis for electro-oxidation of methanol and ethanol. *Electrochim. Acta* **2013**, *109*, 570–576. [CrossRef]

25. Liu, Y.; Liu, L.; Shan, J.; Zhang, J. Electrodeposition of palladium and reduced graphene oxide nanocomposites on foam-nickel electrode for electrocatalytic hydrodechlorination of 4-chlorophenol. *J. Hazard. Mater.* **2015**, *290*, 1–8. [CrossRef] [PubMed]

26. Konda, S.K.; Chen, A. One-step synthesis of Pd and reduced graphene oxide nanocomposites for enhanced hydrogen sorption and storage. *Electrochem. Commun.* **2015**, *60*, 148–152. [CrossRef]

27. Li, Y.; Fan, X.; Qi, J.; Ji, J.; Wang, S.; Zhang, G.; Zhang, F. Palladium nanoparticle-graphene hybrids as active catalysts for the Suzuki reaction. *Nano Res.* **2010**, *3*, 429–437. [CrossRef]

28. Fu, L.; Lai, G.; Zhu, D.; Jia, B.; Malherbe, F.; Yu, A. Advanced catalytic and electrocatalytic performances of polydopamine-functionalized reduced graphene oxide-palladium nanocomposites. *ChemCatChem* **2016**, *8*, 2975–2980. [CrossRef]

29. Yan, H.; Cheng, H.; Yi, H.; Lin, Y.; Yao, T.; Wang, C.; Li, J.; Wei, S.; Lu, J. Single-Atom Pd$_1$/Graphene Catalyst Achieved by Atomic Layer Deposition: Remarkable Performance in Selective Hydrogenation of 1,3-Butadiene. *J. Am. Chem. Soc.* **2015**, *137*, 10484–10487. [CrossRef] [PubMed]

30. Van Bui, H.; Grillo, F.; Helmer, R.; Goulas, A.; van Ommen, J.R. Controlled growth of palladium nanoparticles on graphene nanoplatelets via scalable atmospheric pressure atomic layer deposition. *J. Phys. Chem. C* **2016**, *120*, 8832–8840. [CrossRef]

31. Narayanan, R. Synthesis of green nanocatalysts and industrially important green reactions. *Green Chem. Lett. Rev.* **2012**, *5*, 707–725. [CrossRef]

32. Majeed, M.I.; Lu, Q.; Yan, W.; Li, Z.; Hussain, I.; Tahir, M.N.; Tremel, W.; Tan, B. Highly water-soluble magnetic iron oxide (Fe$_3$O$_4$) nanoparticles for drug delivery: Enhanced in vitro therapeutic efficacy of doxorubicin and MION conjugates. *J. Mater. Chem. B* **2013**, *1*, 2874–2884. [CrossRef]

33. Polshettiwar, V.; Varma, R.S. Green chemistry by nano-catalysis. *Green Chem.* **2010**, *12*, 743–754. [CrossRef]

34. Adil, S.F.; Assal, M.E.; Khan, M.; Al-Warthan, A.; Siddiqui, M.R.H.; Liz-Marzán, L.M. Biogenic synthesis of metallic nanoparticles and prospects toward green chemistry. *Dalton Trans.* **2015**, *44*, 9709–9717. [CrossRef] [PubMed]

35. Alam, M.N.; Roy, N.; Mandal, D.; Begum, N.A. Green chemistry for nanochemistry: Exploring medicinal plants for the biogenic synthesis of metal NPs with fine-tuned properties. *RSC Adv.* **2013**, *3*, 11935–11956. [CrossRef]

36. Peralta-Videa, J.R.; Huang, Y.; Parsons, J.G.; Zhao, L.; Lopez-Moreno, L.; Hernandez-Viezcas, J.A.; Gardea-Torresdey, J.L. Plant-based green synthesis of metallic nanoparticles: Scientific curiosity or a realistic alternative to chemical synthesis? *Nanotechnol. Environ. Eng.* **2016**, *1*, 4. [CrossRef]

37. Kharissova, O.V.; Dias, H.R.; Kharisov, B.I.; Pérez, B.O.; Pérez, V.M.J. The greener synthesis of nanoparticles. *Trends Biotechnol.* **2013**, *31*, 240–248. [CrossRef] [PubMed]

38. Al-Marri, A.H.; Khan, M.; Shaik, M.R.; Mohri, N.; Adil, S.F.; Kuniyil, M.; Alkhathlan, H.Z.; Al-Warthan, A.; Tremel, W.; Tahir, M.N. Green synthesis of Pd@graphene nanocomposite: Catalyst for the selective oxidation of alcohols. *Arab. J. Chem.* **2016**, *9*, 835–845. [CrossRef]

39. Nasrollahzadeh, M.; Maham, M.; Rostami-Vartooni, A.; Bagherzadeh, M.; Sajadi, S.M. Barberry fruit extract assisted in situ green synthesis of Cu nanoparticles supported on a reduced graphene oxide–Fe₃O₄ nanocomposite as a magnetically separable and reusable catalyst for the *O*-arylation of phenols with aryl halides under ligand-free conditions. *RSC Adv.* **2015**, *5*, 64769–64780.
40. Khan, M.; Khan, M.; Kuniyil, M.; Adil, S.F.; Al-Warthan, A.; Alkhathlan, H.Z.; Tremel, W.; Tahir, M.N.; Siddiqui, M.R.H. Biogenic synthesis of palladium nanoparticles using *Pulicaria glutinosa* extract and their catalytic activity towards the Suzuki coupling reaction. *Dalton Trans.* **2014**, *43*, 9026–9031. [CrossRef] [PubMed]
41. Khan, M.; Al-Marri, A.H.; Khan, M.; Mohri, N.; Adil, S.F.; Al-Warthan, A.; Siddiqui, M.R.H.; Alkhathlan, H.Z.; Berger, R.; Tremel, W. *Pulicaria glutinosa* plant extract: A green and eco-friendly reducing agent for the preparation of highly reduced graphene oxide. *RSC Adv.* **2014**, *4*, 24119–24125. [CrossRef]
42. Movahed, S.K.; Dabiri, M.; Bazgir, A. Palladium nanoparticle decorated high nitrogen-doped graphene with high catalytic activity for Suzuki-Miyaura and Ullmann-type coupling reactions in aqueous media. *Appl. Catal. A Gen.* **2014**, *488*, 265–274. [CrossRef]
43. Polshettiwar, V.; Decottignies, A.; Len, C.; Fihri, A. Suzuki-Miyaura Cross-Coupling reactions in aqueous media: Green and sustainable syntheses of biaryls. *ChemSusChem* **2010**, *3*, 502–522. [CrossRef] [PubMed]
44. Dhakshinamoorthy, A.; Asiri, A.M.; Garcia, H. Metal–organic frameworks catalyzed C–C and C–heteroatom coupling reactions. *Chem. Soc. Rev.* **2015**, *44*, 1922–1947. [CrossRef] [PubMed]
45. Littke, A.F.; Dai, C.; Fu, G.C. Versatile catalysts for the Suzuki Cross-Coupling of arylboronic acids with aryl and vinyl halides and triflates under mild conditions. *J. Am. Chem. Soc.* **2000**, *122*, 4020–4028. [CrossRef]
46. Martin, R.; Stephen, L.B. Palladium-catalyzed Suzuki–Miyaura Cross-Coupling reactions employing dialkylbiaryl phosphine ligands. *Acc. Chem. Res.* **2008**, *41*, 1461–1473. [CrossRef] [PubMed]
47. Handa, S.; Wang, Y.; Gallou, F.; Lipshutz, B.H. Sustainable Fe–ppm Pd nanoparticle catalysis of Suzuki-Miyaura Cross-Couplings in water. *Science* **2015**, *349*, 1087–1091. [CrossRef] [PubMed]
48. Khan, M.; Albalawi, G.H.; Shaik, M.R.; Khan, M.; Adil, S.F.; Kuniyil, M.; Alkhathlan, H.Z.; Al-Warthan, A.; Siddiqui, M.R.H. Miswak mediated green synthesized palladium nanoparticles as effective catalysts for the Suzuki coupling reactions in aqueous media. *J. Saudi Chem. Soc.* **2016**, in press. [CrossRef]

catalysts

MDPI

Article

Preparation of Pd-Diimine@SBA-15 and Its Catalytic Performance for the Suzuki Coupling Reaction

Jiahuan Yu [1,†], An Shen [2,†], Yucai Cao [2,*] and Guanzhong Lu [1,3,*]

1   Research Institute of Applied Catalysis, School of Chemical and Environmental Engineering,
     Shanghai Institute of Technology, Shanghai 200235, China; hotta725@163.com
2   State Key Laboratory of Polyolefins and Catalysis, Shanghai Key Laboratory of Catalysis Technology for
     Polyolefins and Organic Chemistry Division, Shanghai Research Institute of Chemical Industry,
     345 East Yunling Road, Shanghai 200062, China; 15921959294@126.com
3   Key Laboratory for Advanced Materials and Research Institute of Industrial Catalysis,
     East China University of Science and Technology, Shanghai 200237, China
*   Correspondence: caoyc@srici.cn (Y.C.); gzhlu@ecust.edu.cn (G.L.);
     Tel.: +86-139-1872-5152 (Y.C.); +86-21-64252827 (G.L.)
†   These authors contributed equally to this work.

Academic Editor: Ioannis D. Kostas
Received: 1 October 2016; Accepted: 17 November 2016; Published: 24 November 2016

Abstract: A highly efficient and stable Pd-diimine@SBA-15 catalyst was successfully prepared by immobilizing Pd onto diimine-functionalized mesoporous silica SBA-15. With the help of diimine functional groups grafted onto the SBA-15, Pd could be anchored on a support with high dispersion. Pd-diimine@SBA-15 catalyst exhibited excellent catalytic performance for the Suzuki coupling reaction of electronically diverse aryl halides and phenylboronic acid under mild conditions with an ultralow amount of Pd (0.05 mol % Pd). When the catalyst amount was increased, it could catalyze the coupling reaction of chlorinated aromatics with phenylboronic acid. Compared with the catalytic performances of Pd/SBA-15 and Pd-diimine@SiO$_2$ catalysts, the Pd-diimine@SBA-15 catalyst exhibited higher hydrothermal stability and could be repeatedly used four times without a significant decrease of its catalytic activity.

Keywords: Pd-diimine@SBA-15; catalyst preparation; functionalization of SBA-15; Suzuki coupling reaction; manufacture of biphenyl compounds

1. Introduction

Palladium is one of the most versatile and widely applied catalysts for the construction of carbon–carbon bonds in Heck, Suzuki, Sonogashira, or Stille coupling reaction. As a homogeneous catalyst, Pd with phosphine ligands, carbene ligands, and other coordinates have often exhibited a higher turnover number (TON) and better catalytic activity for inactive chloride substances, especially in the Suzuki coupling reaction [1–7]. However, the homogeneous catalytic system may cause greater difficulties, including purification of final products and recycling of the expensive catalysts in large-scale applications. Fortunately, the heterogeneous catalytic system could be an alternative strategy to overcome the above difficulties [8].

The palladium nanoparticle supported on polymeric organic [9–13] or inorganic [14–18] supports is a high-efficient heterogeneous catalyst, and its catalytic activity strongly depends on the size of Pd particles. Some heterogeneous Pd catalysts showed lower catalytic activity due to leaching of Pd species or aggregation of nanoparticles. To disperse and stabilize the supported palladium nanoparticles, the nature of supports, advanced preparation methods, and essential functionalization should be taken into proper consideration [19,20].

Palladium nanoparticles immobilized on inorganic materials such as silica [21–23], carbon [24], and metal oxides [25] could be one promising solution. Owing to highly ordered mesoporous structures with regular channels, larger surface areas and pore size, thicker walls, and higher hydrothermal stability, mesoporous silica SBA-15 is a desirable solid support for Pd nanoparticles [26]. In addition, palladium nanoparticles supported on the channels of SBA-15 with two-dimensional (2D) hexagonal structures can be ideally dispersed, to effectively prevent the aggregation of palladium nanoparticles [27,28].

Moreover, the functional groups (such as amino, thiol, and vinyl groups) can usually be incorporated on the mesoporous walls [29]. Some of functional groups have been shown to act as anchoring sites for palladium species, thus providing additional stabilization. For example, Crudden et al. [30,31] anchored Pd on mercaptopropyl-modified mesoporous SBA-15. Their leaching study illustrated the importance of the thiol ligand to retain Pd on the support surface. An efficient and reusable catalyst SBA-15/CCPy/Pd(II) could also be synthesized by grafting melamine-bearing pyridine groups onto SBA-15 [32]. Undoubtedly, the imine groups possess good coordinating ability in a catalytic reaction, and the diimine groups might intensify its coordinating ability.

The Suzuki coupling reaction is a typical carbon–carbon bond-forming process, and has become one of the most powerful and convenient means in the fields of agrochemistry, pharmaceutical chemistry, materials, and synthetic chemistry [33]. Despite the successful application of the Suzuki coupling reaction in homogeneous catalytic systems, the highly efficient and stable heterogeneous catalysts are still in great demand, and the activities and stabilities of heterogeneous catalysts for the Suzuki coupling reaction require be improvement.

Herein, we describe our design and preparation of the Pd-diimine@SBA-15 catalyst for the Suzuki coupling reaction by immobilizing Pd species onto diimine-functionalized mesoporous SBA-15 silica (Scheme 1), in which Pd ions were dispersed atomically onto the functionalized mesoporous SBA-15. The scope and limitations of Pd-diimine@SBA-15 for Suzuki coupling reactions of electronically diverse aryl halides and phenylboronic acid were also evaluated.

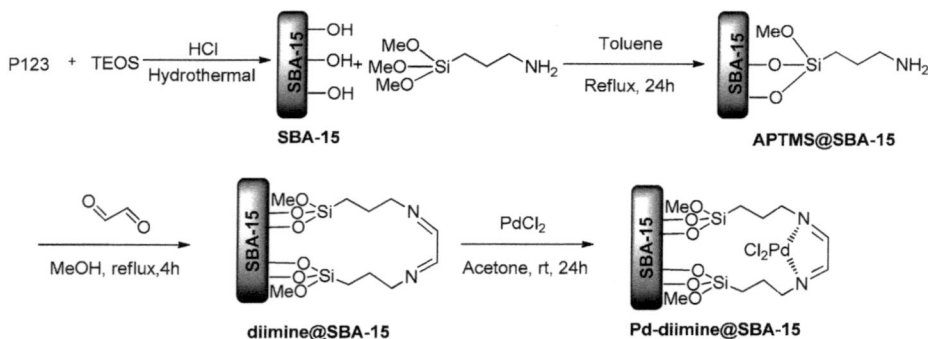

Scheme 1. Schematic diagram of Pd-diimine@SBA-15 fabrication.

2. Results and Discussions

2.1. Structure and Textural Properties of Samples

The Fourier-transform infrared (FT-IR) spectra of SBA-15, 3-aminopropyl trimethoxysilane (APTMS)@SBA-15, diimine@SBA-15, and Pd-diimine@SBA-15 are shown in Figure 1. The peaks at 600–1200 cm^{-1} can be attributed to the vibration of Si–O groups in the mesoporous silica framework. The absorption bands of SBA-15-based materials at 1083, 801, and 466 cm^{-1} are attributed to the Si–O–Si anti-stretching vibration, the Si–O–Si stretching vibration, and the bending vibration of Si–O, respectively [34]. Meanwhile, the band around 956 cm^{-1} should be attributed to the bending vibration of Si–O–H in SBA-15 [35]. In the FT-IR spectrum of APTMS@SBA-15, there are the characteristic bands of –NH_2 at 1645 and 1538 cm^{-1}. For the diimine@SBA-15 sample, the absorption peak at 1656 cm^{-1} is observed and attributed to the characteristic peaks of diimine, due to the presence of the C=N bond. After diimine@SBA-15 coordinated with Pd, this IR absorption peak shifted from 1656 to 1622 cm^{-1}, which is indicative for the formation of a Pd–ligand bond [22,36]. The results above imply the presence of diimine bonded on the surface of SBA-15, and that the molecular structure of these functional moieties can be perfectly retained in the complex of Pd-diimine@SBA-15.

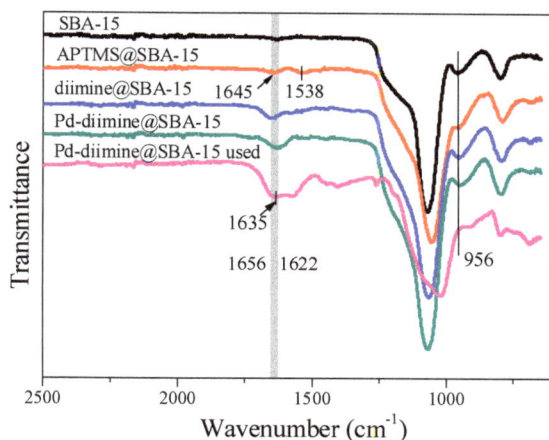

Figure 1. Fourier-transform infrared (FT-IR) spectra of SBA-15, 3-aminopropyl trimethoxysilane (APTMS)@SBA-15, diimine@SBA-15, and Pd-diimine@SBA-15 (fresh and used repeatedly four times).

The Pd-diimine@SBA-15 catalyst before and after reactions were tested by ^{13}C cross-polarization magic-angle spinning (CPMAS) solid-state NMR, and the results are shown in Figure 2. ^{13}C {^{1}H} CPMAS NMR spectrum of fresh Pd-diimine@SBA-15 catalyst exhibits peaks at 10.1 (CH_2Si groups), 22.4 (CH_2 groups), and 50.6 ppm (CH_2 groups in α-position of C=N bond). The other broad peak at 169.8 ppm could be attributed to C=N function groups. After this catalyst was repeatedly used four times, the ^{13}C {^{1}H} CPAMS NMR spectrum of Pd-diimine@SBA-15 catalyst is hardly changed. The additional singlet at 128.5 ppm might correspond to the presence of a biphenyl product. This shows that the diimine ligands remain on the catalyst and were not hydrolyzed after the catalytic reactions.

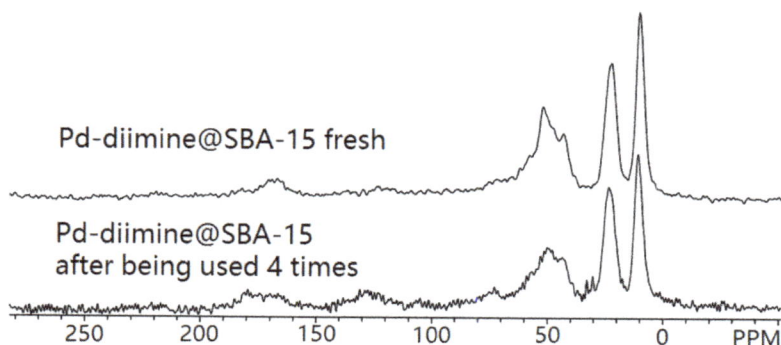

Figure 2. ^{13}C cross-polarization magic-angle spinning (CPMAS) NMR spectra of Pd-diimine@SBA-15 before and after being repeatedly used four times.

The N contents in APTMS@SBA-15 and diimine@SBA-15 were tested by element analysis; the former was 2.05 mmol/g and the latter was 1.92 mmol/g. Thus, the loadings of APTMS and diimine were 2.05 and 0.96 mmol/g, respectively. The Pd content in the Pd-diimine@SBA-15 sample was 5.8 wt %, determined by inductively coupled plasma optical emission spectrometry (ICP-OES). The oxidation state of Pd in the Pd-diimine@SBA-15 catalyst was investigated by the X-ray photoelectron spectroscopy (XPS) technique, and the results are shown in Figure 3. In the fresh Pd-diimine@SBA-15 sample, the peak at $3d_{3/2}$ BE = 337.8 eV (or $3d_{5/2}$ BE = 343.0 eV) corresponds to the Pd(II) ions [18], and the peak at 336.5 eV (or $3d_{5/2}$ BE = 341.7 eV) corresponds to the Pd(II) ions in the diimine-Pd(II) complex [27]. This shows that the Pd(II) ions existed in two kinds of chemical environments, and Pd(II) was not reduced to Pd(0) during the synthesis.

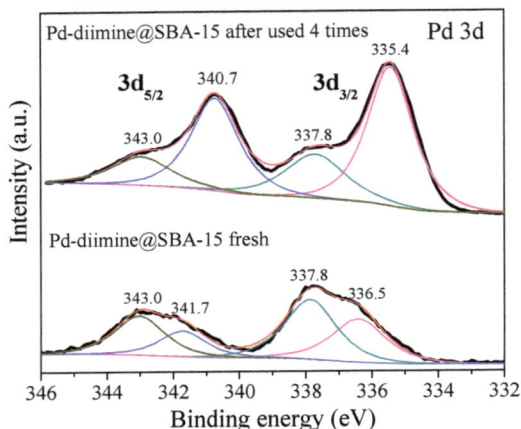

Figure 3. X-ray photoelectron spectroscopy (XPS) spectra of Pd-diimine@SBA-15 (fresh and after being used four times).

The N_2 sorption isotherms and pore-size distribution curves of the samples are displayed in Figure 4, and their textural parameters (Brunauer–Emmett–Teller (BET) surface areas, pore volumes, and pore diameters) are listed in Table 1. As shown in Figure 4, all the isotherms exhibited a typical type IV isotherm with an H_1 hysteresis loop starting from $P/P_0 = 0.6$. This is the characteristic of mesoporous SBA-15 with ordered pore structures [37], which is quite important to disperse and stabilize the supported palladium species. Compared with the BET surface area (641 m^2/g) of SBA-15, the surface area of Pd-diimine@SBA-15 was decreased to 351 m^2/g after SBA-15 was functionalized. Meanwhile, its pore volume was decreased from 0.96 to 0.48 cm^3/g, average pore diameter was decreased from 6.35 to 5.50 nm, and most probable pore diameter was decreased from 10.5 to 6.90 nm. These results are in good agreement with the fact that the surface of mesoporous SBA-15 has been modified successfully by diimine groups. Pd species have entered into the channels of the SBA-15 materials, resulting in the decrease in its pore size.

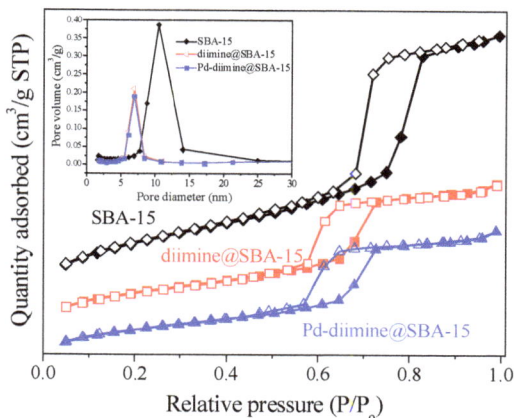

Figure 4. N_2 sorption isotherms and pore-size distribution curves of SBA-15, diimine@SBA-15, and Pd-diimine@SBA-15.

Table 1. Textural parameters of SBA-15, diimine@SBA-15, and Pd-diimine@SBA-15.

Sample	BET Surface Area (m^2/g)	Pore Volume (cm^3/g)	Pore Diameter (nm)	
			Average	Most Probable
SBA-15	641	0.96	6.35	10.5
diimine@SBA-15	398	0.53	5.58	6.90
Pd-diimine@SBA-15	351	0.48	5.50	6.90

The X-ray diffraction (XRD) patterns of SBA-15 and Pd-diimine@SBA-15 are shown in Figure 5. The SBA-15 sample exhibits three peaks at $2\theta = 0.9°$, $1.6°$, and $1.9°$, which correspond to (100), (110), and (200) facets, the characteristic diffraction peaks of mesoporous SBA-15 with a hexagonal geometry [38]. After SBA-15 was modified with diimine, its peak of (110) facet shifted to a lower angle and its intensity was obviously weakened. Compared with SBA-15, the diffraction peak intensities of (110) and (200) facets for Pd-diimine@SBA-15 were greatly decreased. These results are consistent with the fact that Pd-diimine@SBA-15 has smaller surface area, pore volume, and pore size than the SBA-15 sample. The reduced ordering degree of the mesoporous structure is mainly attributed to the occupation of Pd species in the pore channels [39]. Thus, it is reasonable to conclude that Pd species have been dispersed well in the pore channels of SBA-15.

In the XRD patterns of SBA-15, Pd/SBA-15, and Pd-diimine@SBA-15 (Figure 5 inset), a very broad and typical diffraction peak of silica is slightly visible around $2\theta = 25°$. That is, there is no major change in the crystallinity of SBA-15 after diimine-functionalization and Pd immobilization. Furthermore, the diffraction peaks of Pd species cannot be detected, which also shows that the Pd species were immobilized into the pore channels of SBA-15 in the atom dispersion [9], and no crystal Pd species existed in the sample.

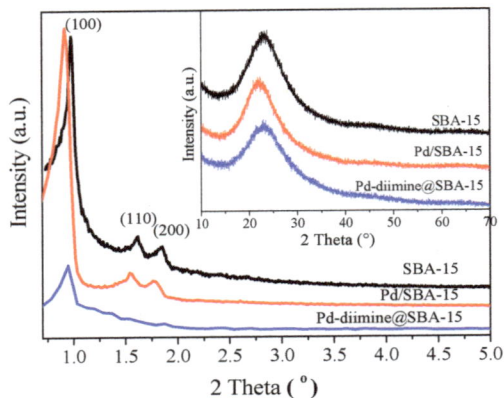

Figure 5. Small-angle X-ray diffraction (SA-XRD) and XRD patterns of SBA-15, Pd/SBA-15, and Pd-diimine@SBA-15.

As shown in the SEM images of Figure 6A, the SBA-15 sample is that with the bagel-shaped particles with relatively uniform sizes. After being functionalized with diimine and Pd, the shape of SBA-15 is unchanged (Figure 6B). To investigate the role of diimine groups in the catalyst, we prepared the Pd/SBA-15 sample as a comparison. The Pd/SBA-15 sample was synthesized by a method similar to Pd-dimine@SBA-15, but without using 3-aminopropyl trimethoxysilane and glyoxal. The TEM image of the Pd-diimine@SBA-15 sample (Figure 6C) reveals that no palladium nanoparticles can be observed in the pores, which shows the Pd^{2+} ions have coordinated with two N atoms in diimine@SBA-15. Unlike the Pd-diimine@SBA-15 sample, more Pd nanoparticles in the pores of SBA-15 can be observed in the TEM image of Pd/SBA-15 (Figure 6D), which shows that the growth of Pd nanoparticles can be inhibited by Pd ions coordinating with diimine anchored on SBA-15.

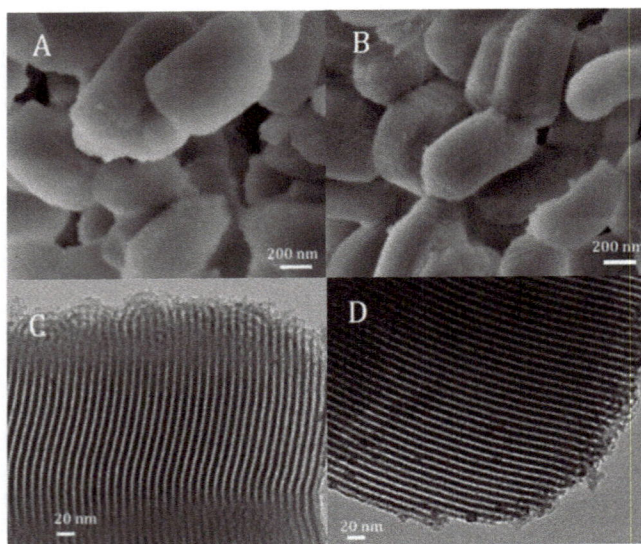

Figure 6. SEM images of (**A**) SBA-15 and (**B**) Pd-diimine@SBA-15, and TEM images of (**C**) Pd-diimine@SBA-15 and (**D**) Pd/SBA-15.

2.2. Catalytic Activities of Pd-Diimine@SBA-15 for Suzuki Coupling Reactions

The catalytic activity of Pd-diimine@SBA-15, SBA-15, and PdCl$_2$ were tested for the Suzuki coupling reaction of para-bromotoluene with phenylboronic acid at 80 °C, and the results are listed in Table 2. The reaction could not occur over SBA-15 (Table 2, entry 1), and 85% yield could be obtained on the PdCl$_2$ homogenous catalyst (Table 2, entry 2). Using the Pd-diimine@SBA-15 catalyst, almost the same yield as that on the PdCl$_2$ catalyst was achieved. The effect of alkalinity in the reaction system on the stability of mesoporous silica SBA-15 and the catalytic activity of the Pd-diimine@SBA-15 catalyst was tested. The results show that stronger alkalinity could increase the product yield. For instance, 91% yield was obtained by adding KOH after 4 h, while only 84% yield could be reached by adding K$_2$CO$_3$ after 24 h (Table 2, entries 3–5).

Table 2. Effect of base on the Suzuki coupling reaction of para-bromotoluene with phenylboronic acid over the Pd-diimine@SBA-15 catalyst at 80 °C (a).

Cat, in N$_2$ / base, solvent, 80 °C

Entry	Catalyst	Base	Time (h)	Conversion (%)	Selectivity (%)	Yield (%) (b)
1	SBA-15	K$_2$CO$_3$	12	C	0	0
2	PdCl$_2$ (0.05 mol% Pd)	K$_2$CO$_3$	12	85	100	85
3	Pd-diimine@SBA-15	K$_2$CO$_3$	24	84	100	84
4	Pd-diimine@SBA-15	K$_3$PO$_4$	12	88	100	88
5	Pd-diimine@SBA-15	KOH	4	91	100	91

(a) Reaction condition: 3 mmol 4-bromotoluene, 3.3 mmol phenylboronic acid, 4.5 mmol base, solvent (i-PrOH/H$_2$O = 3 mL/3 mL), 3 mg catalyst (0.05 mol % Pd); (b) gas chromatography (GC) yield based on 4-bromotoluene.

However, the alkali solution might damage the structure of mesoporous SBA-15 during the Suzuki coupling reaction. The TEM, SEM, and XRD analyses for Pd-diimine@SBA-15 treated with KOH and K$_3$PO$_4$ during the Suzuki coupling reaction were carried out, and their results are shown in Figure 7. The TEM and SEM pictures of Pd-diimine@SBA-15 used in the KOH solution display irregular structures and Pd crystallites (Figure 7A,C). However, Pd-diimine@SBA-15 used in the K$_3$PO$_4$ solution still retained the basic shape of mesoporous SBA-15 (Figure 7D) and abundant wormlike mesoporous structures can be clearly observed (Figure 7B inset). The Pd nanoparticles can be observed in Figure 7A (inset) but not in Figure 7B (inset). As shown in Figure 7, after the Pd-diimine@SBA-15 catalyst was treated in the alkali (KOH or K$_3$PO$_4$) solution, the dispersing diffraction peaks of Pd° at 2θ = ~40° can be observed. Unlike the XRD pattern of the sample treated in the K$_3$PO$_4$ solution, the sample treated in the KOH solution has three weak diffraction peaks at 2θ = ~40°, ~46°, and ~68°, which shows that the sample treated with the KOH solution has the larger size (2.6 nm, estimated by Scherrer equation) of Pd° crystallites than that (2.1 nm) of the sample treated with the K$_3$PO$_4$ solution. These results show that that active Pd species were dispersed quite well on the support, though the mesoporous structure of SBA-15 had been damaged after being treated in the KOH solution. Thus, it is reasonable that the Pd-diimine@SBA-15 catalyst could maintain its catalytic activity due to the high dispersion of Pd.

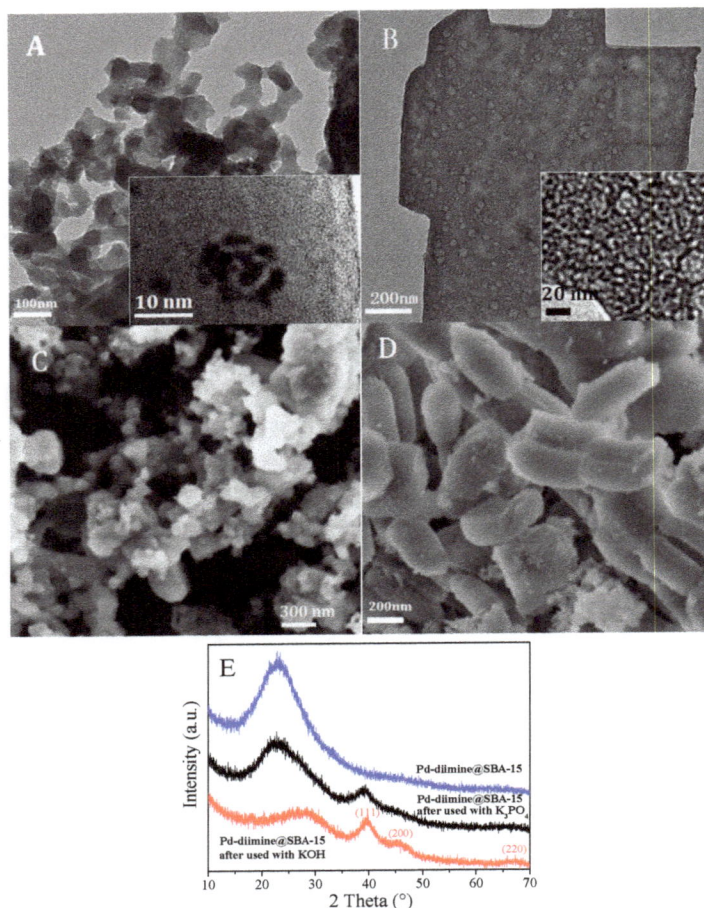

Figure 7. TEM images of Pd-diimine@SBA-15 after use with (**A**) KOH and (**B**) K_3PO_4; SEM image of Pd-diimine@SBA-15 after use with (**C**) KOH and (**D**) K_3PO_4, and (**E**) their XRD patterns.

To evaluate the scope and limitation of the Pd-diimine@SBA-15 catalyst for Suzuki coupling reactions, the reactions of electronically diverse aryl halides and phenylboronic acid were further examined under optimized condition (Table 3). High yields can be obtained for the coupling reactions of bromobenzene with electron-withdrawing and -donating substituents and phenylboronic acid (Table 3, entries 1–8). The better catalytic activity of Pd-diimine@SBA-15 might be contributed to the good dispersion of Pd in the pore channels and high surface area of the SBA-15 support. For instance, the biphenyl yield reached 87% for the reaction of bromobenzene and phenylboronic acid at 80 °C for 12 h (Table 3, entry 1). When an aryl bromide including an electron-withdrawing groups (as bromopentafluorobenzene) was used as a reactant, only 27% yield was obtained at 80 °C for 12 h (Table 3, entry 9) due to the homocoupling of bromopentafluorobenzene, and only 33% selectivity for product. When 2-bromopyridine was used as a reactant, only 13% yield was achieved (Table 3, entry 10). These results indicate that the Pd-diimine@SBA-15 catalyst can easily catalyze the Suzuki coupling reaction of the electron-rich Br-including aromatic compounds with phenylboronic acid. When the amount of catalyst was increased to 60 mg (1 mol % Pd), aryl chlorides became applicable substrates (Table 3, entries 11 and 12); for instance, 20% yield was obtained for the coupling of activated (electron-deficient) 4-nitrochlorobenzene and phenylboronic acid (Table 3, entry 12).

Table 3. Suzuki coupling reactions of various aryl halides (X) and phenylboronic acid over Pd-diimine@SBA-15 catalyst (a).

$$Ar^1X \;+\; Ar^2B(OH)_2 \xrightarrow[\text{K}_3\text{PO}_4(1.5\ \text{equiv}),\ \text{i-PrOH/H}_2\text{O}=1/1]{\text{Pd-diimine@SBA-15, 80}^{\circ}\text{C, in N}_2} Ar^1\text{-}Ar^2$$

X = Br or Cl

Entry	Aryl X	Boronic acid	Time (h)	Conversion (%)	Selectivity (%)	Yield (%) (b)
1	⟨phenyl⟩—Br	⟨phenyl⟩—B(OH)₂	12	87	100	87 (87)
2	Me—⟨phenyl⟩—Br	⟨phenyl⟩—B(OH)₂	12	88	100	88 (84)
3	2-Me-⟨phenyl⟩—Br	⟨phenyl⟩—B(OH)₂	8	91	100	91 (87)
4	Me-⟨phenyl⟩—Br	⟨phenyl⟩—B(OH)₂	8	91	100	91 (88)
5	MeO—⟨phenyl⟩—Br	⟨phenyl⟩—B(OH)₂	6	87	100	87 (79)
6	ᵗBu—⟨phenyl⟩—Br	⟨phenyl⟩—B(OH)₂	10	84	100	84 (77)
7	ᵗBu,ᵗBu-⟨phenyl⟩—Br	⟨phenyl⟩—B(OH)₂	3	88	100	88 (81)
8	F₃C,F₃C-⟨phenyl⟩—Br	⟨phenyl⟩—B(OH)₂	4	99	100	99 (96)
9	F,F,F,F,F-⟨phenyl⟩—Br	⟨phenyl⟩—B(OH)₂	12	83	33	27
10	⟨2-pyridyl⟩—Br	⟨phenyl⟩—B(OH)₂	4	13	100	13
11	2-Me-⟨phenyl⟩—Cl	⟨phenyl⟩—B(OH)₂	4	2	100	2 (c)
12	O₂N—⟨phenyl⟩—Cl	⟨phenyl⟩—B(OH)₂	4	20	100	20 (c)

(a) Reaction condition: 3 mg catalyst (0.05 mol % Pd), ArX (3.0 mmol), ArB(OH)₂ (3.3 mmol, 1.1 equiv), K₃PO₄ (4.5 mmol, 1.5 equiv), i-PrOH/H₂O (3 mL/3 mL); (b) GC yield based on ArX and isolated yield shown in brackets; (c) 60 mg catalyst (1 mol % Pd).

Based on the research results above, the possible mechanism of the Suzuki coupling reaction of aryl halide and phenylboronic acid over Pd-diimine@SBA-15 can be described as Scheme 2 [32,40]. In Scheme 2, there are three main stages in the mechanism of the Suzuki cross-coupling reaction over Pd-diimine@SBA-15: the oxidative addition of substrate, transmetalation, and reductive elimination to produce the final product, in which the catalytic cycle of Pd(0)/Pd(II) occurred.

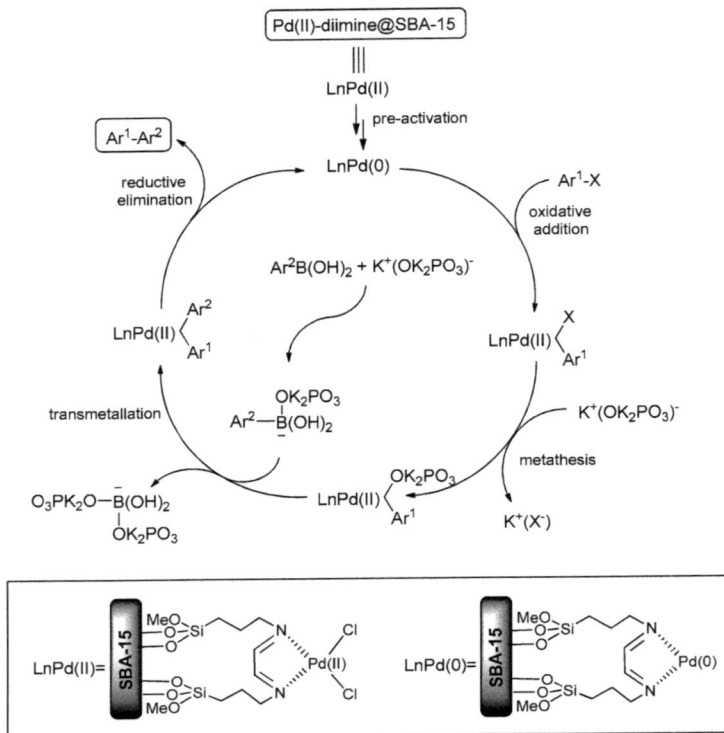

Scheme 2. Possible mechanism of the Suzuki coupling reaction over Pd-diimine@SBA-15.

2.3. Repeated Use of the Pd-Diimine@SBA-15 Catalyst

The catalyst was separated immediately by filtration after the reaction was finished, and was washed by deionized water, ethanol, and diethyl ether successively. After drying in air, the recovered catalyst could be used repeatedly in the Suzuki coupling reaction. To facilitate the recycling of the catalyst, the reaction scale was magnified to 50 mg catalyst. As shown in Table 4 and Figure 8, after the Pd-diimine@SBA-15 catalyst was repeatedly reused four times, its catalytic performance had not decreased based on the yield obtained (first 87%, second 87%, third 86%, and fourth 85%), in which the slight reduction of yield should be ascribed to the loss of the catalyst in the process of recovery and washing. As a comparison, the performances of Pd/SBA-15 and Pd-diimine@SiO$_2$ catalysts after repeated use were also measured. With an increase in recycle times, the catalytic activities of Pd/SBA-15 and Pd-diimine@SiO$_2$ catalysts were reduced gradually, in which the stability of Pd-diimine@SiO$_2$ was higher than that of Pd/SBA-15. For the Pd/SBA-15 catalyst, the yield of first run was 84%, and after the fourth run its yield fell to 19%.

Table 4. Recyclability test of the Pd catalyst for the coupling reaction of bromobenzene and phenylboronic acid at 80 °C for 12 h (a).

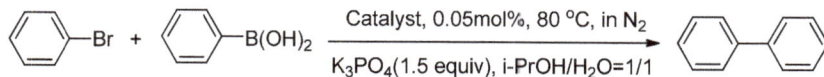

Catalyst, 0.05mol%, 80 °C, in N$_2$

K$_3$PO$_4$(1.5 equiv), i-PrOH/H$_2$O=1/1

Catalyst	Cycle Times	Yield (b) (%)	Pd Leaching (ppm)	Pd leaching/Pd Loading (%)
Pd-diimine@SBA-15 (5.8wt % Pd)	1	87	3.7	1.6
	2	87	3.0	1.3
	3	86	2.2	0.9
	4	85	1.9	0.8
Pd/SBA-15 (4.7wt % Pd)	1	84	23	12.2
	2	73	13	6.9
	3	56	23	12.2
	4	19	9.1	4.8
Pd-diimine@SiO$_2$ (5.2wt % Pd)	1	88	-	-
	2	87	-	-
	3	74	-	-
	4	73	-	-

(a) Reaction condition: 50 mg catalyst (0.05 mol % Pd), ArX (50 mmol), ArB(OH)$_2$ (55 mmol, 1.1 equiv), K$_3$PO$_4$ (75 mmol, 1.5 equiv), i-PrOH/H$_2$O (50 mL/50 mL); (b) GC yield based on 4-bromotoluene.

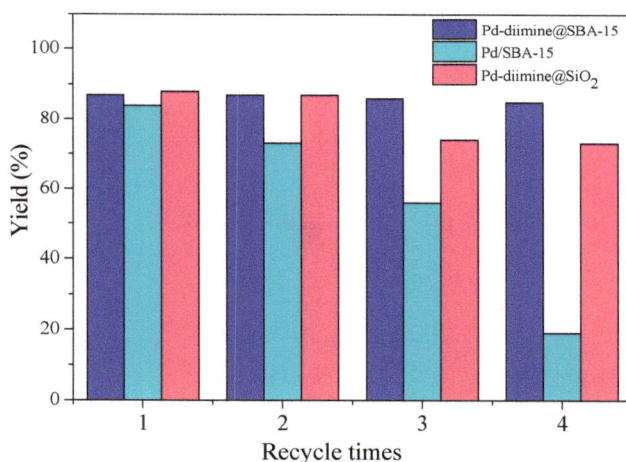

Figure 8. Repeated-use testing of the Pd catalysts for the coupling reaction of bromobenzene and phenylboronic acid at 80 °C for 12 h. (Reaction condition: 50 mg catalyst (0.05 mol% Pd), ArX (50 mmol), ArB(OH)$_2$ (55 mmol, 1.1 equiv), K$_3$PO$_4$ (75 mmol, 1.5 equiv), i-PrOH/H$_2$O (50 mL/50 mL)).

The TEM image, N$_2$ sorption isotherm curve, and the small-angle (SA)-XRD pattern were employed to evaluate the reused catalyst (Figure 9). After Pd-diimine@SBA-15 was repeatedly used four times, the TEM image (Figure 9A) still showed basic shape of SBA-15, which meant that the Pd species were still highly dispersed with some nanoparticles of 2~3 nm. The N$_2$ sorption isotherm and the SA-XRD pattern suggested that the 1D mesoporous structure of Pd-diimine@SBA-15 cannot be observed, but abundant wormlike mesoporous structures still exists obviously (Figure 9C). However, the TEM image of reused Pd/SBA-15 catalyst after four runs (Figure 9B) showed that Pd nanoparticles were aggregated on the surface of the support, and their size was 5~7 nm. In the XPS Pd 3d spectrum of Pd-diimine@SBA-15 after being used four times (Figure 3), the small peak at 337.8 eV (or 3d$_{5/2}$ BE = 343.0 eV) corresponds to the Pd(II) ions, and the larger peak at 335.4 eV (or $3d_{5/2}$ BE = 340.7 eV) corresponds to Pd(0), which shows that a majority of Pd(II) ions were reduced to Pd(0) after the reaction. The FT-IR spectrum of Pd-diimine@SBA-15 after being used four times is shown in Figure 1. The absorption band at 1622~1656 (the top at 1635 cm^{-1}) was observed, which shows the existence of diimine@SBA-15 and Pd-diimine@SBA-15; that is to say, some Pd-diimine@SBA-15 was damaged and turned to supported Pd(0) and diimine@SBA-15 after being used four times. These are evidences accounting for the effect of diimine-functionalization on silica. With the help of diimine groups, Pd species could be anchored on the support tightly with high dispersion and stability, which greatly enhanced the catalytic activity and recyclability [41]. Thus, the mesoporous structure and organic functional groups grafted on the SBA-15 simultaneously improved the catalytic performance of Pd-diimine@SBA-15.

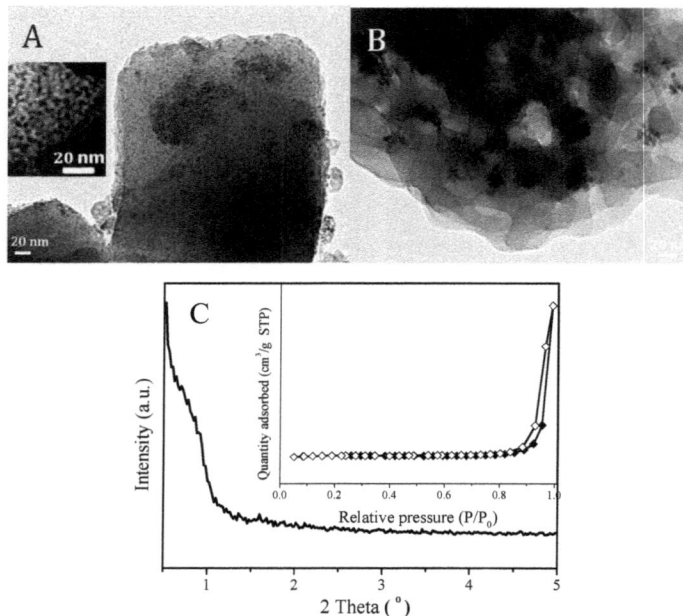

Figure 9. TEM images of (**A**) Pd-diimine@SBA-15 and (**B**) Pd/SBA-15 repeatedly used four times; (**C**) SA-XRD pattern and N_2 sorption isotherm of Pd-diimine@SBA-15 repeatedly used four times.

Palladium leaching was also taken into consideration. After the reaction, the solid catalyst was filtrated, and the filtrate was measured by ICP; the results are shown in Table 4. The amount of leaching Pd in the reaction solution using Pd-diimine@SBA-15 catalyst (1.9–3.7 ppm, Pd leaching/Pd loading of ~1%) was lower than that using the Pd/SBA-15 catalysts (9.1–22.5 ppm, Pd leaching/Pd loading of 5%~12%). This result suggests that diimine groups can anchor palladium ions effectively.

The hot filtration test was done to check the reaction process. The test entails filtering a portion of the reaction solution to remove the catalyst before the reactants are added and the reaction is initiated. The results show that unfiltered portion still proceeded up to 87% yield, while the filtered portion could not react any more. When the reaction mixture after 2 h of the reaction proceeding at 80 °C was filtered with 38% yield, the filtrate was further heated at 80 °C for 2 h and 8% yield was added on the basis of original yield. This shows that Pd leaching can catalyze this coupling reaction. Nevertheless, this test could lead to inaccurate conclusions if fast redeposition of soluble species occurs; if an induction period can be observed, due to an activation of the catalyst, the immobilized species is definitely not the active catalytic species [42].

3. Experimental Section

Triblock copolymer Pluronic P123 ($EO_{20}PO_{70}EO_{20}$) was purchased from Sigma-Aldrich (St. Louis, MO, USA) and 3-aminopropyl trimethoxysilane and others chemicals were obtained from Aladdin (Los Angeles, CA, USA). All chemicals were purchased as reagent grade from commercial suppliers and used without further purification, unless otherwise noted.

3.1. Synthesis of Materials

Synthesis of SBA-15. In a 250 mL round-bottomed flask, Pluronic P123 (4 g) was dissolved in the mixed solution of water (30 mL) and 2 M HCl (120 mL) under stirring at room temperature for 2 h. Then, the mixed solution was heated to 40 °C, and 8.50 g tetraethyl orthosilicate (TEOS) was added under stirring. After this synthesis solution was stirred at 40 °C for 20 h, it was aged at 100 °C for 24 h under static condition. The solid formed was filtered and dried at room temperature overnight. The template in as-synthesized sample was removed by washing twice with the mixed solution of hydrochloric acid and ethanol (1.5 g hydrochloric acid per 200 mL ethanol) under reflux for 6 h and designated as SBA-15 [43,44].

Synthesis of APTMS@SBA-15. In a 100 mL three-necked round-bottomed flask, 3-aminopropyl trimethoxysilane (APTMS) (8 mL, 45 mmol) was added dropwise to a suspension of SBA-15 (5 g) in dry toluene (30 mL) under a N_2 atmosphere. This mixture solution was refluxed for 24 h. After that, the solid was filtered and washed repeatedly with dichloromethane to remove the unreacted starting material, and dried in a vacuum oven at 120 °C for 8 h. The white powder obtained was designated as APTMS@SBA-15 [45].

Synthesis of diimine@SBA-15. In a 250 mL round-bottomed flask, glyoxal (4 mmol) and formic acid (4 drops) was added to the suspension of APTMS@SBA-15 (4 g) in MeOH (100 mL). The mixture solution was refluxed for 4 h. Then, the solid was filtered and washed repeatedly with ice-cold MeOH and dried at room temperature by infrared radiation. The yellow solid obtained was designated as diimine@SBA-15.

Synthesis of Pd-diimine@SBA-15. In a 100 mL round-bottomed flask, diimine@SBA-15 (0.4 g) was added to the mixture solution of $PdCl_2$ (35.4 mg, 0.2 mmol) in acetone (40 mL). After this mixture solution was stirred at room temperature for 24 h, the solid was filtered and washed repeatedly with acetone until the eluate became colorless, and then dried at room temperature by infrared radiation. The light-orange solid obtained was designated as Pd-diimine@SBA-15. The Pd content in the Pd-diimine@SBA-15 catalyst was 5.8 wt % Pd, determined by ICP-OES. The samples above were prepared by the strategy shown in Scheme 1.

Synthesis of Pd/SBA-15. As a comparison sample, the Pd/SBA-15 catalyst was prepared. In a 100 mL round-bottomed flask, SBA-15 (0.4 g) was added to the mixed solution of $PdCl_2$ (35.4 mg, 0.2 mmol) in acetone (40 mL). After this mixed solution was stirred at room temperature for 24 h, the solid was filtered and washed repeatedly with acetone until the eluate became colorless, and then dried at room temperature by infrared radiation. The Pd content in the Pd/SBA-15 catalyst was 4.7 wt % Pd, determined by ICP-OES.

Synthesis of Pd-diimine@SiO$_2$. As a comparison example, the Pd-diimine@SiO$_2$ catalyst was prepared by using the same method as Pd-diimine@SBA-15. In a 100 mL round-bottomed flask, diimine@SiO$_2$ (0.4 g) was added to the mixture solution of $PdCl_2$ (35.4 mg, 0.2 mmol) in acetone (40 mL). After this mixture solution was stirred at room temperature for 24 h, the solid was filtered and washed repeatedly with acetone until the eluate became colorless, and then dried at room temperature by infrared radiation. The Pd content in the Pd-diimine@SiO$_2$ catalyst was 5.2 wt % Pd, determined by ICP-OES.

3.2. Characterization of Sample

The Pd content of the catalyst was determined with an inductively coupled plasma optical emission spectroscopy (ICP-OES, Optima 7000DV, Perkin-Elmmer Co., Waltham, MA, USA) after the sample was dissolved in the HF solution. The FT-IR spectra were recorded on a Nicolet NEXUS 670 FT-IR spectrometer (Thermo Scientific Co., Madison, WI, USA), and the sample to be measured was ground with KBr and pressed into thin wafer. The N_2 adsorption–desorption isotherms of samples were performed on a Micromeritics ASAP 2020 Sorptometer (Micromeritics Instrument Co., Norcross, GA, USA) using static adsorption procedures, and the surface areas of catalysts were calculated by the BET method. The pore-size distribution was calculated by Barrett–Joyner–Halanda (BJH) method. The single-point pore volume (Vp) of sample was estimated based on the amount adsorbed at the relative pressure of $P/P_0 = 0.989$. The X-ray diffractions (XRD) patterns of catalysts were performed on a PANalytical PW3040/60 X'pert PRO diffractometer (PANalytical, Almelo, Netherlands) with Cu $K\alpha$ radiation. Scanning electron microscopy (SEM) images of samples were taken on a Hitachi S4800 scanning electron microscope. Transmission electron microscopy (TEM) images were obtained on a JEOL JEM-2100 microscope (JEOL, Tokyo, Japan), and the sample to be measured was first dispersed in ethanol and then collected on a copper grids covered with carbon film.

3.3. Suzuki Coupling Reactions of Aryl Halides over Supported Pd Catalyst

Aryl halide (3 mmol), arylboronic acid (3.3 mmol), K_3PO_4 (4.5 mmol), catalyst (appropriate quantity), and solvent (6 mL) were added in a 50 mL two-necked round-bottomed flask under a N_2 atmosphere and stirring. After this mixture solution was heated to 80 °C in an oil bath under stirring, the Suzuki coupling reaction occurred at 80 °C for 3–12 h. After the reaction was finished, the catalyst was separated by filtration immediately, and washed by deionized water, ethanol, and diethyl ether, successively. After drying in air, the recovered catalyst could be used repeatedly in the Suzuki coupling reaction. The filtrate was extracted with ethyl acetate, which was uncolored and clear, and no catalyst particles could be observed. The solvent was evaporated and the resultant crude products were purified by silica column chromatography with hexane as eluent. The reactant conversion and product selectivity were determined by gas chromatograph (GC-950, Shanghai, China) equipped with a flame ionization detector (FID) and the method of correction area.

4. Conclusions

In summary, the highly efficient Pd-diimine@SBA-15 catalyst for Suzuki coupling reactions has been successfully prepared by immobilizing Pd species onto diimine-functionalized mesoporous SBA-15, in which the Pd species were atomically dispersed on support. The Pd-diimine@SBA-15 catalyst exhibited high-efficient catalytic performance for the Suzuki coupling reaction of electronically diverse aryl halides and phenylboronic acid under mild conditions with an ultralow amount of Pd (0.05 mol % Pd). High yield of product can be obtained for almost all coupling reactions of bromobenzene with electron-withdrawing and -donating substituents and phenylboronic acid. When the amount of catalyst was increased to 1 mol % Pd, this catalyst can catalyze the coupling reaction of chlorinated aromatics and phenylboronic acid. Compared with the catalytic performances of Pd/SBA-15 and Pd-diimine@SiO_2 catalysts, the Pd-diimine@SBA-15 catalyst exhibited higher hydrothermal stability and could be repeatedly used four times without significant decrease of its catalytic activity. This study might offer a new strategy for synthesis of supported metal catalysts on different types of mesoporous silica.

Acknowledgments: Financial supports from Shanghai Municipal Science and Technology Commission (12NM0504500 and 13XD1421700) are gratefully acknowledged.

Author Contributions: All authors conceived the idea for the work and designed the experiments. G.L. and Y.C. directed the experiments; J.Y. and A.S. performed the experiments; J.Y. and G.L. wrote the manuscript. All authors contributed to the analysis and interpretation of the results.

Conflicts of Interest: The authors declare no conflict of interest.

References

1. Miyaura, N.; Yamada, K.; Suzuki, A. A new stereospecific cross-coupling by the palladium-catalyzed reaction of 1-alkenylboranes with 1-alkenyl or 1-alkynyl halides. *Tetrahedron Lett.* **1979**, *20*, 3437–3440. [CrossRef]
2. Miyaura, N.; Suzuki, A. Palladium-catalyzed cross-coupling reactions of organoboron compounds. *Chem. Rev.* **1995**, *95*, 2457–2483. [CrossRef]
3. Navarro, O.; Kaur, H.; Mahjoor, P.; Nolan, S.P. Cross-coupling and dehalogenation reactions catalyzed by (*N*-heterocyclic carbene) Pd (allyl) Cl complexes. *J. Org. Chem* **2004**, *69*, 3173–3180. [CrossRef] [PubMed]
4. Braga, A.A.C.; Morgon, N.H.; Ujaque, G.; Maseras, F. Computational characterization of the role of the base in the Suzuki-Miyauracross-Coupling Reaction. *J. Am. Chem. Soc.* **2005**, *127*, 9298–9307. [CrossRef] [PubMed]
5. Bonnet, S.; Lutz, M.; Spek, A.L.; Koten, G.V.; KleinGebbink, R.J.M. Bimetallic η6,η1-and PCP-pincer Ruthenium Palladium complexes: Synthesis, structure, and catalytic activity. *Organometallics* **2010**, *29*, 1157–1167. [CrossRef]
6. Sabater, S.; Mata, J.A.; Peris, E. Coordination singularities of a bis(p-xylyl)bis(benzimidazolylidene) ligand and the bis-iridium and rhodium-related complexes. *Organometallics* **2013**, *32*, 1112–1120. [CrossRef]
7. Shen, A.; Ni, C.; Cao, Y.C.; Zhou, H.; Song, G.H.; Ye, X.F. Novel monoligated imine-Pd–NHC complexes: Extremely active pre-catalysts for Suzuki-Miyaura coupling of aryl chlorides. *Tetrahedron Lett.* **2014**, *55*, 3278–3282. [CrossRef]
8. Molnár, Á. Efficient, Selective, and recyclable palladium catalysts in carbon-carbon coupling reactions. *Chem. Rev.* **2011**, *111*, 2251–2320.
9. Narayanan, R.; El-Sayed, M.A. Effect of catalysis on the stability of metallic nanoparticles: Suzuki reaction catalyzed by PVP-palladium nanoparticles. *J. Am. Soc. Chem.* **2003**, *125*, 8340–8347. [CrossRef] [PubMed]
10. Narayanan, R.; El-Sayed, M.A. Effect of colloidal catalysis on the nanoparticle size distribution: Dendrimer-Pd vs. PVP-Pd nanoparticles catalyzing the Suzuki coupling reaction. *J. Phys. Chem. B* **2004**, *108*, 8572–8580. [CrossRef]
11. Bakherad, M.; Keivanloo, A.; Bahramian, B.; Jajarmi, S. Suzuki, Heck, and copper-free Sonogashira reactions catalyzed by 4-amino-5-methyl-3-thio-1, 2, 4-triazole-functionalized polystyrene resin-supported Pd (II) under aerobic conditions in water. *J. Org. Chem.* **2013**, *724*, 206–212. [CrossRef]
12. Kim, J.H.; Kim, J.W.; Shokouhimehr, M.; Lee, Y.S. Polymer-Supported *N*-Heterocyclic Carbene-Palladium Complex for Heterogeneous Suzuki Cross-Coupling Reaction. *J. Org. Chem.* **2005**, *70*, 6714–6720. [CrossRef] [PubMed]
13. Shokouhimehr, M.; Kim, J.H.; Lee, Y.S. Heterogeneous Heck Reaction Catalyzed by Recyclable Polymer-Supported *N*-Heterocyclic Carbene-Palladium Complex. *Synlett* **2006**, *37*, 618–620. [CrossRef]
14. Corma, A.; Garcia, H.; Leyva, A. Catalytic activity of palladium supported on single wall carbon nanotubes compared to palladium supported on activated carbon: Study of the Heck and Suzuki couplings, aerobic alcohol oxidation and selective hydrogenation. *J. Mol. Catal. A Chem.* **2005**, *230*, 97–105. [CrossRef]
15. Jun, S.W.; Shokouhimehr, M.; Lee, D.J.; Jang, Y.; Park, J.; Hyeon, T. One-pot synthesis of magnetically recyclable mesoporous silica supported acid–base catalysts for tandem reactions. *Chem. Commun.* **2013**, *49*, 7821–7823. [CrossRef] [PubMed]
16. Kim, A.; Rafiaei, S.M.; Abolhosseini, S.; Shokouhimehr, M. Palladium Nanocatalysts Confined in Mesoporous Silica for Heterogeneous Reduction of Nitroaromatics. *Energy Environ. Focus* **2015**, *4*, 18–23. [CrossRef]
17. Shokouhimehr, M. Magnetically Separable and Sustainable Nanostructured Catalysts for Heterogeneous Reduction of Nitroaromatics. *Catalysts* **2015**, *5*, 534–560. [CrossRef]
18. Gruttadauria, M.; Liotta, L.F.; Salvo, A.M.P.; Giacalone, F.; Parola, V.L.; Aprile, C.; Noto, R. Multi-Layered, covalently supported ionic liquid phase (mlc-SILP) as highly cross-linked support for recyclable palladium catalysts for the Suzuki reaction in aqueous medium. *Adv. Synth. Catal.* **2011**, *353*, 2119–2130. [CrossRef]
19. Wang, P.Y.; Wang, Z.Y.; Li, J.G.; Bai, Y.X. Preparation, characterization, and catalytic characteristic of Pdnanoparticles encapsulated in mesoporoussilica. *Microp. Mesop. Mater.* **2008**, *116*, 400–405. [CrossRef]
20. Ma, C.Y.; Dou, B.J.; Li, J.J.; Cheng, J.; Hu, Q.; Hao, Z.P.; Qiao, S.Z. Catalytic oxidation of benzyl alcohol on Au or Au-Pdnanoparticles confined in mesoporoussilica. *Appl. Catal. B Environ.* **2009**, *92*, 202–208. [CrossRef]

21. Mubofu, E.B.; Clark, J.H.; Macquarrie, D.J. A novel Suzuki reaction system based on a supported palladium catalyst. *Green Chem.* **2001**, *3*, 23–25. [CrossRef]
22. Paul, S.; Clark, J.H. Highly active and reusable heterogeneous catalyst for the Suzuki reaction: Synthesis of biaryls and polyaryls. *Green Chem.* **2003**, *5*, 635–638. [CrossRef]
23. Sarmah, C.; Sahu, D.; Das, P. Anchoring palladium acetate onto imine-functionalized silica gel through coordinative attachment: An effective recyclable catalyst for the Suzuki-Miyaura reaction in aqueous-isopropanol. *Catal. Today* **2012**, *198*, 197–203. [CrossRef]
24. Mahouche-Chergui, S.; Ledebt, A.; Mammeri, F.; Herbst, F.; Carbonnier, B.; Ben Romdhane, H.; Delamar, M.; Chehimi, M.M. Hairy carbon nanotube@nano-Pd heterostructures: Design, characterization, and application in Suzuki C-C coupling reaction. *Langmuir* **2010**, *26*, 16115–16121. [CrossRef] [PubMed]
25. Kong, G.Q.; Ou, S.; Zou, C.; Wu, C.D. Assembly and post-modification of a metal–organic nanotube for highly efficient catalysis. *J. Am. Chem. Soc.* **2012**, *134*, 19851–19857. [CrossRef] [PubMed]
26. Huang, J.L.; Yin, J.W.; Chai, W.; Liang, C.; Shen, J.; Zhang, F. Multifunctional mesoporous silica supported palladium nanoparticles as efficient and reusable for water-medium Ullmann reaction. *New J. Chem.* **2012**, *36*, 1378–1384. [CrossRef]
27. Yang, H.Q.; Han, X.J.; Li, G.; Wang, Y.W. N-Heterocyclic carbenepalladium complex supported on ionic liquid-modified SBA-16: An efficient and highly recyclable catalyst for the Suzuki and Heck reactions. *Green Chem.* **2009**, *11*, 1184–1193. [CrossRef]
28. Yang, H.Q.; Han, X.J.; Li, G.; Ma, Z.C.; Hao, Y.J. Mesoporous ethane-silicas functionalized with a Bulky N-Heterocyclic carbene for Suzuki-Miyaura coupling of aryl chlorides and benzyl chlorides. *J. Phys. Chem. C* **2010**, *114*, 22221–22229. [CrossRef]
29. Ghorbani-Vaghei, R.; Hemmati, S.; Veisi, H. Pd Immobilized on amidioxime- functionalized mesoporous SBA-15: A novel and highly active heterogeneous catalyst for Suzuki-Miyauracoupling reactions. *J. Mol. Catal. A Chem.* **2014**, *393*, 240–247. [CrossRef]
30. Crudden, C.M.; Sateesh, M.; Lewis, R. Mercaptopropyl-modified mesoporoussilica: A remarkable support for the preparation of a reusable, heterogeneous palladium catalyst for coupling reactions. *J. Am. Chem. Soc.* **2005**, *127*, 10045–10050. [CrossRef] [PubMed]
31. Webb, J.D.; MacQuarrie, S.; McEleney, K.; Crudden, C.M. Mesoporous Silica-Supported Pd Catalysts: An Investigation into Structure, Activity, Leaching and Heterogeneity. *J. Catal.* **2007**, *252*, 97–109. [CrossRef]
32. Veisi, H.; Hamelian, M.; Hemmati, S. Palladium anchored to SBA-15 functionalized with melamine-pyridine groups as anovel and efficient heterogeneous nanocatalyst for Suzuki-Miyaura coupling reactions. *J. Mol. Catal. A Chem.* **2014**, *395*, 25–33. [CrossRef]
33. Alonso, F.; Beletskaya, P.; Yus, M. Non-conventional methodologies for transition-metal catalysed carbon-carbon coupling: A critical overview. Part 2: The Suzuki reaction. *Tetrahedron* **2008**, *64*, 3047–3101. [CrossRef]
34. Bass, J.D.; Solovyov, A.; Pascall, A.J.; Katz, A. Acid-basebifunctional and dielectric outer-sphere effects in heterogeneous catalysis: A comparative investigation of model primary amine catalysts. *J. Am. Chem. Soc.* **2006**, *128*, 3737–3747. [CrossRef] [PubMed]
35. Jiang, Y.; Gao, Q.; Yu, H.; Chen, Y.; Deng, F. Intensively competitive adsorption for heavy metal ions by PAMAM-SBA-15 and EDTA-PAMAM-SBA-15 inorganic–organic hybrid materials. *Microp. Mesop. Mater.* **2007**, *103*, 316–324. [CrossRef]
36. Poel, H.V.D.; Koten, G.V.; Vrieze, K. Novel bonding modes of α-diimines. Synthesis and characterization of [MCl$_2$L(α-diimine)] and [MCl$_2$(α-diimine)n] (M = Pd, Pt; L = phosphine, arsine; n = 1, 2) containing σ,σ-N,N', σ-N, or σ-N ↔ σ-N' bonded α-diimines. *Inorg. Chem.* **1980**, *19*, 1145–1151. [CrossRef]
37. Zhang, G.H.; Wang, P.Y.; Wei, X.F. Palladium supported on functionalized mesoporous silica as an efficient catalyst for Suzuki—Miyauracoupling reaction. *Catal. Lett.* **2013**, *143*, 1188–1194. [CrossRef]
38. Hu, Q.Y.; Hampsey, J.E.; Jiang, N.; Li, C.J.; Lu, Y.F. Surfactant-Templated organic functionalized mesoporous silica with phosphinoligands. *Chem. Mater.* **2005**, *17*, 1561–1569. [CrossRef]
39. Shi, X.J.; Ji, S.F.; Wang, K.; Li, C.Y. Oxidative Dehydrogenation of Ethane with CO$_2$ over Novel Cr/SBA-15/Al$_2$O$_3$/FeCrAl Monolithic Catalysts. *Energy Fuels* **2008**, *22*, 3631–3638. [CrossRef]
40. Navarro, O.; Marion, N.; Oonishi, Y.; Kelly, R.A.; Nolan, S.P. Suzuki-Miyaura, α-Ketone Arylation and Dehalogenation Reactions Catalyzed by a Versatile N-Heterocyclic Carbene-Palladacycle Complex. *J. Org. Chem.* **2006**, *71*, 685–692. [CrossRef] [PubMed]

41. Li, P.; Liu, H.; Yu, Y.; Cao, C.Y.; Song, W.G. One-Pot multistep cascade reactions over multifunctional nanocomposites with Pd nanoparticles supported on amine-modified mesoporous silica. *Chem. Asian J.* **2013**, *8*, 2459–2465. [CrossRef] [PubMed]

42. Taladriz-Blanco, P.; Hervés, P.; Pérez-Juste, J. Supported Pd Nanoparticles for Carbon-Carbon Coupling Reactions. *Top. Catal.* **2013**, *56*, 1154–1170.

43. Zhao, D.Y.; Huo, Q.S.; Feng, J.L.; Chmelka, B.F.; Stucky, G.D. Nonionic triblock and star diblock copolymer and oligomeric surfactant syntheses of highly ordered, hydrothermally stable, mesoporous silica structures. *J. Am. Chem. Soc.* **1998**, *120*, 6024–6036. [CrossRef]

44. Melero, J.A.; Stucky, G.D.; Grieken, R.V.; Morales, G. Direct syntheses of ordered SBA-15 mesoporous materials containing arenesulfonic acid groups. *J. Mater. Chem.* **2002**, *12*, 1664–1670. [CrossRef]

45. Isfahani, A.L.; Mohammadpoor-Baltork, I.; Mirkhani, V.; Khosropour, A.R.; Moghadam, M.; Tangestaninejad, S.; Kia, R. Palladium nanoparticles immobilized on nano-Silica triazinedendritic polymer (Pdnp-nSTDP): An efficient and reusable catalyst for Suzuki-Miyauracross-coupling and Heck reactions. *Adv. Synth. Catal.* **2013**, *355*, 957–972. [CrossRef]

catalysts

[MDPI]

Article

Palladium Nanoparticles Tethered in Amine-Functionalized Hypercrosslinked Organic Tubes as an Efficient Catalyst for Suzuki Coupling in Water

Arindam Modak [1,2], Jing Sun [1], Wenjun Qiu [1] and Xiao Liu [1,*]

[1] Key Laboratory for Green Chemical Technology of Ministry of Education, School of Chemical Engineering and Technology, Tianjin University, Tianjin 300072, China; arindam_modak_2006@yahoo.co.in (A.M.); dugufanying@163.com (J.S.); wingqiu85@163.com (W.Q.)

[2] Indian Association for the Cultivation of Science, Jadavpur, Kolkata-700032, India

* Correspondence: liuxiao71@tju.edu.cn; Tel./Fax: +86-22-2789-0859

Academic Editor: Ioannis D. Kostas

Received: 7 September 2016; Accepted: 14 October 2016; Published: 20 October 2016

Abstract: It is highly desirable to design functionalized supports in heterogeneous catalysis regarding the stabilization of active sites. Pd immobilization in porous polymers and henceforth its application is a rapidly growing field. In virtue of its' scalable synthesis and high stability in reaction conditions, amorphous polymers are considered an excellent scaffold for metal mediated catalysis, but the majority of them are found as either agglomerated particles or composed of rough spheres. Owing to several important applications of hollow organic tubes in diverse research areas, we aimed to utilize them as support for the immobilization of Pd nanoparticles. Pd immobilization in nanoporous polymer tubes shows high activity in Suzuki cross coupling reactions between aryl halides and sodium phenyl trihydroxyborate in water, which deserves environmental merit.

Keywords: porous organic tubes; heterogeneous catalysis; Suzuki coupling in water

1. Introduction

Porous organic polymers (POPs) are emerging as next-generation support materials for heterogeneous catalysis [1–6]. Cheap and readily available organic precursors, tailorable functionality arising from diverse building blocks, are generally advantageous for making high surface area POPs, which are not only used as "support" for metal-mediated catalysis, but also possess significant applications in diverse research, owing to the advantages of densely packed organic groups [7–9]. Therefore, it is customary to mention that the inherent advantages of having organic units in POPs is tremendous, as a support for metal complexes/nanoparticles or as a catalyst because of the virtue of having an electronic interaction between the organic units in POPs and metal nanoparticles. In fact, the main advantage of POPs with its competitive porous support viz. metal organic frameworks (MOFs) [10] and periodic mesoporous organosilica (PMOs) [11,12] is its stability in drastic reaction conditions, which has been immensely highlighted as POPs have shown usability in water medium for catalysis.

Considering adverse environmental impact, using volatile organic solvent for catalysis is a serious concern, which should be replaced by water as an environmentally more demanding. However, reactions using water as the only solvent is frequently encountered as fatal because of the moisture sensitive organic precursors, catalysts, and intermediates. Therefore, it is highly challenging to develop water-compatible catalysts that could be stable, active, and reusable for a number of times without being deteriorated [13,14]. In this context, a Suzuki–Miyaura cross coupling reaction in water as an eco-friendly solvent is meritorious because of its high stability and good solubility of phenyl boronic acid/salt [15]. Pd-catalyzed Suzuki–Miyaura cross coupling reactions have significant importance in organic chemistry, pharmaceuticals research, drug discovery, and development as an elegant tool for C–C bond formation reactions, mainly because of the wide availability of starting materials and relatively mild reaction conditions [16,17]. Although a majority of research has been done in homogeneous conditions using either organic/water as solvent, it suffers from the formation of Pd black as an inactive catalytic species. Again, the use of triphenyl phosphine for stabilization of Pd intermediates under homogeneous conditions often encounters toxicity/poison [18]. In this regard, high surface area heterogeneous catalysts possess tremendous applications in fine chemical industries, owing to its repetitive usability, tailorability of surface modification for stabilization of active sites, and so on. Pd-grafted heterogeneous catalysts comprising porous silica, zeolite, MOFs, and polymers are hereby reported as solid-supported catalysts for Suzuki reaction, but few show considerable recyclability and stability in a water medium [19–21]. On the other hand, microporous POPs are generally formed through precipitation polymerization as agglomerated solid particles/irregular flakes, instead of any well defined nanostructure [22]. Nevertheless, POPs having uniform morphology of hollow tubes/fibers are still scarce; moreover, their application is merely limited to device manufacturing [23]. Recently, Modak and Bhaumik reported interesting microporous polymer tubes (PP-1, PP-2, PP-3; PPs) that are uniform and show high heterogeneity for one-pot tandem catalysis [24]. However, PPs could also be interesting as support for metal nanoparticles' owing to their amine functionality; therefore, we investigated its activity for Suzuki–Miyaura cross coupling reactions. The catalysis in water might be advantageous, since the hydrophobic 4-tritylaniline-based tubes are stable and prevent the active sites from agglomeration and inactivation. Hopefully, this research can provide tremendous scientific interest in the utilization of highly functionalized porous organic tubes/fibers as support of nanoparticles.

2. Results and Discussion

2.1. Synthesis and Characterization of Pd/PP-3 Tubes

The amine-functionalized hypercrosslinked polymer (PP-3) is formed through a one-pot polymerization condensation using 4-Tritylaniline as starting precursor, dimethoxymethane (DMM) as linker and FeCl$_3$ as catalyst/mediator (Scheme 1).

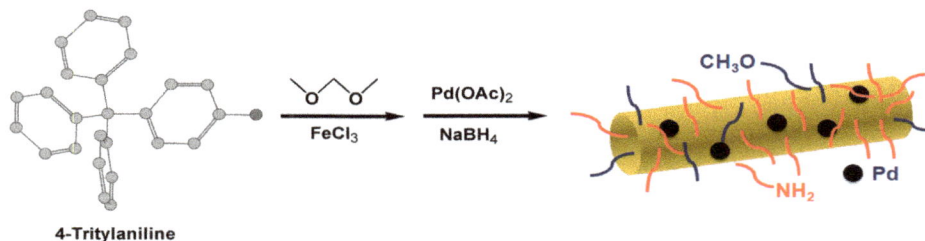

4-Tritylaniline

Scheme 1. Schematic representation for the formation of Pd/PP-3 from a 4-tritylaniline precursor.

The resulting light brown precipitate shows a unique hollow tube shaped morphology as shown in both scanning electron microscopy (SEM) and transmission electron microscopy (TEM) images (Figure 1). Unlike other tube-shaped porous polymers, synthesis of PP-3 is performed in relatively mild conditions utilizing FeCl$_3$ as a cheap and non-harmful chemical [23,25]. The size of the organic tubes are found to be ~5–7 μm in length, and the inner hollow diameter is ~80–100 nm, together with ~300–350 nm is the wall thickness. This thickness accounts for excessive polymerization/non-covalent interaction, which is due to the addition of a large quantity of DMM linker during the synthesis [24].

Pd immobilization to PP-3 was performed by loading with Pd(OAc)$_2$, followed by a reduction with aqueous NaBH$_4$ under mild conditions. It was observed in TEM images (Figure 2) that Pd nanoparticles with dimensions of 5–9 nm were distributed throughout PP-3, which is possibly because of the stabilization by built-in amine sites (Figure 2) [26–29].

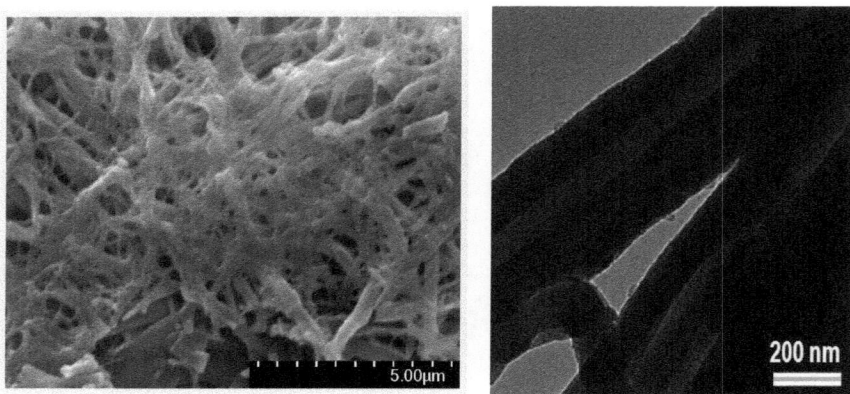

Figure 1. Scanning electron microscopy (SEM) (**left**) and transmission electron microscopy (TEM) (**right**) images of PP-3 tubes (Scale 200 nm for TEM).

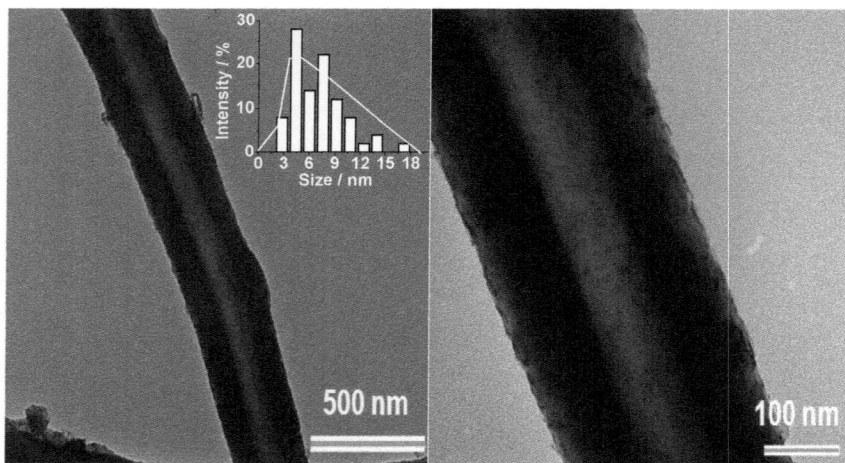

Figure 2. TEM images of Pd/PP-3.

Size distribution of Pd nanoparticles is shown in the inset of Figure 2, suggesting a broad distribution pattern (3–9 nm). Further characterization of Pd/PP-3 was achieved through powder X-ray diffraction (XRD), N$_2$-sorption, X-ray photoelectron spectroscopy (XPS), and inductively coupled plasma atomic emission spectroscopy (ICP) analysis and provides good justification for the presence of Pd nanoparticles in our hypercrosslinked PP-3 tubes.

PXRD of Pd/PP-3 is given in Figure 3a, which shows sharp diffraction at the 40.2°, 43.9°, and 47.3° regions, corresponding to different facets of Pd crystal particles. In comparison with Pd/PP-3, only PP-3 shows a broad peak because of the presence of an amorphous pore wall [30]. Porous properties of Pd/PP-3 were measured from the N_2 sorption isotherm, as shown in Figure 3b, which preferentially suggests Type I characteristics of the isotherm. Like other microporous materials, a high uptake at a low P/P_0 is observed, followed by a flat extrapolation at 0.2–0.8 P/P_0, along with a step uptake at 0.9 P/P_0. The increase of the isotherm at 0.9 P/P_0 depicts the inter-particle mesoporosity. The Brunauer–Emmett–Teller (BET) surface area of Pd/PP-3 was calculated to be 420 $m^2 \cdot g^{-1}$, lower than 530 $m^2 \cdot g^{-1}$ for PP-3 tubes, which is basically due to pore blocking by Pd nanoparticles. In Figure 4a, we provide a survey XPS spectrum of Pd/PP-3, which shows the presence of N, O, and Pd. Figure 4b shows the XPS of Pd 3d electrons, which indicates that Pd sites were reduced to a metallic state in PP-3 with the presence of $NaBH_4$. The confirmation of Pd(0) has been characterized at 334.4–334.8 eV and 339.8–340.2 eV, which were assigned to the Pd $3d_{5/2}$ and Pd $3d_{3/2}$ electrons, respectively [31,32]. Traces of a PdO peak at 341 eV have been observed [33]. All these results, however, demonstrate almost a complete conversion of $Pd(OAc)_2$ into Pd nanoparticles with the utilization of hollow PP-3 tubes.

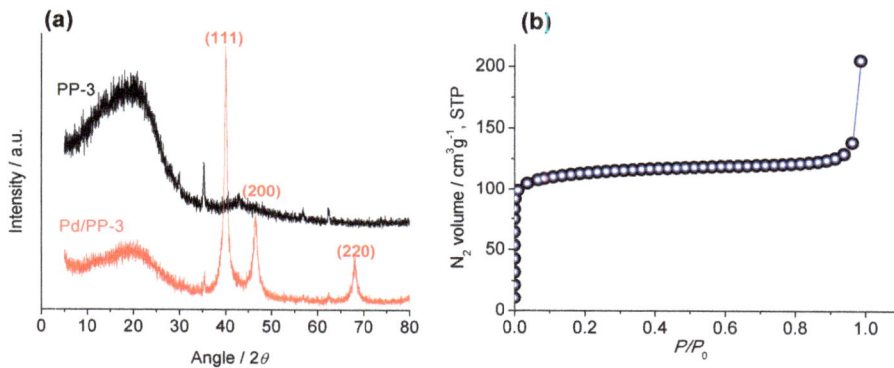

Figure 3. (a) Powder X-ray diffraction pattern; (b) N_2 adsorption-desorption isotherm of Pd/PP-3.

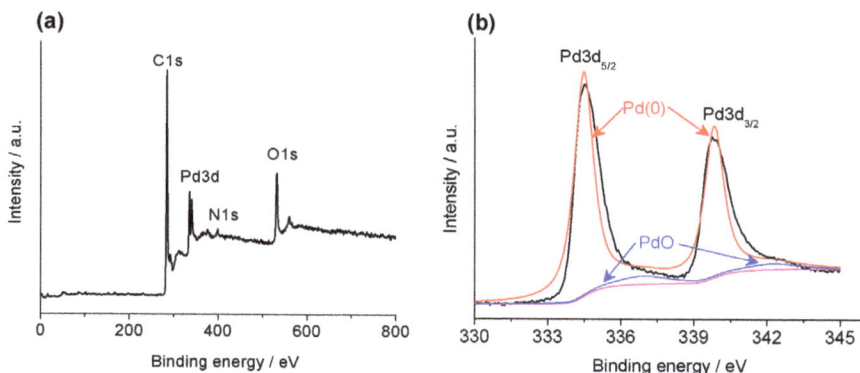

Figure 4. (a) Survey XPS (X-ray photoelectron spectroscopy) spectra of Pd/PP-3; (b) Deconvoluted XPS spectra of Pd 3d electrons in Pd/PP-3, showing the presence of Pd nanoparticles.

2.2. Heterogeneous Catalysis for Suzuki–Miyaura Cross Coupling Reaction in Water

Based on this perspective, the benchmark Suzuki reaction between bromobenzene and sodium phenyltrihydroxyborate was tested in water as a solvent at 100 °C. Because of the better solubility of sodium phenyltrihydroxyborate (PHB) in water compared with phenylboronic acid, the use of PHB for Suzuki coupling without the aid of an additional base was considered advantageous. Owing to the easy preparation and highly stable PHB as organoboron salt, Pd-mediated C–C bond formation reactions are quite meritorious. [34]. Initially, we investigated the scope of Suzuki catalysis in several solvents such as toluene, dicholomethane, dioxane, water, and dimethylformamide (DMF) (Table 1). We observed that the reaction was very sluggish in non-polar, aprotic solvents such as toluene and dicholomethane, partly successful in 1,4-dioxane, and takes place quite efficiently in DMF and DMF/water mixtures (Table 1). Reaction in pure water shows only a 60% yield of biphenyl at 24 h, which considerably improves to >90% at 24 h upon the addition of tetrabutylammonium bromide (TBAB) as phase transfer catalyst.

Table 1. Optimization in reaction conditions for Suzuki coupling reactions between 1 equiv bromobenzene and 1.2 equiv sodium phenyltrihydorxyborate, catalyzed by Pd/PP-3 support.

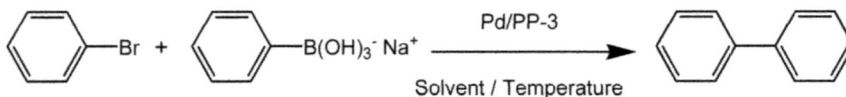

Entry	Solvent	Temperature/°C	Time/h	Yield [a]/%
1	Toluene	100	24	Traces
2	1,4-dioxane	20	24	35
3	DMF	100	24	>90
4	Dichloromethane	20	24	<10
5	Acetone	20	24	50
6	Water/DMF	25	24	>90
7	Water [b]	100	24	60
8	Water [c]	100	24	>90
9	Water [d]	25	24	50

[a] Yield refers to isolated products after purification; [b] Without using TBAB; [c, d] 1 equiv of TBAB.

Using TBAB during the reaction is essential in order to increase the solubility of organic precursors in water. However, upon the addition of TBAB, the reaction can also take place at room temperature conditions, as shown in Table 1. In this regard, we show a temperature-dependent conversion of bromobenzene to biphenyl in Figure 5a, with or without TBAB. In the case of adding TBAB, our model reaction shows much faster kinetics than the reaction devoid of TBAB, demonstrating that phase transfer catalyst is indeed essential for reactions in water. Furthermore, the optimum loading of Pd(OAc)$_2$ and consequently the generation of Pd nanoparticles in PP-3 influences catalytic activity, and we observe the highest catalytic activity/reaction rate (12.5 h^{-1}) with 2 wt % Pd/PP-3, which shows the highest yield of biphenyl from bromobenzene (Figure 5b).

It is worthy to mention that the reaction was carried out in aerial conditions without using degassed water/co-solvent, which could partially lead to the formation of agglomerated Pd/Pd black, as this phenomenon is quite common with several known palladacycles [35]. Therefore, the problem of using Pd/PP-3 in recycling studies was outperformed when choosing degassed water under N$_2$ prior to adding substrates and catalysts, which worked out well in our study.

Next, the efficiency of Pd/PP-3 was tested by encountering a broad substrate scope ranging from electron rich to electron poor aromatic halides, as given in Table 2. Aromatic iodo and/or bromo compounds worked efficiently for the formation of substituted biphenyl; however, for chloro derivates, a partially sluggish reaction was observed. Nonetheless, there are many reports for heterogeneous catalysts such as Suzuki coupling reactions, which can work efficiently with expensive aromatic iodo/bromo derivatives, [36], but very few catalysts can activate aromatic chloro compounds, which are cheap and have more economically feasible applications [31].

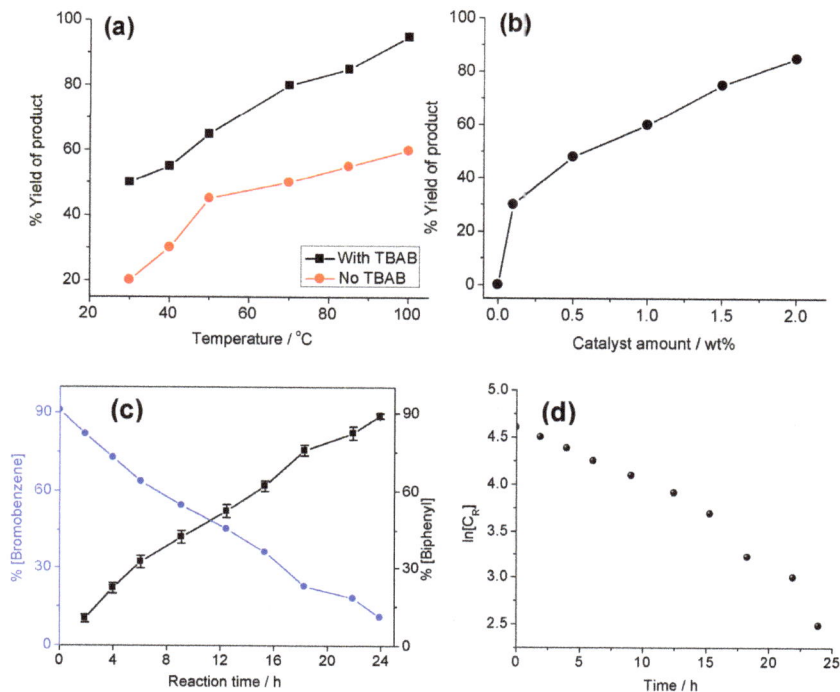

Figure 5. (**a**) Influence of reaction temperature; (**b**) Amount of catalyst on yield of biphenyl from bromobenzene; (**c**) Reaction profile/kinetic plot for bromoenzene conversion to biphenyl; (**d**) Plot of natural logarithm of the remained bromobenzene conc. during reaction with time.

Table 2. Substrate scope of Suzuki coupling reaction by Pd/PP-3.[a]

Entry	Ar-X	Time/h	% Conv. of Ar-X [b]	% Selectivity	Rate [c]/h^{-1}
1	Iodo benzene	20	90	>99	12.5
2	Bromobenzene	24	86	>99	11
3	Chlorobenzene	24	48	>80	6
4	4-bromo nitrobenzene	18	88	>95	13.8
5	4-chloro nitrobenzene	24	40	>78	8
6	4-bromo benzaldehyde	24	75	-	5.3
7	4-bromo acetophenone	20	76	>80	12.5
8	4-iodo acetophenone	15	88	>98	16.6
9	4-iodo anisole	19	85	>98	13
10	4-bromo anisole	24	79	>90	10
11	4-bromo benzonitrile	24	70	>90	9.2
12	4-chloro benzonitrile	24	30	>70	4.7
13	4-bromo toluene	20	70	>85	11
14	4-bromobenzoic acid	18	60	>80	12.8
15	Brombenzene [d, e]	24	70 [d], 55 [e]	>90 [d], >86 [e]	-

[a] 1 mmol aryl halide, 1.2 mmol sodium phenyltrihydroxyborate, 5 mL water, 100 °C, 0.02 g Pd/PP-3 (2 wt % Pd); [b] Yield refer to the isolated product; [c] Reaction rate (mol product per mol of total Pd per time) at 10 min; [d] Reaction was carried out with Pd/MCM-41; [e] Pd/C catalyzed reaction.

It is worthy to mention that the Suzuki coupling reaction of aromatic chloride by Pd/PP-3 showed almost a 40% yield of biphenyl in water, which indicates that Pd/PP-3 also has a potential for utilizing much cheaper chloroaromatics for making biphenyl in a cost-effective way. All products of the Suzuki reaction, given in Table 2, were characterized by ^{1}H NMR and ^{13}C NMR (see Supplementary Materials). It is pertinent to mention that the model Suzuki reaction, when compared with Pd/MCM-41 (2 wt %·Pd; 5–8 nm) and Pd/C (2 wt % Pd; 10–15 nm) in water, shows much lower activity than Pd/PP-3, signifying the importance of amine functionality decorated in hydrophobic PP-3 for the stabilization of active sites, as well as organometallic intermediates. Similarly, Pd nanoparticles stabilized by phenol resign also show comparable catalytic activity as that of Pd/PP-3, which again proves that the surface functional sites could endow a significant binding interaction with guest metal particles for stabilization and improvement in its catalytic property [37,38].

In Figure 5c,d, we present the kinetic aspect of the Pd/PP-3 catalyzed Suzuki reaction, which shows that the reaction is devoid of any long induction time. The reaction rate shows first order dependency with respect to bromobenzene; the rate constant is calculated to be $k_{obs} = 0.081 \pm 0.006$ h^{-1}, and the half life period, $t_{1/2}$, is 6.93 h. The first order reaction rate can be explained on the basis of an initial oxidative addition by bromobenzene with surface-exposed Pd nanoparticles in a slow step, accompanied by rapid elimination of the cross coupled product.

Finally, the heterogeneity of Pd/PP-3 was proved through a hot filtration test. In this regard, we initially separate the catalyst after the 10 h reaction is over. The hot filtrate without any Pd/PP-3 was then investigated for the Suzuki reaction, which did not produce any cross coupling products, signifying that the filtrate is devoid of any Pd-based impurity (Figure 6a). Again, ICP measurements of the filtrate solution did not detect any Pd content, i.e., the Pd amount in the filtrate is beyond the scope of its detection limit. All these results clearly suggest that Pd sites in Pd/PP-3 are stable for liquid-phase catalytic reactions.

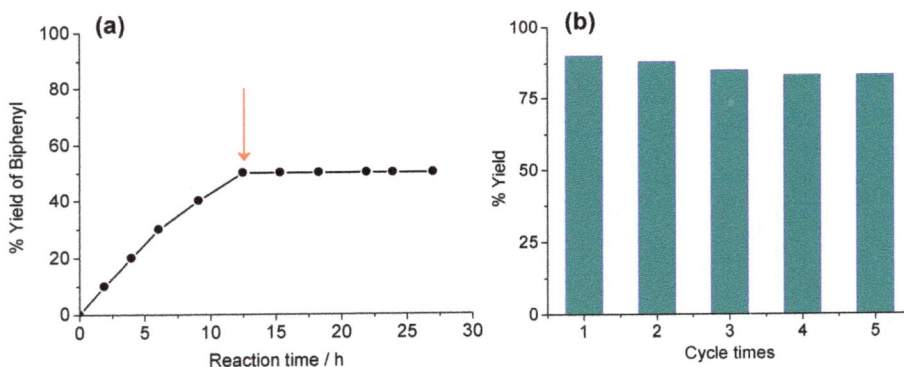

Figure 6. (**a**) Percent yield of biphenyl versus time during a leaching experiment, where the red arrow indicates the time when catalyst was separated and the supernatant was run afterward; (**b**) Recycling study of Pd/PP-3 for Suzuki coupling reaction between iodobenzene and sodium salt of phenyltrihydroxyborate.

Furthermore, we studied poisoning experiments in order to check if the reaction is still catalyzed by leached Pd nanoparticles or not. In this regard, we added 2–3 drops of metallic Hg at the middle of the reaction, and the addition of Hg hardly affects the rate as well as the overall yield of final product, demonstrating that Pd/PP-3 has good heterogeneous characteristics for successive reactions.

Later, while investigating the recycling experiments, we found that Pd/PP-3 retains its catalytic activity for five cycles without much deterioration in its activity (Figure 6b). All these results, however, demonstrate that Pd/PP-3 is a stable heterogeneous catalyst for long-term applications. Since Pd/PP-3 was used to repeat Suzuki coupling experiments in a water medium, there might be another possibility of damage in either morphology or active Pd sites. In this regard, we investigated the stability of reused Pd/PP-3 through TEM and XPS analysis. XPS investigation of the 5th reused Pd/PP-3 shows marginal change in Pd 3d electrons, possibly because of the partial oxidation or continuous exposure in water (Figure 7). On the other hand, TEM analysis of the 3rd reused Pd/PP-3 suggests no collapse of hollow tube morphology and almost no change in the Pd size distribution (3–6 nm particles), as shown in Figure 8.

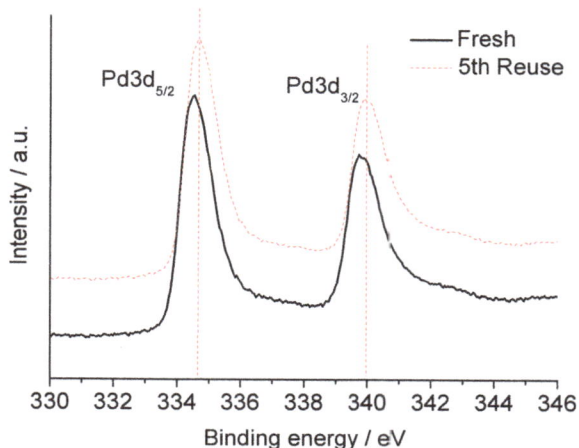

Figure 7. XPS spectra of 5th recycled Pd/PP-3 catalyst.

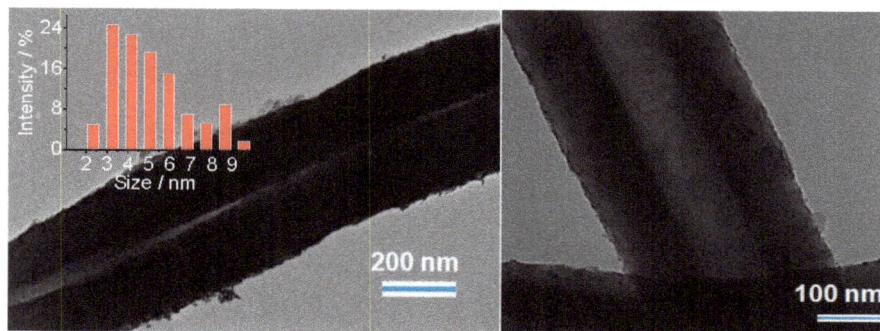

Figure 8. TEM of Pd/PP-3 after third catalytic cycle is over.

3. Materials and Methods

3.1. Instrumentation

Porosity was measured at 77 K using a Quantachrome Instrument (Quantachrome Instrument; Boynton Beach, FL, USA), Autosorb-1, where all samples was degassed at 100 °C for 4 h before the measurement. The Brauner–Emmett–Teller (BET) surface area was calculated over the entire pressure region from ~0.05 to ~0.18 P/P$_0$. Transmission electron microscopy (TEM) was obtained from Hitachi HT-7700 (Nishi-shimbashi, Minato-Ku, Tokyo, Japan) with an acceleration voltage at 100 kV after the samples were dispersed in ethanol via sonication and placed onto an ultrathin carbon film supported on a copper grid. Powder X-ray diffraction (PXRD) was performed with a Rigaku D/Max2500 PC diffractometer with CuKα radiation (λ = 1.5418 Å) over the 2θ range of 5°–70° at a scan speed of 5° per min at room temperature. A Bruker DPX-300 NMR spectrometer was used to measure the ^{1}H and ^{13}C NMR of catalytic products in a liquid state. X-ray photoelectron spectroscopy (XPS) was recorded on a VG ESCALAB MK2 apparatus using AlKα ($h\nu$ = 1486.6 eV) as the excitation light source. Pd content was determined via PLASAM-SPEC-II inductively coupled plasma atomic emission spectrometry (ICP). Particle sizes were determined from Nano Measurer 1.2 software, 2008 by Jie Xu, Fudan University, China.

3.2. Methods

Synthesis of PP-3 was followed in accordance with a previously reported procedure [24]. For the preparation of Pd/PP-3, we initially dissolved 0.010 g of palladium acetate in a 20 mL glass vial containing 10 mL of distilled water and stirred until the solution became yellow. Next, 0.05 g of PP-3 was mixed, and the solution was stirred for 7 h. The mixture was centrifuged several times and washed with H$_2$O and ethanol followed by drying at 60 °C for 12 h, which was denoted as Pd/PP-3. Pd loading was found to be ~2 wt %, as confirmed by ICP analysis.

3.3. General Procedure for Suzuki–Miyaura Coupling Reaction

In the typical synthesis condition, a mixture of Arylhalide (1 mmol), sodium phenyltrihydroxyborate (1.2 mmol), 0.02 g Pd/PP-3 (2 wt % Pd), and TBAB (1 mmol) was stirred in water (5 mL). The mixture was refluxed with magnetic stirring for several hours, as shown in Table 2. After the reaction was complete (monitored by thin layer chromatography technique), the mixture was filtered to separate the catalyst, and the filtrate was subjected to extract (10–20 min) with diethyl ether (20 mL). The combined organic layers were then washed with brine (10 mL), dried by anhydrous Na$_2$SO$_4$, and evaporated. The residue was purified on a short column of silica using petroleum ether as the eluent to afford the desired substituted biphenyl as pure product.

4. Conclusions

Herein, we report on a Suzuki–Miyaura cross coupling reaction in water with Pd-grafted PP-3 as a porous polymer support, which is thought to be promising and environmentally appealing. High catalytic activity, good stability, and reusability of Pd/PP-3 essentially signify its advantages as a heterogeneous catalyst for the liquid phase synthesis of fine chemicals. Moreover, the hollow tube geometry of PP-3 is suitable for anchoring Pd sites by exploiting both the inner and outer hollow spaces, providing an enormous stabilization of Pd and preventing the formation of inactive Pd clusters. Owing to the importance of particle morphologies in solid catalytic research, we believe our efforts could motivate others' for developing organic nanotubes/carbon tubes and utilization of its hollow space for developing nanoreactors, which could ultimately lead to a sustainable and environmentally benign solid catalysis research.

Supplementary Materials: The following are available online at www.mdpi.com/2073-4344/6/10/161/s1.

Acknowledgments: We acknowledge the National Natural Science Foundation of China (No. 21276191), the Specialized Research Fund for the Doctoral Program of Higher Education of China (No. 20120032120083), the Natural Science Foundation of Tianjin, China (No. 16JCQNJC06200) for financial support.

Author Contributions: A.M. designed scheme and experiments; A.M., J.S. & W.Q. performed all experiments and collected data; X.L. & A.M. analyzed the data and finally wrote the manuscript.

Conflicts of Interest: The authors declare no conflict of interest.

References

1. Kaur, P.; Hupp, J.T.; Nguyen, S.T. Porous Organic Polymers in Catalysis: Opportunities and Challenges. *ACS Catal.* **2011**, *1*, 819–835. [CrossRef]
2. Modak, A.; Pramanik, M.; Inagaki, S.; Bhaumik, A. A triazine functionalized porous organic polymer: excellent CO_2 storage material and support for designing Pd nanocatalyst for C–C cross-coupling reactions. *J. Mater. Chem. A* **2014**, *2*, 11642–11650. [CrossRef]
3. Sun, Q.; Dai, Z.; Meng, X.; Wang, L.; Xiao, F.S. Task-Specific Design of Porous Polymer Heterogeneous Catalysts beyond Homogeneous Counterparts. *ACS Catal.* **2015**, *5*, 4556–4567. [CrossRef]
4. Zhang, P.; Weng, Z.; Guo, J.; Wang, C. Solution-Dispersible, Colloidal, Conjugated Porous Polymer Networks with Entrapped Palladium Nanocrystals for Heterogeneous Catalysis of the Suzuki–Miyaura Coupling Reaction. *Chem. Mater.* **2011**, *23*, 5243–5249. [CrossRef]
5. Modak, A.; Mondal, J.; Bhaumik, A. Highly Porous Organic Polymer containing Free -CO_2H Groups: A Convenient Carbocatalyst for Indole C-H Activation at Room Temperature. *ChemCatChem* **2013**, *5*, 1749–1753. [CrossRef]
6. Zhang, Y.; Riduan, S.N. Functional porous organic polymers for heterogeneous catalysis. *Chem. Soc. Rev.* **2012**, *41*, 2083–2094. [CrossRef] [PubMed]
7. Sprick, R.S.; Jiang, J.X.; Bonillo, B.; Ren, S.; Ratvijitvech, T.; Guiglion, P.; Zwijnenburg, M.A.; Adams, D.J.; Cooper, A.I. Tunable Organic Photocatalysts for Visible-Light-Driven Hydrogen Evolution. *J. Am. Chem. Soc.* **2015**, *137*, 3265. [CrossRef] [PubMed]
8. Modak, A.; Yamanaka, K.I.; Goto, Y.; Inagaki, S. Photocatalytic H_2 Evolution by Pt-Loaded 9,9'-Spirobifluorene-Based Conjugated Microporous Polymers under Visible-Light Irradiation. *Bull. Chem. Soc. Jpn.* **2016**, *89*, 887–891. [CrossRef]
9. Liras, M.; Marta Iglesias, M.; Félix Sánchez, F. Conjugated Microporous Polymers Incorporating BODIPY Moieties as Light-Emitting Materials and Recyclable Visible-Light Photocatalysts. *Macromolecules* **2016**, *49*, 1666–1673. [CrossRef]
10. Zhou, H.C.J.; Kitagawa, S. Metal-Organic Frameworks (MOFs). *Chem. Soc. Rev.* **2014**, *43*, 5415–5418. [CrossRef] [PubMed]
11. Liu, X.; Maegawa, Y.; Goto, Y.; Hara, K.; Inagaki, S. Heterogeneous Catalysis for Water Oxidation by an Iridium Complex Immobilized on Bipyridine-Periodic Mesoporous Organosilica. *Angew. Chem. Int. Ed.* **2016**, *55*, 7943–7947. [CrossRef] [PubMed]
12. Modak, A.; Mondal, J.; Aswal, V.K.; Bhaumik, A. A new periodic mesoporous organosilica containing diimine-phloroglucinol, Pd(II)-grafting and its excellent catalytic activity and trans-selectivity in C–C coupling reactions. *J. Mater. Chem.* **2010**, *20*, 8099–8106. [CrossRef]
13. Jagtap, S.; Deshpande, R. True water soluble palladium-catalyzed Heck reactions in aqueous–organic biphasic media. *Tetrahedron Lett.* **2013**, *54*, 2733–2736. [CrossRef]
14. Gilbert, L.; Mercier, C. Solvent effects in heterogeneous catalysis: Application to the synthesis of fine chemicals. *Stud. Surface Sci. Catal.* **1993**, *78*, 51–66.
15. Shaughnessy, K.H.; DeVasher, R.B. Palladium-Catalyzed Cross-Coupling in Aqueous Media: Recent Progress and Current Applications. *Curr. Org. Chem.* **2005**, *9*, 595. [CrossRef]
16. Selander, N.; Szabó, K.J. Catalysis by Palladium Pincer Complexes. *Chem. Rev.* **2011**, *111*, 2048–2076. [CrossRef] [PubMed]
17. Miyaura, N.M.; Suzuki, A. Palladium-Catalyzed Cross-Coupling Reactions of Organoboron Compounds. *Chem. Rev.* **1995**, *95*, 2457–2483. [CrossRef]
18. Masjedi, M.; Demiralp, T.; Özkar, S. Testing catalytic activity of ruthenium(III) acetylacetonate in the presence of trialkylphosphite or trialkylphosphine in hydrogen generation from the hydrolysis of sodium borohydride. *J. Mol. Catal. A Chem.* **2009**, *310*, 59–63. [CrossRef]
19. Chen, L.; Gao, Z.; Li, Y. Immobilization of Pd(II) on MOFs as a highly active heterogeneous catalyst for Suzuki–Miyaura and Ullmann-type coupling reactions. *Catal. Today* **2015**, *245*, 122–128. [CrossRef]
20. Li, B.; Guan, Z.; Wang, W.; Yang, X.; Hu, J.; Tan, B.; Li, T. Highly Dispersed Pd Catalyst Locked in Knitting Aryl Network Polymers for Suzuki–Miyaura Coupling Reactions of Aryl Chlorides in Aqueous Media. *Adv. Mater.* **2012**, *24*, 3390–3395. [CrossRef] [PubMed]

21. Kumbhar, A.; Kamble, S.; Mane, A.; Jha, R.; Salunkhe, R. Modified zeolite immobilized palladium for ligand-free Suzuki–Miyaura cross-coupling reaction. *J. Organomet. Chem.* **2013**, *738*, 29–34. [CrossRef]

22. Gokmen, M.T.; Prez, F.E.D. Porous polymer particles—A comprehensive guide to synthesis, characterization, functionalization and applications. *Prog. Polym. Sci.* **2012**, *37*, 365–405. [CrossRef]

23. Feng, X.; Liang, Y.; Zhi, L.; Thomas, A.; Wu, D.; Lieberwirth, I.; Kolb, U.; Mullen, K. Synthesis of Microporous Carbon Nanofibers and Nanotubes from Conjugated Polymer Network and Evaluation in Electrochemical Capacitor. *Adv. Funct. Mater.* **2009**, *19*, 2125–2129. [CrossRef]

24. Modak, A.; Bhaumik, A. High-throughput Acid-Base Tandem Organocatalysis over Hollow Tube-Shaped Porous Polymers and Carbons. *ChemistrySelect* **2016**, *6*, 1192–1200. [CrossRef]

25. Kang, N.; Hon Park, J.; Choi, J.; Jin, J.; Chun, J.; Jung, I.G.; Jeong, J.; Geun Park, J.; Lee, S.M.; Kim, H.J.; et al. Nanoparticulate Iron Oxide Tubes from Microporous Organic Nanotubes as Stable Anode Materials for Lithium Ion Batteries. *Angew. Chem. Int. Ed.* **2012**, *51*, 6626–6630. [CrossRef] [PubMed]

26. Mukhopadhyay, K.; Phadtare, S.; Vinod, V.P.; Kumar, A.; Rao, M.; Chaudhari, R.V.; Sastry, M. Gold Nanoparticles Assembled on Amine-Functionalized Na-Y Zeolite: A Biocompatible Surface for Enzyme immobilizaiton. *Langmuir* **2003**, *19*, 3858–3863. [CrossRef]

27. Mandal, S.; Roy, D.; Chaudhari, R.V.; Sastry, M. Pt and Pd Nanoparticles Immobilized on Amine-Functionalized Zeolite: Excellent Catalysts for Hydrogenation and Heck Reactions. *Chem. Mater.* **2004**, *16*, 3714–3724. [CrossRef]

28. Huang, Y.; Zheng, Z.; Liu, T.; Lü, J.; Lin, Z.; Li, H.; Cao, R. Palladium nanoparticles supported on amino functionalized metal-organic frameworks as highly active catalysts for the Suzuki–Miyaura cross coupling reactions. *Catal. Commun.* **2011**, *14*, 27–31. [CrossRef]

29. Dai, B.; Wen, B.; Zhu, M.; Kanga, L.; Yu, F. Nickel catalysts supported on amino functionalized MCM-41 for syngas methanation. *RSC Adv.* **2016**, *6*, 66957–66962. [CrossRef]

30. Zhu, Q.L.; Tsumori, N.; Xu, Q. Immobilizing Extremely Catalytically Active Palladium Nanoparticles to Carbon Nanospheres: A Weakly-Capping Growth Approach. *J. Am. Chem. Soc.* **2015**, *137*, 11743–11748. [CrossRef] [PubMed]

31. Huang, N.; Xu, Y.; Jiang, D. High-performance heterogeneous catalysis with surface-exposed stable metal nanoparticles. *Sci. Rep.* **2014**, *4*, 7228. [CrossRef] [PubMed]

32. Ohtaka, A.; Okagaki, T.; Hamasaka, G.; Uozumi, Y.; Shinagawa, T.; Shimomura, O.; Nomura, R. Application of "Boomerang" Linear Polystyrene-Stabilized Pd Nanoparticles to a Series of C–C Coupling Reactions in Water. *Catalysts* **2015**, *5*, 106–118. [CrossRef]

33. Wang, C.; Yang, F.; Yang, W.; Ren, L.; Zhang, Y.; Jia, X.; Zhang, L.; Li, Y.F. PdO nanoparticles enhancing the catalytic activity of Pd/carbon nanotubes for 4-nitrophenol reduction. *RSC Adv.* **2015**, *5*, 27526–27532. [CrossRef]

34. Cammidge, A.N.; Goddard, V.H.M.; Gopee, H.; Harrison, N.L.; Hughes, D.L.; Schubert, C.J.; Sutton, B.M.; Watts, G.L.; Whitehead, A.J. Aryl Trihydroxyborates: Easily Isolated Discrete Species Convenient for Direct Application in Coupling Reactions. *Org. Lett.* **2006**, *8*, 4071. [CrossRef] [PubMed]

35. Phan, T.S.; Van Der Sluys, M.; Jones, C.W. On the Nature of the Active Species in Palladium Catalyzed Mizoroki–Heck and Suzuki–Miyaura Couplings—Homogeneous or Heterogeneous Catalysis, A Critical Review. *Adv. Synth. Catal.* **2006**, *348*, 609. [CrossRef]

36. Rostamnia, S.; Alamgholiloo, H.; Liu, X. Pd-grafted open metal site copper-benzene-1,4-dicarboxylate metal organic frameworks (Cu-BDC MOF's) as promising interfacial catalysts for sustainable Suzuki coupling. *J. Colloid Interface Sci.* **2016**, *469*, 310–317. [CrossRef] [PubMed]

37. Xu, T.Y.; Zhang, Q.F.; Yang, H.F.; Li, X.N.; Wang, J.G. Role of Phenolic Groups in the Stabilization of Palladium Nanoparticles. *Ind. Eng. Chem. Res.* **2013**, *52*, 9783–9789. [CrossRef]

38. Nishiwaki, N.; Hamada, S.; Watanabe, T.; Hirao, S.; Jun Sawayama, J.; Asahara, H.; Saigo, K.; Kamata, T.; Funabashi, M. Development of a new palladium catalyst supported on phenolic resin. *RSC Adv.* **2015**, *5*, 4463–4467. [CrossRef]

catalysts

Article

Synthesis, Structural Characterization and Catalytic Evaluation of Anionic Phosphinoferrocene Amidosulfonate Ligands

Jiří Schulz, Filip Horký, Ivana Císařová and Petr Štěpnička *

Department of Inorganic Chemistry, Faculty of Science, Charles University, Hlavova 2030, 128 40 Prague, Czech Republic; jiri.schulz@natur.cuni.cz (J.S.); Filip.blud@seznam.cz (F.H.); ivana.cisarova@natur.cuni.cz (I.C.)
* Correspondence: petr.stepnicka@natur.cuni.cz; Tel.: +420-221-951-260

Academic Editor: Ioannis D. Kostas
Received: 20 April 2017; Accepted: 19 May 2017; Published: 24 May 2017

Abstract: Triethylammonium salts of phosphinoferrocene amidosulfonates with electron-rich dialkyphosphino substituents, $R_2PfcCONHCH_2SO_3(HNEt_3)$ (**4a–c**), where fc = ferrocene-1,1'-diyl, and R = i-Pr (**a**), cyclohexyl (Cy; **b**), and t-butyl (**c**), were synthesized from the corresponding phosphinocarboxylic acids-borane adducts, $R_2PfcCO_2H \cdot BH_3$ (**1a–c**), via esters $R_2PfcCO_2C_6F_5 \cdot BH_3$ (**2a–c**) and adducts $R_2PfcCONHCH_2SO_3(HNEt_3) \cdot BH_3$ (**3a–c**). Compound **4b** was shown to react with $[Pd(\mu\text{-}Cl)(\eta\text{-}C_3H_5)]_2$ and $AgClO_4$ to afford the zwitterionic complex $[Pd(\eta^3\text{-}C_3H_5)(Cy_2PfcCONHCH_2SO_3\text{-}\kappa^2O,P)]$ (**5b**), in which the amidosulfonate ligand coordinates as a chelating donor making use of its phosphine moiety and amide oxygen. The structures of **3b**·CH_2Cl_2, **4b** and **5b**·CH_2Cl_2 were determined by single-crystal X-ray diffraction analysis. Compounds **4a–c** and their known diphenylphosphino analogue, $Ph_2PfcCONHCH_2SO_3(HNEt_3)$ (**4d**), were studied as supporting ligands in Pd-catalyzed cyanation of aryl bromides with $K_4[Fe(CN)_6]$ and in Suzuki–Miyaura biaryl cross-coupling performed in aqueous reaction media under mild reaction conditions. In the former reaction, the best results were achieved with a catalyst generated from $[PdCl_2(cod)]$ (cod = $\eta^2\text{:}\eta^2$-cyclooocta-1,5-diene) and 2 equiv. of the least electron-rich ligand **4d** in dioxane–water as a solvent. In contrast, the biaryl coupling was advantageously performed with a catalyst resulting from palladium(II) acetate and ligand **4a** (1 equiv.) in the same solvent.

Keywords: ferrocene ligands; phosphines; sulfonates; aqueous catalysis; Suzuki–Miyaura reaction; cyanation; palladium

1. Introduction

Sulfonation of phosphines represents an efficient and time-tested approach toward hydrophilic ligands. Introduction of a single sulfonate moiety into a molecule of a phosphine donor usually increases polarity and hydrophilicity to such an extent that the resulting derivatives can be used as donors to prepare highly hydrophilic coordination compounds as well as supporting ligands for transition metal-catalyzed reactions performed in aqueous and water-organic biphase media [1–3]. Phosphinosulfonate Donors including the archetypal and widely used trisodium tris(sulfonatophenyl)phosphine (TPPTS) [4–6] are typically obtained by direct sulfonation of the parent phosphines, which somewhat limits the scope of the accessible compounds, mainly due to problems associated with a high reactivity of the sulfonation agents and their compatibility with other functional groups.

We have recently demonstrated that phosphinosulfonate donors can also be accessed via amidation of phosphinocarboxylic acids with ω-aminosulfonic acids [7]. This synthetic approach, which inherently leads to a simultaneous incorporation of the polar amide linking group, is modular and allows the synthesis of libraries of donors with modified structures and also eliminates some limitations of the direct-sulfonation approach. So far, we have utilized this method to prepare a series of phosphinoferrocene amidosulfonates differing in the length of the aliphatic spacer between the amide and sulfonate groups, compounds **A** (*n* = 1–3) [7], and a pair of isomeric phosphinobiphenyl donors **B** and **C** [8,9] (Scheme 1). These compounds were successfully tested as supporting ligands in Pd-catalyzed cyanation of aryl bromides and in Suzuki–Miyaura cross-coupling [7–9].

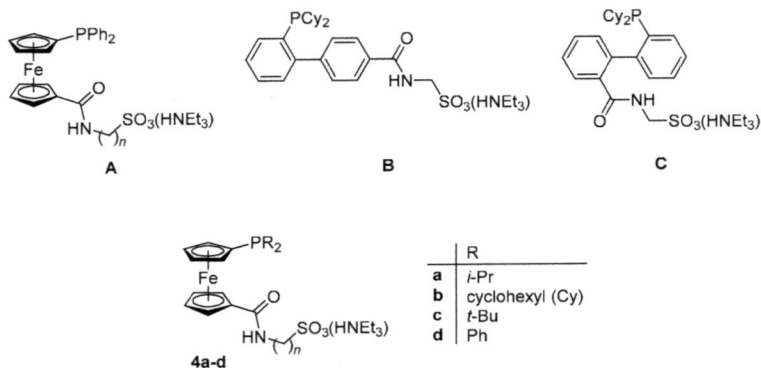

Scheme 1. Examples of phosphino-amidosulfonate donors.

Having recently established a reliable synthetic route to new (dialkylphosphino)ferrocenecarboxylic acids in their stable, phosphine-protected form (viz. compounds **1a–c** [10], we decided to use these compounds further in the preparation of phosphinoferrocene amidosulfonate donors with varied, electron-rich phosphine substituents, compounds **4a–c** (Scheme 1). The synthesis, structural characterization and an evaluation of these compounds as supporting ligands in Pd-catalyzed cyanation of aryl bromides to the corresponding nitriles [11,12] and Suzuki–Miyaura cross-coupling of aryl bromides with arylboronic acids [13–17] performed in aqueous reaction media are reported in this contribution.

2. Results and Discussion

2.1. Synthesis of the Ligands

Compounds **1a–c** were prepared in analogy to the amidosulfonate donors **A**–**C** mentioned in the Introduction except that their oxidation-sensitive phosphine moieties were protected in the form of BH$_3$ adducts [18] during the synthesis (Scheme 2). In the first step, phosphinocarboxylic acid-borane adducts **1a–c** were reacted with pentafluorophenol in the presence of 1-ethyl–3-[3-(dimethylamino)propyl]carbodiimide (EDC) and 4-(dimethylamino)pyridine (DMAP) to afford the corresponding esters **2a–c**. These active esters were, in turn, reacted with aminomethanesulfonic acid in a mixture of *N,N*-dimethylformamide (DMF) and triethylamine to give the target ligands in their P-protected form, compound **3a–c**. In the last step, the borane protecting group was removed by heating these borane adducts in neat morpholine (65 °C/16 h) and the resulting free phosphines **4a–c** were isolated by column chromatography using a NEt$_3$-pretreated silica gel column to avoid cation exchange. The target ligands **4a–c** as well as all reaction intermediates were characterized by NMR and IR spectroscopy, electrospray ionization (ESI) mass spectrometry and by elemental analysis.

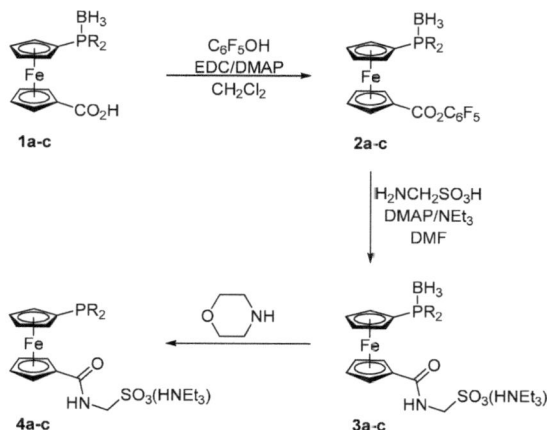

Scheme 2. Synthesis of ligands **4a–c**. Legend: R = *i*-Pr (**a**), cyclohexyl (Cy; **b**), and *t*-Bu (**c**); EDC = 1-ethyl-3-[3-(dimethylamino)propyl]carbodiimide, DMAP = 4-(dimethylamino)pyridine, and DMF = *N,N*-dimethylformamide.

In addition, solid-state structures of **3b·CH₂Cl₂** and **4b** were determined by single-crystal X-ray diffraction analysis. Views of the molecular structures are shown in Figure 1 and the relevant structural parameters for the phosphinoferrocene amidosulfonate anions are presented in Table 1. The ferrocene units in both structures adopt their usual geometry with similar Fe–C distances (cf. Fe–C(1–10): 2.036(2)–2.066(2) Å for **3b·CH₂Cl₂** and 2.032(2)–2.070(1) Å for **4b**) and tilt angles not exceeding 5°. The substituents attached in the positions 1 and 1′ are diverted from each other, so that the ferrocene cyclopentadienyls assume conformations near to anticlinal eclipsed (compare τ angles with the ideal value of 144° [19]). Generally, the individual geometric parameters compare well with the values reported previously for compound **4d** [7] and acid **1b** [10]. In the pair, the structures of the anions differ mainly by the orientation of their CH₂SO₃ pendant groups. In the structure of **3b·CH₂Cl₂**, this moiety extends above the amide plane toward the ferrocene unit, whereas in **4b**, it is directed away from the ferrocene moiety (compare the dihedral angles C11-N1-C24-S of −88.5(2)° and 105.3(1)° in **3b·CH₂Cl₂** and **4b**, respectively). Slight differences are observed also in the overall molecular conformation (compare angles τ and φ in Table 1) and in the lengths of the P–C bonds, which are longer in the free phosphine **4b** than in the corresponding BH₃ adduct **3b·CH₂Cl₂** [**3b·CH₂Cl₂**: P–C6 1.798(2) Å, P–C12 1.843(2) Å, and P–C18 1.840(2) Å; **4b**: P–C6 1.825(1) Å, P–C12 1.873(1) Å, and P–C18 1.862(1) Å]. The cyclohexyl substituents assume chair conformations, which is clearly indicated by the ring puckering parameter θ being 0.0(2)° [2.2(2)] and 178.0(2)° [175.3(2)°] for the rings C(12–17) and C(18–23) in **3b·CH₂Cl₂** [**4b**], respectively (N.B. ideal chair requires θ = 0/180°, see ref. [20]). In all rings, the P–C bonds occupy equatorial positions.

Despite their obvious structural similarity, compounds **3b·CH₂Cl₂** and **4b** constitute different solid-state assemblies in their crystals (Figure 2). The amidosulfonate anions in the structure of **3b·CH₂Cl₂** interact with their inversion-related counterparts through a pair of N1–H1N···O4 hydrogen bonds (N1···O4 = 2.811(2) Å) to form a closed dimeric motif, which further binds two adjacent Et₃NH⁺ cations via N2–H2N···O1 hydrogen bonds involving the amide oxygen (N2···O1 = 2.794(2) Å). A similar central unit formed from a pair of inversion-related amidosulfonate anions is found also in the structure of **4b** (N1···O2 = and 2.813(2) Å) and this dimeric moiety even interacts with a pair of the triethylammonium cations though via hydrogen bonds toward the sulfonate oxygen O3 (N2–H2N ··· O3 = 2.717(2) Å).

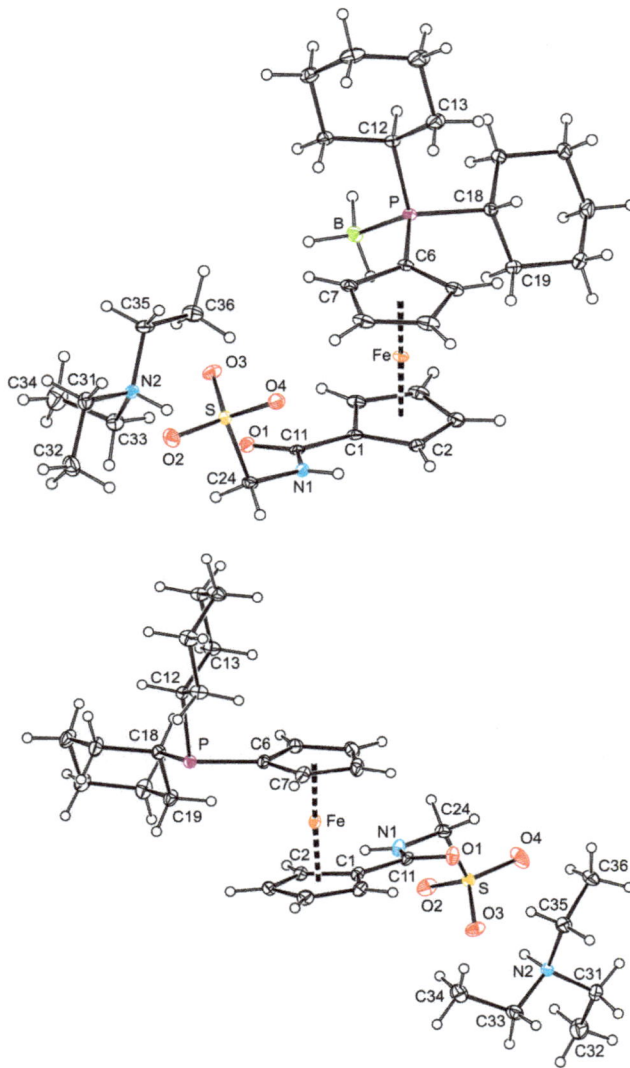

Figure 1. PLATON plots of the molecular structures of **3b·CH₂Cl₂** (**top**) and **4b** (**bottom**). The displacement ellipsoids are scaled to the 30% probability level. Note: all rings are numbered consecutively and, hence, only the labels of the pivotal and the adjacent atom are shown to avoid complicating the Figure. The molecule of solvent is also omitted.

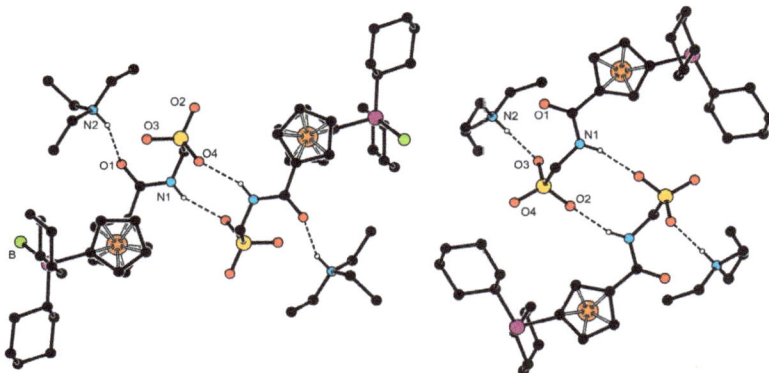

Figure 2. Simplified packing diagrams for **3b·CH₂Cl₂** (**left**) and **4b** (**right**).

Table 1. Selected distances and angles for the anions in the structures of **3b·CH₂Cl₂** and **4b** (in Å and deg) [a].

Parameter	3b·CH$_2$Cl$_2$	4b
Fe–Cg1	1.6506(8)	1.6452(7)
Fe–Cg2	1.6490(8)	1.6490(8)
∠Cp1, Cp2	4.8(1)	4.50(9)
τ	−130.3(1)	146.1(1)
P–C6	1.798(2)	1.825(1)
P–B	1.931(2)	n.a.
C1–C11	1.483(2)	1.477(2)
C11–O1	1.239(2)	1.230(2)
C11–N1	1.346(2)	1.353(2)
O–C11–N1	123.0(1)	122.6(1)
φ	11.9(2)	5.9(2)
N1–C24	1.443(2)	1.434(2)
C24–S	1.804(2)	1.791(2)
S–O2	1.453(1)	1.445(1)
S–O3	1.454(1)	1.458(1)
S–O4	1.460(1)	1.442(1)
N1–C24-S	112.3(1)	115.0(1)
C24–S–O [b]	104.59(7)–106.20(7)	103.06(8)–107.16(7)
O2–S–O3	114.07(7)	110.89(7)
O2–S–O4	113.27(8)	114.73(8)
O3–S–O4	112.16(7)	113.57(8)

[a] Definition of the parameters: Cp1 and Cp2 are the cyclopentadienyl rings C(1–5) and C(6–10), respectively. Cg1 and Cg2 are the respective centroids. τ denotes the dihedral angle C1–Cg1–Cg2–Cg, and φ is the angle subtended by the planes of the ring Cp1 and the amide group {C11, O1 N}. n.a. = not applicable. [b] The range of C24–S–O(2,3,4) angles.

The reaction of [Pd(µ-Cl)(η-C₃H₅)]₂ with ligand **4b** chosen as a representative and then with silver(I) perchlorate proceeded under cleavage of the chloride bridges in the dimeric Pd(II) precursor and removal of the Pd-bound chloride to afford the zwitterionic (η³-C₃H₅)Pd(II) complex **5b** wherein the phosphinosulfonate anion Cy₂PfcCONHCH₂SO₃⁻ coordinates through its phosphine substituent and the amide oxygen, forming an O,P-chelate ring (Scheme 3). The coordination of the phosphine group in **5b** was manifested through a shift of the ³¹P NMR resonance to a lower field (the ³¹P NMR coordination shift, Δ$_P$ = δ$_P$(complex) − δ$_P$(free ligand), was 36.0 ppm), whereas the shift of the amide C=O vibration in the IR spectrum by 65 cm⁻¹ to lower energies suggested that the amide oxygen is involved in coordination to the (η³-C₃H₅)Pd fragment rather that the anionic sulfonate moiety [7].

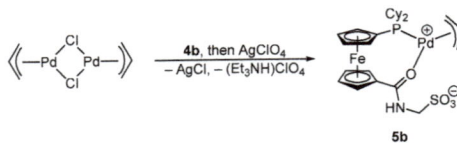

Scheme 3. Synthesis of complex **5b**.

The formulation of **5b** was unequivocally corroborated by X-ray diffraction analysis on the stoichiometric solvate **5b**·CH$_2$Cl$_2$. A view of the complex molecule is shown in Figure 3, and the selected geometric parameters are given in Table 2. The allyl moiety {C31, C32, C33} in the structure of **5b** is rotated by 69.7(4)° with respect to the plane defined by the palladium atom and the remaining ligating atoms P and O1, so that the carbon atom C32 in the *meso* position is diverted from the ferrocene ligand [21]. The individual Pd–C(allyl) distances decrease gradually from C31 to C33, reflecting the *trans*-influence of the donor atoms located opposite the allyl moiety (P > O) [22,23].

Table 2. Selected distances and angles for **5b**·CH$_2$Cl$_2$ (in Å and deg) [a].

Pd–P	2.3161(7)	P–Pd–O1	105.07(5)
Pd–O1	2.131(2)	C31–Pd–C33	67.1(1)
Pd–C31	2.225(3)	P–Pd–C33	97.00(9)
Pd–C32	2.140(3)	O1–Pd–C31	91.1(1)
Pd–C33	2.082(3)	C31–C32–C33	121.4(3)
Fe–Cg1	1.650(1)	∠Cp1,Cp2	6.5(2)
Fe–Cg2	1.652(1)	τ	−65.1(2)
C1–C11	1.467(3)	O1–C11–N	120.6(2)
C11–O	1.249(3)	φ	15.3(3)
C11–N	1.342(3)	C11–N–C24	123.2(2)
N–C24	1.436(4)	N–C24–S	112.5(2)
C24–S	1.793(3)	C24–S–O [b]	103.0(2)–106.8(2)
S–O2	1.442(3)	O2–S–O3	111.1(2)
S–O3	1.460(2)	O2–S–O4	116.9(2)
S–O4	1.443(3)	O3–S–O4	111.8(2)
C6–P	1.813(3)	C6–P–C12	104.6(1)
P–C12	1.844(3)	C6–P–C18	101.2(1)
P–C18	1.854(3)	C12–P–C18	104.9(1)

[a] Definition of the parameters: Cp1 and Cp2 are the cyclopentadienyl rings C(1–5) and C(6–10), respectively. Cg1 and Cg2 denote their respective centroids. τ stands for the dihedral angle C1–Cg1–Cg2–Cg, and φ is the angle subtended by the planes of the ring Cp1 and the amide moiety {C11, O1 N}. [b] The range of C24–S–O(2,3,4) angles.

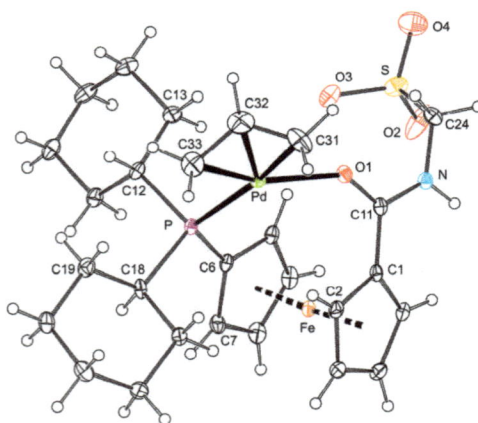

Figure 3. PLATON plot of the complex molecule in the structure of **5b**·CH$_2$Cl$_2$ showing displacement ellipsoids at the 30% probability level.

Apparently because of chelate coordination, the 1,1'-disubtituted ferrocene unit is less open than in the structure of **4b**, adopting a near synclinal eclipsed conformation as evidenced by the torsion angle C1–Cg1–Cg2–C6 (τ) of $-65.1(2)°$ (cf. the ideal value: 72°). The ferrocene moiety is somewhat tilted (dihedral angle: 6.5(2)°) with the C1 and C10 atoms that reside above each other in the ferrocene unit forming the shortest bonds toward the central iron atom (N.B. the individual Fe–C bonds span the range 2.011(2)–2.074(2) Å). More importantly, the distortion propagates to the amide unit, which appears twisted by 15.3(3)° with respect to its parent cyclopentadienyl ring Cp1 and the pivotal atom C11 is displaced from the Cp1 plane inward of the ferrocene unit by as much as 0.214(2) Å. All this aids in bringing the amide oxygen O1 into a position suitable for chelate ring formation. The CH_2SO_3 arm is oriented above the amide unit and to the side of the ferrocene unit so that the angle subtended by the C24–S bond and the axis of the ferrocene unit (Cg1···Cg2) is 21.75(6)°. In contrast, the phosphine phosphorus lies in the plane of its bonding cyclopentadienyl ring (perpendicular distance from the Cp2 plane is only 0.047(1) Å). However, because of the proximity of one of the cyclohexyl substituents, the C7–C6–P angle is 3.6° less acute than the C10–C6–P angle opening to the less sterically encumbered side of the ferrocene moiety (N.B. the difference between the C2–C1–C11 and C5–C1–C11 angles is only 2.0°).

In the crystal, the molecules of complex **5b** associate into dimers via hydrogen bonds between their amide NH groups and sulfonate oxygen O2 in a proximal molecule, N–H1N···O2 (N1···O2 = 2.779(3) Å, angle at H1N = 155°). Additional soft C–H···O_3S interactions interconnect these dimers into a three-dimensional array.

2.2. Catalytic Experiments

The series of phosphinoferrocene amidosulfonates **4a–d** bearing different substituents at the phosphorus atom, which can be regarded as the primary coordination site for the catalytically active soft metal ions, was firstly evaluated in Pd-catalyzed cyanation of aryl bromides leading to the corresponding nitriles [11,12] using potassium hexacyanoferrate(II) as a practically non-toxic and environmentally benign cyanide source [24]. In particular, we chose the cyanation of N-Boc protected 4-bromophenylalanine **6** (Scheme 4) performed in dioxane–water (1:1) mixture at 100 °C in the presence of potassium carbonate as a base. The results collected in Table 3 indicate that the yield of the coupling product **7** depends strongly on the Pd source. At a Pd loading of 2 mol. % and with the model ligand **4d**, the best catalyst performance resulted from [PdCl$_2$(cod)] (cod = η^2:η^2-cycloocta-1,5-diene) and 2 equiv. of the phosphine ligand, which reached full conversion of **6** to **7** within 3 h. An analogous catalyst prepared at a 1:1 [PdCl$_2$(cod)]:**4d** molar ratio ensued in only 30% conversion. Similar catalysts resulting from **4d** and palladium(II) acetate, which is commonly used as a Pd-precursor in cross-coupling reactions [25] and even afforded very good yields of the coupling products in similar cyanation reactions [7,26], performed considerably worse (conversions <5%). Poor results were also obtained when [Pd(μ-Cl)(η-C$_3$H$_5$)]$_2$ was employed as the Pd source, whereas the reaction performed in the presence of [PdCl$_2$(cod)] (2 mol. %) without any supporting ligand did not proceed in any appreciable extent. Notably, the yields of the coupling product **7** markedly decreased when the supporting ligand **4d**, used during the screening experiments, was replaced with its more electron-rich dialkylphosphino analogues **4a–c** (see Table 3).

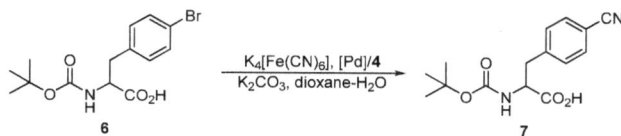

Scheme 4. Pd-catalyzed cyanation of *N*-protected (4-bromophenyl)alanine **6**.

Table 3. Summary of the catalytic results in Pd-catalyzed cyanation of substrate **6** [a].

Entry	Pd Source (Loading)	Ligand (Amount)	NMR Yield of 7 [%]
1	[PdCl$_2$(cod)] (2 mol. %)	**4d** (4 mol. %)	100 (95 [c])
2	[PdCl$_2$(cod)] (2 mol. %)	**4d** (2 mol. %)	30
3	Pd(OAc)$_2$ (2 mol. %)	**4d** (4 mol. %)	<5%
4	Pd(OAc)$_2$ (2 mol. %)	**4d** (2 mol. %)	<5%
5	[Pd(μ-Cl)(η-C$_3$H$_5$)]$_2$ (2 mol. %) [b]	**4d** (2 mol. %)	<5%
6	[PdCl$_2$(cod)] (2 mol. %)	none	0
7	[PdCl$_2$(cod)] (2 mol. %)	**4a** (2 mol. %)	6
8	[PdCl$_2$(cod)] (2 mol. %)	**4b** (2 mol. %)	14
9	[PdCl$_2$(cod)] (2 mol. %)	**4c** (2 mol. %)	12

[a] Conditions: Substrate **6** (0.50 mmol), K$_2$CO$_3$ (1 mmol) and K$_4$[Fe(CN)$_6$] (0.25 mmol) were reacted in the presence of in situ generated catalysts in dioxane–water (1:1, 4 mL) at 100 °C for 3 h. [b] 2 mol. % of Pd. [c] Isolated yield.

Next, we turned to Suzuki–Miyaura biaryl coupling, which is one of the most widely utilized cross-coupling reactions [14–17]. We chose the reactions of bromobenzoic (**8o/8m/8p**) and (bromophenyl)acetic (**9o/9m/9p**) acids with 4-fluorophenyl (**10a**) and 4-tolylboronic (**10b**) acids (Scheme 5) that can be advantageously performed in aqueous solvents, and used the most stable and accessible ligand **4d** for the initial screening experiments. The reaction conditions were optimized for the coupling of **9p** and **10a** yielding biphenyl **12pa**, which can be easily monitored by ^{1}H and ^{19}F NMR spectroscopy. The results are summarized in Table 4.

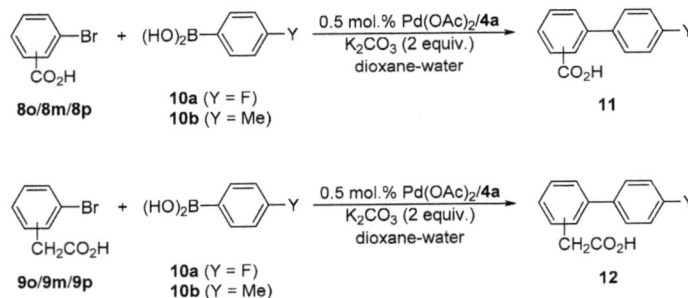

Scheme 5. Pd-catalyzed Suzuki–Miyaura cross-coupling of aromatic acids **8/9** with boronic acids **10a** and **10b**.

The first reaction tests were aimed at finding a suitable solvent (Table 4, entries 1–5). When performed in water at 40 °C and with a catalyst formed from [PdCl$_2$(cod)] and **4d** (1:2 ratio, 1 mol. %), the model reaction produced the coupling product **12pa** in a decent 60% yield after 2 h and a 72% yield after 6 h. Better results were obtained in ethanol and, in particular, an ethanol–water 1:1 mixture, where the yield was 81% after 6 h. Reaction in pure dioxane proceeded to only a negligible extent (10% yield after 6 h), presumably for solubility reasons. However, the addition of water to the system markedly improved the reaction outcome. Thus, the yields of **12pa** achieved in a 1:1 dioxane–water mixture were 82% after 2 h and quantitative after 6 h at 40 °C.

Table 4. Summary of the optimization experiments for the model coupling of **8p** and **10a**.

Entry	Pd Source (Loading)	Ligand	Solvent	NMR Yield of 12pa [%] after	
				2 h	6 h
1	[PdCl₂(cod)] (1 mol. %)	**4d** (2 mol. %)	water	60	72
2	[PdCl₂(cod)] (1 mol. %)	**4d** (2 mol. %)	ethanol	72	75
3	[PdCl₂(cod)] (1 mol. %)	**4d** (2 mol. %)	ethanol–water (1:1)	72	81
4	[PdCl₂(cod)] (1 mol. %)	**4d** (2 mol. %)	dioxane	0	10
5	[PdCl₂(cod)] (1 mol. %)	**4d** (2 mol. %)	dioxane–water (1:1)	82	quant.
6	[PdCl₂(cod)] (1 mol. %)	**4d** (1 mol. %)	dioxane–water (1:1)	41	n.a.
7	Pd(OAc)₂ (1 mol. %)	**4d** (2 mol. %)	dioxane–water (1:1)	78	n.a.
8	Pd(OAc)₂ (1 mol. %)	**4d** (1 mol. %)	dioxane–water (1:1)	85	n.a.
9	[Pd(μ-Cl)(η-C₃H₅)]₂ (1 mol. %) [b]	**4d** (1 mol. %)	dioxane–water (1:1)	0	n.a.
10	Pd(OAc)₂ (1 mol. %)	none	dioxane–water (1:1)	56	n.a.
11	Pd(OAc)₂ (0.5 mol. %)	**4a** (0.5 mol. %)	dioxane–water (1:1)	79	98
12	Pd(OAc)₂ (0.5 mol. %)	**4b** (0.5 mol. %)	dioxane–water (1:1)	74	88
13	Pd(OAc)₂ (0.5 mol. %)	**4c** (0.5 mol. %)	dioxane–water (1:1)	43	67
14	Pd(OAc)₂ (0.5 mol. %)	**4d** (0.5 mol. %)	dioxane–water (1:1)	85	89

[a] Conditions: substrates **8p** (1.0 mmol) and **10a** (1.15 mmol) were reacted in the presence of in situ generated catalysts and K₂CO₃ (2.0 mmol) in 4 mL of the respective solvent at 40 °C. n.a. = not available. [b] 1 mol. % of Pd.

Evaluation of different Pd(II) precursors performed next (Table 4, entries 5–13) revealed that the best yield of the coupling product (85%) is obtained with a catalyst resulting in situ from palladium(II) acetate and 1 equiv. of the amidosulfonate ligand **4d**. Addition of another equivalent of **4d** slightly reduced the yield of **12pa**. A different trend was noted for [PdCl₂(cod)] in which case the catalyst obtained after the addition of 2 equiv. of **4d** achieved a higher yield than the analogous catalyst after the addition of only 1 equiv. of the supporting ligand. The catalyst generated from [Pd(μ-Cl)(η-C₃H₅)]₂ and **4d** (1 equiv. per Pd atom) proved to be inactive. It is also noteworthy that unsupported palladium(II) acetate also catalyzed the reaction but the yield was substantially lower than for both tested Pd(OAc)₂/**4d** catalysts.

In the last step, we have compared catalysts resulting from palladium(II) acetate and different ligands **4**. For this purpose, the metal loading was reduced to 0.5 mol. % and the reaction was monitored after 2 h and 6 h to check whether the catalysts retain their activity. In the case of the Pd(OAc)₂/**4d** catalyst, the yields achieved after 2 and 6 h were 85% and 89%, respectively. Analogous catalyst resulting from **4b** showed a similar yield after 6 h (namely 88%) but a significantly lower yield after 2 h (only 74%), which may suggest a slower catalyst activation and/or slower reaction rate [6]. A similar situation was noted in the case of ligand **4a** possessing diisopropylphosphino substituent, except that the yield of **12pa** after 6 h was the highest among the catalysts tested (nearly quantitative). In contrast, catalysts resulting from the most bulky and electron-rich ligand **4c** acquired the lowest conversions in the entire series, presumably due to rapid deactivation [10]. Based on these results, the Pd(OAc)₂/**4a** catalyst (0.5 mol. % Pd) was chosen for the following reaction scope tests, which were limited to aryl bromides because the reaction with (4-chlorophenyl)acetic acid and **10a** did not yield any coupling product (0.5 mol. % Pd(OAc)₂/**4a**, dioxane–water, 40 or 80 °C/6 h).

The results collected in Table 5 reveal several trends. First of all, both 2-bromobenzoic and (2-bromophenyl)acetic acid reacted only sluggishly with arylboronic acids **10a/b**, which can be ascribed to a steric hindrance. All other isomeric aryl bromide substrates reacted well, achieving very good to practically complete conversions. In all cases, the reactions with **10a** bearing the electron-withdrawing substituent proceeded with higher conversions than those with **10b**. The differences in pairs of the analogous reactions involving **10a** and **10b** were considerably larger for *ortho-* and *para-*substituted aryl bromides than for their *meta-*substituted counterparts, which in turn points to a dominant electronic influence (*I-* and *M-*effect).

Table 5. Summary of the reaction scope tests [a].

Entry	Aryl Bromide	Boronic Acid	Product	NMR Yield [%]	Isolated Yield [%]
1	8o	10a	11oa	19	n.d.
2	8m	10a	11ma	89	67
3	8p	10a	11pa	92	77
4	8o	10b	11ob	24	n.d.
5	8m	10b	11mb	88	77
6	8p	10b	11pb	84	64
7	9o	10a	12oa	<5	n.d.
8	9m	10a	12ma	97	78
9	9p	10a	12pa	98	78
10	9o	10b	12ob	12	n.d.
11	9m	10b	12mb	93	90
12	9p	10b	12pb	80	71

[a] Conditions: the respective substrates **8/9** (1.0 mmol) and **10** (1.15 mmol) were reacted in the presence of a catalyst generated from palladium(II) acetate (0.5 mol.%) and ligand **4a** (1 equiv. with respect to Pd) and K_2CO_3 (2.0 mmol) as the base in 4 mL of dioxane–water (1:1) at 40 °C. n.d. = not determined.

Attempted coupling of 4-bromophenyl alanine substrate **6** with **10a** and **10b** (Scheme 6) under similar conditions proceeded satisfactorily, producing the biphenyl amino acids **13a** and **13b**, respectively, in approximately 90% yields (Table 6). Unfortunately, these products could not be separated from unreacted **6a** either by chromatography or crystallization due to a similar retention characteristic and reluctance to crystallize, respectively. Upon increasing the amount of the catalyst to 1 mol. %, the reaction achieved complete conversions within 6 h at 40 °C, which in turn allowed for the isolation of the pure products **13** in good yields. Notably, NMR analysis of the crude reaction mixtures revealed that the Boc-protecting group is stable under the reaction conditions, which is indeed manifested in the good yields.

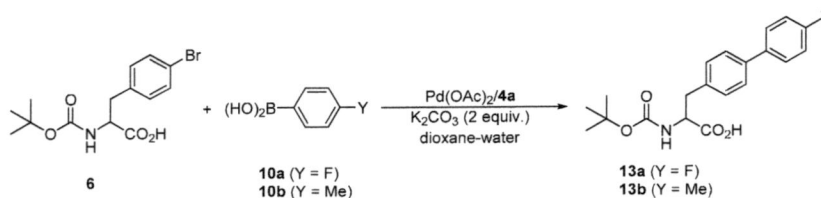

Scheme 6. Pd-catalyzed Suzuki–Miyaura biaryl coupling involving substrate **6**.

Table 6. Reactions of **10a/b** with amino acid **6** [a].

Entry	Boronic Acid	Product	Pd Loading	NMR Yield [%]	Isolated Yield [%]
1	10a	13a	0.5 mol. %	88	n.d.
2	10b	13b	0.5 mol. %	91	n.d.
3	10a	13a	1.0 mol. %	100	76
4	10b	13b	1.0 mol. %	100	83

[a] Conditions: compound **6** (1.0 mmol) and **10a** or **10b** (1.15 mmol) were reacted in the presence of a catalyst generated from palladium(II) acetate (0.5 or 1.0 mol. %) and ligand **4a** (1 equiv.) and K_2CO_3 (2.0 mmol) as the base in 4 mL of dioxane–water (1:1) at 40 °C for 6 h. n.d. = not determined.

3. Experimental

3.1. Materials and Methods

Compounds **1a–c** [10] and **4d** [7] were prepared according to the literature methods. Anhydrous dichloromethane was obtained from a PureSolv MD5 Solvent Purification System (Innovative Technology Inc., Amesbury, MA, USA). Triethylamine and dioxane were distilled from sodium metal. Other chemicals (Alfa-Aesar or Sigma-Aldrich, Ward Hill, MA, USA; Saint Louis, MO, USA), anhydrous *N,N*-dimethylformamide over molecular sieves (Sigma-Aldrich), anhydrous ethanol (Penta, Prague, Czech Republic) and all solvents (reagent grade from Lach-Ner, Neratovice, Czech Republic) used for workup, column chromatography and crystallizations were without any additional purification.

NMR spectra were recorded at 298 K on a Varian Unity Inova 400 spectrometer (^{1}H, 399.95 MHz; ^{13}C, 100.58 MHz; and ^{31}P, 161.90 MHz) or a Bruker AVANCE III 400 spectrometer (^{1}H, 400.13 MHz; ^{13}C{^{1}H}, 100.62 MHz; ^{19}F, 376.46 MHz; and ^{31}P, 161.97 MHz). Chemical shifts (δ/ppm) are given relative to internal tetramethylsilane (^{1}H and ^{13}C NMR), to external 85% aqueous H_3PO_4 (^{31}P NMR), and to external neat $CFCl_3$ (^{19}F NMR), respectively. In addition to the standard notation of NMR signals, vt and vq are used to distinguish virtual triplets and quartets arising from the C=O and phosphine-substituted cyclopentadienyl rings, respectively. Conventional low-resolution electrospray ionization mass spectra (ESI MS) were recorded with an Escuire 3000 (Bruker). High-resolution (HR) analyses were performed with a compact Q-TOF (Bruker Daltonik) instrument. The samples were dissolved in HPLC-grade methanol. Infrared spectra were collected in Nujol mulls on a FTIR Thermo Fisher Nicolet 760 instrument in the range 400–4000 cm^{-1}.

3.2. General Procedure for the Synthesis of Esters $H_3B \cdot R_2PfcCO_2C_6F_5$ (2a–c)

The respective acids $H_3B \cdot R_2PfcCO_2H$ (1a–c, 5.0 mmol), 1-ethyl-3-[3-(dimethylamino)propyl]-carbodiimide hydrochloride (1.15 g, 6.0 mmol), pentafluorophenol (1.10 g, 6.0 mmol) and 4-(dimethylamino)pyridine (122 mg, 1.0 mmol) were dissolved in dichloromethane (50 mL). After stirring overnight, brine (50 mL) was added to the reaction mixture and stirring was continued for another 10 min. Then, the organic layer was separated and washed with brine (50 mL). The combined aqueous phases were back-extracted with dichloromethane (20 mL). The combined organic layers were dried over anhydrous $MgSO_4$ and evaporated to dryness. The crude product was purified by flash column chromatography over silica gel as described below.

3.2.1. Preparation of $iPr_2PfcCO_2C_6F_5 \cdot BH_3$ (2a)

Ester **2a** was synthesized from acid **1a** (1.80 g, 5.0 mmol) following the general procedure and isolated as an orange solid. Ethyl acetate-hexane (1:8) was used during the chromatography. Only the first band from the product was collected. Yield: 2.46 g (94%).

^{1}H NMR (CDCl$_3$): δ 0.15–1.10 (br m, 3 H, BH$_3$), 1.16 (dd, $^3J_{HH}$ = 7.1 Hz, $^3J_{PH}$ = 2.7 Hz, 6 H, CHMe$_2$), 1.19 (dd, $^3J_{HH}$ = 7.1 Hz, $^3J_{PH}$ = 3.5 Hz, 6 H, CHMe$_2$), 2.16 (d of sept, $^2J_{PH}$ = 10.1 Hz, $^3J_{HH}$ = 7.1 Hz, 2 H, CHMe$_2$), 4.52 (vq, J' = 1.8 Hz, 2 H, fc), 4.61 (d of vt, $J' \approx 0.9$, 1.9 Hz, 2 H, fc), 4.80 (vt, J' = 2.0 Hz, 2 H, fc), 5.06 (vt, J' = 2.0 Hz, 2 H, fc). ^{13}C{^{1}H} NMR (CDCl$_3$): δ 17.16 (s, 2 C, CHMe$_2$), 17.47 (d, $^2J_{PC}$ = 2 Hz, 2 C, CHMe$_2$), 22.58 (d, $^1J_{PC}$ = 35 Hz, 2 C, CHMe$_2$), 68.05 (s, C$_{ipso}$-CO of fc), 70.65 (d, $^1J_{PC}$ = 53 Hz, C$_{ipso}$-P of fc), 72.32 (s, 2 C, CH of fc), 73.44 (d, J_{PC} = 7 Hz, 2 C, CH of fc), 73.59 (d, J_{PC} = 6 Hz, 2 C, CH of fc), 75.43 (s, 2 C, CH of fc), 125.16 (s, C$_{ipso}$ of C$_6$F$_5$), 137.96 (dm, $^1J_{FC}$ = 253 Hz, C$_{meta}$ of C$_6$F$_5$), 139.45 (dm, $^1J_{FC}$ = 253 Hz, C$_{para}$ of C$_6$F$_5$), 141.48 (dm, $^1J_{FC}$ = 250 Hz, C$_{ortho}$ of C$_6$F$_5$), 167.45 (s, C=O). ^{19}F{^{1}H} NMR (CDCl$_3$): δ = −162.60 (m, F$_{meta}$ of C$_6$F$_5$), −158.34 (t, $^3J_{FF}$ = 22 Hz, F$_{para}$ of C$_6$F$_5$), −153.08 (m, F$_{ortho}$ of C$_6$F$_5$) ppm. ^{31}P{^{1}H} NMR (CDCl$_3$): δ 31.9 (br d). IR (Nujol): 2373 m, 2342 m, 1768 s, 1522 s, 1305 w, 1261 (s), 1203 w, 1169 w, 1142 w, 1072 s, 1058 m, 1023 m, 1011 m, 974 m, 885 m, 870 w, 852 w, 833 w, 753 w, 639 w, 611 w, 531 w, 505 w cm^{-1}. MS (ESI+): m/z 549.1 ([M + Na]$^+$), 565.1 ([M + K]$^+$). Anal. Calc. for C$_{23}$H$_{25}$BF$_5$FeO$_2$P (526.06): C 52.51, H 4.79%. Found: C 52.50, H 4.69%.

3.2.2. Preparation of $Cy_2PfcCO_2C_6F_5 \cdot BH_3$ (2b)

Ester **2b** was obtained from acid **1b** (2.20 g, 5.0 mmol) as described above and isolated as an orange solid. Ethyl acetate-hexane (1:10) was used during the chromatography. Only the first band from the product was collected. Yield: 2.54 g (84%).

^{1}H NMR (CDCl$_3$): δ 0.15–1.05 (br m, 3 H, BH$_3$), 1.10–1.40 (m, 10 H, Cy), 1.65–1.73 (m, 2 H, Cy), 1.75–2.01 (m, 10 H, Cy), 4.49 (vq, J' = 1.8 Hz, 2 H, fc), 4.60 (d of vt, J' ≈ 0.9, 1.8 Hz, 2 H, fc), 4.78 (vt, J' = 2.0 Hz, 2 H, fc), 5.04 (vt, J' = 2.0 Hz, 2 H, fc). ^{13}C{^{1}H} NMR (CDCl$_3$): δ 25.89 (d, J_{PC} = 1 Hz, 2 C, CH$_2$ of Cy), 26.77 (d, J_{PC} = 2 Hz, 2 C, CH$_2$ of Cy), 26.88 (d, J_{PC} = 1 Hz, 2 C, CH$_2$ of Cy), 26.95 (s, 2 C, CH$_2$ of Cy), 27.27 (d, J_{PC} = 2 Hz, 2 C, CH$_2$ of Cy), 32.33 (d, $^1J_{PC}$ = 34 Hz, 2 C, CH of Cy), 68.03 (s, C$_{ipso}$-CO of fc), 71.31 (d, $^1J_{PC}$ = 53 Hz, C$_{ipso}$-P of fc), 72.34 (s, 2 C, CH of fc), 73.42 (d, J_{PC} = 6 Hz, 2 C, CH of fc), 73.58 (d, J_{PC} = 7 Hz, 2 C, CH of fc), 75.39 (s, 2 C, CH of fc), 125.18 (m, C$_{ipso}$ of C$_6$F$_5$), 137.95 (dm, $^1J_{FC}$ = 252 Hz, C$_{meta}$ of C$_6$F$_5$), 139.44 (dm, $^1J_{FC}$ = 253 Hz, C$_{para}$ of C$_6$F$_5$), 141.38 (dm, $^1J_{FC}$ = 251 Hz, C$_{ortho}$ of C$_6$F$_5$), 167.46 (s, C=O). ^{19}F{^{1}H} NMR (CDCl$_3$): δ = −162.62 (m, F$_{meta}$ of C$_6$F$_5$), −158.33 (t, $^3J_{FF}$ = 22 Hz, F$_{para}$ of C$_6$F$_5$), −153.10 (m, F$_{ortho}$ of C$_6$F$_5$) ppm. ^{31}P{^{1}H} NMR (CDCl$_3$): δ 24.3 (br d). IR (Nujol): 2373 m, 2342 m, 1768 s, 1522 s, 1305 w, 1261 m, 1203 w, 1169 w, 1142 w, 1072 s, 1058 m, 1023 m, 1011 s, 974 m, 885 m, 870 w, 852 w, 833 w, 753 w, 639 w, 611 w, 531 w, 505 w cm^{-1}. MS (ESI+): m/z 629.2 ([M + Na]$^+$), 645.1 ([M + K]$^+$). Anal. Calc. for C$_{29}$H$_{33}$BF$_5$FeO$_2$P (606.18): C 57.50, H 5.50%. Found: C 57.46, H 5.49%.

3.2.3. Preparation of H$_3$B·tBu$_2$PfcCO$_2$C$_6$F$_5$ (2c)

Ester **2c** was synthesized from **1c** (1.94 g, 5.0 mmol) by using the general procedure and isolated as an orange solid. Ethyl acetate-hexane (1:8) was used during the column chromatography. Only the first band from the product was collected. Yield: 2.06 g (93%).

^{1}H NMR (CDCl$_3$): δ 0.20–1.20 (br m, 3 H, BH$_3$), 1.30 (d, $^3J_{PH}$ = 12.9 Hz, 18 H, CMe$_3$), 4.62–4.65 (m, 4 H, fc), 4.80 (vt, J' = 2.0 Hz, 2 H, fc), 5.03 (vt, J' = 2.0 Hz, 2 H, fc). ^{13}C{^{1}H} NMR (CDCl$_3$): δ 28.64 (d, $^2J_{PC}$ = 2 Hz, 6 C, CMe$_3$), 33.42 (d, $^1J_{PC}$ = 27 Hz, 2 C, CMe$_3$), 67.95 (s, C$_{ipso}$-CO of fc), 72.46 (s, 2 C, CH of fc), 72.76 (d, $^1J_{PC}$ = 48 Hz, C$_{ipso}$-P of fc), 73.51 (d, J_{PC} = 6 Hz, 2 C, CH of fc), 74.91 (d, J_{PC} = 6 Hz, 2 C, CH of fc), 75.92 (s, 2 C, CH of fc), 125.18 (m, C$_{ipso}$ of C$_6$F$_5$), 137.92 (dm, $^1J_{FC}$ = 252 Hz, C$_{meta}$ of C$_6$F$_5$), 139.44 (dm, $^1J_{FC}$ = 253 Hz, C$_{para}$ of C$_6$F$_5$), 143.37 (dm, $^1J_{FC}$ = 251 Hz, C$_{ortho}$ of C$_6$F$_5$), 167.45 (s, C=O). ^{19}F{^{1}H} NMR (CDCl$_3$): δ = −162.60 (m, F$_{meta}$ of C$_6$F$_5$), −158.35 (t, $^3J_{FF}$ = 22 Hz, F$_{para}$ of C$_6$F$_5$), −153.01 (m, F$_{ortho}$ of C$_6$F$_5$) ppm. ^{31}P{^{1}H} NMR (CDCl$_3$): δ 45.5 (br d). IR (Nujol): 2391 m, 2363 m, 2344 m, 1752 s, 1521 s, 1304 w, 1267 s, 1162 w, 1144 w, 1080 s, 1057 w, 1026 m, 1006 w, 995 m, 980 w, 910 w, 895 m, 874 w, 852 w, 833 m, 815 w, 755 w, 647 w, 626 w, 573 w, 531 w, 514 w cm^{-1}. MS (ESI+): m/z 577.2 ([M + Na]$^+$), 593.1 ([M + K]$^+$). Anal. Calc. for C$_{25}$H$_{29}$BF$_5$FeO$_2$P (554.11): C 54.19, H 5.28%. Found: C 54.18, H 5.22%.

3.3. General Procedure for the Synthesis of Amides R$_2$PfcCONHCH$_2$SO$_3$(HNEt$_3$)·BH$_3$ (3a–c)

The respective ester **1** (3.0 mmol), aminomethanesulfonic acid (0.51 g, 4.5 mmol) and 4-(dimethylamino)pyridine (122 mg, 1.0 mmol) were dissolved in a mixture of *N*,*N*-dimethylformamide (10 mL) and triethylamine (2 mL). After stirring overnight, the solvents were evaporated under vacuum and the residual DMF was removed by trituration with diethyl ether (50 mL). The crude product was purified by flash column chromatography over silica gel. The column was packed in CH$_2$Cl$_2$–MeOH–Et$_3$N (100:5:5) and the product was eluted with CH$_2$Cl$_2$–MeOH (100:5). The first orange band was collected and evaporated. The solid residue was triturated with diethyl ether (3 × 20 mL) to remove residual triethylamine and, finally, dried under vacuum.

3.3.1. Preparation of iPr$_2$PfcCONHCH$_2$SO$_3$(HNEt$_3$)·BH$_3$ (3a)

Amide **3a** was synthesized from ester **2a** (1.58 g, 3.0 mmol) following the general procedure described above and isolated as an orange crystalline solid after crystallization from hot ethyl acetate. Yield: 1.32 g (79%).

^{1}H NMR (CD$_2$Cl$_2$): δ 0.10–1.10 (br m, 3 H, BH$_3$), 1.11 (dd, $^3J_{HH}$ = 7.1 Hz, $^3J_{PH}$ = 1.4 Hz, 6 H, CHMe$_2$), 1.15 (d, $^3J_{HH}$ = 7.1 Hz, 6 H, CHMe$_2$), 1.31 (t, $^3J_{HH}$ = 7.3 Hz, 9 H, CH$_3$ of Et$_3$NH$^+$), 2.13 (d of sept, $^2J_{PH}$ = 10.1 Hz, $^3J_{HH}$ = 7.1 Hz, 2 H, CHMe$_2$), 3.10 (q, $^3J_{HH}$ = 7.3 Hz, 6 H, CH$_2$ of Et$_3$NH$^+$), 4.40 (vq, J' = 1.7 Hz, 2 H, fc), 4.43 (d, $^3J_{HH}$ = 6.6 Hz, 2 H, CH$_2$N), 4.50 (vt, J' = 2.0 Hz, 2 H, fc), 4.65 (d of vt,

$J' \approx 0.9$, 1.9 Hz, 2 H, fc), 4.83 (vt, $J' = 2.0$ Hz, 2 H, fc), 6.92 (t, $^3J_{HH} = 6.5$ Hz, 1 H, NH). ^{13}C{^{1}H} NMR (CD$_2$Cl$_2$): δ 8.85 (s, 3 C, CH$_3$ of Et$_3$NH$^+$), 17.30 (s, 2 C, CHMe_2), 17.62 (d, $^2J_{PC} = 2$ Hz, 2 C, CHMe_2), 22.82 (d, $^1J_{PC} = 35$ Hz, 2 C, CHMe_2), 46.46 (s, 3 C, CH$_2$ of Et$_3$NH$^+$), 56.05 (s, CH$_2$N), 69.36 (d, $^1J_{PC} = 55$ Hz, C$_{ipso}$-P of fc), 70.09 (s, 2 C, CH of fc), 73.35 (s, 2 C, CH of fc), 73.59 (s, 2 C, CH of fc), 73.66 (d, $J_{PC} = 1$ Hz, 2 C, CH of fc), 76.96 (s, C$_{ipso}$-CO of fc), 169.54 (s, C=O). ^{31}P{^{1}H} NMR (CD$_2$Cl$_2$): δ 31.6 (br d). IR (Nujol): 3312 m, 2376 w, 2360 m, 2332 m, 1654 s, 1533 s, 1499 m, 1366 m, 1313 w, 1285 w, 1245 m, 1216 m, 1165 s, 1070 m, 1035 m, 1012 m, 980 m, 899 w, 888 w, 868 w, 852 w, 840 m, 831 m, 812 w, 797 w, 754 m, 692 w, 667 w, 632 w, 615 w, 585 w, 535 w, 515 m cm^{-1}. MS (ESI+): m/z 555.2 ([M + H]$^+$), 656.4 ([M + HNEt$_3$]$^+$); MS (ESI$-$): m/z 451.9 ([M $-$ HNEt$_3$]$^+$). Anal. Calc. for C$_{24}$H$_{44}$BFeN$_2$O$_4$PS (554.30): C 52.00, H 8.00, N 5.06%. Found: C 51.90, H 8.03, N 4.93%.

3.3.2. Preparation of Cy$_2$PfcCONHCH$_2$SO$_3$(HNEt$_3$)·BH$_3$ (3b)

Amide **3b** was synthesized from **2b** (1.82 g, 3.0 mmol) according to the general procedure and was isolated as an orange crystalline solid after crystallization from hot ethyl acetate. Yield: 1.67 g (88%).

^{1}H NMR (CD$_2$Cl$_2$): δ 0.04–1.04 (br m, 3 H, BH$_3$), 1.08–1.38 (m, 10 H, Cy), 1.31 (t, $^3J_{HH} = 7.3$ Hz, 9 H, CH$_3$ of Et$_3$NH$^+$), 1.64–1.71 (m, 2 H, Cy), 1.74–1.99 (m, 10 H, Cy), 3.11 (q, $^3J_{HH} = 7.3$ Hz, 6 H, CH$_2$ of Et$_3$NH$^+$), 4.37 (vq, $J' = 1.8$ Hz, 2 H, fc), 4.42 (d, $^3J_{HH} = 6.6$ Hz, 2 H, CH$_2$N), 4.47 (vt, $J' = 2.0$ Hz, 2 H, fc), 4.64 (d of vt, $J' \approx 0.9$, 1.8 Hz, 2 H, fc), 4.80 (vt, $J' = 1.9$ Hz, 2 H, fc), 6.85 (t, $^3J_{HH} = 6.6$ Hz, 1 H, NH). ^{13}C{^{1}H} NMR (CD$_2$Cl$_2$): δ 8.87 (s, 3 C, CH$_3$ of Et$_3$NH$^+$), 26.39 (d, $J_{PC} = 2$ Hz, 2 C, CH$_2$ of Cy), 27.15–27.30 (m, 6 C, CH$_2$ of Cy), 27.57 (d, $J_{PC} = 2$ Hz, 2 C, CH$_2$ of Cy), 32.49 (d, $^1J_{PC} = 35$ Hz, 2 C, CH of Cy), 46.50 (s, 3 C, CH$_2$ of Et$_3$NH$^+$), 56.02 (s, CH$_2$N), 69.99 (d, $^1J_{PC} = 55$ Hz, C$_{ipso}$-P of fc), 70.04 (s, 2 C, CH of fc), 73.35 (s, 2 C, CH of fc), 73.48 (d, $J_{PC} = 7$ Hz, 2 C, CH of fc), 73.70 (d, $J_{PC} = 7$ Hz, 2 C, CH of fc), 76.98 (s, C$_{ipso}$-CO of fc), 169.44 (s, C=O). ^{31}P{^{1}H} NMR (CD$_2$Cl$_2$): δ 24.1 (br d). IR (Nujol): 3531 m, 3456 m, 3337 m, 2378 m, 2334 m, 1648 s, 1523 m, 1296 w, 1253 w, 1211 m, 1168 s, 1053 m, 890 w, 853 w, 826 w, 765 w, 635 w, 616 w, 528 w, 506 w cm^{-1}. MS (ESI+): m/z 635.2 ([M + H]$^+$), 736.5 ([M + HNEt$_3$]$^+$); MS (ESI$-$): m/z 532.0 ([M $-$ HNEt$_3$]$^+$). Anal. Calc. for C$_{30}$H$_{52}$BFeN$_2$O$_4$PS (634.43): C 56.79, H 8.26, N 4.42%. Found: C 56.64, H 8.31, N 4.29%.

3.3.3. Preparation of *t*Bu$_2$PfcCONHCH$_2$SO$_3$(HNEt$_3$)·BH$_3$ (3c)

Amide **3c** was synthesized from **2c** (1.66 g, 3.0 mmol) according to the general procedure and isolated as an orange solid. The compound decomposes upon attempted crystallization. Yield: 1.62 g (92%).

^{1}H NMR (CD$_2$Cl$_2$): δ 0.15–1.15 (br m, 3 H, BH$_3$), 1.27 (d, $^3J_{PH} = 12.8$ Hz, 18 H, CMe_3), 1.31 (t, $^3J_{HH} = 7.3$ Hz, 9 H, CH$_3$ of Et$_3$NH$^+$), 3.10 (q, $^3J_{HH} = 7.3$ Hz, 6 H, CH$_2$ of Et$_3$NH$^+$), 4.43 (d, $^3J_{HH} = 6.6$ Hz, 2 H, CH$_2$N), 4.48 (vt, $J' = 2.0$ Hz, 2 H, fc), 4.50 (vq, $J' = 1.7$ Hz, 2 H, fc), 4.69 (d of vt, $J' \approx 0.9$, 1.9 Hz, 2 H, fc), 4.80 (vt, $J' = 2.0$ Hz, 2 H, fc), 6.91 (t, $^3J_{HH} = 6.5$ Hz, 1 H, NH). ^{13}C{^{1}H} NMR (CD$_2$Cl$_2$): δ 8.86 (s, 3 C, CH$_3$ of Et$_3$NH$^+$), 28.84 (d, $^2J_{PC} = 2$ Hz, 6 C, CMe_3), 33.60 (d, $^1J_{PC} = 28$ Hz, 2 C, CMe_3), 46.44 (s, 3 C, CH$_2$ of Et$_3$NH$^+$), 56.03 (s, CH$_2$N), 70.23 (s, 2 C, CH of fc), 71.66 (d, $^1J_{PC} = 49$ Hz, C$_{ipso}$-P of fc), 73.49 (d, $J_{PC} = 6$ Hz, 2 C, CH of fc), 73.85 (s, 2 C, CH of fc), 75.16 (d, $J_{PC} = 7$ Hz, 2 C, CH of fc), 76.93 (s, C$_{ipso}$-CO of fc), 169.42 (s, C=O). ^{31}P{^{1}H} NMR (CD$_2$Cl$_2$): δ 45.1 (br m). IR (Nujol): 3296 m, 2382 m, 2363 m, 2343 m, 1656 s, 1517 m, 1498 m, 1316 w, 1299 w, 1251 w, 1211 m, 1161 s, 1068 w, 1039 s, 1014 m, 983 m, 895 w, 858 w, 838 w, 814 w, 751 w, 647 w, 613 w, 598 w, 516 w cm^{-1}. MS (ESI+): m/z 684.4 ([M + HNEt$_3$]$^+$), 605.3 ([M + Na]$^+$); MS (ESI$-$): m/z 479.9 ([M $-$ HNEt$_3$]$^+$). Anal. Calc. for C$_{26}$H$_{48}$BFeN$_2$O$_4$PS·0.1CH$_2$Cl$_2$ (590.84): C 53.05, H 8.22, N 4.74%. Found: C 52.98, H 7.88, N 4.49%.

3.4. General Procedure for Deprotection of Borane Adducts 3

A Schlenk flask was charged with a borane adduct **3** (1.5 mmol) and freshly distilled morpholine (10 mL). The reaction mixture was thoroughly degassed by five freeze–pump–thaw cycles and then heated at 65 °C for 16 h. Next, the morpholine was removed under vacuum and the crude product was purified by flash column chromatography over silica gel. The chromatographic column was packed in CH$_2$Cl$_2$–MeOH–Et$_3$N (100:5:5) and the product was eluted with CH$_2$Cl$_2$–MeOH (100:5). The first orange band was collected and evaporated. The residue was triturated with diethyl ether (3 × 20 mL) to remove residual triethylamine.

3.4.1. Preparation of *i*Pr$_2$PfcCONHCH$_2$SO$_3$(HNEt$_3$) (**4a**)

Amide **4a** was prepared from **3a** (0.831 g, 1.5 mmol) as outlined above and isolated as an orange crystalline solid after crystallization from hot ethyl acetate. Yield: 0.657 g (81%).

^{1}H NMR (CD$_2$Cl$_2$): δ 1.05 (dd, $^3J_{HH}$ = 7.0 Hz, $^3J_{PH}$ = 2.1 Hz, 6 H, CH*Me*$_2$), 1.08 (dd, $^3J_{HH}$ = 7.0 Hz, $^3J_{PH}$ = 4.2 Hz, 6 H, CH*Me*$_2$), 1.32 (t, $^3J_{HH}$ = 7.3 Hz, 9 H, CH$_3$ of Et$_3$NH$^+$), 1.91 (d of sept, $^2J_{PH}$ = 2.0 Hz, $^3J_{HH}$ = 7.3 Hz, 2 H, C*H*Me$_2$), 3.12 (q, $^3J_{HH}$ = 7.3 Hz, 6 H, CH$_2$ of Et$_3$NH$^+$), 4.25 (vq, J' = 1.6 Hz, 2 H, fc), 4.31 (vt, J' = 1.9 Hz, 2 H, fc), 4.41 (d, $^3J_{HH}$ = 6.5 Hz, 2 H, CH$_2$N), 4.47 (vt, J' = 1.8 Hz, 2 H, fc), 4.68 (vt, J' = 1.9 Hz, 2 H, fc), 6.77 (t, $^3J_{HH}$ = 5.8 Hz, 1 H, NH). ^{13}C{^{1}H} NMR (CD$_2$Cl$_2$): δ 8.90 (s, 3 C, CH$_3$ of Et$_3$NH$^+$), 20.03 (d, $^2J_{PC}$ = 11 Hz, 2 C, CH*Me*$_2$), 20.25 (d, $^2J_{PC}$ = 15 Hz, 2 C, CH*Me*$_2$), 23.76 (d, $^1J_{PC}$ = 12 Hz, 2 C, *C*HMe$_2$), 46.61 (s, 2 C, CH$_2$ of Et$_3$NH$^+$), 56.08 (s, CH$_2$N), 69.66 (s, 2 C, CH of fc), 72.21 (d, J_{PC} = 2 Hz, 2 C, CH of fc), 72.75 (s, 2 C, CH of fc), 73.13 (d, J_{PC} = 10 Hz, 2 C, CH of fc), 76.28 (s, C$_{ipso}$-CO of fc), 76.28 (d, $^1J_{PC}$ = 10 Hz, C$_{ipso}$-P of fc), 170.09 (s, C=O). ^{31}P{^{1}H} NMR (CD$_2$Cl$_2$): δ − 0.1 (s). IR (Nujol): 3509 s, 3448 s, 3285 s, 1659 s, 1637 s, 1540 s, 1342 w, 1315 m, 1286 m, 1243 m, 1209 m, 1157 s, 1073 w, 1038 s, 964 w, 893 w, 834 m, 769 w, 658 w, 635 w, 616 m, 595 w, 541 w, 525 w, 517 w cm^{-1}. MS (ESI+): *m/z* 440.0 ([M + H − NEt$_3$]$^+$), 541.2 ([M + H]$^+$); MS (ESI−): *m/z* 437.9 ([M − HNEt$_3$]$^+$). Anal. Calc. for C$_{24}$H$_{41}$BFeN$_2$O$_4$PS (540.47): C 53.33, H 7.65, N 5.18%. Found: C 53.07, H 7.69, 5.24%.

3.4.2. Preparation of Cy$_2$PfcCONHCH$_2$SO$_3$(HNEt$_3$) (**4b**)

Amide **4b** was synthesized from **3b** (0.952 g, 1.5 mmol) by following the general procedure and isolated as an orange crystalline solid after crystallization from hot ethyl acetate. Yield: 0.78 g (84%).

^{1}H NMR (CD$_2$Cl$_2$): δ 0.96–1.37 (m, 10 H, Cy), 1.32 (t, $^3J_{HH}$ = 7.3 Hz, 9 H, CH$_3$ of Et$_3$NH$^+$), 1.61–1.82 (m, 10 H, Cy), 1.86–1.95 (m, 2 H, Cy), 3.12 (q, $^3J_{HH}$ = 7.3 Hz, 6 H, CH$_2$ of Et$_3$NH$^+$), 4.22 (vq, J' = 1.6 Hz, 2 H, fc), 4.29 (vt, J' = 1.9 Hz, 2 H, fc), 4.42 (d, $^3J_{HH}$ = 6.5 Hz, 2 H, CH$_2$N), 4.46 (vt, J' = 1.8 Hz, 2 H, fc), 4.67 (vt, J' = 1.9 Hz, 2 H, fc), 6.86 (t, $^3J_{HH}$ = 6.3 Hz, 1 H, NH). ^{13}C{^{1}H} NMR (CD$_2$Cl$_2$): δ 8.91 (s, 3 C, CH$_3$ of Et$_3$NH$^+$), 26.87 (d, J_{PC} = 1 Hz, 2 C, CH$_2$ of Cy), 27.65 (d, J_{PC} = 9 Hz, 2 C, CH$_2$ of Cy), 27.78 (d, J_{PC} = 11 Hz, 2 C, CH$_2$ of Cy), 30.48 (d, J_{PC} = 2 Hz, 2 C, CH$_2$ of Cy), 30.60 (s, 2 C, CH$_2$ of Cy), 33.73 (d, J_{PC} = 12 Hz, 2 C, CH of Cy), 46.63 (s, 3 C, CH$_2$ of Et$_3$NH$^+$), 56.09 (s, CH$_2$N), 69.67 (s, 2 C, CH of fc), 72.16 (d, J_{PC} = 3 Hz, 2 C, CH of fc), 72.76 (d, J_{PC} = 1 Hz, 2 C, CH of fc), 73.32 (d, J_{PC} = 10 Hz, 2 C, CH of fc), 76.23 (s, C$_{ipso}$-CO of fc), 170.21 (s, C=O). One ferrocene resonance (C$_{ipso}$-P) was not identified, presumably due to overlaps. ^{31}P{^{1}H} NMR (CD$_2$Cl$_2$): δ − 8.3 (s). IR (Nujol): 3309 s, 1653 s, 1533 s, 1314 m, 1284 m, 1247 w, 1217 m, 1169 s, 1077 w, 1044 s, 966 w, 891 w, 847 w, 831 w, 771 w, 613 m, 547 w, 533 w, 516 w cm^{-1}. MS (ESI+): *m/z* 621.3 ([M + H]$^+$); MS (ESI−): *m/z* 518.0 ([M − HNEt$_3$]$^+$). Anal. Calc. for C$_{30}$H$_{49}$BFeN$_2$O$_4$PS (620.59): C 58.06, H 7.96, N 4.52%. Found: C 57.81, H 7.82, N 4.50%.

3.4.3. Preparation of *t*Bu$_2$PfcCONHCH$_2$SO$_3$(HNEt$_3$) (**4c**)

Amide **4c** was synthesized from **3c** (0.874 g, 1.5 mmol) as described above and isolated as an orange solid. Yield: 0.605 g (71%).

^{1}H NMR (CD$_2$Cl$_2$): δ 1.18 (d, $^3J_{PH}$ = 11.2 Hz, 18 H, C*Me*$_3$), 1.32 (t, $^3J_{HH}$ = 6.0 Hz, 9 H, CH$_3$ of Et$_3$NH$^+$), 3.12 (q, $^3J_{HH}$ = 6.4 Hz, 6 H, CH$_2$ of Et$_3$NH$^+$), 4.32–4.45 (m, 4 H, fc), 4.42 (d, $^3J_{HH}$ = 6.5 Hz, 2 H, CH$_2$N), 4.54 (vt, J' = 1.8 Hz, 2 H, fc), 4.71 (vt, J' = 1.9 Hz, 2 H, fc), 6.76 (t, $^3J_{HH}$ = 5.4 Hz, 1 H, NH). ^{13}C{^{1}H} NMR (CD$_2$Cl$_2$): δ 8.92 (s, 3 C, CH$_3$ of Et$_3$NH$^+$), 30.92 (d, $^2J_{PC}$ = 13 Hz, 6 C, C*Me*$_3$), 33.00 (d, $^1J_{PC}$ = 20 Hz, 2 C, *C*Me$_3$), 46.58 (s, 3 C, CH$_2$ of Et$_3$NH$^+$), 56.10 (s, CH$_2$N), 69.72 (s, 2 C, CH of fc), 72.17 (d, J_{PC} = 3 Hz, 2 C, CH of fc), 73.34 (s, 2 C, CH of fc), 74.65 (d, J_{PC} = 12 Hz, 2 C, CH of fc), 76.11 (s, C$_{ipso}$-CO of fc), 170.00 (s,

C=O). One ferrocene resonance (C$_{ipso}$-P) was not identified, presumably due to overlaps. ^{31}P{^{1}H} NMR (CD$_2$Cl$_2$): δ 27.3 (s). IR (Nujol): 3448 s, 3324 s, 1644 s, 1541 s, 1317 m, 1292 m, 1186 s, 1072 w, 1039 s, 936 w, 896 w, 837 m, 812 m, 756 w, 734 w, 617 m, 522 m cm^{-1} MS (ESI+): *m/z* 468.1 ([M + H − NEt$_3$]$^+$), 569.2 ([M + H]$^+$); MS (ESI−): *m/z* 465.9 ([M − HNEt$_3$]$^+$). Anal. Calc. for C$_{26}$H$_{45}$FeN$_2$O$_4$PS·0.3CH$_2$Cl$_2$ (594.00): C 53.18, H 7.74, N 4.72%. Found: C 53.16, H 7.75, N 4.53%.

3.4.4. Preparation of [Pd(Cy$_2$PfcCONHCH$_2$SO$_3$-κ^{2}O,P)(η^3-C$_3$H$_5$)] (**5b**)

Solid [Pd(μ-Cl)(η^3-C$_3$H$_5$)]$_2$ (18.3 mg, 0.05 mmol) was added to a solution of **4b** (62.1 mg, 0.1 mmol) in dichloromethane (5 mL). After stirring for 30 min, a solution of AgClO$_4$ (21.0 mg, 0.1 mmol) in benzene (1 mL) was added causing immediate precipitation of AgCl. The resulting mixture was stirred for 1 h and filtered through a syringe filter (PTFE, 0.45 μm pore size). The filtrate was evaporated under vacuum to afford a crude product, which was purified by two subsequent crystallizations by liquid-phase diffusion of diethyl ether into a chloroform solution of the complex to fully remove triethylammonium perchlorate. Yield after two crystallizations: 41 mg (62%).

^{1}H NMR (CDCl$_3$): δ 1.11–1.44 (m, 11 H, Cy), 1.67–2.09 (m, 11 H, Cy), 4.41 (br m, 2 H, fc), 4.53 (br m, 2 H, fc), 4.60 (br d, $^3J_{HH}$ = 5.4 Hz, 2 H, CH$_2$N), 4.76 (br m, 2 H, fc), 4.87 (br m, 2 H, fc), 5.67 (quint, $^3J_{HH}$ = 9.9 Hz, 1 H, CH of C$_3$H$_5$), 7.14 (br s, 1 H, NH). ^{31}P{^{1}H} NMR (CDCl$_3$): δ 27.7 (s). IR (Nujol): 3240 s, 1588 s, 1566 s, 1329 w, 1297 m, 1258 m, 1249 w, 1226 w, 1204 s, 1175 s, 1091 s, 1039 s, 1007 m, 976 w, 959 w, 937 w, 921 w, 846 m, 836 w, 771 w, 743 m, 619 m, 604 w, 594 w, 548 w, 529 w, 500 w. MS (ESI+): *m/z* 688.1 ([M + Na]$^+$). Anal. Calc. for C$_{27}$H$_{38}$FeNO$_4$PPdS (665.86): C 48.70, H 5.75, N 2.10%. Found: C 48.52, H 5.68, N 2.04%.

3.5. Catalytic Tests in Pd-Catalyzed Cyanation

A solution of ligand (2, 4 or 8 mol. % with respect to the aryl bromide **6**) in dry dichloromethane (3 mL) was added to the respective palladium source (2 or 4 mol. %) placed in the reaction vessel; the obtained mixture was stirred for 5 min and then evaporated under vacuum. Anhydrous potassium hexacyanoferrate(II) (184 mg, 0.5 mmol), potassium carbonate (138 mg, 1.0 mmol) and **6** (172 mg, 0.5 mmol) were added to the pre-formed catalyst. The flask was equipped with a magnetic stirring bar, flushed with argon, and sealed with a septum. Solvent (1,4-dioxane/water, 1:1; 4 mL) was introduced, the septum was replaced with a glass stopper, and the reaction flask was transferred to an oil bath maintained at 100 °C. After stirring for 3 h, the reaction mixture was cooled to room temperature and diluted with water (10 mL), ethyl acetate (5 mL) and 3 M HCl (3 mL). The organic layer was separated and washed with brine (10 mL). The aqueous layer was back-extracted with ethyl acetate (3 × 5 mL). The combined organic layers were dried over anhydrous magnesium sulfate and evaporated under reduced pressure. The conversion was determined by integration of ^{1}H NMR spectrum.

Characterization data for the coupling product **7**. ^{1}H NMR (CDCl$_3$): δ 1.35 (s, 9 H, CH$_3$), 2.95–3.38 (m, 2 H, CH$_2$), 4.25–5.10 (m, 2 H, CH and NH), 7.32 and 7.60 (2 × d, J_{HH} = 8.0 Hz, 2 H, C$_6$H$_4$). The NMR data are in accordance with the literature [27].

3.6. Catalytic Tests in Pd-Catalyzed Suzuki–Miyaura Cross-Coupling

The procedure was modified from ref. [28]. Thus, palladium(II) acetate (1.0 or 0.5 mol. %) and the respective ligand **4** (1 or 2 equiv. with respect to Pd) were placed into a Schlenk tube and dissolved in dichloromethane (2 mL) under argon. The mixture was stirred for 5 min and then evaporated under vacuum. Next, aryl halide (1.00 mmol), boronic acid (1.15 mmol) and K$_2$CO$_3$ (2.00 mmol) were added to the Schlenk tube and the reaction vessel was filled with argon and sealed with a rubber septum. Degassed water (2 mL) and dioxane (2 mL) were introduced and the reaction flask was placed into a preheated oil bath (40 °C). After stirring for 6 h, the reaction was terminated by cooling on ice and the simultaneous addition of 3 M aqueous HCl (3 mL), water (7 mL) and ethyl acetate (10 mL). The aqueous layer was separated and extracted with ethyl acetate (3 × 15 mL). The organic layers were combined, washed with brine (30 mL), dried over MgSO$_4$ and evaporated under reduced pressure. Conversion was determined by ^{1}H NMR spectroscopy.

If appropriate, the coupling products were isolated as follows. K$_2$CO$_3$ (ca. 300 mg) was added to the crude product and the mixture was dissolved in water (10 mL). Brine (10 mL) and NaCl (ca. 300 mg) were added to the solution and the resulting precipitate was filtered. The collected precipitate was dissolved in boiling water (100 mL) and filtered immediately by suction to remove the by-product resulting from self-coupling of aryl boronic acids. The filtrate was cooled to laboratory temperature and acidified with dilute hydrochloric acid until pH 3–4 was reached whereupon a white precipitate formed. The separated product was filtered off, washed with distilled water (20 mL) and then taken up with ethyl acetate (30 mL). The solution was dried over MgSO$_4$ and evaporated under reduced pressure to afford pure coupling product, which was analyzed by ^{1}H NMR spectroscopy. Characterization data of the coupling products were as follows.

2-(4-Fluorophenyl)benzoic acid (**11oa**). ^{1}H NMR (CDCl$_3$): δ 7.13–7.01 (m, 2 H), 7.31–7.25 (m, 2 H), 7.33 (dd, J = 7.8, 1.1 Hz, 1H), 7.43 (td, J = 7.6, 1.4 Hz, 1H), 7.57 (td, J = 7.6, 1.4 Hz), 7.96 (dd, J = 7.8, 1.1 Hz, 1H), (all aromatics). The collected data correspond to those in the literature; see ref. [29].

3-(4-Fluorophenyl)benzoic acid (**11ma**). ^{1}H NMR (dmso-d_6): δ 7.28–8.17 (m, 8 H, aromatics), 13.16 (br s, 1 H, CO$_2$H). The NMR data are in line with those in the literature (see ref. [27]).

4-(4-Fluorophenyl)benzoic acid (**11pa**). ^{1}H NMR (dmso-d_6): δ 7.30–7.37 (m, 2 H, aromatics), 7.75–7.83 (m, 4 H, aromatics), 7.98–8.04 (m, 2 H, aromatics), 13.02 (br s, 1 H, CO$_2$H). The data are in accordance with those in the literature; see ref. [27].

2-(4-Tolyl)benzoic acid (**11ob**). ^{1}H NMR (CDCl$_3$): δ 2.38 (s, 3 H, CH$_3$), 7.26–7.14 (m, 4 H, aromatics), 7.35 (dd, J = 7.7, 0.9 Hz, 1 H, aromatics), 7.39 (td, J = 7.6, 1.4 Hz, 1 H, aromatics), 7.53 (td, J = 7.6, 1.4 Hz, 1 H, aromatics), 7.92 (dd, J = 7.8, 1.1 Hz, 1 H, aromatics). The analytical data are in agreement with those in the literature, see ref. [28].

3-(4-Tolyl)benzoic acid (**11mb**). ^{1}H NMR (dmso-d_6): δ 2.35 (s, 3 H, CH$_3$), 7.27–8.19 (m, 8 H, aromatics). The NMR data are in agreement with those in the literature; see ref. [27].

4-(4-Tolyl)benzoic acid (**11pb**). ^{1}H NMR (CDCl$_3$): δ 2.42 (s, 3 H, CH$_3$) 7.27–7.32 (m, 2 H, aromatics), 7.52–7.57 (m, 2 H, aromatics), 7.66–7.71 (m, 2 H, aromatics), 8.13–8.18 (m, 2 H, aromatics). These data correspond with those in the literature; see ref. [27].

2-(4-Fluorophenyl)phenylacetic acid (**12oa**). ^{1}H NMR (dmso-d_6): δ 3.50 (s, 2 H, CH$_2$), 7.21–7.38 (m, 8 H, aromatics), 12.27 (br s, 1 H, CO$_2$H). ^{13}C{^{1}H} NMR (dmso-d_6): δ 38.49 (s, CH$_2$), 115.06 (d, $^2J_{CF}$ = 21 Hz, 2 C, aromatic CH), 126.91 (s, CH), 127.45 (s, CH) 129.76 (s, aromatic CH), 130.79 (s, CH), 130.83 (d, $^3J_{CF}$ = 7.9 Hz, 2 C, aromatic CH), 132.53 (s, aromatic C), 137.06 (d, $^4J_{CF}$ = 3 Hz, aromatic C), 140.76 (s, aromatic C), 161.42 (d, $^1J_{CF}$ = 244 Hz, aromatic C), 172.67 (s, CO$_2$H). HRMS (ESI−) calc. for C$_{14}$H$_{10}$FO$_2$ ([M − H]$^-$): 229.0665, found 229.0670.

3-(4-Fluorophenyl)phenylacetic acid (**12ma**). ^{1}H NMR (dmso-d_6): δ 3.65 (s, 2 H, CH$_2$), 2.24–2.33 (m, 3 H, aromatics), 7.37–7.43 (m, 1 H, aromatics), 7.50–7.55 (m, 2 H, aromatics), 7.65–7.71 (m, 2 H, aromatics), 12.35 (br s, 1 H, CO$_2$H). ^{13}C{^{1}H} NMR (dmso-d_6): δ 40.61 (s, CH$_2$), 115.68 (d, $^2J_{CF}$ = 21 Hz, 2 C aromatic CH), 124.88 (s, aromatic CH), 127.79 (s, aromatic CH), 128.56 (s, aromatic CH), 128.60 (d, $^3J_{CF}$ = 7.9 Hz, 2 C aromatic CH) 128.83 (s, aromatic CH) 135.72 (s, aromatic C), 136.50 (d, $^4J_{CF}$ = 3 Hz, aromatic C), 139.04 (s, aromatic C), 161.82 (d, $^1J_{CF}$ = 244 Hz, aromatic C), 172.62 (s, CO$_2$H). HRMS (ESI−) calc. for C$_{14}$H$_{10}$FO$_2$ ([M − H]$^-$): 229.0665, found 229.0669.

4-(4-Fluorophenyl)phenylacetic acid (**12pa**). ^{1}H NMR (dmso-d_6): δ 3.61 (s, 2 H, CH$_2$), 7.24–7.74 (m, 8 H, aromatics), 12.38 (br s, 1 H, CO$_2$H). The data are in accordance with those in the literature; see ref. [30].

2-(4-Tolyl)phenylacetic acid (**12ob**). ^{1}H NMR (CDCl$_3$): δ 2.39 (s, 3 H, CH$_3$), 3.59 (s, 2 H, CH$_2$), 7.25–7.55 (m, 8 H, aromatics). The NMR parameters are in agreement with those in the literature; see ref. [31].

3-(4-Tolyl)phenylacetic acid (**12mb**). ^{1}H NMR (dmso-d_6): δ 2.34 (s, 3 H, CH$_3$), 3.64 (s, 2 H, CH$_2$), 7.20–7.56 (m, 8 H, aromatics), 12.35 (br s, 1 H, CO$_2$H). ^{13}C{^{1}H} NMR (dmso-d_6): δ 20.67 (s, CH$_3$), 40.66 (s, CH$_2$), 124.68 (s, aromatic CH), 126.45 (s, 2 C, aromatic CH), 127.56 (s, aromatic CH), 128.17(s, aromatic CH), 128.78(s, aromatic CH), 129.51 (s, 2 C, aromatic CH), 135.62 (s, aromatic C), 136.72 (s, aromatic C), 137.15(s, aromatic C), 140.013 (s, aromatic C), 172.69 (s, CO$_2$H). HRMS (ESI−) calc. for C$_{15}$H$_{13}$O$_2$ ([M − H]$^-$): 225.0916, found 225.0957.

4-(4-Tolyl)phenylacetic acid (**12pb**). ^{1}H NMR (CDCl$_3$): δ 2.39 (s, 3 H, CH$_3$), 3.59 (s, 2 H, CH$_2$), 7.25–7.55 (m, 8 H, aromatics). The data correspond with those in the literature (ref. [27]).

Rac-2-{[(*tert*-butyloxy)carbonyl]amino}-3-(4′-fluorobiphenyl-4-yl)propionic acid (**13a**). ^{1}H NMR (dmso-d_6): δ 1.32 (s, 9 H, CMe$_3$) 2.83–2.89 (m, 1 H, CH$_2$), 3.03–3.08 (m, 1 H, CH$_2$), 4.01–4.15 (m, 1 H, CH), 7.16–7.09 (m, 1 H), 7.13 (d, $^3J_{HH}$ = 8.3 Hz, 1 H, NH), 7.24–7.30 (m, 2 H, aromatics), 7.33 (d, *J* = 8.1 Hz, 2 H, aromatics), 7.56 (d, *J* = 8.2 Hz, 2 H, aromatics), 12.65 (br s, 1 H, CO$_2$H). The data are in accordance with those reported in ref. [32].

Rac-2-{[(*tert*-butyloxy)carbonyl]amino}-3-(4′-methylbiphenyl-4-yl)propionic acid (**13b**). ^{1}H NMR (dmso-d_6): δ 1.32 (s, 9 H, CMe$_3$), 2.33 (s, 3 H, CH$_3$), 2.75–3.08 (m, 2 H, CH$_2$), 4.01-4.16 (m, 1 H, CH), 7.12 (d,$^3J_{HH}$ = 8.3 Hz, 1 H, NH), 7.24–7.26 (m, 2 H, aromatics), 7.30–7.32 (m, 2 H, aromatics), 7.51–7.56 (m, 4 H, aromatics), 12.63 (br s, 1 H, CO$_2$H). The analytical data are in agreement with those in the literature (see ref. [8]).

3.7. X-ray Crystallography

The crystals of **3b**·CH$_2$Cl$_2$ (orange prism, 0.17 × 0.20 × 0.25 mm^3) and **4b** (orange plate, 0.07 × 0.17 × 0.23 mm^3) were grown from dichloromethane–hexane and ethyl acetate–heptane, respectively. Full-set diffraction data for both compounds (±*h* ± *k* ± *l*, θ$_{max}$ = 27.5°) were collected on a Bruker D8 VENTURE Kappa Duo PHOTON100 diffractometer equipped with IμS micro-focus sealed tube (MoKα radiation, λ = 0.71073 Å) and a Cryostream cooling device (Oxford Cryosystems) at 150(2) K.

Selected crystallographic data for **3b**·CH$_2$Cl$_2$: C$_{31}$H$_{54}$BCl$_2$FeN$_2$O$_4$PS·CH$_2$Cl$_2$ (*M* = 719.35 g mol^{-1}), triclinic, space group *P* − 1 (no. 2); *a* = 8.3583(3) Å, *b* = 11.0812(4) Å, *c* = 20.6164(8) Å, α = 99.276(1)°, β = 100.551(1)°, γ = 102.947(1)°; *V* = 1788.3(1) Å^3, *Z* = 2, *D*$_{calc}$ = 1.336 g mL^{-1}, *F*(000) = 764, μ(MoKα) = 0.711 mm^{-1}. A total of 37,660 diffractions were collected, of which were 8211 unique (*R*$_{int}$ = 2.93%) and 7117 observed according to the $I_o > 2σ(I_o)$ criterion.

Selected crystallographic data for **4b**: C$_{30}$H$_{49}$FeN$_2$O$_4$PS (*M* = 620.59 g mol^{-1}), monoclinic, space group P2$_1$/*c* (no. 14); *a* = 17.3904(6) Å, *b* = 9.1114(3) Å, *c* = 19.7014(7) Å, β = 97.552(1)°; *V* = 3094.6(2) Å^3, *Z* = 4, *D*$_{calc}$ = 1.332 g mL^{-1}, *F*(000) = 1328 μ(MoKα) = 0.643 mm^{-1}. A total of 82,146 diffractions were collected, of which were 7125 unique (*R*$_{irt}$ = 4.12%) and 3275 observed according to the $I_o > 2σ(I_o)$ criterion.

The structures of **3b**·CH$_2$Cl$_2$ and **4b** were solved by direct methods (XT2014 [33]) and refined by unrestricted least-squares against F^2 (SHELXL-97 [34] or SHELXL-2014 [35]). All non-hydrogen atoms were refined with anisotropic displacement parameters. The amide hydrogens H1N were identified on an electron density map and refined as riding atom with U_{iso}(H1N) set to 1.2U_{eq}(N). Hydrogen atoms in the CH$_n$ groups were included in their theoretical positions using the standard HFIX instructions in SHELXL-2014 and refined as riding atoms. A recent version of the PLATON program [36] was used to perform all geometric calculations and prepare the structural diagrams.

In the case of **3b**·CH$_2$Cl$_2$, the refinement converged (Δ/σ ≤ 0.002, 391 parameters) to *R* = 2.96% and *wR* = 7.02% for the observed diffractions and to *R* = 3.84% and *wR* = 7.54% for all diffractions. Extremes on the final difference electron density map were: Δρ$_{max}$ = 0.72, Δρ$_{min}$ = −0.79 e Å^{-3}. CCDC deposition number: 1545049.

For **4b**, the refinement converged (Δ/σ ≤ 0.002, 355 parameters) to *R* = 2.68% and *wR* = 6.53% for the observed diffractions and to *R* = 3.31% and *wR* = 6.83% for all diffractions. Extremes on the final difference electron density map were: $\Delta\rho_{max} = 0.36$, $\Delta\rho_{min} = -0.36$ e Å^{-3}. CCDC deposition number: 1545048.

An orange, bar-like crystal of **5b**·CH$_2$Cl$_2$ with approximate dimensions of $0.14 \times 0.25 \times 0.36$ mm^3 was obtained by recrystallization of the complex from dichloromethane–diethyl ether. Full-diffraction data ($\pm h \pm k \pm l$, $\theta_{max} = 27.56°$) were collected at 150(2) K with a Nonius Kappa CCD diffractometer equipped with a Bruker APEX-II image plate detector and a Cryostream cooling device (Oxford Cryosystems) using MoKα radiation ($\lambda = 0.71073$ Å). A total of 26,856 diffractions were collected, of which 7049 were independent ($R_{int} = 2.98\%$) and 5970 observed according to the $I_o > 2\sigma(I_o)$ criterion. Selected crystallographic data: C$_{27}$H$_{38}$FeNO$_4$PPdS·CH$_2$Cl$_2$ ($M = 750.79$ g mol^{-1}), monoclinic, space group $P2/c$ (no. 13); $a = 17.257(1)$ Å, $b = 10.4304(6)$ Å, $c = 17.2618(8)$ Å, $\beta = 100.882(2)°$; $V = 3051.2(3)$ Å^3, $Z = 4$, $D_{calc} = 1.634$ g mL^{-1}, $F(000) = 1535$, μ(MoKα) = 1.395 mm^{-1}.

The structure was solved and refined as described above for **3b**·CH$_2$Cl$_2$ and **4b**. The refinement converged ($\Delta/\sigma \leq 0.002$, 352 parameters) to $R = 3.30\%$ and $wR = 7.74\%$ for the observed diffractions and to $R = 4.26\%$ and $wR = 8.22\%$ for all diffractions. The final difference electron density map showed peaks of no chemical significance ($\Delta\rho_{max} = 1.51$, $\Delta\rho_{min} = -0.75$ e Å^{-3}). CCDC deposition number: 1544852.

4. Conclusions

In summary, several new phosphinoferrocene amidosulfonate donors with different dialkylphosphino substituents were synthesized and characterized. Together with their diphenylphosphino- substituted counterpart, these compounds were used as a ligand in Pd-mediated cyanation of aryl bromides and Suzuki–Miyaura cross-coupling of bromo-substituted benzoic and phenylacetic acids with boronic acids to the corresponding biphenyls. The testing reactions—aimed mainly at comparing different ligands and catalysts resulting thereof, rather than at obtaining the highest yields—revealed differences between the two types of reactions. Namely, ligand **4d** bearing the least electron-donating phosphine moiety and [PdCl$_2$(cod)] as a palladium source provided the best results in the cyanation reaction, while the biaryl coupling reactions provided the best yields with a catalyst resulting from the electron-rich donor **4a** and palladium(II) acetate. In a wider perspective, the collected results confirm that careful optimization of the catalytic system and reaction conditions is needed in each particular case to obtain good results in Pd-catalyzed C–C bond forming reactions.

Acknowledgments: The research leading to these results has received funding from the Norwegian Financial Mechanism 2009–2014 and the Ministry of Education, Youth and Sports of the Czech Republic under Project Contract no. MSMT-23681/2015-2.

Author Contributions: Jiří Schulz prepared and characterized the ligands evaluated in this study and Pd-complex **5b**; Filip Horký performed all catalytic tests; Ivana Císařová collected the diffraction data; Petr Štěpnička conceived the experiments and analyzed the collected results. All co-authors participated in writing the paper.

Conflicts of Interest: The authors declare no conflict of interest. The funding agency had no role in the design of the present study; in the collection, analyses, or interpretation of the data; in the preparation of the manuscript and in the decision to publish the results.

References

1. Herrmann, W.A.; Kohlpainter, C.W. Water-soluble ligands, metal complexes, and catalysts: Synergism of homogeneous and heterogeneous catalysis. *Angew. Chem. Int. Ed. Engl.* **1993**, *32*, 1524–1544. [CrossRef]
2. Pinault, N.; Bruce, D.W. Homogeneous catalysts based on water-soluble phosphines. *Coord. Chem. Rev.* **2003**, *241*, 1–25. [CrossRef]
3. Herrmann, W.A.; Cornils, B. *Aqueous-Phase Organometallic Catalysis*, 2nd ed.; Wiley-VCH: Weinheim, Geramny, 2004.
4. Herrmann, W.A.; Kulpe, J.A.; Kellner, J.; Riepl, H.; Bahrmann, H.; Konkol, W. Water-Soluble Metal Complexes of the Sulfonated Triphenylphosphane TPPTS: Preparation of the Pure Compounds and their Use in Catalysis. *Angew. Chem. Int. Ed. Engl.* **1990**, *29*, 391–393. [CrossRef]

5. Cornils, B.; Kuntz, E.G. Introducing TPPTS and related ligands for industrial biphasic processes. *J. Organomet. Chem.* **1995**, *502*, 177–186. [CrossRef]

6. Herrmann, W.A.; Albanese, G.P.; Manetsberger, R.B.; Lappe, P.; Bahrmann, H. New Process for the Sulfonation of Phosphane Ligands for Catalysts. *Angew. Chem. Int. Ed. Engl.* **1995**, *34*, 811–813. [CrossRef]

7. Schulz, J.; Císařová, I.; Štěpnička, P. Phosphinoferrocene Amidosulfonates: Synthesis, Palladium Complexes, and Catalytic Use in Pd-Catalyzed Cyanation of Aryl Bromides in an Aqueous Reaction Medium. *Organometallics* **2012**, *31*, 729–738. [CrossRef]

8. Schulz, J.; Císařová, I.; Štěpnička, P. Synthesis of an amidosulfonate-tagged biphenyl phosphine and its application in the Suzuki-Miyaura reaction affording biphenyl-substituted amino acids in water. *J. Organomet. Chem.* **2015**, *796*, 65–72. [CrossRef]

9. Schulz, J.; Horký, F.; Štěpnička, P. Different Performance of Two Isomeric Phosphinobiphenyl Amidosulfonates in Pd-Catalyzed Cyanation of Aryl Bromides. *Catalysts* **2016**, *6*, 182. [CrossRef]

10. Schulz, J.; Vosáhlo, P.; Uhlík, F.; Císařová, I.; Štěpnička, P. Probing the Influence of Phosphine Substituents on the Donor and Catalytic Properties of Phosphinoferrocene Carboxamides: A Combined Experimental and Theoretical Study. *Organometallics* **2017**, *36*, 1828–1841. [CrossRef]

11. Anbarasan, P.; Schareina, T.; Beller, M. Recent developments and perspectives in palladium-catalyzed cyanation of aryl halides: Synthesis of benzonitriles. *Chem. Soc. Rev.* **2011**, *40*, 5049–5067. [CrossRef] [PubMed]

12. Vafaeezadeh, M.; Hashemi, M.M.; Karbalaie-Reza, M. The possibilities of palladium-catalyzed aromatic cyanation in aqueous media. *Inorg. Chem. Commun.* **2016**, *72*, 86–90. [CrossRef]

13. Miyaura, N.; Suzuki, A. Palladium-Catalyzed Cross-Coupling Reactions of Organoboron Compounds. *Chem. Rev.* **1995**, *95*, 2457–2483. [CrossRef]

14. Miyaura, N. Organoboron compounds. *Top. Curr. Chem.* **2002**, *219*, 11–59.

15. Miyaura, N. *Metal-Catalyzed Cross-Coupling Reactions*, 2nd ed.; De Meijere, A., Diederich, F., Eds.; Wiley-VCH: Weinheim, Germany, 2004; Volume 1, Chapter 2; pp. 41–123.

16. Molnár, Á. *Palladium-Catalyzed Coupling Reactions*; Wiley-VCH: Weinheim, Germany, 2013.

17. Maluenda, I.; Navarro, O. Recent developments in the Suzuki-Miyaura reaction: 2010–2014. *Molecules* **2015**, *20*, 7528–7557. [CrossRef] [PubMed]

18. Brunel, J.M.; Faure, B.; Maffei, M. Phosphane–boranes: Synthesis, characterization and synthetic applications. *Coord. Chem. Rev.* **1998**, *178–180*, 665–698. [CrossRef]

19. Gan, K.-S.; Hor, T.S.A. *Ferrocenes: Homogeneous Catalysis, Organic Synthesis, Materials Science*; Togni, A., Hayashi, T., Eds.; VCH: Weinheim, Germany, 1995; Chapter 1; pp. 18–35.

20. Cremer, D.; Pople, J.A. General definition of ring puckering coordinates. *J. Am. Chem. Soc.* **1975**, *97*, 1354–1358. [CrossRef]

21. Redhouse, A.D. The chemistry of the metal-carbon bond. In *The Structure, Preparation, Thermochemistry and Characterization of Organometallic Compounds*; Hartley, F.R., Patai, S., Eds.; John Wiley: New York, NY, USA, 1982; Volume 1, Chapter 1; pp. 20–22.

22. Appleton, T.G.; Clark, H.C.; Manzer, L.E. The trans-Influence. Its measurement and significance. *Coord. Chem. Rev.* **1973**, *10*, 335–422. [CrossRef]

23. Hartley, F.R. Cis- and trans-effects of ligands. *Chem. Soc. Rev.* **1973**, *2*, 163–179. [CrossRef]

24. Schareina, T.; Zapf, A.; Beller, M. Potassium hexacyanoferrate(II)—A new cyanating agent for the palladium-catalyzed cyanation of aryl halides. *Chem. Commun.* **2004**, 1388–1389. [CrossRef] [PubMed]

25. Carole, W.A.; Colacot, T.J. Understanding Palladium Acetate from a User Perspective. *Chem. Eur. J.* **2016**, *22*, 7686–7695. [CrossRef] [PubMed]

26. Škoch, K.; Císařová, I.; Štěpnička, P. Phosphinoferrocene ureas: Synthesis, structural characterization and catalytic use in Pd-catalyzed cyanation of aryl bromides. *Organometallics* **2015**, *34*, 1942–1956. [CrossRef]

27. Kuroki, Y.; Ueno, H.; Tanaka, M.; Takata, K.; Motoyama, T.; Baba, K. *N*-acylamino Acid Amide Compounds and Intermediates for Preparation Thereof. U.S. Patent 6265418, 24 July 2001.

28. Shiwen, L.; Meiyun, L.; Daoan, X.; Xiaogang, L.; Xiuling, Z.; Mengping, G. A highly efficient catalyst of a nitrogen-based ligand for the Suzuki coupling reaction at room temperature under air in neat water. *Org. Biomol. Chem.* **2014**, *12*, 4511–4516.

29. Ramirez, N.P.; Bosque, I.; Gonzalez-Gomez, J.C. Photocatalytic dehydrogenative lactonization of 2-arylbenzoic acids. *Org. Lett.* **2015**, *17*, 4550–4553. [CrossRef] [PubMed]

30. Capparelli, E.; Zinzi, L.; Cantore, M.; Contino, M.; Perrone, M.G.; Luurtsema, G.; Berardi, F.; Perrone, R.; Colabufo, N.A. SAR studies on tetrahydroisoquinoline derivatives: The Role of flexibility and bioisosterism to raise potency and selectivity toward P-glycoprotein. *J. Med. Chem.* **2014**, *57*, 9983–9994. [CrossRef] [PubMed]

31. Dastbaravardeh, N.; Toba, T.; Farmer, M.E.; Jin-Quan, Y. Monoselective o-C-H functionalizations of mandelic acid and α-phenylglycine. *J. Am. Chem. Soc.* **2015**, *137*, 9877–9884. [CrossRef] [PubMed]

32. Ahmed, S.T.; Parmeggiani, F.; Weise, N.J.; Flitsch, S.L.; Turner, N.J. Chemoenzymatic synthesis of optically pure and L- and D-biarylalanines through biocatalytic asymmetric amination and palladium-catalyzed arylation. *ACS Catal.* **2015**, *5*, 5410–5413. [CrossRef]

33. Sheldrick, G.M. SHELXT—Integrated space-group and crystal-structure determination. *Acta Crystallogr. Sect. A Found. Adv.* **2015**, *71*, 3–8. [CrossRef] [PubMed]

34. Sheldrick, G.M. A short history of SHELX. *Acta Crystallogr. Sect. A Found. Crystallogr.* **2008**, *64*, 112–122. [CrossRef] [PubMed]

35. Sheldrick, G.M. Crystal structure refinement with SHELXL. *Acta Crystallogr. Sect. C Struct. Chem.* **2015**, *71*, 3–8. [CrossRef] [PubMed]

36. Spek, A.L. Structure validation in chemical crystallography. *Acta Crystallogr. Sect. D Biol. Crystallogr.* **2009**, *65*, 148–155. [CrossRef] [PubMed]

MDPI AG

St. Alban-Anlage 66

4052 Basel, Switzerland

Tel. +41 61 683 77 34

Fax +41 61 302 89 18

http://www.mdpi.com

Catalysts Editorial Office

E-mail: catalysts@mdpi.com

http://www.mdpi.com/journal/catalysts

www.ingramcontent.com/pod-product-compliance
Lightning Source LLC
Chambersburg PA
CBHW051723210326
41597CB00032B/5586